Robot Technology and Applications

MANUFACTURING ENGINEERING AND MATERIALS PROCESSING

A Series of Reference Books and Textbooks

SERIES EDITORS

Geoffrey Boothroyd

*Chairman, Department of Industrial
and Manufacturing Engineering
University of Rhode Island
Kingston, Rhode Island*

George E. Dieter

*Dean, College of Engineering
University of Maryland
College Park, Maryland*

OTHER VOLUMES IN PREPARATION

Robot Technology and Applications

edited by

Ulrich Rembold
*University of Karlsruhe
Karlsruhe, Federal Republic of Germany*

CRC Press
Taylor & Francis Group
Boca Raton London New York

CRC Press is an imprint of the
Taylor & Francis Group, an **informa** business

CRC Press
Taylor & Francis Group
6000 Broken Sound Parkway NW, Suite 300
Boca Raton, FL 33487-2742

First issued in paperback 2019

ISBN-13: 978-0-8247-8206-1 (hbk)
ISBN-13: 978-0-367-40319-5 (pbk)

Library of Congress Cataloging-in-Publication Data

Robot technology and applications / edited by Ulrich Rembold
        p.   cm.--(Manufacturing engineering and materials
        processing; v.34)
        Includes bibliographical references
        ISBN 0-8247-8206-2
        1. Robots, Industrial.   2. Computer integrated manufacturing
        systems.   3. Flexible manufacturing systems.   I. Title.
        TS191.8.R46   1990
        670.42'72--dc20   90-35116
                                                                CIP

Visit the Taylor & Francis Web site at
http://www.taylorandfrancis.com

and the CRC Press Web site at
http://www.crcpress.com

Preface

This book is a multi-authored effort to discuss the new generation of
robots to be employed by the factories of the 1990s. These devices
will be capable of planning and supervising their own work assignments
with information obtained from engineering. The execution of their
work will be supported by intelligent sensors and expert knowledge
about manufacturing. Additional features of this new generation of
robots are a high degree of versatility and the ability to recognize
errors and take corrective actions. Flexible manufacturing systems
will be supported by autonomous mobile robots which can freely move
about the plant floor and transport materials and tools from one work
station to another.

Within recent years, the robot has become an indispensable tool for
manufacturing. It has found its place in material handling, spot welding, seam welding, spray painting, machine-tool tending, and numerous other applications. Initially, it was anticipated that the robot
would find it place in medium-sized production runs where the products to be manufactured are often changed several times a day. However, this type of production requires the availability of robots with a
limited amount of intelligence. Since very few intelligent sensors are
available, most of the current assignments of robots are rather simple and repetitive. The robots are employed by mass producers of
consumer goods such as the automotive and appliance companies. Another area where the robot has had very little impact so far is assembly. Most assembly operations are being conceived for human assem-
blers who have two dextrous hands, vast assembly experience and an

intricate sensory system. Without such features, robots can only perform very simple stack type assemblies.

Various kinematic principles are being used to construct industrial robots. The underlying design and the number of degrees of freedom of the robot determine the versatility and the applicability of the device. In order to reach a point in a three-dimensional space, the robot must have three degrees of freedom. If, however, the robot wants to do useful work at an object located at this point, the effector must have another three degrees of freedom. For this reason, most industrial robots have six or more degrees of freedom. The present robot has a development history of about three decades, and during this time, several design principles have become an industry standard. There are designs available which can be used for various tasks; however, there are also specific designs that are best used for dedicated applications. The kinematic, dynamic and control principles for the present industrial robots are amply covered in other literature and are not discussed in this book.

The greatest potential of advanced industrial robots lies in medium-sized production runs and in assembly. For such applications, the robot must be provided with an adequate amount of intelligence, complex sensors, and, if necessary, mobility. A plan of action for the robot may be obtained from the description of the workpiece. For example, the assembly drawing along with a model of the robot world environment and expert knowledge of manufacturing processes should contain all information necessary to plan the assembly system and assembly operation. Such a planning system will also furnish the sensor hypotheses and control information needed for assembly. The control module of the assembly operation must be able to anticipate and detect difficulties. In case of a problem, a new plan of action has to be initiated and executed. An advanced robot should also have some capabilities of deducing from a present assignment basic operations needed for solving future similar tasks.

In most applications, the industrial robot has an integrating function in the factory environment; for example, it closes the chain of the material flow between a workpiece buffer and a machine tool or it interconnects several machine tools to a work center. Planning such a complex, flexible integrated manufacturing system is a very involved task, and conventional trial-and-error approaches for the plant layout will lead to an inefficient facility. For this reason, simulation tools have been developed which can help the plant designer conceive an optimal solution.

Within the simulation, it is possible to display a pictorial image of the plant on a graphic display, and the manufacturing engineer can observe the creation of a workpiece through its different production stages. Once the layout of the manufacturing floor and equipment has been determined, programming of the machine tools, robots and other facilities can commence.

There are explicit, implicit, and graphical programming tools available for robots. A user-oriented approach for programming is to direct the motions of the robot with task-oriented instructions. This method is of particular interest for planning and programming of assembly work. However, assembly tasks are usually complex enough that the robot must be equipped with a comprehensive sensor system consisting of, for example, vision, approach, touch, and force torque sensors. The interpretation of the implicit robot instructions and the coordination of the robot control and sensors must be done by an expert module which is an essential part of the programming system. The user or an operator may communicate with an implicit programming system by speech or formalized input in a task-oriented mode.

Expert systems will play an important role in the plant of the future. For example, when the user programs the robot by natural language, an expert system is necessary to translate the human instructions into robot specific commands. In case of autonomous planning, the system must know the configuration of the robot, its workplace and the workpiece. A knowledge-based module is necessary to automatically plan the assembly sequences and operations. The robots of an assembly cell must know how to cooperate with other robots, machine tools and peripheral production equipment. There is also expert knowledge necessary to recognize and solve conflicts. The coordination of the work of the different experts is done via a common blackboard system.

For planning and controlling the work of robot-based manufacturing cell, a centralized data management system must be provided. It is the data repository for modeling and programming the robot and supervising its actions. Such a facility may be conceived according to a non-normalized relational data model. Hierarchical relationships among data objects can easily be described with it by deep nesting of relations. For this purpose, it is necessary to define new data types and operations on these systems.

Often, the versatility of production equipment can be enhanced when it is placed on an autonomous mobile platform. This enables easier material distribution, machine tool tending, low piece rate production and flexible assembly. The platform can serve as a carrier for robots, part containers, loading and unloading devices, etc. Such a mobile vehicle may consist of a basic universal drive platform with three degrees of freedom. The mobility can be directed by a navigator and a comprehensive sensory and control system. No external lead or guide systems are needed. An autonomous intelligent platform should be able to plan, execute and supervise a mission along a route of a manufacturing floor. If a conflict occurs, it must recognize it and independently try to find a solution. The major components of the mobile platform are the mechanics and drive system, sensor system, controller, computer architecture, planning and nagivation system, world model and knowledge acquisition and in turn world modeling modules.

The design of these components involves research knowledge from a variety of disciplines, including physics, electrical engineering, computer science, and mechanical engineering. It is important to coordinate the cooperation of the different disciplines for the design, the construction and the interfaces of the overall concept of the autonomous system. This book will focus on all these aspects of advanced robot applications to inform the reader about the state of the art and future developments.

Ulrich Rembold

Contents

Contributors

JIM BROWNE   University College Galway, Republic of Ireland

C. BUCKLEY   Integrated Systems Laboratory, Federal Institute of Technology at Zürich, Switzerland

DETLEF CLASSE   Robert Bosch GmbH, Nürnberg, Federal Republic of Germany

RÜDIGER DILLMANN   University of Karlsruhe, Karlsruhe, Federal Republic of Germany

KLAUS R. DITTRICH   University of Karlsruhe, Karlsruhe, Federal Republic of Germany

THEO DOLL   University of Karlsruhe, Karlsruhe, Federal Republic of Germany

GERHARD DRUNK   Fraunhofer Institute of Technical Production and Automation, Stuttgart, Federal Republic of Germany

KLAUS FELDMANN   Friedrich Alexander University, Erlangen, Federal Republic of Germany

BRUNO FRANKENHAUSER   Robert Bosch GmbH, Reutlingen, Federal Republic of Germany

MARIA GINI   University of Minnesota, Minneapolis, Minnesota

PAUL LEVI*   University of Karlsruhe, Karlsruhe, Federal Republic of Germany

KLAUS HÖRMANN   University of Karlsruhe, Karlsruhe, Federal Republic of Germany

ALFONS KEMPER   University of Karlsruhe, Karlsruhe, Federal Republic of Germany

PETER C. LOCKEMANN   University of Karlsruhe, Karlsruhe, Federal Republic of Germany

ULRICH REMBOLD   University of Karlsruhe, Karlsruhe, Federal Republic of Germany

VOLKER TURAU†   University of Karlsruhe, Karlsruhe, Federal Republic of Germany

SUBHASH WADHWA   University College Galway, Republic of Ireland

HANS-JÜRGEN WARNECKE   Fraunhofer Institute of Technical Production and Automation, Stuttgart, Federal Republic of Germany

Present affiliations:
*Technical University, Munich, Federal Republic of Germany.
†Mathematic and Data Processing Services, Darmstadt, Federal Republic of Germany.

1

Types of Robots and Their Integration into Computer-Integrated Manufacturing Systems

RÜDIGER DILLMANN *University of Karlsruhe, Karlsruhe, Federal Republic of Germany*

1.1 INTRODUCTION

Robotics is an applied engineering discipline and has been a subject of research and development for about 20 years. This new technology has found entrance to many industries. It still is a very difficult and often controversial subject because numerous new spectacular robot applications are announced every year. The range of possible applications is very wide. Robots are employed in the nuclear industry, civil engineering, marine work, orbital space missions, ship building, agriculture, household work, and security supervision. A new robot type, the third-generation robot, will be of strategic importance. Third-generation robots are autonomous systems that can perform their task in an unknown environment. They will be the basic building blocks of the factory of the future. The related technologies of robotics and automation, which are used for factory automation, integrate mechanics, electronics, controls, sensor perception, and computers into a complex control system. Usually robots are operating in conjunction with material flow systems, numerically controlled machine tools, fixtures, and material processing machines. The concepts of fixed automation, programmable automation, and especially flexible automation are strongly based on robotics.

Robots occupied man's mind for a long time before becoming an integral part of the industrial manufacturing world. Science fiction writers have dreamed and written about armies of good-natured slave robots performing hard or hazardous work, and bad dangerous robots which, when out of control, were trying to rule human beings. Often,

robots are seen as a substitute for man. This opinion of robots influences the discussions about their use, benefit, and acceptance.
Different opinions about the capabilities of robots exist among engineers, technical publishers, managers, workers, and unions. Indeed, today's robots can be visualized as poor duplicates of man, capable of doing simple and in rare occasions sophisticated work of a
repetitive nature. Humanlike behavior is still missing, and it seems
that enormous research activities are necessary to realize even primitive intelligence. Robotics is not a research subject on its own, but
a conglomerate of various disciplines [mechanics, electronics, informatics, artificial intelligence (AI), computer-integrated manufacturing (CIM), etc.]. To design a robot and enable it to do a task requires a strong interaction among disciplines. This is a complex
task for research laboratories, industrial companies, and robot manufacturers. The exchange of expertise among researchers, robot designers, and users is absolutely necessary to obtain application-oriented robots.

In Europe, enormous research activities at universities and industrial laboratories have been started. In many countries, large-scale
national and international research and development programs have
been launched and are still going on. ESPRIT, BRITE, and EUREKA
are European research programs that have robotics as a key topic.
The United States, Canada, and Japan work on third-generation robots. NASA and the European Space Agency do robotics research
for teleoperations and manipulations in space.

In the industrial area several hundred companies are manufacturing robots of various types, as well as support systems and peripherals. Only a few are really successful and have a large share of
the market. The most important buyers of robots are the large automotive companies, which have the capability and potential of applying
the robotics technology. The automotive industries have advanced
robot applications in various areas. Smaller companies often do not
have the manufacturing requirements that support robot applications.
This is because today's general-purpose robots are too expensive and
inflexible to replace man in jobs like assembly, which call for skill,
dexterity, and intelligence. Further developments are necessary to
exploit the skills of robots. For the factory of the future, where the
integration of robots into CIM is necessary, tools must be developed
to help the designer conceive a product for manufacturability. The
era of unmanned factories, when we can claim "We are free because
we have robots" seems to be far away. In the near future robots may
create some unemployment but they will increase productivity, quality,
and flexibility. The wide spectrum of robot applications is summarized
in Table 1.1.

This first chapter introduces basic approaches of robotics, explains the state of the art, and shows how to get started with a robot-

TABLE 1.1 Typical Robot Application Areas

Industrial robots	Special-purpose robots	Autonomous robots
Industrial manufac-turing	Nuclear material han-dling	Industrial manufactur-ing
Pick-and-place oper-ations	Underwater handling and inspection	Transport and han-dling of tools and workpieces
Loading/unloading	Space servicing, re-pair, assembly, and exploration	Exploration of unknown terrain
Assembly operations		Inspection
Spot welding	Civil engineering	Automatic traffic
Arc welding	Telerobotics	Military and security areas
Spray painting	Personal service and others	
Inspection		
Features	*Features*	*Features*
Repetitive operation	No repetitive tasks	No repetitive tasks
Rigid	Mobile, multiarms	Mobile
Preprogrammed (fixed)	Flexible behavior	Flexible behavior
Structured environ-ment	Multisensor system	Learning capabilities
Sensor guidance	Unstructured en-vironment	Complex sensor sys-tem
In some cases, mo-bile		Unstructured environ-ment

based CIM system. It discusses basic aspects of robot components, architectures, design, control methods, and applications, outlining and discussing the following areas:

Components and architecture of robots
Activities of robot applications
Information processing for robot applications and robot operational control

A survey of recent robot developments and research in leading re-search centers illustrates how many basic problems are solved. We begin with some introductory remarks and offer a short survey of the state of the art on robot technology.

1.2 THE ROBOT SYSTEM

An industrial robot is a general-purpose machine that is programmable
and performs skillfully manipulation tasks. It is designed as a programmable, multifunctional manipulator to handle materials, parts,
tools, or special devices in various manufacturing operations. In some
cases a robot may have locomotive capabilities. Figure 1.1 shows several robot configurations conceived for manipulation, locomotion, or
both. A robot consists of the kinematic and mechanical system, motor drives, axis servo control, sensors, and end-effector, workcell control, and the programming software. In CIM a more generalized view
of a robot system is used. Several levels of control processing data
with different degrees of abstraction are part of the generalized robot system, namely the manipulator, robot, cell, shop floor, and in
some cases the factory control levels. Product design, manufacturing planning, and programming must consider each control level. The
various supporting tools of these activities and their integration into
an overall system are described in the chapters of this book.

1.2.1 Basic Components of a Robot System

A robot system comprises the functional elements and components that
are necessary to prepare and execute a manipulation and/or locomotion task. Each component influences the flexibility and the integration of robots into a manufacturing system. A robot system is characterized by the following:

Kinematic structure
Process control architecture
Sensor environment
Safety capabilities
Programming methodology
Interfaces to the environment
Data management and data presentation

We now proceed to a brief analysis and characterization of the most
significant system subcomponents of robots.
    A robot manipulator consists of a mechanical structure with de-
fined kinematics (open-loop or closed-loop kinematic chain architecture), which allows a motion of its end-effector along trajectories
with several degrees of freedom. If the robot is provided with locomotion, it can be divided into three components: the vehicle, the
arm, and the end-effector; each may be provided with several degrees of freedom (Fig. 1.2). The vehicle gives the robot its mobility, the arm moves the end-effector along a trajectory, and the end-
effector interacts directly with the environment.

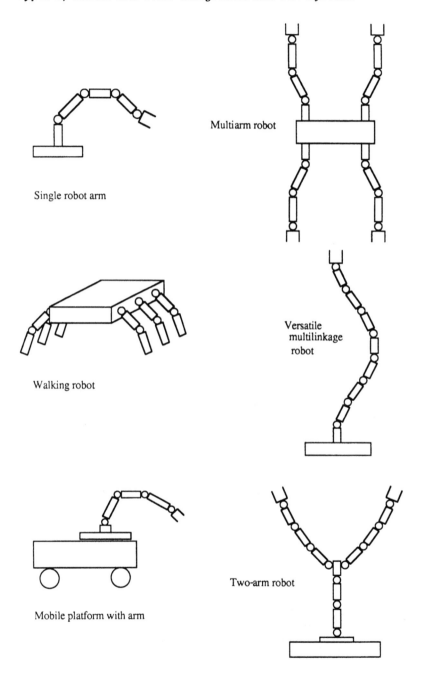

FIGURE 1.1 Various robot configurations and robot designs.

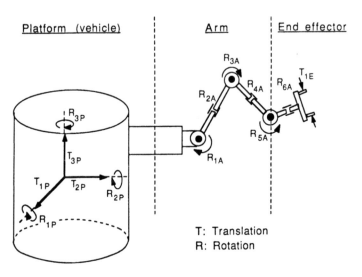

FIGURE 1.2 General kinematic configuration of a robot system.

Sensor systems, tactile and nontactile, internal or external, per-
ceive the interaction between the robot and the environment. A sensor processing system interpreting sensor data from the actual operation enhances the autonomy of the robot. The robot control system
translates the sensor data and motion commands into signals for the
servo controllers and the electromechanical transducers, to instruct
the mechanical system to follow the desired trajectory. The control
system tries to minimize the positioning or path errors. The programming system allows the specification of a given task with high level
language instructions, graphic test facilities, and decision support
(in many cases, with the aid of a user-friendly function menu). Programming may be done interactively using simulation techniques or
automatically with the aid of an action sequence planner. On this
level, the use of artificial intelligence techniques is of great advantage. Advanced autonomous robot systems perform action sequence
planning, program execution, mission supervision, and on-line error
recovery. Thus, goal-oriented behavior, reaction to unexpected
events, and handling of exceptions are expected. Passive and active
learning strategies enable the robot to acquire skills using experience
from the past. Building such a system requires a strong interaction
among all components and a clear separation of the control levels that
are performing their specific tasks.
    Interesting and important problem areas are multiarm control, co-
operating robots, and the interaction between robots and peripherals

within a manufacturing cell. Thus, a robot and its peripherals may be combined into a complex functional unit, designed to perform complex manufacturing tasks.

1.2.2 The Kinematic Problems

Most manipulator arms are configured from three major linkages, which produce the gross motion of the end-effector. The purpose of the lower linkages is to orient the end-effector. The design of the kinematic architecture is done under the following presumptions:

The type of the required motion (planar, linear, circular, general path) of the end-effector
The degree of freedom of the end-effector
The required forces of the application
The computational calculations necessary for executing the coordinate transformations, motion equations, and control algorithms
The Coriolis effects caused by the acceleration and velocity of the end-effector
The costs of the motor drives and gears
The required accuracy
The flexibility considerations of the kinematic elements
The compliant robot structures
The modular robot components

Mobile robots will be applied in the future for industrial production. Inductively guided transport vehicles are under development. Often such systems are equipped with simple manipulators for workpiece handling. Research work is being carried out for various locomotion principles of robots (legged, wheeled, and caterpillar driven).

There exists a vast amount of experience in solving the kinematics of robot configurations. In general, solutions are available for handling the direct and inverse kinematic calculations of kinematic chains (open or closed loop) for most 6 degree of freedom robots. For kinematically overdefined robots, approximation methods or additional programmer instructions ("left shoulder," "right shoulder," "elbow high," "elbow low," etc.) are required to solve the inverse kinematic problem. The kinematic solutions were derived for cooperating multiarm robots, dextrous multijoint, multifinger systems, and walking machines.

Homogeneous vectors and matrices were generalized to obtain a standardized representation method for kinematic problems. Kinematic structures (serial links and joints) are analyzed by defining the position and orientation of one link with respect to another. A widely used method consists of assigning coordinate frames that are fixed to their respective links. The position of the consecutive links is then defined by a homogeneous transformation matrix—for example, transforming a frame attached to link $n - 1$ to the frame attached to link n (Fig. 1.3). A manipulator with n links and n joints can be described by n homogeneous matrices. They are generalized as unit vector base frames, which are used to define the geometric relationship between any kind of reference coordinate system and robot coordination base. The Jacobian matrix is used to describe the transformation of a global kinematic chain in terms of changes of the individual joint angles. It can be applied to the calculation of joint movements to achieve small, but directionally defined changes of the tool or end-effector. The Jacobian is also useful for the calculation of joint movement to achieve a specific tool velocity or a path. By using the virtual work principle, it can relate forces, applied by or to the end-effector, to the resultant forces and torques generated at each joint.

With the availability of fast 32-bit microcomputers and transputers, the numerical methods for solving and calculating the kinematic equations and interpolation routines for continuous trajectories were greatly simplified.

1.2.3 The Mechanical Structure of a Robot and the Axis Drive System

Often, robot manufacturers have tried to increase the low positioning accuracy of industrial robots by improving the mechanical design. The requirements for a mechanical axis are as follows:

Fast response to the variation of the reference input
Good immunity to noise
Freedom from dynamic noise resulting from varying torques
Accuracy of periodic transfer motion
Optimal path accuracy without time delay
Low friction

Various robot configurations are available using Cartesian axes, such as the Scara, universal spherical joint, and gantry designs. High accuracy can be achieved using a heavy rigid structure, which is in contradiction to light and flexible robots with good dynamic response. Research work on new lightweight arms, improved robot configurations, and new axis drives is currently in progress. The results will improve the repeatability, velocity, and accuracy of robots. The direct-drive robot is another popular subject of investigation. The integration of the stator and rotor directly into the joint axis allows a reduction of backlash, hysteresis, and weight of the drive system; thereby a higher accuracy is achieved. The use of samarium cobalt

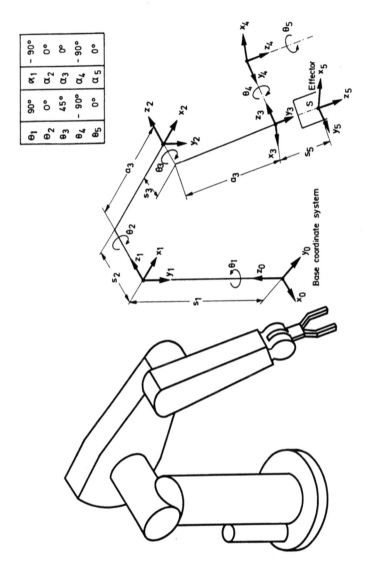

θ_1	90°	α_1	-90°
θ_2	0°	α_2	0°
θ_3	45°	α_3	0°
θ_4	-90°	α_4	-90°
θ_5	0°	α_5	0°

FIGURE 1.3 Coordinate frame relations between robot links defining a Puma robot.

based materials for the direct-drive motor helps in improving the acceleration and speed characteristics of the robot.

For controlling the high speed and acceleration of the direct-drive servo, new control algorithms must be developed. The reduction of the undesirable dynamic effects is a key topic of new development in robot control theory. The formulation and solution of the equations of motion and the control of various robot structures are research topics, worldwide. Various suboptimal solutions for handling direct and indirect dynamic problems are used for robot axis control. Advances in wrist sensor development allow force—torque control of robots. Hybrid controls now under development combine position with force control principles. Today's robots use the following control strategies:

Closed-loop servo controls
Cascade control
State space control
Dynamic control based on dynamic models
Decoupling algorithms for fast movements

Dynamic controls are computation extensive and therefore are used only in advanced pilot robot installations. In the future it is expected that they will become standard controls. Advanced pneumatic and hydraulic servo drive technology promises economical alternatives for electric robot drives. New developments for modular axis drives will allow a flexible mechanical structuring of the robot configuration.

1.2.4 Trajectory Planning

Trajectory planning is a basic on-line task of the robot control system to determine the motion parameters of the manipulator under sensor control. The following type of motion are typical for industrial robots:

Point to point
Straight line
Defined curves
Active compliance motion for assembly
Sensor-guided motion
Relative motion (tracking)
Movements to avoid collision

The calculation and optimization of the motions are performed by algorithms that use such constraint criteria as work space, velocity, acceleration, and vibrations. The amount of computation can be reduced if trajectory planning is divided into separate programs for trajectory preparation (parametric), interpolation, and execution.

The planning program determines the parameters, constraints, sampling frequency, and method for trajectory smoothing. The interpolator calculates the coordinate values of the points for the given path. Forces and torques necessary for the application are also calculated on this level. The coordinate transformation is the arithmetic interface between the Cartesian and the robot-specific state space. Most of the currently available robots are capable of performing the transformations both from Cartesian space into robot space and from robot space into Cartesian space. The latter is necessary for teach-in programming, sensor data processing, and path control algorithms.

Coordinate transformation makes extensive use of trigonometric calculations and matrix operations. The control software for trajectory planning consists of task-specific modules necessary for special-purpose applications. Examples are:

Search functions to find and to reach a workpiece (fine and gross motion)
Trajectories for tool changes
Weaving functions
Palletizing applications
Conveyor tracking operations
Sensor path correction

Usually, they are implemented as modules, subroutines, or macros in hardware, software, or firmware.

1.2.5 Locomotion and Mobile Robots

Mobile vehicles using manipulators were already developed in the early 1970s for application in risky environments (e.g., nuclear power plants; mobile underwater and space applications). In space, mobile telemanipulators are applied to explore unknown surfaces of planets. Mobile telemanipulators in nuclear power plants may have to work under high radiation within an unstructured environment. New applications of mobile robots for civil engineering, agriculture, fire fighting, personal services, and rescue operations as a result of nuclear or chemical accidents are being designed and developed.

In the industrial environment, autonomous mobile robots work in conjunction with material conveyors, part buffers, automated guided vehicles (AGVs), stationary robots, and production machines like machine tools. Their task is to transport workpieces, parts, and tools for machining. They also serve as devices for loading of production machines, administration of surface treatment, welding of large product components, and performance of assembly operations. Locomotion is not limited to classical AGVs, which are guided by inductive wires buried in the ground or markers on the floor. Free-moving robots contain for navigation an internal map of the world that represents

their known environment. Changes in the environment and the encounter of obstacles can be handled by the route planner. Locomotion in the industrial environment may be done in three modes:

Locomotion in a manufacturing microworld
Locomotion between manufacturing cells
Locomotion in the overall factory macroworld

Autonomous mobile robots are being developed for applications in a
natural unstructured environment and in a man-made structured industrial environment. These robots use wheels or caterpillar treads
or are designed as walking machines with legs. Walking machines
have one-, two-, four-, and six-legged configurations, usually with
several joints. Legged locomotion is still in the early experimental
stage. In the United States and especially in Japan, projects on
legged locomotion are sponsored by governmental funding agencies.
The locomotion of various animals is being studied intensively and
has resulted in prototypes of moving robots resembling crabs, snakes,
or other creatures.

Usually, the industrial mobile robots have a platform with wheels
operating with 2—3 degrees of freedom (Fig. 1.4). The wheels may
be passive or active, fixed, or steerable. Hovercraft or Mecanum-
type wheel-based robots are under research to realize omnidirectional
locomotion. Figure 1.5 shows an experimental omnidirectional mobile
robot equipped with two manipulator arms. This system is being developed at the University of Karlsruhe.

Outer space and underwater robots use pulsating jets and propel-
lers for propulsion.

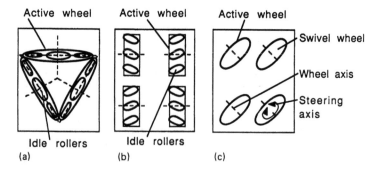

FIGURE 1.4 Basic wheel drive systems for mobile platforms.

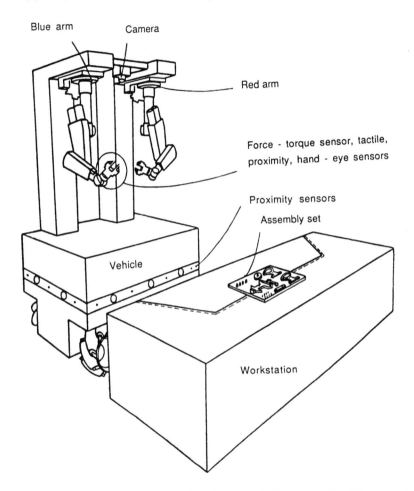

Blue arm   Camera

Red arm

Force - torque sensor, tactile, proximity, hand - eye sensors

Proximity sensors
Assembly set

Vehicle

Workstation

FIGURE 1.5 The KAMRO (Karlsruhe Autonomous Mobile Robot) system using Mecanum wheels for omnidirectional traveling.

1.2.6 End-Effector and Gripper Systems

The end-effector is a critical component of the robot system; it interacts directly with the world. The simple two-finger grippers used by most industrial robots are not flexible enough to handle a wide spectrum of workpieces. For this reason, industrial gripper design is concerned with dedicated applications. Depending on the actual task to be performed, fast gripper exchange mechanisms allow a changeover to various gripper types. An interface in the wrist has to supply the end-effector with electric, hydraulic, or pneumatic power as well as with control signals. For about 5 years there has been quite a lot of research on flexible soft grippers, multipurpose fingers, and dextrous hands. Figure 1.6 shows several flexible configurations of wrists for end-effectors with highly dextrous capabilities.

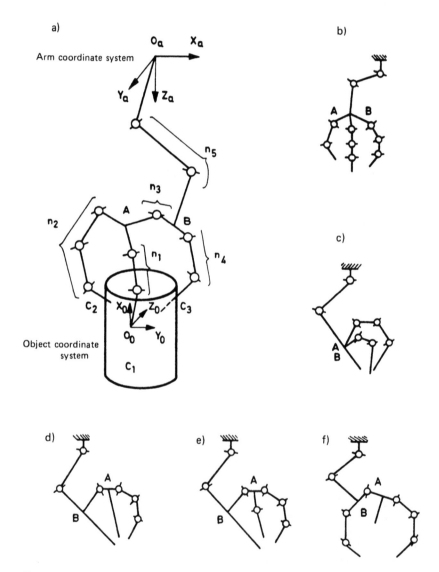

FIGURE 1.6 Various configurations of dextrous end-effector/wrist principles.

University laboratories are trying to emulate the capabilities of the human hand. Efforts are made to equip grippers with an artificial skin for static or dynamic tactile sensing. The three-, four-, and five-finger systems, each finger with two or three links, require very complex controls. A multiple-joint five-finger gripper can be compared with five small robots trying to handle one workpiece. Special problem areas are the miniaturization of the actuator and drive systems. Small dc motors, magnetic solenoids, or tendons are used to solve the problem. Controls for dextrous hands need hybrid components to obtain accurate positions and forces. Force—torque sensors, tactile sensor arrays, and grasp force controls will improve the skills of a robot hand. With such a hand, fine motion and active compliant operation are possible. The two-hand robot of the Waseda University in Japan, can play an organ, demonstrating that multifinger systems can have a very quick time response.

1.2.7 Sensors and Sensor Fusion

To perform its task, an autonomous robot needs real-world data, which must be acquired by sensors. Research on robot sensing proceeds along three different lines. First, various physical sensor principles are investigated for touch, force—torque, proximity, and precision positioning. Second, interest is focused on the complex problems of sensor data reduction and sensor data fusion, with the purpose of interpreting sensor data and obtaining meaningful information. The use of multiple sensors is also a topic of intensive research. Aspects of sensor compatibility, confidence in data, and the use of complementary and combined sensors are of interest. Third, the use of intelligent sensors for robots is investigated. Adaptive controls, decision strategies, and behavioral controls need reliable sensor data as input.

In general, sensors improve robot flexibility, adaptability, and intelligence. Today, there exist no general sensor models and interfaces to integrate sensors into various robot controls. Modeling of sensors with CAD systems to improve the design and use of complex systems is being investigated. The processing of sensor information depends on the sensor and the control level where the information is being needed for the robot control (Figure 1.7).

Tactile Sensing

Tactile sensors are used to identify and to control directly the interaction between the robot end-effector and the environment. The interaction with the object may occur by direct contact, reaction force, compliance, gravity, and temperature and optical patterns, to name a few.

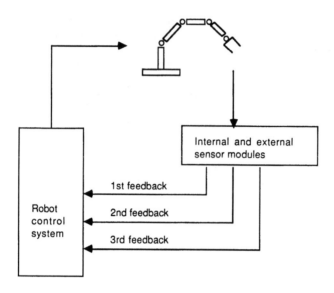

FIGURE 1.7 Different levels of sensory control.

Tactile sensors find their use in various locations of a robot system, including:

The joints between the robot links
The wrist between the arm and the end-effector
The finger joints
The fingertips
In the robot environment

These sensors are used to detect contact and threshold forces, sliding objects, and contact patterns between objects and the end-effector. Three groups of tactile sensors can be identified. With the first group, the object to be detected is much smaller than the sensor. In this case, one measuring cycle is necessary to acquire the information about the object. The quality of the information obtained from the sensor depends on the density and resolution of the tactile array. With the second group, the gripper is equal in size to the object. Information about the object is obtained from a tactile array and a finger position sensor. The quality of the sensor information depends on the resolution of the tactile arrays, the finger position sensors, and the number of sensors used. With the third group, the object is much larger than the end-effector. A sensing operation consists of a series of single measurements that must be coordinated by moving the end-effector and the robot arm. The robot movement is controlled by a series of single measurement. Much attention is devoted to the development of universal force--torque sensors for the effector wrist. Usually, the sensor is part of a closed-loop force control system. Active compliance operations are based on force—torque sensors. Tactile sensors are also applied in sensitive multifinger hands and in adaptable feet of walking machines. Sensor fusion and interpretation of multitactile sensor information are topics of research in artificial intelligence.

Nontactile Sensors and Vision

Nontactile sensors are a basic device of robots of the third generation. Topics of research include:

Inspection and location of objects
Detection of object surfaces (features like edges, holes, and relations
    between objects)
Inspection of objects for quality control
Scene analysis and navigation
Interpretation of movements in the robots work environment
Navigation of mobile robots
Stereo vision

The nontactile sensors are used for one-, two-, and three-dimensional applications. A typical 1D problem is solved by a proximity sensor, which detects distances between an end-effector and an object. The technology used is based on inductive, capacitive, and optoelectronic principles. Sensory problems involving two and three dimensions can be solved by two different approaches:

1. Use of multiple sensors with defined geometric locations
2. The use of sensors capable of taking multidimensional measurements

For the first approach, inductive and capacitive sensors may be applied. Optoelectronic sensors (e.g., silicon photodiodes and infrared diodes) are suitable for multidimensional measurements. The most frequently used silicon-based sensors for 2D vision are the charge-coupled device (CCD) elements, which have a photosensitive surface where electric charges proportional to the incoming light are collected. A transfer gate controls the flow of the charges from the photosensitive surface to a transfer register. From the transfer register the single charges are transmitted to the output port of the sensor, where they are converted for further processing. There are linear and array CCDs available. Linear CCDs have a resolution of 256—4096 pixels. The clock rate of such a sensor is 20 MHz and the dynamic resolution is about 2500:1−5000:1, giving a gray-scale resolution of 11−12 bits. Linear sensors with color processing capabilities are under development. Some matrix sensors (arrays) have electronic exposure control, which allows the adaption to different light conditions.

Three-dimensional vision systems consist of an illumination source and a detector. Depending on the use of the light source, 3D sensor systems may be classified as passive or active. With the passive principle, the illumination is independent of the system. Pictures are taken by two cameras and one uncontrolled light source with defined topology. The distance is determined by applying triangulation to the points of interest of the image. The solution of the correspondence problem requires a large amount of computing capacity. Active systems solve the correspondence problem with an exact control of the light source. Various structured light methods are applied. Laser techniques can be employed with triangulation and flight time of light measurement principles. However, at short distances the flight time of light is difficult to measure. Useful 3D vision systems are not yet available on the market. There is much research done on stereo vision and analysis of image sequences. Despite all these endeavors, the recognition of overlapping objects, visual servoing, and guiding of mobile robots by cameras represent unsolved problems.

Numerous fast, dedicated vision systems are already in industrial use and represent an interesting market. High level vision, where knowledge about the object and the robot task is to be processed, is supported by CAD and AI techniques. The interpretation of images, reasoning about spatial relationships, stereo vision, and processing of image sequences are being explored by many researchers. The fusion of data from sonic, infrared, and laser-ranging sensors is studied in the context with collision avoidance algorithms, docking procedures, and free navigation of mobile robots.

Sensors for Vehicle Guidance

For autonomous mobile robots, sensors for navigation and vehicle guidance are of interest. The type of sensor to be used is determined by the type of guidance method selected. There are two basic navigation principles. With the first principle, external fixed points or markers for the determination of the actual position are referenced. With the second principle, independent on-board sensor systems are used to identify the vehicle's location and orientation. Several guidance principles are discussed briefly.

Electromagnetic guidance of vehicles: This sensor guidance method is very popular and is being successfully used for AGVs. A frequency-modulated signal is applied to a cable, which is embedded in the floor. It induces a voltage in a magnetic pickup coil mounted on the vehicle.

This induced signal is used for guiding the vehicle along the route. Often, multiple or parallel cable systems are used for flexibility. Different frequency signals can be applied to each cable to outline a specific route; in addition, communication signals are applied to each cable for station-to-vehicle communication and for checking and avoiding collision.

Optical tape guidance systems: With this method reflective tape (white tape, metal tape, etc.) is fastened to the floor to mark the route. An optical sensor measures the deviation from the course and guides the ground vehicle. Light emitters of various types are fastened to the bottom side of the vehicle. The light is reflected and sensed by a photosensor under the vehicle.

Often, tape guidance systems offer a better solution than electromagnetic ones; they are more flexible and easy to install.

Guidance with the help of magnetic paint strips: Strips of paint loaded with ferrite magnetic powder are applied to the floor to outline the desired path of the vehicle. Magnetic sensors are installed on both ends of the vehicle to monitor the course. The sensors detect the deviation of the vehicle from the predefined course by monitoring the strength of the signal.

Laser guidance systems: A laser is mounted on the ceiling of the work area and scans the shop floor. The vehicle uses the light to find its desired path. Via a photosensor array, the vehicle detects any deviation from its course and transmits this information to the vehicle controller. With this type of guidance system the vehicle can be sent along a fixed or random course.

Optical guidance system: Numerous vision systems for vehicles have been studied, developed, and reported. The scope of the methods used ranges from simple marker systems to complicated computerized three-dimensional vision systems. They are usually based on video cameras, which are able to detect bar code patterns or to spot markers on the floor to obtain information about the actual position of the vehicle and the route.

On-board position sensing systems: On-board positioning sensing systems are based on the gyrocompass, sonic or laser beacons, or combinations of several methods. Fiber optical gyrocompasses are becoming competitive with mechanical gyrocompasses; they are a special topic of research.

Collision avoidance sensors: Here, proximity, touch, and vision sensors are successfully combined and applied to control the navigation of an AGV. Sonic, optical range (laser, infrared), and video image sensors are the basic components of collision avoidance systems. Collision avoidance is performed in three steps. If an obstacle is detected and identified, the normal control of the vehicle is interrupted. A collision avoidance algorithm is activated, and it modifies the route or selects an alternative path. As soon as the obstacle has been passed, a search algorithm is activated to bring the vehicle back to the original path. A new research topic is the control of a traffic scene in which multiple vehicles are on a possible collision course.

1.2.8 Robot Control Systems

The majority of today's robot controllers are designed for position and trajectory control and in some cases support the interaction with external sensors for adaptive control. They may contain the following modules:

Trajectory planning
Trajectory interpolation
Coordinate transformation (forward and backward transformation)
Dynamic control
Sensor control
End-effector control
Interpretation of the program
Real-time control systems

The task of the control system is to execute the movements and actions of the manipulator according to the robot program and to coordinate the interaction with the manufacturing environment. Design, installation, and maintenance of robot controls are realized under the aspect of functionality and modularity. Currently, most robot controls perform point-to-point and continuous path control. Active compliant motions are realized in some robot-operated assembly installations.

Advanced controllers are hierarchically organized in several control levels. Multiprocessor architectures with parallel processors and real-time capabilities are topics of intensive research. There are two basic control tiers to be distinguished: the task planning and control level and the subtask interpretation and primitive operations level. The algorithms for the time-critical calculation of the robot dynamics and force control can be implemented in dedicated VLSI chips.

The general implementation languages for robot control sofware are Pascal, Modula 2, C, and often Assembler. Logic programming languages like Prolog and logic control structures are increasingly being used to implement planning and control systems for intelligent robots. Various hierarchically structured control system architectures have been proposed. The real-time requirements to be fulfilled for fast information processing in autonomous robot systems necessitate the development of innovative hardware architectures. Currently, neural networks with parallel hardware architectures have been proposed for robots. They have a control structure that allows a fast response to exceptions.

1.2.9 Robot-Based Manufacturing Cells

There is an increasing need in industry for manufacturing cells consisting of multiple robots, machine tools, and peripherals that cooperate and are integrated as a functional unit performing a complex manufacturing task. Different cell types (e.g., machining, welding, or assembly cells) can be configured depending on the application. The workstations are structured from functional combinations of robots, machine tools, part buffers, peripherals, etc. A machining cell, for example, can consist of a milling machine, an inspection station, and a material handling system. Other cells may be configured from a number of assembly devices that work in parallel and can be used for precision assembly.

Vision systems in combination with other sensors are basic elements of a workstation to perform surface machining, complex assemblies, and workpiece inspection. When programming and controlling a robot-based manufacturing cell, each device gets its dedicated task, and the assignments are synchronized by an overall control structure. Petri nets and Prolog-based predicate transition nets are excellent tools for programming. Distributed processing systems are readily programmed with high level cell languages, supported by real-time kernel systems. Hierarchical control architectures for manufacturing cells are the topic of numerous research activities. The techniques and learning strategies of artificial intelligence are used for diagnosis and for preventive maintenance of manufacturing cells.

1.2.10 Robot Safety

In 1979 the German Engineering Society (VDI) drafted a safety standard for industrial robots, which was published. In addition, there is an effort by the International Organization for Standardization (ISO), to set standards for robots. The most important areas discussed by ISO are:

Installation and operation of robots
Protection of dangerous areas
Protection facilities for parts feeding
Setup, repair, and maintenance
Emergency shutdown provisions
Disruption of the energy supply
Collision

Man/machine interfaces
Instruction of personnel

Safety measures are concerned with hardware and software, fault tolerance, robust controls, and collision avoidance. The basic components for safety monitoring are illustrated in Figure 1.8.

1.3 PROGRAMMING OF ROBOTS

The key for the integration of industrial robots into existing manufacturing systems is the availability of efficient software tools for the
development of application and control software. Over the past 15
years, robot programming methods have been rapidly changed. Most
robot applications are carried out in an industrial environment where
robots, simple positioning devices, sensors, peripherals, and machine
tools perform discrete repetitive operations. For this purpose, the
following programming methods are available (Fig. 1.9):

Manual lead-through, teach-in
Manual programming
Tactile and optical sequence programming

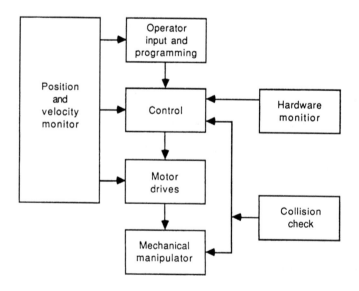

FIGURE 1.8 System components for safety monitoring.

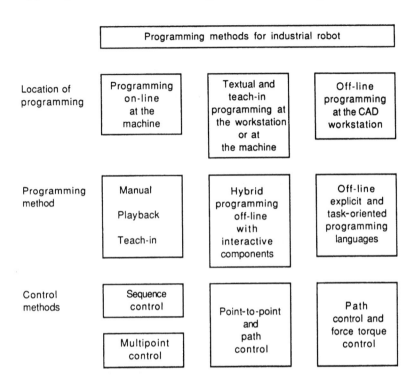

FIGURE 1.9 Classification of programming methods for industrial robots.

Master—slave programming
Textual programming
Pictorial programming
Acoustic programming
Explicit programming
Implicit programming

Several scientific programming languages were extended with movement instructions, sensor control statements, and data types (frames, vectors, matrices, etc.). Various dedicated languages were developed with robot-specific commands derived from other automation languages, including APT.

Modern programming systems consist of software development aids, problem- or application-oriented languages, and a graphical simulator for testing the program. The use of a world model is suggested in

AL, SRL, and RAPT. Pure CAD-oriented world models are not yet
available because robot-specific instructions for the description of
trajectories, grasping points, force, and other variables, have not
been defined adequately. Also, useful world models for sensors do
not exist today. Several graphic aids are available to visualize the
world geometry and to simulate the execution of robot program. Database systems specifically designed for robots are under development.

1.3.1 Graphical Simulation

The use of graphical simulation tools for robot programming is being
tested by many researchers. Systems like MC Autoplace, ROBCAD,
CATIA, and ADAMS are already available on the market. The description of the geometric object in these systems is done with 2D or 3D
CAD modeling packages based on solid models or surface representations. The systems contain specific tools for:

Robot modeling
World modeling
Description of motion
Collision detection
Control code generation

Current research efforts are focused on the simulation of sensors, robot dynamics, and collision detection. There is ongoing work on the
development of interactive dialogue systems and debugging facilities
as well as for standardized interfaces for robot databases and programming system (Fig. 1.10). Standards for the graphic animation
of robots on the basis of graphical kernel systems (GKS) or Programmer's Hierarchical Interactive Graphics System (PHIGS) are still
missing.

1.3.2 Interfaces Between the Programming System and Robot

Most programming systems are designed for robots of a specific type.
Thus, the user who intends to apply different types of robots has to
use dedicated vendor-made programming systems for each robot. To
establish a robot-independent programming system, it is necessary to
develop a standardized interface similar to CLDATA, which was conceived for numerically controlled (NC) languages for machine tools.
Figure 1.11 illustrates several interface levels that can be used for
standardization. VDI is engaged in defining the IRDATA code as a
standardized robot-independent interface (Fig. 1.12). In addition,
national and international standardizing committees are defining hardware interfaces and communication and transport protocols.

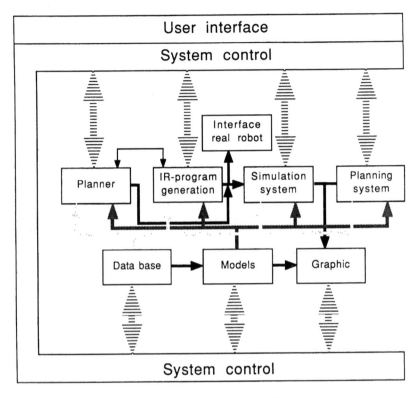

FIGURE 1.10 Information flow between different robot programming system components.

1.3.3 Robot Database Technology

Operational control of robot applications can be efficiently supported by a technical database management system. A robot database supports the following activities:

Task and product description
Robot application planning and layout
Description of cell components, robots, and the overall manufacturing
    cell
Robot programming
Control of robots in a runtime environment

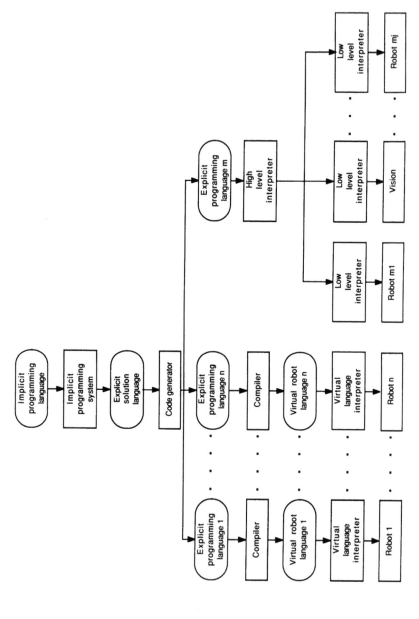

FIGURE 1.11 Hierarchy of robot programming language levels.

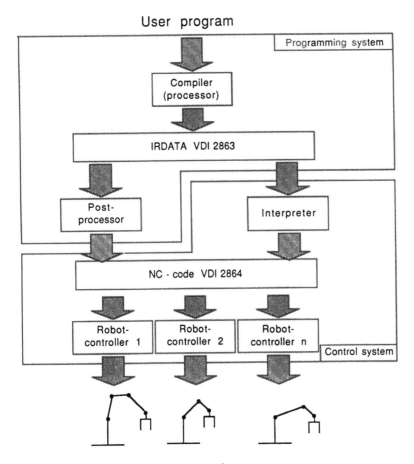

FIGURE 1.12 The IRDATA interface.

The data to be represented and processed are of five basic types:

1. Technical data of the robot, including the kinematic, dynamic, and geometric description of the robot, tools, and grippers, as well as a description of robot workspace, accuracy, and safety
2. The robot work environment, including a description of the manipulated objects, cell components, layout of the workstation, collision space, free space, and maps of the environment
3. Material characteristics, including the physical parameters of the materials (weight, centerpoint of mass, color, elasticity, etc.)
4. Geometric models of the robot world in CAD-specific representation

5. Representation of trajectories and program control structures for specific tasks

For defining a robot in a database system, hierarchical network and relational models are used, as discussed in other chapters of the book.

1.4 ROBOT APPLICATIONS

This chapter discusses the basic requirements of robots in industrial applications. The major tasks of industrial robots are spot welding, arc welding, surface coating, and machine tool servicing (including tending of die casting, injection molding, and forging machines and assembly stations). The basic areas of robot application are shown in Figure 1.13. The requirements for an industrial robot may vary depending on the skills needed to perform the manufacturing task. The user installs a robot primarily to increase productivity and flexibility, to perform hazardous tasks, or to improve product quality.

1.4.1 Material Handling

Material handling was a typical assignment of a first-generation robot. Often these robots perform only pick-and-place operations. A

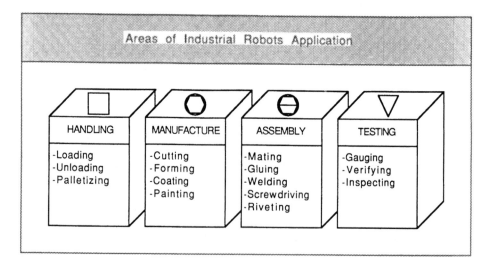

FIGURE 1.13 Areas of industrial robot application.

manipulation cycle may consist of loading of raw material into a machine and removing the finished part or both. Robots are used for loading and unloading of the following machines:

1. Die casting, injection molding, and vacuum forming machines
2. Forging machines, including drop hammers, forging presses, and sheet metal forming machines.
3. Metal cutting machines (lathes; milling, drilling, and deburring machines, etc.)
4. Heat treatment systems
5. Miscellaneous machines (glass processing and inspection machines, etc.)

The machine loading tasks are carried out by stationary stand-alone robots with 4—6 degrees of freedom. Usually they operate in a point-to-point control mode.

For material handling, the robot axis control system may only need point-to-point (PTP) control and can be programmed by the teach-in method. The payload is typically less than 40 kg and the maximum path velocity is less than 1 m/s. The majority of these loading tasks are realized using robots without external sensors. Other robot tasks are coating, cutting, and forming operations. Whereas cutting and forming entail the generation of a new part geometry, coating involves processing of an existing geometry. In cutting, a robot may both generate a new geometry (e.g., milling) and finish an existing geometry (e.g., deburring). Figure 1.14 presents an overview of various industrial robot applications in manufacturing.

1.4.2 Coating by Industrial Robots

One of the first manufacturing processes to be automated by robots, coating today is the second most important application for these devices. Coating includes spray painting, application of enamel and glue, and thermal spraying. A typical coating robot is a stand-alone unit with five or six rotational axes. It is designed to operate within a large work area and can handle a payload of 10—40 kg. Because the robot geometry is often similar to that of the human arm, programming is carried out using the lead-through method, whereby a skilled worker "leads" the robot through its work cycle and coordinates are read into memory for playback. Applications of coating robots are found mostly in the automotive industry. Typical assignments include spraying of PVC undercoats, filler materials, protective coatings, and surface finishes. Coating robots are also employed by the aviation industry for gluing and sealing of large aircraft parts. The polymer industry is another important user of coating robots.

| Requirements for further Application of IR | | | | | | | | | | | | |
| Requirements of — ● major ◐ moderate ○ minor importance | Handling | | | | Assembly | | | Machining | | | | | |
	Paletizing	Machine loading	Interlinking	Other handling	Small part ass.	Spot welding	Arc welding	Milling, grinding	Reforming	Spraypainting	Other coating	Inspection	All applications
Robot capabilities — Low cost, reliable and effective vision sensing	◐	○	○	●	●	○	●	◐	◐	◐	◐	●	◐
Easier, standardized programming	◐	◐	◐	○	●	○	◐	●	◐	●	●	●	●
Improved gripper dexterity	◐	◐	○	●	●	◐	○	◐	●	○	○	◐	◐
Greater flexibility for different applications	○	◐	◐	●	◐	○	○	◐	◐	○	○	○	○
Low cost, effective force sensing	○	○	○	○	◐	○	○	●	●	○	○	◐	○
Lighter, smaller robots	○	◐	◐	◐	●	○	○	○	○	◐	◐	○	○
Improved control systems	○	○	○	○	◐	○	◐	●	◐	◐	◐	◐	○
Robot technology — Greater speed	●	●	●	●	●	◐	○	○	●	○	◐	○	◐
Improved positioning accuracy	○	◐	○	◐	●	◐	◐	◐	○	○	○	●	○
Improved repeatability	◐	◐	◐	◐	●	◐	◐	◐	○	○	○	●	◐
Improved reliability	◐	●	◐	●	◐	◐	◐	●	◐	◐	◐	◐	●
Planning needs — Reduced Robot costs	●	●	●	●	●	◐	◐	◐	●	◐	◐	◐	●
Improved ability to interface with existing equipment	○	◐	●	◐	◐	○	◐	○	◐	◐	○	○	○
Product design/redesign	○	◐	◐	◐	●	●	●	◐	◐	◐	◐	◐	○
Turnkey CIM-systems	◐	◐	●	◐	●	◐	◐	●	◐	◐	◐	◐	●
All needs	○	◐	◐	◐	●	○	◐	●	◐	○	◐	◐	

FIGURE 1.14 Typical requirements for industrial robot applications in manufacturing.

1.4.3 Cutting Operation

In contrast to conventional machine tools, the robot can cover a relatively large workspace and has a high flexibility. Its flexibility, however, imparts to the device a limited positional accuracy, and robots cannot be used for cutting operations requiring the accuracy of a machine tool.

Typical applications are:

Milling of metal, wood, and plastic parts with low accuracy requirements
Drilling of sheet metal with the aid of templates
Water jet, torch, and laser beam cutting of complicated contours of
 flat sheets and plates
Finishing operations such as belt grinding, deburring, brushing, pol-
 ishing, and buffing

1.4.4 Product Assembly

The assembly of discrete components is an area of manufacturing in
which productivity is small in comparison with other processes such
as machining and inspection. Here, the increase in productivity generally is realized by special-purpose machines and improved assembly
procedures. The development of programmable automatic assembly
machines is different from that of NC machines. The automation of
machining processes was mainly done by automating existing machines,
whereas in programmable assembly the product often must be redesigned
for automatic assembly and new assembly systems must be conceived.
  Two assembly tasks for industrial robots can be distinguished:

Assembly of small discrete parts using active or passive compliance
 devices
Assembly using joining operations including spot, seam, and arc welding

There are two approaches for building flexible assembly systems using
robots:

To design sophisticated equipment and robots to handle the parts and
 subassemblies, whereby the human assembler is emulated
To redesign the product and to take into consideration the requirements
 and constraints of the industrial robot

The key to successful automatic assembly by robots is the redesign of
the product. Robot-based assembly systems are conceived according
to two principles:

1. Conventional manual assembly systems are copied, whereby the total assembly process is subdivided into small subtasks. For each
   subtask a special workstation is designed and the assembly of the
   product is done sequentially according to the flow line method.
   Examples are found in the assembly of printed circuit boards
   (PCB), and small electric motors.

2. When the principle of a single assembly station is used, an industrial robot with a high degree of freedom and flexibility (or dexterity) performs all assembly operations in sequential order at one location. Thus, the robot and its peripherals act as an assembly cell. For this form of assembly, gantry-type robots with one or more arms are applied successively. These manipulators can perform translational motions in a Cartesian workspace and are often equipped with "wrists" with three additional secondary rotational axes. Other types of assembly robot have a spherical arm configuration with five or six axes and a kinematic structure similar to that of the human arm.

Joining by Welding

Welding tasks are usually carried out by stationary stand-alone robots. Frequently arc welding robots have three rotational axes as the main axes and operate within a torus-shaped workspace. For most arc welding applications it is necessary to add peripherals, jigs, and fixtures with rotational, swivel, or moving functions to improve the accessibility to the weld seam within the workcell.

Point-to-point control is mainly used for spot welding and stud welding. Cycle times for spot welding with weld distances of approximately 20 mm may be 0.8 second. Arc welding requires continuous path (CP) control, whereby the center point of the tool is moved with a defined speed and orientation along the programmed path. Trajectories for welding applications can be programmed with the aid of the 3D-CAD model of the product. Advanced robot control systems contain a number of special functions that simplify programming and help to increase flexibility. For example, programmable weaving motions can be offered to the user. Some robots use sensor information for closed-loop control to improve the accuracy of the path coordinate and to support testing and diagnosis functions.

Most of the user programs for welding robots are generated with classical methods by teach-in with a handheld programming pendum. To obtain efficient cycle times, spot welding robots move their welding tools with a velocity of 1.5—4.0 m/s along the work path. Because of spatial and physical process restrictions, such a high speed is not practical for arc welding. In practice, most industrial robots have an operation time approximately 10% lower than that of an experienced human welder. However, the robot guarantees higher quality welds.

Application of Spot Welding Robots

The main area of robot applications in the automotive industry is spot welding of car bodies and body parts. Often several robots weld in sequence in a production line, or in parallel in a cell. Typically, a car body contains 5000–6000 welding points. Special lightweight and compact welding tools were developed for robot applications. A number of welding cell configurations were developed for robots. They may be classified as follows:

Welding cells
Dedicated welding lines configured by interconnecting several stations
    into one welding line
Flexible welding lines configured by interconnecting several independent welding stations

Single welding cells are often used for manufacturing of products with a low number of components. The transportation of workpieces between welding stations in a plant may be done by continuous or paced conveyors. Two transportation principles are used: moving the workpiece on mobile pallets between cells and moving the workpiece on a conveyor line.

In the first case the fastening and positioning of the workpiece is done on the pallet and the pallet is brought into a fixed position in the welding cell. For a continuously moving workpiece, there are two methods of realizing the synchronous motion between the welding tool and the workpiece. First, the robot is mobile; during the welding process it moves parallel to the workpiece. Upon completion of welding, it returns to its initial position and waits for the next work cycle. Second, the robot remains stationary and performs its work cycle. The workpiece is clamped in a stationary fixture and the robot, which is also stationary, performs the welding. After the operation the workpiece is released from the fixture and a new one is brought in by a feeding or indexing device. In the majority of manufacturing systems for spot welding, multiple robots are used in parallel. Sometimes, where there is a difficult spatial access to the welding points, the robots have to work from an overhead position.

Application of Arc Welding Robots

Many parts used in the automotive industry require continuous seam welding operations. Depending on the welding beam, the welding gun may have to be programmed for straight-line trajectory, trajectory following a simple curve, and trajectory following a spatial curve.

The analysis of existing applications shows that the straight-line trajectory can be used for approximately 80% of all arc welding applications. In other words, most arc welding applications are relatively simple.

For arc welding requiring complex motions along a workpiece surface and where major deviations from the path are possible, a sensor-based control is necessary. Sensors are used to search the work position and to lead the welding tool along the desired seam. For advanced robots with continuous path control, straight-line and circular trajectories can be defined by two or three points in the coordinate space. For weaving motions additional points at both sides of the desired trajectory must be specified to determine the weaving plane of the oscillating motion. The weaving frequency must be defined by the program. With increasing complexity of arc welding applications, powerful off-line robot programming tools are needed.

For arc welding, mostly stand-alone robots are used. For these applications electrically powered robots with revolute joints are preferred because they have an excellent dynamic performance and high repeatability in comparison to hydraulic systems. Figure 1.15 shows the requirements of robots for welding applications.

The future of robots for arc welding depends on the availability of versatile and accurate sensors. The recognition of the start and other welding points and the compensation of tolerances in space still require intensive research efforts.

1.5 SELECTION OF ROBOTS FOR VARIOUS MANUFACTURING APPLICATIONS

Thus far we have surveyed the state of the art in robot control and robot applications. Based on the need, a general implementation methodology for the design of robot application is required. Such a methodology necessitates a structured decision process for specifying the requirements of the manufacturing task. To introduce industrial robots to a wider spectrum of users, the robot technology must be better understood and incorporated into the phases of the manufacturing planning process, including the design of the product and the conception of the manufacturing system. In the sections that follow, several of these requirements are briefly explained for the automotive, electrical equipment, and aerospace industries.

1.5.1 Decision Process for the Design of Robot Applications

For planning robot applications, a sequence of planning activities and decisions must be performed. Figure 1.16 illustrates a decision schema that consists of a set of decision levels to guide the system designer from the initial manufacturing concept down to the detailed robot configuration needed for manufacturing. Econimic and technologic criteria are taken into consideration. In this example, six hierarchically ordered decision levels are outlined. In analog to software engineering cycles, the specification of the robot application is refined from the highest level down to the lowest level.

	Application ⟍ Specification	Spot welding	Arc welding	Stud welding
Kinematics	stationary, stand-alone	●	●	●
	traversing, stand-alone	◐	◐	◐
	gantry	◐	◐	○
Axes	≤ 4	○	○	◐
	5 - 6	●	●	●
	≥ 7	◐	◐	○
Drives	electric	●	●	●
	hydraulic	●	●	○
Control	PTP	●	○	●
	MP	◐	◐	◐
	CP	◐	●	○
Programming	teach-in	●	●	●
	play-back	○	◐	○
	language	⊘	◐	◐
Max-speed	> 1.5 m/s	◐	○	◐
	0,5 - 1.5 m/s	●	●	●
	< 0.5 m/s	○	◐	○
Load capacity	15 - 100 kg	●	○	○
	2 - 15 kg	○	●	●
	≤ 2 kg	○	◐	◐
Repeatability	≥ 1.0 mm	●	○	○
	1 - 0.5 mm	◐	●	●
	≤ 0.5 mm	○	●	●
	Sensors	○	●	○
	DNC	●	◐	○

● = Standard required   ○ = Non standard

◐ = Parity standard required   ⊘ = Under development

FIGURE 1.15 Specification of robot requirements for welding applications.

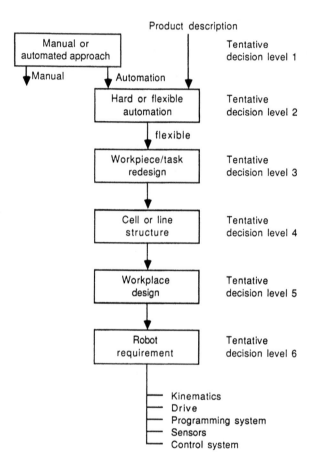

FIGURE 1.16 General design procedure for industrial robot application.

Decision Level 1: Manual or Automatic Assembly?

It is decided to use a manual and/or an automated system. It is important to distinguish between "automatic" and "automated." In an automatic system the task is done by automatic equipment without any human intervention. In an automated system the task is performed in part by human operators and in part by automatic machines.

Decision level 2: Hard or Flexible Automation?

It is decided whether special-purpose hard or fixed-type of automation tools should be used or programmable automation is required

(Figure 1.17). For flexible automation, it is important to distinguish between flexibility in the product design and flexibility in the manufacturing process. A robot, for example, can handle workpieces of various types. But a robot can also use various tools, and thus be able to do various machining operations.

Decision Level 3: Workpiece/Task Redesign?

It is decided whether the product, hence the manufacturing task, must be redesigned to optimize the operation of the robot. Criteria include time, length of the trajectories, limitations in speed and accuracy, collision space, robot tolerances, possible compliance operations, and cost of sensors needed.

Decision Level 4: Cell or Line Structure?

There are two options to be considered. The first option is a cell structure in which the robot does a multitude of operations. With the second option the work is divided along the manufacturing line, where every robot has a specific task. The decision of which option to select is influenced strongly by the overall manufacturing system for which the robot application is planned.

Decision Level 5: Workplace Design and Layout

This decision task includes the specification of the material handling system, the types of robot to be used, and the placement of the

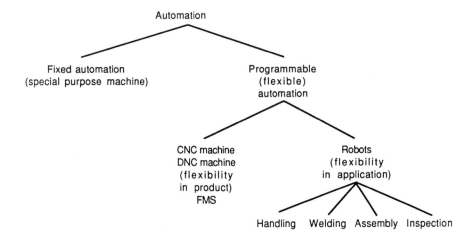

FIGURE 1.17 Various automation principles.

peripherals and manufacturing machines in the workstation. The other equipment (machine tools, magazines, conveyors, positioners, buffers, transport vehicles, etc.) are selected, as well. This decision results in the specification of the components of the workstation and the description of the topology and the workspace layout. The final design is influenced by the degree of intelligence of the robot controls and the sensors used (Fig. 1.18).

Decision Level 6: Detailed Requirements for Each Robot

At this decision level the final robot configuration to be used at each workstation is specified. In addition, the effector, sensors, and other peripherals are selected.

In the past, planning of robots was done specifically for every application because the production processes in which the robots were applied were often poorly described. There was insufficient information about the various possibilities of integrating robots into existing production processes. An automatic procedure for planning and selecting of robots as mentioned above is required to identify economical robot applications. One important parameter of such a selection scheme is the description of the task to which a robot can be applied. With such an aid, existing and future robot installations can be analyzed and planned (Fig. 1.19). Planning requires access to data about

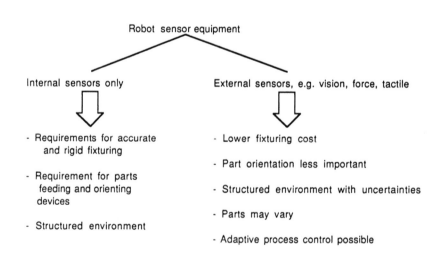

FIGURE 1.18 Requirements for sensor-guided robots.

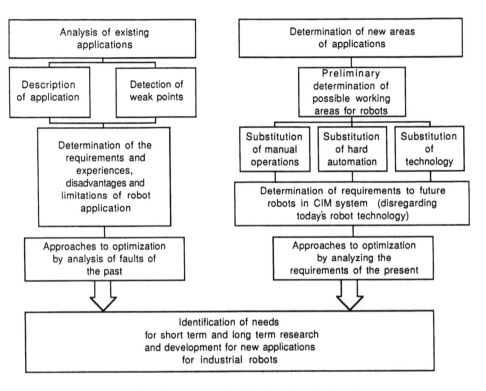

FIGURE 1.19 Schema for the analysis of robot technology.

the geometry and functionality of the product, the technology of the manufacturing process, and the robots and other devices needed by the manufacturing cell. The planning process can be supported by interactive or automatic rule-based tools. Expert systems for automatic planning are under development. Simulation techniques may be used to verify the layout of a planned manufacturing cell. They are discussed elsewhere in this book.

1.5.2 Robot Application in Various Industries

Aspects to be considered for planning of robot applications in typical manufacturing industries are briefly analyzed. The four selected examples are from the automotive, machine tool, aeronautical, and electrical equipment industries. The manufacturing processes considered are tending of machine tools, assembly of small discrete parts, welding, and machining.

The Automotive Industry

The most important robot application in the automotive industry is spot welding. Planning and programming of a spot welding station can be supported by the computer to assist the automatic selection of the robot and the welding tools. It also allows the generation of the program and the calculation of the tool trajectory. The second typs of welding application is arc welding. The most important parameters to be considered are the properties of the material to be joined and the geometry of the welding beam. Intelligent weld path guidance systems are required because it is difficult to follow the weld seam automatically.

Other important applications in the automotive industry are coating and assembly. The greatest potential for advanced robots is in assembly of components. Coating includes spray painting, underbody sealing, and bonding. Assembly is divided into small part and large part assembly. Typical operations are window, tire, gear, and battery assembly. Important requirements for future robot systems in car manufacturing include:

Methods for accurate and stable positioning of the gripper and the workpiece, particularly for tracking applications on a moving conveyor

Methods to assure a smooth material flow through manufacturing operations

The need for good "design for automation" methods (in general, existing product designs consider only assembly by hand)

The integration of CAD systems for programming

The extension of the geometric product model to the technological and task-related models

Conventional flow-line-oriented manufacturing strategies must be compared with flexible cell-oriented and distributed strategies. With distributed strategies robots may be assigned to various tasks.

Machine Tool Industry

A new area of application of the industrial robot in the machine tool industry is automatic tending of machine tools and handling of workpieces. A survey of the use of robots in current machine tool applications suggests the following system improvements:

Integration of cell programming languages into the planning system

Integration of task-related sensors into the cell components

Higher acceleration and deceleration capabilities of axes
Increased positional accuracy and repeatability of the robot
Reduction in time and effort for programming and system setup
Development of mobile robots
Development of space- and energy-saving devices

A significant increase in the reliability of the complete machining cell can be achieved by providing the robot with intelligence (AI). AI techniques can play an important role in:

Programming, where frequent program modifications are necessary
The handling of nonpredictable system errors and equipment problems such as chip blockage
The identification and sorting of defective workpieces
Increasing the safety of man—machine systems

Aerospace Industry

Robots are used in the aerospace industry to obtain a uniform manufacturing process and high product quality. Typical applications are riveting, sealing, spray painting, and metal routing. Problems to be solved for future applications include:

Increasing the accuracy for machining robots
Increasing the flexibility of the robot to handle complex assemblies and parts of small manufacturing lots, which are typical in the aerospace industry
Developing high performance and rapid programming methods for robots

Future development will be related to machining and assembly. This may require the design of new robots that can work in a small space. The robots for future aerospace applications may have features similar to available painting robots (seven axes) but with improved accuracy. Other possible applications are:

Painting of the interior of a fuselage or subassembly of an aircraft
Making of fuselage parts in preparation for chemical milling
Contouring and drilling of large fuselage panels
Processing of composite materials
Circuit board assembly

Because of the small lot sizes and the complexity of the parts, off-line programming is an important requirement for all these applications.

Electrical Equipment Industry

The main applications in the electrical equipment industry are part handling and assembly. Because of high production rates and short product life cycles, the production system must be very reliable. Possible new applications are:

Palletizing of parts (plastic products, keyboards, terminal assemblies, electric motor assemblies, etc.)
Feeding of special-purpose machines
Precision assembly of small parts or products
Assembly (final) of complex products (washing machines, dishwashers, vacuum cleaners, etc.)
Packaging of products

Many applications in food processing, agriculture, ship building, and other industries have specific handling requirements. For most applications off-line programming capabilities are required to assure the integration of the robot into the plant. In the future, robot planning and programming will be simplified by integrating of the database management system, CAD modeler, and simulator into a general open system architecture. Thus product design and manufacturing planning and control can be combined into one system.

1.6 THE INTEGRATION OF THE INDUSTRIAL ROBOT INTO A CIM SYSTEM

The basic activities of an operational robot control system, illustrated in Figure 1.20, are as follows:

The design of a product for robot-oriented manufacture
The planning for the installation of robot-based manufacturing system
The programming of the robot
The programming of the robot control and sensory operation supporting the robot's skill

1.6.1 Product Design for Automatic Manufacture by Robots

Until very recently, the application of robots in various manufacturing areas was restricted to relatively primitive tasks. The development of new robot applications, however, can be supported by the application of advanced hardware and software or by an improved product design that considers the restrictions of the robot. Two research areas can be distinguished: the development of advanced assembly robots and the redesign of the product and its components for assembly by robots.

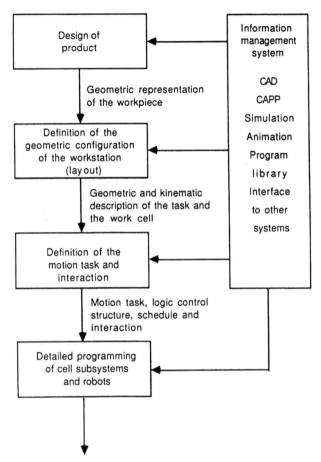

FIGURE 1.20 Information exchange between system components supporting robot programming in a CIM environment.

Both of these approaches are important, and they may complement each other. Till today, the first approach was the one mainly pursued by industry. However, the experience of companies indicates that much greater effort is required to design a product for assembly by robots. Flexible assembly stations using robots require a standardization of the product or part of it and a unique method for the description of the product assembly system. Several rules for the design of a product for assembly by robots are shown in Figure 1.21.

1.6.2 Planning a Robot Application

There is a need for planning tools to support the design and implementation of robot applications. Computer-based planning systems for various robot tasks in a CIM environment are under development. For planning, the following steps are necessary:

1. System analysis
2. Performance specification
3. Layout planning
4. Robot selection

Various expert systems are presently being developed to support the planning step. In addition, a database management system is needed,

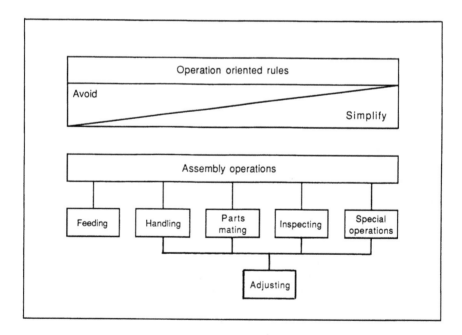

FIGURE 1.21 Basic scheme for assembly-oriented product design.

and methods must be found to coordinate and control cooperating planning modules of such a complex system.

1.6.3 The Robot Programming Environment

Today, robots are mainly used in high volume production. Powerful and user-friendly programming systems are not available yet, and this is one reason for the infrequent use of robots for small and medium-sized production runs. With the classical on-line programming methods a robot cannot be used for production during the programming. For this reason, off-line programming systems are needed as a linkage to production planning. Further improvement in efficiency can be obtained by connecting the CAD system to planning and programming. Thus, a higher flexibility and short manufacturing planning cycles can be realized.

Operational control systems for robots must be integrated into existing CAD/CAM systems as it was done for NC machine tools.

The currently available robot programming languages cannot be used universally; they are specific to the type of robot selected for an application. The programmer must specify trajectories explicitly for every robot. An explicit language forces the user to think in terms of procedures in which all possible situations that might occur in an application are considered. A task-oriented language must specify actions, control structures, and exceptions.

Future on-line programming languages for robots should support:

1. The specification of the program flow control
2. The capability of writing subroutines for exception handling
3. The sensor interaction and the rules for the sensor-guided path control

In addition, they should emphasize the task to be performed rather than the specific robot motion, and they must offer an interface to off-line programming.

Off-line programming languages should have the following capabilities:

1. User-definable tasks and subroutines
2. User-definable end-effectors and robot arms
3. Complex data structures and predefined state variables
4. Coordinate transformation between frames
5. Runtime definition of variables
6. High level instructions for tactile sensors and vision
7. Decision-making capabilities allowing the robot to recover from an unexpected event
8. Use of CAD data

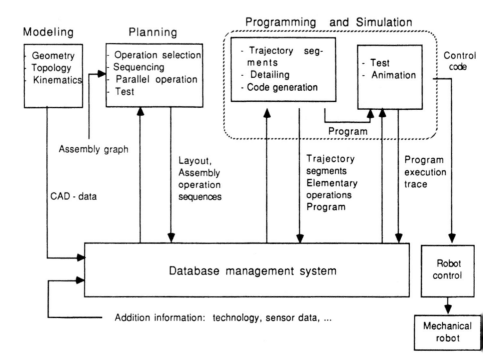

FIGURE 1.22 Structure of a typical integrated programming system for robot applications.

An advanced integrated off-line programming system includes a CAD modeler and many contain the components for a geometric modeler and graphic animation system and an off-line programming language and simulator, as well as an interface to the target robot system. Figure 1.22 shows the principal structure of such a system.

Programming of complex motion sequences requires the programmer to think in a three-dimensional workspace. Algorithms must be provided to test the off-line control data generated for efficient work cycles, trajectories, and collision avoidance. Suitable software tools are simulation programs with graphical capabilities. These systems are based on three-dimensional geometric and kinematic models. The dynamic models of robots require the formulation and solution of the motion equations in real time. A simulator consists of a package of emulated robot control functions. The basic data generated by production planning, the geometric model to the working space, the robot model, and the product geometry are the input data of the simulation system. The objective of linking the CAD system to the robot planner is to increase the efficiency of the robot programming task by integration of design data and production planning data. In general, the following benefits can be expected with an off-line programming system:

Elimination of downtime inherent to normal on-line programming
Cost-efficient redesign of the cell layout
Elimination of possible danger to the operator during teach-in
Availability of application programs for robots of various types
Less duplication in production planning and programming
An increase of the performance of the manufacturing system due to
   the generation of control data with the aid of the computer
Improvement of the editing of control functions
Improvement of the cooperation between design and manufacturing

The geometric modeling usually is done with a 3D-CAD system with a solid modeler or a surface representation scheme. Some of the available systems support modeling of kinematic chains and the definition of trajectory frames. For the direct generation of the position reference frames, geometric models of the raw, intermediate, and final product must exist. Collision check and the optimization of the trajectory are done with a simulation system. The calculation of the collision space and of the free space requires fast geometric computation methods and the application of geometric reasoning.

Geometric modeling of the robots and workspace can be done by approximating objects with cylinders, cubes, and spheres. The robot workspace may contain fixed and mobile objects.

The database technology required for the exchange of information beween product design, planning, and programming is discussed in a later chapter.

1.6.4 CAD-Oriented Robot Programming

A manufacturing cell consists of the robot, peripherals, and machine tools. They may perform one or several manufacturing tasks. An information exchange between the different components of a cell is necessary to synchronize the processes. The efficiency of planning the workcell can be increased if programming is done with an integrated planning system. Such a system must support the functional specification of the global cell and the local tasks. The global plan is broken down into subtasks that must be performed by the components of the cell. The subtasks are interconnected by a synchronization mechanism. Robot and cell programming systems can be divided into three related parts. Figure 1.23 shows the components of this hierarchy, namely:

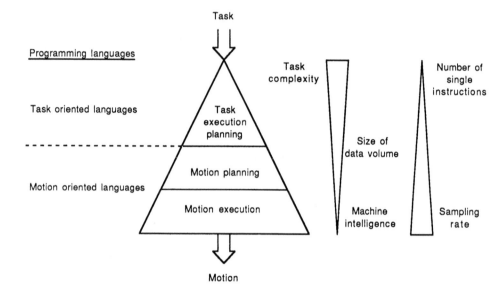

FIGURE 1.23 Multilevel approach for the description of robot motion process.

Planning of the global task sequences and of the task execution
Motion planning and synchronization
Detailing of the motion and execution planning

The planning of the task sequencing and execution is very complex and must be supported by a descriptive language and required data for the global operation. The lower levels are concerned with the sub-tasks of the cell components and with the detailed description of the motion and the interaction between the cell components.

Components of an Integrated Robot Cell Programming System

The off-line generation for a program of a robot cell requires access to geometric, technological, logic, and economic data (Fig. 1.24). Geometric data are available from a 3D-CAD modeler. Technological data, process knowledge, and the description of the robot tasks are stored in the database of the management system. For the generation and use of the different sources, various standardized interfaces are needed.

For the development of the program, specific data must be accessed from the database of the management system and transferred to the programming system (Fig. 1.25). The verification of the cell

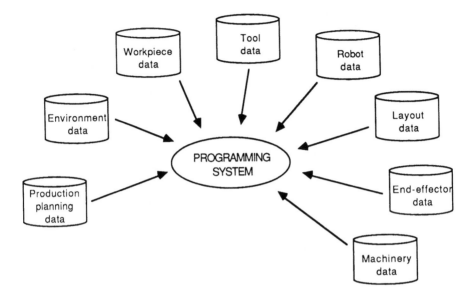

FIGURE 1.24 Required data for programming.

program and the programs for the cell components are carried out with the help of the simulation tool, which is an integral part of the programming system. Collision detection, exception handling, and verification of the functions of the control loops can be performed off-line.

The output of the programming system is a correct and optimized control program that defines the control structure and the data required by the cell and its components. A standard programming language for a cell as interface between the programming system and the target system is needed. Languages for programming of direct NC systems, which consist of multiple NC machine tools and robots, are considered for cell programming. Capabilities of changing the cell program on-line are desirable. A program modification may include a change of data, adjustment of offsets, adaptive control, or altering parts of the user program. For the documentation of the modifications, an interface to the programming system must be provided.

Programming Activities

A flowchart for the off-line cell programming activities appears in Figure 1.27. Before the program development is started, the global planning of the task of the cell is necessary, referencing task-specific data and the cell description as input. The result is a sequence of complex operations of the cell represented in terms of a formal plan.

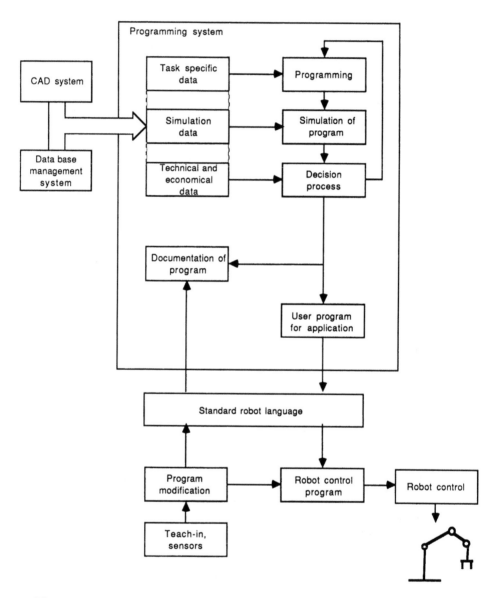

FIGURE 1.25 Structure of an off-line programming system.

In the next step the complex task of the cell is decomposed into several subtasks to be performed by the components of the cell. This requires dispatching of the subtasks to the cell devices, scheduling of the subtasks, and defining of the synchronization rules between

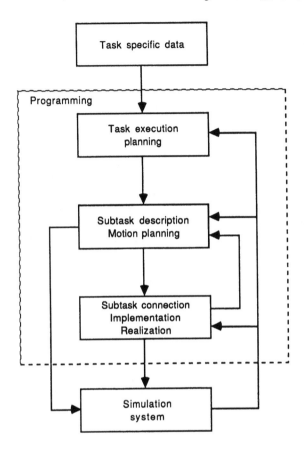

FIGURE 1.26 Information flow during the programming process.

the subtasks. After the sequence of the subtasks and the task-related control structure have been described, the motion must be specified. Each subtask is related to trajectories, which must consider
given geometric constraints, task frames, sensor data, and time intervals. The next step of the trajectory planning is to determine
whether any collision may occur in the cell and to plan alternative
trajectories for collision avoidance. In case of uncertainties, sensor
subroutines may have to be activated to evaluate the situation and
generate a corrective measure. Trajectory planning depends on the
type and complexity of the robot control system.

For PTP control, the motion of the end-effector along a trajectory
can be specified by a sequence of spatial points; for continuous path,
a sequence of polynomial path segments can be used. The interaction

with the robot world along these trajectories (open/close gripper, tool interactions, forces, torques, etc.) must be specified.

The programs of the cell subtasks should be modular and decomposed into elementary operations to ensure that they can be tested and simulated independently. An important aspect of the program development and simulation is the possibility of controlling the program flow by sensor information. This requires the specification of the sensor functions and the control algorithm of the sensors in the closed-loop control. Work is under way on sensor simulation systems to support the development of sensor-guided subtasks. Other programs can simulate noise and are used to evaluate and test the programmed reactions of the system to disturbances. After programming and testing of each one, the subtasks are linked together for configuring the entire cell program. The position-independent parameters of the subtasks are transformed into absolute coordinates (e.g., into the coordinate frames of the robot, vision system, and sensors).

The resulting user program can then be tested directly on the target system. With the aid of such software tools as debugger, tracer, and monitor, the control behavior of the system can be evaluated in the real manufacturing environment.

Programming System Modules

Planning of the cell tasks or subtasks can be carried out either by the robot control system (in the case of an intelligent control) or by an off-line programming system. In the general case, where robot controls have a limited machine intelligence, this task is done by the programmer with the aid of a programming system. Advanced robot controls are equipped with planning capabilities and will be able to perform complex functions with increasing degrees of machine intelligence.

Thus, the user can specify the cell task without considering all details and possible exceptions. Task programming can be performed with the help of implicit and task-oriented programming languages. The decomposition of the implicit program into task sequences, subtasks, and motion sequences is done on-line by the intelligent control system in real time. System programming of such an intelligent robot system requires a language that allows the implementation of a task decomposition module, local planner, monitor, diagnosis, and other intelligent control functions. Such a programming language needs an instruction set that is specifically designed for describing operations on different hierarchical control levels. The programming system must support all robot control levels and should allow the specification of robot operations with varying degrees of detail and abstractions. Thus, different levels of robot planning capabilities must be supported by one programming system (Fig. 1.27). This allows programming of

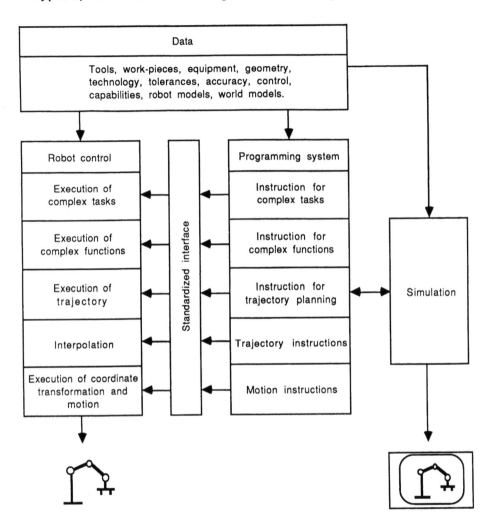

FIGURE 1.27 Instruction/execution of multistage conversion process for robot programming.

complex tasks, subtasks, or elementary operations, as well as description of trajectories or basic motions and sensor controls.

Hierarchically structured programming systems are being developed that support programming of the cell control level, of the cell component control level, and of the elementary robot control levels. The decomposition of a cell task into a sequence of subtasks is done in

several stages. An instruction on the cell level is mapped onto an instruction sequence of the next lower subtask level with the help of a
task assign module and local planner (Fig. 1.28). The semantics of
the instructions of a higher level must be maintained during the decomposition into the instruction sequences of lower levels. A basic
schema for a hierarchically organized control system for a robot cell
was developed by the U.S. National Bureau of Standards (NBS); see
Figure 1.29. The first prototypes of such a system are being implemented.

Programming Languages of Industrial Robots

Presently, there are more than 300 robot programming languages.
Most of them are explicit and are based on such classical programming

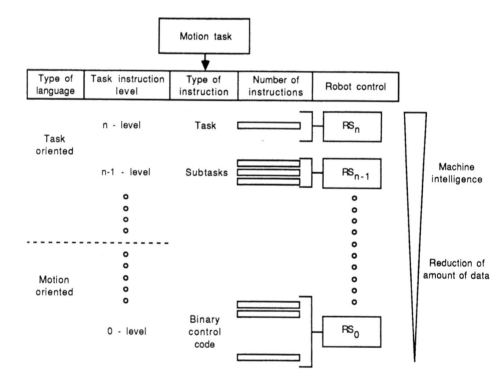

FIGURE 1.28 The various task instruction levels of a robot programming system.

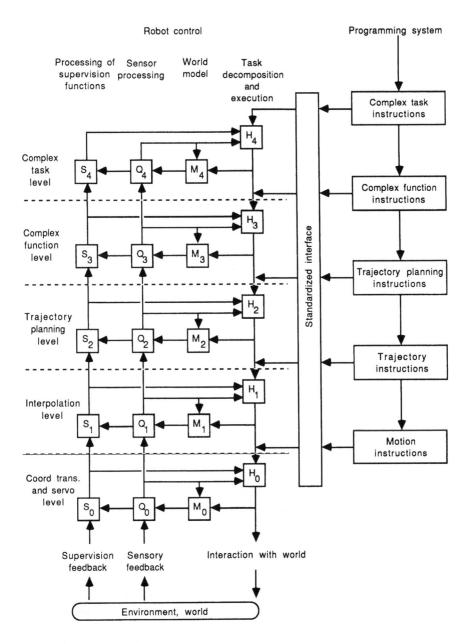

FIGURE 1.29 Linkage robot control/programming system within a hierarchically organized robot system. H-operators, perform the task decomposition and task execution; M-operators, process the world model data; Q-operators process the sensor data; S-operators supervise the robot cell operation.

languages as Pascal, C, Modula-2, BASIC, and Assembler. New object-oriented languages like Smalltalk or C++ are applied in experimental programming systems. There is a worldwide trend to develop implicit robot programming languages. An implicit robot programming system is based on a world model and knowledge about the task to be performed. A world model can be implemented with the help of the frame concept that originates from artificial intelligence. A frame is a logic structure suitable of processing facts and world data; it can be represented with the help of a rational database to obtain direct access to the stored information about the robot world. A relational database is an essential part of an implicit robot programming system. Two types of information are used to describe the world. One contains geometric and the other nongeometric parameters. The programmer must access both types and combine the retrieved information for the application at hand. Relations between objects and the attributes of objects are parameters needed for implicit robot programming. Other data required are sensor parameters, descriptions of the robot (geometric, kinematic, dynamic), and control data (trajectories).

Basic Types of Robot Programming Language

The robot programming languages can be classified according to the robot reference model, the type of control structure used for data, the type of motion specification, the sensors, the interfaces to external machines, and the peripherals used. The following types of robot programming language are available:

Point-to-point motion languages
Basic motion languages at the Assembler level
Nonstructured high level programming languages
Structured high level programming languages
NC-type languages
Object-oriented languages
Task-oriented languages

The dependence of the robot programming language on the robot control system makes it necessary to standardize the interfaces, the test facilities, and other software tools. A complete programming system supports a range of software tools, including the language, the operating system, the device controller, and the debugging and application programs. Most of the advanced software is written in high level languages. An operating system kernels for a task-oriented architecture must include protocols for interprocessor communication, device drivers, interrupt handlers, and a runtime system for peripherals and error deduction and recovery procedures. Often, the absence of reliable hardware can be solved by a hybrid control system—

for example, an application program can be tested on a hybrid hardware—software configuration. New technological developments used
for robot and programming call for the definition of new control functions. These programs must be visible on different levels of the robot programming system with different degrees of abstractions. Another important aspect of robot programming is the specification of
motion. Traditionally, motions are specified by a sequence of special
positions or path segments. With the help of a CAD system, implicit
specifications can be used to calculate the geometric trajectory similar to the APT-like languages.

On-Line and Off-Line Programming Languages

On-line and off-line programming are used on different control
levels of the robot cell. Off-line programming supports the integration of robots into a CIM system. It allows the manipulation of CAD
data to reduce the programming time by eliminating the interaction
with the physical devices. The programming method, however, has
difficulty in handling sensor information, and this obliges the programmer to use simulation, which allows modeling of the sensor input in critical situations and its interpretation. In general, an off-
line programming system needs a software development environment
to specify the application requirements, to analyze the tasks, and
to decompose the global task to subtasks. Program test facilities
are needed, as well. An on-line programming system provides tools
for debugging and testing the program. Here the use of the physical robot system is required. An on-line language is more accessible
to nonexpert users. In the future, on-line languages must have the
following features:

Branching capabilities
Subroutine capabilities
Better sensor interaction with the control
Interface to the off-line programming system
Interface to the on-line simulator and to the debugging facilities

1.6.5 Geometric Requirements to the CAD/Robot Linkage

The off-line programming system provides the possibility of developing
task-oriented and robot-independent user programs. That is, programs can be developed without having a specific robot in mind. When
a task is decomposed into several subtasks, it must be possible to
choose among different coordinate frames according to the specified
task. For general-purpose applications a reference coordinate frame
is used, since it allows the programmer to easily visualize the motion
of the robot in a 3D space. The zero-reference points of the various

coordinate frames are chosen freely by the programmer, who can then use location-independent information from the CAD database, which is accessed by any element of the manufacturing system.

To generate the user program, it is necessary to transform the geometric information of the different coordinates, into a common reference coordinate frame. This can be done automatically by the system. The required information to carry out the transformation is provided by the CAD system. To get an executable user program, a transformation from the Cartesian coordinate to the robot coordinate system must be performed. The transformation is necessary for a translation and rotation. The relative position and orientation between adjacent coordinate frames is calculated with a homogeneous transformation. Mathematically, this transformation is represented by a 4×4 matrix of the following form:

$$\underline{T} = \left[\begin{array}{ccc|c} \underline{O}_1 & \underline{O}_2 & \underline{O}_3 & \underline{P} \\ \hline & 0 & & 1 \end{array} \right]$$

$$\underline{O}_i = (\underline{O}_{ix}, \underline{O}_{iy}, \underline{O}_{iz})^{\underline{T}}$$

To describe the transposition of a frame P1 to a new frame P0, the transformation matrix $\underline{T}SYS0,SYS1$ is calculated. The position and orientation of frame P1 in relation to frame P0 are described by this transformation matrix. This is done by simple multiplication of the matrices (Fig. 1.30). For programming of subtasks, user-oriented reference frames can be chosen. Figure 1.31 shows an example of the position and orientation of frame P1 in the Cartesian robot reference system.

With a freely choosable coordinate system, task-oriented programming is independent of a special robot. The position of the robot in relation to the workspace must not be available during programming. However, the exact position data must be entered before the program is executed either by a simulator or the robot itself. Changes of the geometric relations between robot and its workspace can be simply correlated for. This programming technique is also useful for application to mobile robots.

In an integrated robot programming system, the information generated during the product design phase must be available via an interface to the CAD database. This includes geometry data, information on how to process and assemble the workpiece, and data on technological requirements for programming. For example, a surface representation of a car body must be represented to define spot welding points and collision-free robot trajectories. For assemblies, a solid or a surface representation is required to compute collision-free trajectories and control surfaces. Also required are task frames (e.g.,

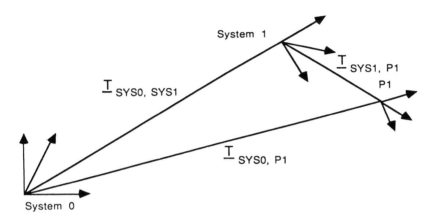

$$\mathrm{I}_{\text{SYS0, P1}} = \mathrm{I}_{\text{SYS1, P1}} \cdot \mathrm{I}_{\text{SYS0, SYS1}}$$

FIGURE 1.30 Relation between different coordinate base systems.

grasp points, approach and deproach direction, and center point of mass). Standard interfaces for the exchange of geometric data are proposed with IGES or with STEP; both are ISO proposals.

1.6.6 Simulation

A graphical simulation system for the validation and specification of the robot program is an integral part of an advanced programming system. It must provide a library of emulated robots, transport devices, and end-effectors to build up a cell model quickly. There must be available modeling software for the robot's environment and of the robot itself. The mathematic and geometric descriptions of the models depend on the desired accuracy of the simulation (2D- or 3D-models wire frame, surface, or CGS representations). The graphic representation enables the operator to check the programmed operation sequence. Today, many PC workstations offer high resolution graphics based on the raster graphic technology. Graphic standards like GKS or PHIGS facilitate the graphic animation of the simulated robot. For program alteration, an interactive interface is necessary, which allows the use of a joystick, mouse, or light pen. A simulation may be discrete and/or continuous. Discrete simulation supports programming of the overall control structure of a cell using predicate transition networks or Petri nets to represent the cell states and their transitions. Continuous simulation is used for testing such details of the program as the robot motion, interaction with other components of the cell, and the dynamic behavior of robots.

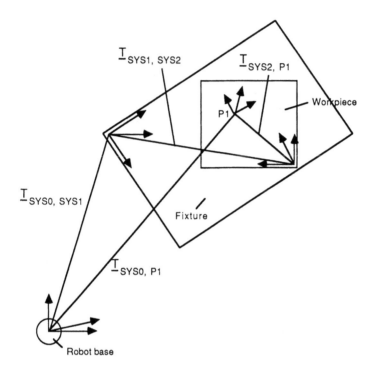

FIGURE 1.31 Examples of possible transformations to describe the position and orientation of a workpiece relative to different coordinate systems.

Program Modification

Program edit facilities must be available for the off-line programming system, for the simulation system, and for the on-line control system of the robot cell. Modifications may be related to the sequence of the program control flow, to the data used by the program (constants or variables), or to uncertainties of the robot environment. The on-line modification of the user program and/or the modification of data that are unknown during the off-line programming phase can be carried out with the help of the real robot. The user program and data can be modified via teach-in or by alphanumeric input. Data can also be changed by information obtained from sensors. For example, a sensor may determine reference points needed for the calibration procedure (Fig. 1.32), a vision system must be focused and calibrated in reference to the robot coordinate base, or spot welding points must be adjusted to tolerances. In any case, to avoid problems caused by different program versions, no unauthorized changes or modifications should be permitted. A standardized software documentation is required.

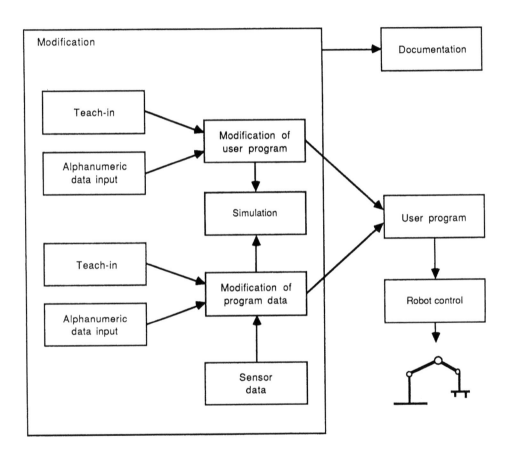

FIGURE 1.32 Modification of a program using teach-in techniques and sensor input.

2

Planning of a Robot Integrated Manufacturing Process

ULRICH REMBOLD   *University of Karlsruhe, Karlsruhe, Federal Republic of Germany*

2.1 INTRODUCTION

Planning is a prerequisite for the conception and operation of an efficient manufacturing process. For every phase of manufacturing, a scheme of planning activities is required to decide how a product should be manufactured and how the production facility should be laid out. Planning starts with the conception of the product and ends with its maintenance. A detailed database must be available for planning, covering the product, manufacturing processes, skills, know-how, and resources, as well as information on costs, product markets, competition, and numerous other factors. Of primary importance is the accessibility of this information, to provide the user with all the data needed for planning. Thus the user must have thorough knowledge of the various tools available, which greatly facilitate planning.

Planning of facilities employing robots is extremely difficult because the robot is a very versatile tool, which often is used for complex handling and assembly operations involving vision and other sensory devices. This chapter emphasizes the description of special tools for planning manufacturing operations using robot work cells. Some of the more conventional planning methods are also mentioned. However, they are not discussed in detail, and the user interested in them should read the literature on planning, which is abundant.

In most cases planning is a recurring process; therefore an attempt should be made to design planning tools in a modular fashion. In addition, many planning problems are similar and can be solved by generalized approaches. Planning aids must be user friendly and accessible from low cost computer workstations.

Basically, there are two types of planning activities: the planning of the manufacturing facility and the planning of the manufacturing operation. The latter consists of process planning and manufacturing scheduling. Several of the tools discussed in this book are used for planning activities of both types.

2.2 PLANNING OF THE MANUFACTURING FACILITY

In general, a planning activity goes through many stages during which manual or automatic tools are applied. For attaining an acceptable solution, it may be necessary to repeat the whole cycle or part of it several times. To avoid the propagation of serious errors through succeeding stages, all planning activities must be carried out very diligently. This requires that the basic data are complete and the planning tools functional. Errors entering at an early stage are in general detrimental if they are not corrected at the source. Usually, they must be removed later at the customer's site, at high cost to the implementer.

A systematic approach to the design and realization of a flexible robot cell requires a good understanding of the design aids and process. There are many experts with different skills involved, and a tight synergy between the participants is necessary. Typically, the planning cycle is divided into the following phases:

1. Manufacturing system analysis
2. Gathering of basic manufacturing data
3. System layout and technical evaluation
4. Return on investment analysis
5. Completion of the final design and system documentation
6. Installation and test in the manufacturing environment

The first four steps are the predesign activities during which all technical and economical information is prepared for the design and realization of the final system, which takes place in the last two steps. These iterative steps will consider many feasible solutions until a technically acceptable plan has been obtained. It is important to eliminate all risky solutions and to assure the practicality of the selected system. Particular attention must be paid to two circumstances: that every planning activity takes time and effort and that the system designer should try to arrive at an optimal solution as quickly as possible. In other words, the effort spent for planning must be justified in terms of time and cost. If possible, the planning tools should follow conventional procedures, and they must be easy to use.

A schema of the predesign phase is shown in Figure 2.1. The upper circular element contains the predesign phases and the lower element the supporting computer-aided design tools. In this predesign environment the human concept of the manufacturing cell is mapped into a computer for specification and presentation. The evaluation and optimization of the concept is done by computer and the results are presented to the system designer in tabulated and graphic form.

For the predesign activities the user should have the following tools available.

1. Semiformal languages for specifying:
   the components of the manufacturing equipment
   the functions of the components and the entire cell
   the layout of the cell
2. A programming environment for:
   explicit programming of robots
   implicit programming of robots
   NC machine tools
   auxiliary manufacturing equipment

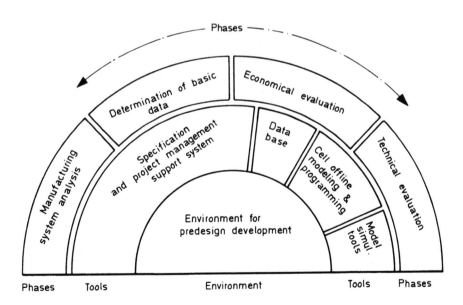

FIGURE 2.1 Computer-aided predesign phases.

3. Modeling and simulation tools for all components of the cell
4. A programming environment for the economic evaluation of the manufacturing system
5. Generation tools for documents and graphs
6. Software and knowledge engineering tools to facilitate the design of the system software
7. A project management support system

In a CAD/CAM environment there will be extensive use of analysis, modeling, and simulation, all trying to access the same database to assure the correctness and integrity of the data. Product design and factory equipment layout concentrate on geometrical aspects, and in particular on solid model presentation. This implies that the same modeling methods will be used and that interfaces will be provided to facilitate the access of data of related activities. Working with nongeometric data is also a vital aspect of any CIM activity (e.g., the bill of materials and manufacturing documents must be generated). Thus, the system must be able to accept these as input and work with them. In the following sections the components of the planning cycle are explained in more detail.

2.2.1 The Predesign Phase

The Manufacturing System Analysis

It is during system analysis that the manufacturing processes for a given product are selected. First, a basic decision will be made to make or buy components for the product. If the components are made in-house they will be produced with existing, modified, or new equipment. Here a crucial decision is made because the setup of a new system or a change in an existing one is not done on the basis of economic considerations alone; it is influenced as well by many other factors, such as management philosophy, manufacturing resources, part families, available component manufacturers, plant location, and material and product distribution systems.

As soon as the manufacturing method has been decided on, the more detailed selection of equipment will commence in a hierarchical fashion. Intuitively, planning from the top down will be the most economical way of performing the system analysis. However, in practice, this method does not always work and the drafting of the plan is also done from the bottom up. Fundamental decisions are made (e.g., to install single, parallel, paced, or continuous assembly lines). With the help of data on the piece rate and geometrical and technical parameters of the workpieces, the manufacturing machines are selected. The machines are interconnected with material handling equipment to form a manufacturing cell. This whole process may be done on the concept of a virtual manufacturing system, whereby the tools for this initial planning phase are independent of the target manufacturing equipment. As soon as the final manufacturing equipment and cell configuration have been selected, the machine-independent planning code will be transformed to machine-dependent code by a postprocessor.

The most important tools for this phase are simulation packages, expert systems, and numerical and statistical tools. Typical data needed are as follows:

Workpiece Data: product model, variant, shape, dimensions, weight, material
Technology: manufacturing methods, processes, and sequences
Customer Order Data: number of product, units ordered, delivery date
Manufacturing Skill: processing know-how, available craftsmen
Component Manufacturer

Gathering of Basic Manufacturing Data

Most of the information required is contained in the drawing, bill of materials, and master files of the products, plus data on production equipment, materials, and manufacturing processes. If properly organized in a database, it can be readily accessed by the user. However, in general there is a considerable amount of additional data to be gathered which are directly related to the customer order or may be needed because there are new manufacturing processes or machines available. User-oriented tools must be available so that this information can be readily entered into the database. To guarantee the success of the later phases, the data gathering must be done meticulously. Typical data gathered are:

Process Planning: raw material, machine tools, machining sequences, machining parameters, machining times
Assembly Planning: components, subassemblies, assembly equipment, tools, fixtures, assembly sequences, assembly times
Process Capabilities: piece rate, quality reliability, cost per unit product
Scheduling: available machine tools, manufacturing sequences, alternative manufacturing routes, due dates
Materials: source, inventory, in-process inventory, vendor performance, unit cost, inventory carrying cost
Labor: labor rate, benefits
Operating Costs: fixed and variable costs, performance of cost centers

System Layout and Technical Evaluation

In the layout and evaluation phase, planning becomes more detailed and is extensively supported by graphic and numeric simulations, expert systems, and knowledge located in databases. The system designer starts with the general layout of the manufacturing facility and defines its components. Whereas in the previous phase the designer was working with a virtual manufacturing cell, in this phase the objects of investigation are virtual or real machines. The following information may be generated:

The physical layout of the workcell
The number of workstations and their location in the workcell
The production capacity at the workstations
The degree of automation of all system components
Inventory and in-process inventory stations
The required quality control stations and their location in the work-
    cell
Maintenance philosophy and provisions

A typical manufacturing cell may be assembled from machine tools, robots, part feeders, fixtures, and auxiliary equipment, the basic functions of all of which must be defined, requiring a thorough knowledge of machines available on the market. If there is no standard equipment, the design of custom-made machines has to be initiated. The designer will consider various manufacturing alternatives.

The information for selecting the machines is contained in the master files for machine tools, robots, fixtures, and auxiliary equipment as well as in catalogs of vendors. If a similar manufacturing system had been operated previously, the new installation can be planned on the basis of available experience.

Of major importance in the selection process are factors such as the number of axes of machine tools, and robots, cycle times, piece rate, redundancy, and possible collision. Attention must be paid to the possibility of equipment failure and the contingency of future expansion.

The last phase in this activity is the selection of a suitable installation from the various alternatives. There are numerous criteria for which this selection can be done, including the operation of similar equipment, expandability, available space, and return on investment.

The data used for system layout and technical evaluation conform to the following types:

Machine Tools: type of machine tool, size, number of axes, piece rate, accuracy, reliability, maintainability, programming method
Robots: payload, reach, number of axes, positional accuracy, repeatability, maintainability, available programming method, work cycle

Material Handling Equipment: operation principle, load, capacity,
space requirement
Fixtures: type of fixture, automatic or manual operation, universal
or special purpose

Return on Investment Analysis

After the technical layout and the equipment have been selected,
the economic benefits of the system must be calculated. Usually, this
is done with readily available return on investment packages. This
analysis calls for fixed and variable cost data. The fixed costs are
easy to determine; however, the assessment of the variable costs may
be very difficult. Because often variable costs are an important factor of the return on investment analysis, however, a substantial effort must be made to obtain this information as precisely and completely as possible. The analysis must be done for all manufacturing
alternatives under consideration.

2.2.2 The Realization Phase

Steps 5 and 6 of the planning of the manufacturing facility constitute
the realization phase. All hardware and software components are
specified and designed in detail, and the system is set up for production.

The Completion of the Final Design and System Documentation

In the penultimate phase the planning documents are finalized for
all subsystems and parts, including those to be purchased and those
to be custom built. Since the manufacturing equipment is computer
controlled, there will be hardware and software components. Computers and programmable machines require a considerable amount of test
and maintenance software, which must be included. For all the components, requests for bids will be sent out to vendors. After comparison of the bids, vendors will be selected on the basis of price, delivery dates, and performance.

Particular attention will be paid to the control system. Because it
contains vital manufacturing know-how, it may not be released to a
vendor, to ensure that it cannot be readily duplicated in a competitor's
facility. This problem can also be resolved with a nondisclosure con-
tract, requiring the vendor to keep the installation confidential for a
specified period.

Complete documentation must be furnished for all components. This
is of utmost importance for the software. The flow of control should
be recorded by graphical aids, such as flowcharts or Petri nets. There
must be source code for all programs to be sure that software problems
can be located easily by operational personnel.

Installation and Test in the Manufacturing Environment

In the final phase of the system realization, all components are installed, tested, and integrated into the manufacturing operation. When
the hardware and software have been checked out properly, the system is ready for activation. The start-up should be done step by step
according to a predetermined procedure. Usually, this phase is the
most frustrating part of the entire planning cycle. Rarely will the installation be on-line after the first start-up trial. In most cases, considerable difficulties are encountered with hardware and software. Control elements may interact with each other, or the computer will suddenly enter an unexpected path, which causes the collapse of the entire system or part of it. Often, numerous controllers, sensors, software, or machine elements do not function according to the specification.

The problems and their remedy should be recorded carefully to be
able to pinpoint the source of the malfunction and to accumulate statistics for recurring breakdowns. Graphical and numerical simulation
tools are excellent means to simulate the operation of a faulty system
and to interpret its behavior [1].

After the system has been operated over a longer period of time, a
postaudit should be performed. Management must assess the performance of the entire installation and its components, to determine its
benefits. Valuable information will be obtained to justify future installations and improve the present system. The performance of each
component must be measured against its specification. When the post-
evaluation is done, the user should be aware that it often takes a year
or longer for a system to reach steady-state condition. The importance
of the evaluation lies in its findings and the recommendations for actions. As a result of this postaudit, many of the planning and programming tools may have to be revised.

2.3 PLANNING OF THE ASSEMBLY

When an order for a product is received, it is placed in a queue with
other orders, competing for the same manufacturing resources. With
assembly operations there are three activities which have to be performed to build the product from parts and subassemblies, namely assembly planning, assembly scheduling, and control of the assembly.

There are numerous factors determining the planning of an assembly. Typically, they include the number of assemblies that must be
performed, the complexity of the task, the similarity of the assembly
to previous work, the delivery dates of the product, and assembly resources.

The planning and scheduling of a flexible robot cell is a very sophisticated task and usually cannot be performed by conventional,
often very costly, trial-and-error methods. To obtain an optimal

manufacturing system, the planning must be done with the aid of the computer, using all its numeric and graphic capabilities to investigate manufacturing alternatives.

For an efficient assembly planning operation, a comprehensive software and modeling system must be available, consisting of tools with various functions. An overall structure of such a system is shown in Figure 2.2. Its basic components are:

1. A workcell and assembly planning system for setting up the workcell for a given product and for planning the assembly process
2. A program planner to support the implicit or explicit programming of the robots and the NC programming of the machine tools
3. A program generator for producing the application code for the robot and the machine tools
4. A simulation system for planning the workcell and performing its animation in real time

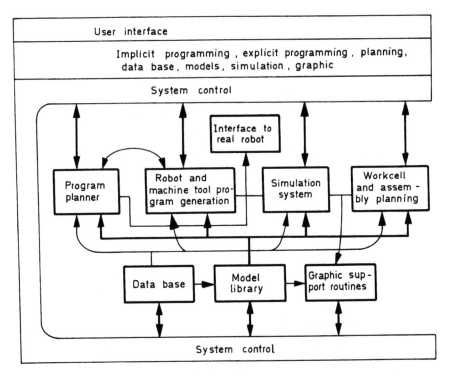

FIGURE 2.2 Concept of a planning and programming system of assembly cells.

5. A database containing all information and knowledge needed for planning
6. A library of models of robots, machine tools, and auxiliary equipment
7. A graphic module containing all graphic support routines
8. A system controller for planning and supervising the functions and cooperation of all components
9. An interface to the user
10. An interface to the robot workcell

Since it is the task of the assembly planner to plan the operation of the entire assembly cell, this software system will be of hierarchical structure. The individual modules of such a planning system may contain numerous submodules. With a well-designed system, the user should be given the capability of configuring a custom-tailored planner.

In the following discussion the overall operation of a hierarchical assembly planner and the task of the individual tiers will be explained with the help of Figure 2.3. The planning operation commences with the entry of an assembly order.

On the first level, the assembly cell to be used is selected. The planner investigates the description of the assembly and consults an expert system that has knowledge about the capabilities of the various assembly cells and their availability. If a cell has been found that can perform the assembly, the information about the cell is obtained from the shop floor data file. Now, the shop floor programming system begins to program the flow of the assembly operation through the plant. Thereafter, the program is executed with the help of the simulation system. A graphical output will help the system engineer to observe the assembly on a display terminal. If a solution has been found, the planning operation will enter the second tier. If there is no solution, the planning is done over again.

On the second tier the machines of the cell are selected. This means that there must be knowledge in the database on the performance of all the equipment available. The knowledge required to solve the problem will be extracted, and the applicable machines of the cell will be selected. Now the flow of the part through the cell is programmed. Again, with this information it is possible to simulate the asembly operation, however, in more detail. If the solution is acceptable, the planning commences on the next tier. If no solution is found, planning is repeated.

The third tier is concerned with the planning of the operation of the individual machine tools and robots used for the assembly. The expert system will provide information about the elementary functions each machine can perform. For example, for the robots it will describe possible trajectories, collision space, the coordination of the work of

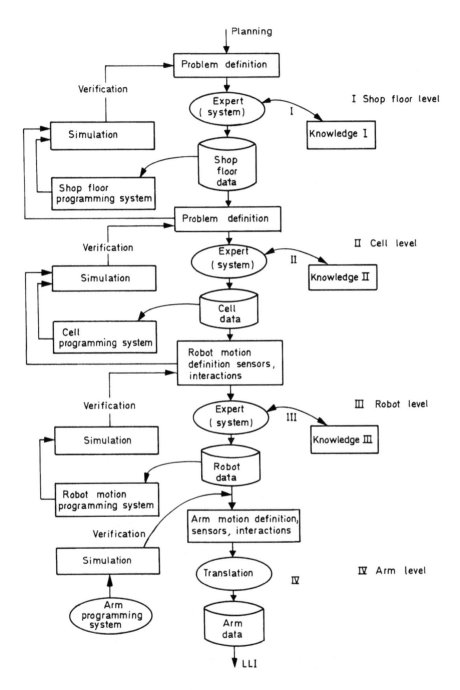

FIGURE 2.3 Hierarchical techniques for the programming of manufacturing processes.

two robot arms, the gripper functions, and so on. The detailed description of the robots selected is obtained from the robot database. Now the programming of the robots can be done, and the results of this phase can be simulated and observed.

On the fourth tier are generated the control commands of all the equipment used for assembly, including robots, sensors, vision systems, machine tools, and auxiliary equipment. The individual functions may also be observed on a graphical display, allowing the operator to make corrections if necessary.

The expert systems, knowledge bases, simulation facilities, and programming aids on every level will contain information and tools of various levels of abstraction, with the highest degree of abstraction on tier 1 and the lowest degree on tier 4.

A synopsis of the function of the components for the assembly planning is given below. The reader will find a more detailed discussion of the most important components in other chapters of the book.

Workcell and Assembly Planning

During workcell layout, the arrangement of the components is determined, and whereas in assembly planning, the sequence of the assembly operation is determined. The layout of the workcell and the observance of its operation are supported by numerical and graphical simulation tools. For the selection of the assembly equipment and for determining the sequence of the operation, planning software is used. A more recent tool employed for various phases of planning is the expert system.

Two assembly planning principles are used: the generative and the variant methods.

The generative planning method starts with the three-dimensional model of the workpiece. To determine the sequence of the assembly operation, the model is disassembled and a precedence or assembly graph is drawn. As the next step, various assembly strategies are considered, and the most promising one is determined. This phase is followed by the selection of the assembly tools from a pool of resources. All three phases mentioned so far are done on the basis of a virtual robot cell. Now the actual equipment can be specified from a catalog of robots, machine tools, conveyors, and grippers. With the knowledge of the robots and the work environment, the fine motion of assembling the product can commence. Thereafter planning the gripping operation is done by determining the best gripping position for the selected effector. The last phase is planning of a collision-free trajectory of the robot arms. This method must be applied for any product for which the assembly is planned the first time. Expert systems are a suitable tool to facilitate the generative planning. For more about assembly planning, see Chapter 7.

The variant planning method is based on the principle that the assembly is known from a similar product. The method provides a non-parameterized variant to the user. When a product is to be assembled, the user consults a catalog to find the variant of his assembly. Thereafter, he enters the values for the product parameters and the system selects the assembly sequences, motion, and equipment. This method works well only if the variants are similar in size. If the dimensional difference is too great, the assembly methods and equipment will be quite different.

Programming of Robots and Machine Tools

The program planner to support the implicit or explicit programming of the robots and NC programming of the machine tools consists of several modules. In general, program preparation for the machine tools is done with APT or an APT-like language, whose principles are well known, hence are not discussed further. For the interconnection of the machine tools with the robot, conveyors, and auxiliary equipment, a cell-level language is needed. This language may be based on the APT or a task-oriented programming principle. For programming of the robots, an implicit or explicit programming system may be used.

With implicit programming the user just specifies the task the robot is to perform (e.g., "take a wrench and fasten a bolt"). The instructions for implicit programming assume that the system knows all objects, their relations to one another, and all the motions to be performed to execute the command. Such a programming environment is supported by a world model and a knowledge-based expert system. The world model contains all information about the robot world, and the knowledge base is familiar with the robot capabilities. In addition, a metaknowledge module is needed, which can infer from the knowledge base the desired robot actions.

Presently, most robot programming systems are based on explicit languages, which means that every instruction must be stated explicitly. Figure 2.4 shows a schema of an explicit programming system for robots [2]. The most important components of such a system are:

1. An interface for the user to communicate with the system components
2. A system control unit that supervises the cooperation of all modules
3. The robot programming module, consisting of the compiler, robot frame handler, editor, and documentation module
4. The simulator, consisting of the simulator for the robots, machine tools, and so on, and a collision detector

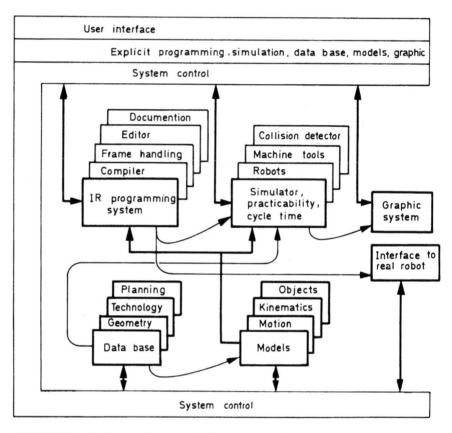

FIGURE 2.4 Schema of an explicit programming system for robots.

5. The models for the kinematics, motion, and objects
6. The database for all the data required for programming

The level of the programming language available for a robot depends on the configuration of the robot and the language designed for it. More detailed aspects of programming are discussed in Chapter 6.

The Robot Program Generator

   The programming system may use languages of different levels depending on the type and make of the robot. If various types of robot are used, compilation of the application program into an intermediate language should be considered. For this purpose, a program generator is provided. With this concept, the robot manufacturer must supply for the robot a controller interface that accepts the code of the intermediate language. Thus, the output of the compiler becomes independent of the robot being used. This output is interpretative code, which is entered into the control computer of the robot for execution.

The Simulation System

At least two distinctly different simulation packages are required for planning. With the numeric or discrete simulation, a mathematical model is built of the process. The model can be tested with various input parameters (number of products to be manufactured on an assembly line, product completion dates, parallel and serial operation of workstations, location of workstations, in-process inventory stations, etc.). The output of the model is information on the performance of an assembly line, possible bottlenecks, in-process inventory, and so on. The model allows investigation of the stochastic behavior of the assembly line, as well as the influence of disturbances. With the iterative use of the model, optimization of the assembly line may be possible. For the layout of the assembly system the output of the numeric simulation may be displayed on a 2D screen. Chapter 4 treats modeling in great detail.

Graphic simulation allows the animation on a display screen of the operation of the assembly cell and that of all its components. Here the user can observe the flow of a product through the assembly system, its various states of assembly, the function of the component of the assembly station, possible collision, and other phenomena.

A schema of a graphical simulation system is shown in Figure 2.5 [3]. Its main components are modules for robot modeling, world modeling, and motion generation. The two modeling modules are accessed by the user via the application program. From the user description, the components of the robots and those of the workcell are selected from the module libraries. With these components, the robots and the equipment of the assembly system are built up and prepared for the simulation. The application program for the performance of the work is entered into the motion generation module; this can be done either with a robot and machine tool programming language or a task description procedure. The outputs of the two modeling modules and that of the motion data module are entered into the emulator and displayed on the graphical screen.

Programming of a robot can also be done by generating the robot trajectory for a specific robot on a display and transferring the trajectory directly to the robot controller. Graphic simulation is discussed later in this chapter.

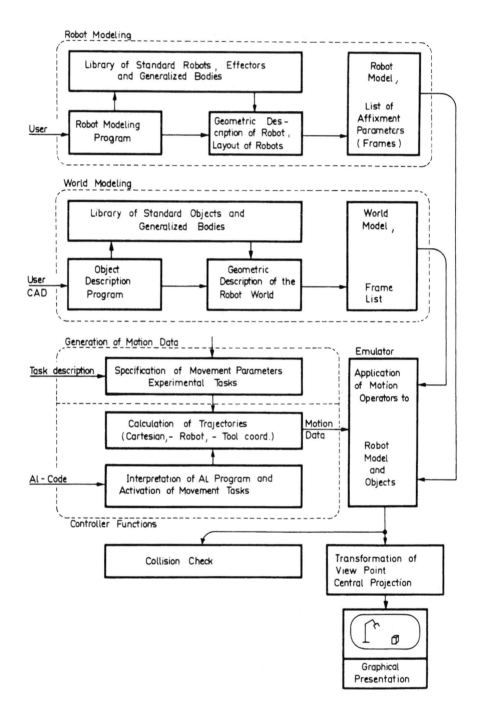

FIGURE 2.5 A graphical simulation system for robots.

The Database for a Workcell

The database is the central repository for all data needed for the planning activities. In it the robot and the other manufacturing equipment can be conceived as a nonnormalized relational data NF2 model. With this concept, hierarchical relationships among robot components are described by arbitrary deep nesting of relations among data objects. Based on the model, an object-oriented user interface can be provided for defining application-specific abstract data types (ADT) that consist of the internal representation method and the operations for manipulating the data objects. The operations are integrated into the data model to allow data manipulation of user defined datatypes and operations.

Figure 2.6 shows a schema of a database for manufacturing applications. Through a programming or a user interface, information about a robot can be accessed at varying levels of detail. For example, if a simulation system wants to place an object from location a to b, the instruction to search the object is given to the object cache module by the interface to the robot application program. It interprets the operation and searches for the object. When the object has been found, the operation is carried out and the object is placed at another location. Likewise, the user can generate or alter an object such as a robot via the interface of the robot user dialogue. This interface also permits the application of read and write operations. The robot ADT manager interprets the ADTs and prepares the manipulations in the database. The query translator transforms the access request to the code of the Heidelberg Databank Language (HDBL). The query processor executes the manipulation, and a search is done for the objects via the tuple; element, segment, and buffer manager. During this process a search path is generated from a complex tuple to subtuple right down to the individual elements. The index manager provides the pointers for this search. The reader will find more about databases in Chapter 9.

The Model Library

For planning the assembly system and assembly operations, the modeler must have access to a library containing the three-dimensional description of all objects located in the workcell, including the workpiece, tools, robots, and machine tools.

The description of the workpiece can be obtained from the CAD databank; thus, it is essential that the objects in the CAD system and those in the system model library be presented by the same modeling technique. With the three-dimensional presentation of the workpiece description, information about assembly sequences and interferences

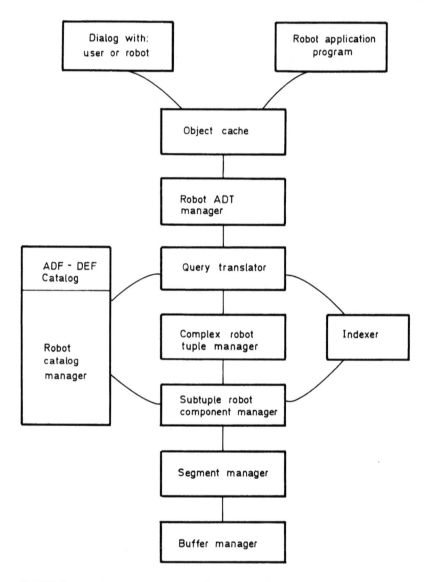

FIGURE 2.6 Access manager of the relational robot database.

can be derived. The workpiece model is also needed for the selection
of the gripper and the gripping surfaces.

The library of the robots, machine tools, fixtures, and tools will
contain a description of all equipment located in the assembly cell.

To permit presentation of machine tools and robots of different configurations, the objects are described in a hierarchical fashion with modules of varying detail. Figure 2.7 shows typical contents of the model library for the example of a robot. The model library is coded and presented in the relational database for the workcell.

Graphic Support Routines

The graphic support routines belong to the system software that is necessary to perform the graphic simulation. Typical routines are for the change of viewpoints and drawing of hidden lines and visible surfaces, as well as for coloring and shading. Usually, this software is provided by the vendor of the graphic system. For frequently used display objects, graphic menus are conceived and implemented by the designer of the software of the graphic simulation system.

2.4 SPECIAL PLANNING TOOLS

The preceding section introduced the planning of assembly cells. It was shown that for efficient planning, many tools must be available. Several of these tools are explained in more detail in dedicated chapters of this book. Very promising planning aids are graphical simulation and expert systems of various kinds.

A good simulation system should provide tools for the entire planning process including global and local planning. In addition, it should be possible to simulate all components of the assembly cell, such as the robots, machine tools, fixtures, sensors, and controllers. Since it is very difficult to visualize physical phenomena (temperature, pressure, etc.), the simulation of sensors imposes a special problem. Part of this section is devoted to the ROSI simulation system for visualizing assembly cells and to a simulator for a camera. Both systems were developed at the University of Karlsruhe.

Expert systems will play an ever-increasing role in planning cycles. Although this technology is very young, many promising expert systems have been built. The reader will find that much attention is devoted to expert systems and that this tool has found a place where general algorithm methods cannot be applied or where they lead to unwieldy software solutions. Two expert systems are introduced at the end of this chapter: one for selecting robot grippers and sensors for a specified assembly application and the other for the configuration of a flexible manufacturing cell. Both systems were developed at the University of Karlsruhe.

2.4.1 The ROSI Simulation System for Simulating Assembly Cells

Interactive planning and programming of a manufacturing process can be conveniently performed with CAD modeling software [4]. In addition

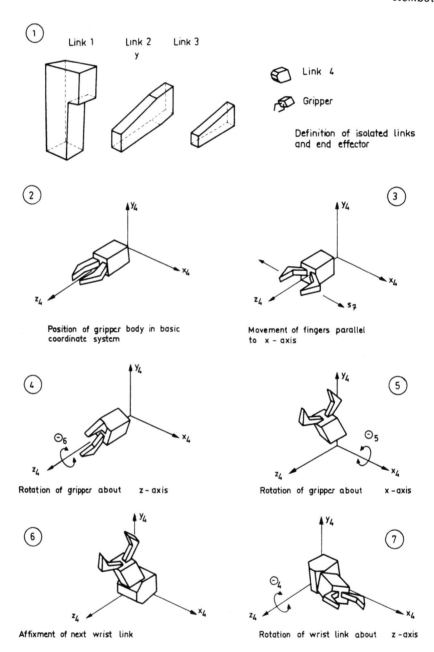

FIGURE 2.7 (a) Modeling and composition of basic arm elements.

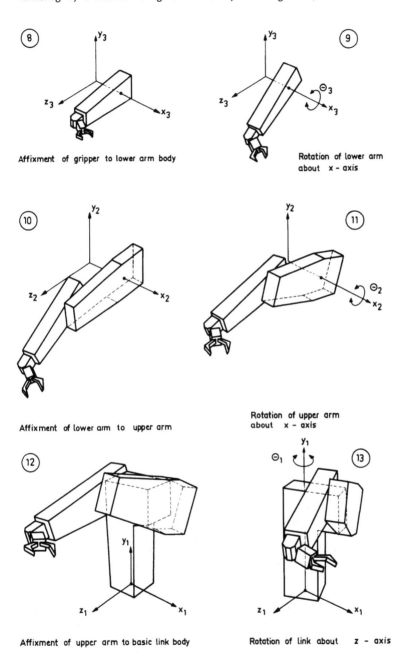

FIGURE 2.7 (b) Successive composition of the PUMA 600.

to CAD data, technical, procedural, and logical data must be processed to solve a manufacturing task. Interactive planning and programming with a graphic simulator supports the layout of a manufacturing cell and the design of the programs that control the manufacturing task. Robot motions and their interaction with the environment, material flow, sensor operation, etc., can be displayed on a screen and evaluated by the programmer. The operator can specify various tasks to identify restrictions, to observe the performance of the operation, and to detect such conflicts as the collision of a robot with objects located in its environment. With the simulator, an interactive planning and programming system can be designed to provide a linkage between CAD/CAM and robotic applications. A CAD modeler is used to define the geometry of the robots, machine tools, transport systems, tools, pieceparts, and the overall cell. In addition, the kinematic and dynamic behavior of the components of a cell, the sensors, and peripherals can be emulated. The result is a model of the cell which is called a virtual cell. The user can apply instructions to the virtual cell or to a virtual robot; they are executed by the simulator and visualized on the graphic screen. It is possible to animate the operation with considerable realism to support programming. Different levels of abstraction can be identified for interactive programming of a robot-based manufacturing cell.

In accordance with the control hierarchy of a manufacturing system, the simulation can be applied to the shop floor level, the cell level, the robot level, and the arm level (Fig. 2.3). The presentation of the robot is different on each level. The degree of detailing increases from the shop floor level toward the robot arm level. The simulation of a robot on the arm level necessitates the use of a microscopic virtual axis control, an end-effector control, and a basic sensor control. In the case of a cell simulation, the robot appears as a functional unit that cooperates with machine tools, peripherals, and conveyors.

To emulate a control function means to imitate in a virtual sense a control system in software. Emulation is an important tool for supporting the design of a manufacturing system [5]. The control system for a manufacturing cell can be emulated in the computer as a virtual process, and no detailed knowledge of the production hardware is needed. With the aid of the emulated control system, user software can be developed for the manufacturing equipment and tested in the different design phases of the installation.

Different concepts of graphical simulation systems for interactive manufacturing cells and robot programming systems are under development worldwide. Most of the known systems use CAD-based geometric modelers [6–9]. Commercially available CAD systems for robots like McAuto-Place (McDonnell-Douglas) [10] or CATIA-Robotic

(IBM/Dasault) [11] contain numerous library functions, such as collision detection, trajectory optimization, and evaluation of robot dynamics. User-friendly and problem-oriented programming can be realized using methods of artificial intelligence [12].

In the sections that follow, we outline the interactive planning and programming system known as ROSI (*RO*bot *SI*mulation), of the University of Karlsruhe (Figure 2.8) [13,14].

Structure of ROSI

The system offers to the user a dialogue to specify the desired interaction. It is possible to communicate with the system in the interactive, textual, and graphic mode. Via a command selection procedure the programmer can use:

The modeler
The emulator
The programming module
The simulator
The graphical animation module

The dialogue is divided into functional groups with specifically defined dialogue modules. All modules that can be accessed by dialogue are of similar design. The kernel of each module is a library of programs and programming elements, which provide the necessary methods for the execution of a module-specific task. In addition, each module has a method monitor that assembles and controls the execution of the tasks specified by dialogue. The method monitor activates the relevant program modules and configures from these the methods. A central data management system assures the separation of methods and data. The basic modules are described below.

The modeler describes the manufacturing cell and the piecepart, including geometric, physical, and functional properties [15]. The CAD system ROMULUS is used as a tool for the geometric presentation of the robots, tools, workpieces, and sensors. The CAD data structures generated can be extended with additional nongeometric data. Kinematic and dynamic attributes of the robots, conveyors, and fixtures, and functional properties of sensors and axis control loops, as well as the relations between objects within the cell, can be defined.

The emulator comprises the methods and functions necessary to plan and animate the robot within a cell and to define its operation. It is also used to build a virtual cell controller to be used for the simulation. Emulated functions are trajectory planning, coordinate transformation, and sensor operations, as well as control and decision

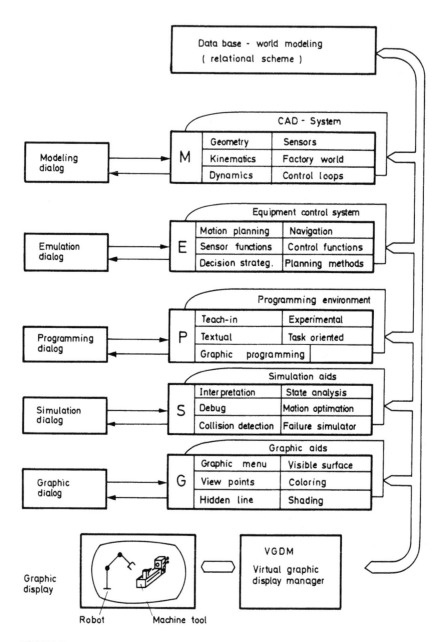

FIGURE 2.8 A comprehensive system simulation module for manufacturing.

functions. The virtual cell controller executes the programs defined by the programming module.

The programming module permits the programmer to develop programs by different modes. The dialogue allows interactive textual and/or graphical program development. The completed program can be down loaded to the robot via an interface with the robot controller. An experimental mode is used for the development and testing of individual trajectory sequences and basic operations. It is also possible to combine these functions in a robot program. The purpose of this mode is to provide a user-friendly, problem-oriented program interface.

The simulator is used for validation and verification of the generated program. The execution of the manufacturing program for the virtual cell is displayed on the graphic screen. A performance index (criterion) and a number of analysis functions (collision check, error recovery procedure, tolerance analysis, etc.) allow the programmer to analyze and validate the program. If the simulator detects errors, sources of errors, or critical states, these are displayed as graphic or alphanumeric information.

The graphic module facilitates the interpretation of the simulated operation. It offers methods for enhancing the information on the screen with the aid of graphic animation subroutines (hidden line, visible surface, coloring, shading, zooming, viewpoint transformation, etc.).

The interface between the simulator and the graphic system is realized via a virtual graphic display manager (VGDM), which makes the system independent of the type of the graphic workstation used. It offers the simulator pseudographic instructions for the construction and manipulation of graphic pictures. With these commands, each module can change the structure of the graphic image. Workstation-dependent software drivers adapt the VGDM pseudodisplay file to a workstation-specific control structure.

Each module has access to the central robot database [16]. Furthermore, the methods and data are clearly separated to assure consistency for the data to be processed. The data management is based on a relational data model, where objects of the same type are grouped together in object classes. Each object has a structure, which is described by attributes. Relations between the objects are indicated by attributes, the values of which present object names.

The Dialogue System

The dialogue component is the communication interface between the user and the system: it performs two essential functions:

It provides the user with textual and graphic communication via an
alphanumeric terminal and a graphic terminal.

It decodes the user input, transfers the function parameters to the
specific method monitor, and activates the monitor.

The syntax and the sequences of the commands are checked upon entry. For decoding of the textual input commands, the dialogue frame
provides a function-independent and table-driven decoding concept
[17]. This makes the execution of the command independent of the
decoding. The syntax of the commands and the parameter sequences
are described by different file entries. During the system start-up
phase, the files are read in and the tables are initialized. This concept allows an easy change and expansion of the system command language.

For graphical interaction, the system offers menu and window techniques and uses graphic input devices such as a dial, mouse, light pen,
and tablet. The following examples demonstrate the use of the graphic
input devices:

With the light pen, the frames of the robot motion and the choice of
objects to be grasped can be selected.

With the dials, the position of the angles of the arm joints of the robot
can be defined.

With the light pen or function keys, a specific type of motion can be
selected from a menu that contains predefined segments for the robot motion. The same technique allows the specification of predefined emulation macros for the execution of elementary or complex robot tasks.

In addition, a formatized dialogue supports the definition and manipulation of object data of the manufacturing cell.

All these communication methods support an application-oriented
user interface. Inexperienced users are instructed by helpful system
information. Table 2.1 shows the assignments of the dialogue mode,
the methods that can be chosen, and the corresponding technique employed.

Modeling System

The model definition system fulfills two basic purposes:

Generation of the geometric layout of the robot configuration of a manufacturing workcell and of an entire manufacturing plant

Definition and computation of physical and functional data, describing
the objects and their structured storage in the database

TABLE 2.1 Methods and Techniques for Five Dialogue Modes

Dialogue mode	Methods	Dialogue techniques
Modeling dialogue	CAD modeling	Command language
	Sensors, relations, functional and technological object data	Menu, formatized dialogue
	Dynamic	
	Kinematic	Formatized dialogue, graphical dialogue
Emulation dialogue	Emulation of functions, chaining of functions, definition of data flow between functions	Command language, graphical dialogue for the chaining of the functions
Programming dialogue	Textual program generation	Textual command language
	Graphical teach-in	Function keys, graphical dialogue (dials, light pen)
	Experimental programming	
	Task-oriented programming (workpiece oriented)	Menu, function keys, graphical dialogue
Simulation dialogue	Selection of analysis and evaluation operations	Command language, menu
	Debugging	Command language
	Error simulator	
Graphic dialogue	Selection of graphic algorithms and transformations	Command language, menu

Physical objects are the robots, part feeders (e.g., conveyors or material movement vehicles), machine tools, fixtures, and workpieces. All physical objects are described by geometric and technological data. In addition, the robot linkages require the definition of kinematic and dynamic functions.

Functional objects are sensors, control loops, and relations between objects of the workplace—for example, the relation between a gripper and a gripped workpiece--whereby sensors can be described physically by their geometry and measuring principle.

For all physical and functional objects, specific object classes are established which in their entity define the world model stored in the database. Figure 2.9 shows the specific structure of the object classes "manipulator," "axis," and "gripper," which allow one to model a robot [14]. Each object class is characterized by number and type of specific attributes. Object classes for feeding devices and workpieces are implemented in the same manner. The description of functional objects is based on object classes like "sensor," "control loop," and "relation." This concept allows easy addition of new object classes.

The CAD system ROMULUS is used for geometric modeling. ROMULUS is a solid modeler that allows modeling of exact geometries and produces a volume-oriented internal representation. In addition, ROMULUS allows the computation of such geometry-dependent properties as volume and center of mass, which are of interest for calculating the dynamics of robot motions. Each object class for the description of a physical object points via the attribute "geometry" to the internal representation stored in the file.

The parameters for the coordinate transformations, the robot joint coordinates to Cartesian coordinates, and the inverse transformation describe the robot kinematics. The system uses the Denavit—Hartenberg method, which solves the direct coordinate transformation for a large number of robot configurations. Figure 2.10 illustrates a method for the parameter definition using a graphic dialogue. The user can specify with the help of four knobs the four parameters, which represent the relations between the coordinate systems of two adjacent arm joints [14]. The inverse coordinate transformation from world to robot coordinates is described by a robot-dependent method. A pointer to this procedure in the robot library completes the kinematic data of the robot description. In addition, data derived from physical parameters are necessary for the dynamic description of the robot. These inertia tensors, friction, force, and torque parameters are added to the world model as parameters of physical objects.

Figure 2.11 shows the internal representation of a Puma 600 model. The model is generated by objects characterized by the structure of the different object classes. The objects are linked together by pointers and define the kinematic chain of the robot. Each object includes a pointer to the geometric description.

Modeling of functional objects is performed by the definition of typical functional parameters. The emulation of control loops requires information on the type of controller, the controller strategy, and controller parameters. These parameters can be entered by a formatized input. Relations between the objects and the world are described by objects of the object class "relation." Such an object is uniquely identified by a name and contains the names of related objects and describes the relation. A relation may describe a temporary connection

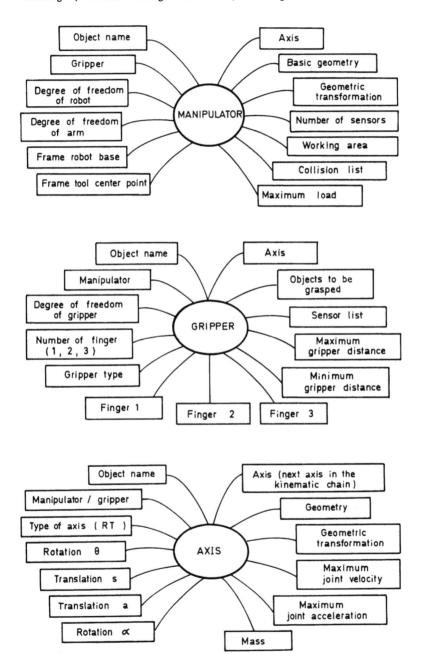

FIGURE 2.9 The structure of the object classes "manipulator," "gripper," "axis."

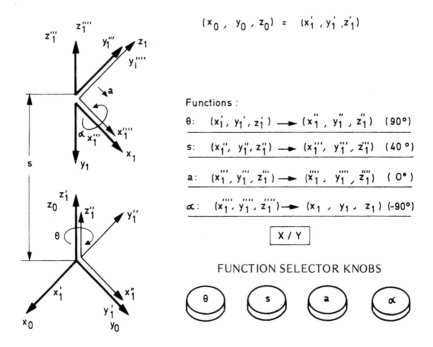

$$(x_0, y_0, z_0) = (x_1', y_1', z_1')$$

Functions :

θ: $(x_1', y_1', z_1') \longrightarrow (x_1'', y_1'', z_1'')$ (90°)

s: $(x_1'', y_1'', z_1'') \longrightarrow (x_1''', y_1''', z_1''')$ (40°)

a: $(x_1''', y_1''', z_1''') \longrightarrow (x_1'''', y_1'''', z_1'''')$ (0°)

α: $(x_1'''', y_1'''', z_1'''') \longrightarrow (x_1, y_1, z_1)$ (-90°)

$$\boxed{X / Y}$$

FUNCTION SELECTOR KNOBS

FIGURE 2.10 Graphical input of transformation parameters.

between two objects that are stacked together, or a permanent connection of objects bolted together. Furthermore, the definition of the position and orientation of an object in relation to the world coordinate system (the coordinate system of the manufacturing cell) is an essential function of the world modeling subsystem.

To integrate the sensor into a complex robot simulation system requires the functional modeling of the sensor. Therefore, the modeling system provides an object class "sensor," which describes the sensors (Fig. 2.12). This description is suited for geometric sensors. In this context geometric sensors are tactile and approach sensors. The sensor functions are modeled by the geometric operations in accordance with the geometric world model. Sensor attributes can be separated into two categories. One group describes the physical sensor properties (e.g., the way the sensor is mounted to the robot, the sensor type, the signal characteristic, and if necessary, the signal threshold). The important attributes for the sensor modeler are represented by the second category. They define the geometric state of the sensor--for example, the relationship between a straight line (representing the sensor signal) and a plane located in the world. The

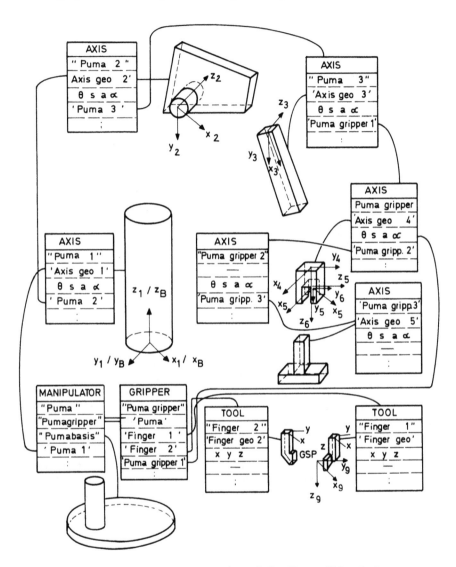

FIGURE 2.11 Internal representation of the Puma 600 robot.

plane may represent the face of a workpiece. The attribute operator describes the result of the modeling—for example, the distance between the position of the sensor and the face of the workpiece. Depending on the modeled sensor signal, a predefined execution sequence can define further actions. For this purpose the simulator

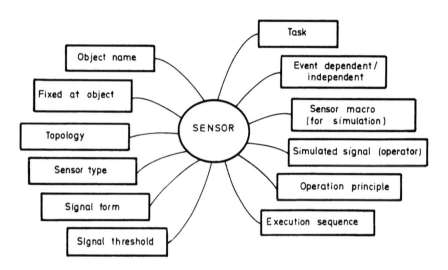

FIGURE 2.12 The structure of the object class "sensor."

uses a sensor macro that selects the corresponding methods and allows the visualization of the sensor signals on the graphic screen.

The modeling of nongeometric sensors needs other methods; for example, it could be accomplished by sensor signal tables, whereby the system determines the access to the sensor table. Figure 2.13 presents the concept for the sensor simulator. All objects are represented in the world. The horizontal line separates the robot and the sensors from the robot peripheral devices. The world without the robot is described by object-specific classes. The robot is modeled accordingly and equipped with sensors. The function of sensors is given by the interaction between the sensors and the environment. These specific functions must be modeled by software, which should assure a realistic sensor emulation.

Emulation of Functions

The control architecture and the behavior of a robot is characterized by a number of hierarchical functions and their integration into a system. The hierarchical control levels may influence one another, vertically and horizontally. The simulation of the behavior of the control system requires the emulation of those functions by software. Components of robot control systems to be emulated are:

Motion planning
Collision avoidance

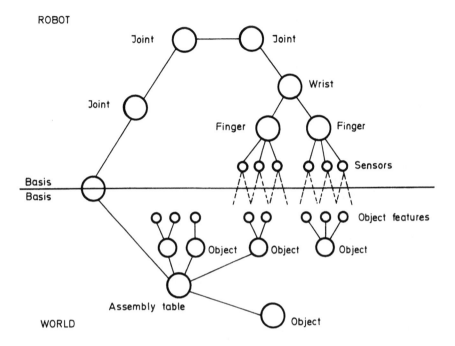

FIGURE 2.13 The concept of sensor emulation according to the world model.

Offset correction
Coordinate transformation (direct, inverse)
Adaptive sensor-guided motion planning
Servo control for motor drives
Gripper control
Sensor functions
Sensor data processing
Logical control functions
Navigation function

A robot model consisting of these emulated functions is called a virtual robot. The formal description of each emulated function is given by the input and output data, and the internal processing of the data, which represents the behavior of the function. During the emulation process the function of an entire assembly cell and of its components can be visualized. For displaying these functions, numerous calculations must be performed by the emulator (e.g., the robot trajectories, the motion of the robot effectors, and the functions of sensor operations must be determined). The supporting tools for these operations must be implemented in the emulator.

It is the task of motion planning to compute a continuous trajectory of the tool center point of a robot. More complex motion trajectories are partitioned into single segments to avoid expensive computation. The simulation system described in this text offers different trajectory types for the specification of the motion within one segment. In addition to Cartesian point-to-point moves, there are disposable straight lines, polynomials, and circle moves with user-defined velocities as initial and/or final condition of each segment. Figure 2.14 shows seven straight-line trajectories with different velocity profiles; Figure 2.15 shows a polynomial trajectory with its specific equation. Work on the integration of object-based trajectories generated from CAD models of workpieces is in progress. The correctness of the segment transitions is verified by the system.

The required Cartesian orientation of the end-effector during the motion is important. The orientation specified by three rotation angles (Euler or Cartesian) can be either maintained for the segment or continuously interpolated from the actual to the desired orientation by two simultaneous rotations. Thus the motion of a robot is specified by a motion command that consists of a unique identifier and the segment parameters, such as end point, end orientation, velocity, and trajectory type. The parameters can be related to different coordinate systems (robot base, effector, environment, workpiece). The necessary conversions are executed by the motion planning module based on the internal world model.

The computation of the discrete control values is performed in time cycles within a defined interpolation interval Δt, which corresponds to the global system time interval and depends on the desired performance of the trajectory. Synchronization operations with other

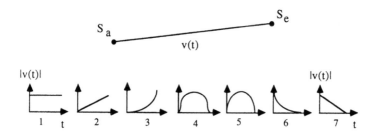

FIGURE 2.14 Velocity profile of seven straight line trajectory segments.

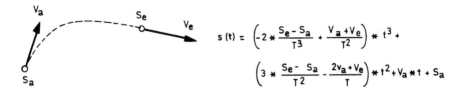

$$s(t) = \left(-2 * \frac{S_e - S_a}{T^3} + \frac{V_a + V_e}{T^2}\right) * t^3 +$$

$$\left(3 * \frac{S_e - S_a}{T^2} - \frac{2V_a + V_e}{T}\right) * t^2 + V_a * t + S_a$$

FIGURE 2.15 Equation of a polynomial trajectory segment.

active components of the workcell (i.e., sensor signals) must be taken into consideration. They are processed like constraints within the control of a real robot.

Cartesian trajectories must be transformed into joint angles for the determination of the control values of the robot. The required coordinate transformation routines depend on the kinematic structure of the robot. They are activated for each interpolation point. Ambiguities in the configuration of the robot must be resolved by additional control information.

The modularization of the system allows such simulation system components as motion execution, trajectory planning, and robot-dependent functions to be implemented independently. Thus the use of a new robot model requires only programming and insertion of new function blocks into the simulation software. It contains all routines for the robot-dependent transformations and work space boundaries to achieve a new application using this robot.

Commands for the manipulation of objects with simple (parallel) grippers contain the specification of the gripper distance, which will be transformed into an internal control value to adjust the gripper fingers.

For more complex sensor-guided grippers, the gripper emulator can by itself plan the grip operation, using the CAD model of the workpiece, and carry out necessary corrections. Therefore it is necessary to have a feedback message from the gripper control to the robot for repositioning of the arm that adjusts the position of the gripper relative to the workpiece. The gripper emulation module provides functions for communication between the robot and the gripper control. Three-dimensional offsets can be computed in end-effector frames, which are transformed into Cartesian robot reference frames during the interpolation of the end-effector operation. Thus the assembly of complex workpieces that are difficult to manipulate can also be studied in the simulation.

For the generation and testing of programs that are controlled by sensors, a sensor emulator should emulate the sensing and sensor processing tasks.

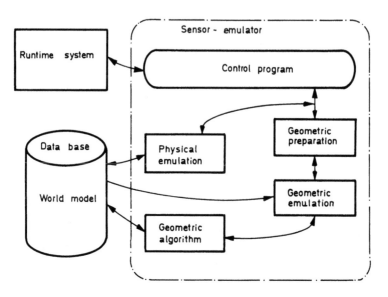

FIGURE 2.16 Block structure of the sensor emulator.

Figure 2.16 shows the structure of the sensor emulator module and the interfaces between the functional blocks inside this module. The emulation process for a specific sensor is activated by the runtime system, which calls the sensor emulator and passes the name of the sensor. As a result of the sensor emulation, the emulator transfers the computed sensor value to the specific program variable. The description of the sensor to be emulated is stored in the form of a sensor model in the database. In a preprocessing step the position of the sensor and its sensing direction are calculated, depending on the actual situation stored in the world model. Thereafter, the objects that lie in the measuring area of the sensor are detected. From these objects the faces identified from the sensor signal are indicated. The methodology of the geometrical calculation of the sensor signal in the emulation is derived from the type of the physical sensor. In the emulation process, basic geometrical algorithms are combined and used subsequently for the calculation of the emulated sensor value.

In the last step the influences of the real sensor in the real world and the effect on the sensor signal can be approximately emulated. The physical emulation unit evaluates the environment conditions described in the world model and accordingly modifies the sensor result calculated in the geometrical emulation unit.

To investigate the dynamic aspects of the kinematic chain, solutions of the motion equation must be available in software. Wittenburg and

Wolz [18] have written a program solving this problem. To control the motion equations, the controller properties must be emulated and described by the dynamic equations. The emulated sensor functions allow the emulation of sensor-guided adaptive control loops or complex multisensor statements. To complete the system, it is useful to add software routines performing logical decisions, learning, and optimization.

The textual and graphical dialogue functions allow the user to configure a specific emulator. The interface to the modeled objects of the cell is defined by the database management system. The emulation of the functions of cell objects as well as the modeling of the objects provide the user with a tool to do realistic, interactive off-line programming.

Programming

Communication between the application-oriented user and the system is performed through functions of the programming module. In this module high level language programs or graphically defined instructional sequences are converted into the system internal command language. The runtime system interprets the semantics of each individual command and activates the specific emulators. The desired operation, which may be a robot move, gripper, peripheral, or synchronization command, is executed under constraint checks. In case of an error, the execution is interrupted with a message about the kind of error. A successful operation is terminated by an actualization of the state variables of the world model. After this stage, the system is ready to interpret the next command.

The ROSI system allows the user to choose from one of the following programming techniques:

Textual programming
Interactive graphical programming
Experimental programming
Debugging

Depending on their complexity, the programming methods are divided into categories that correspond to the architecture of the robot emulator. The following categories of programming levels are implemented:

Complex task description (implicit statements)
Complex function description (explicit statements)
Motion description
Motion condition description
Trajectory statements

Via these programming levels, the corresponding emulated robot function and other manufacturing machine functions are programmed and activated. For planning of the motions emulated, interpolation routines and coordinate transformations are activated. Gripper operations are accomplished by emulated gripper functions. For the simulation of tracking operations (relative motions), the kinematic behavior of conveyors, bowl feeders, and moving objects must be emulated. Emulated sensor functions for adaptive control strategies are obtained by the insertion of sensor and branching statements into the program. The trajectory statements allow the definition of frames and reference points.

A very practical method for the generation of programs is direct interactive programming. In this method a single statement is generated through a dialogue with the system. The user selects the desired command with options from a menu. The system requests all necessary parameters and checks the syntax of these values. Different graphical input capabilities support the specification of parameters (e.g., objects that should be grasped can be selected with a light pen; joint angles of a robot can be controlled by knobs; predefined frame variables can be selected by function keys). A robot teach-box processor, controlled with a keyboard, allows "manual movements" of robot joints, the recording of frame variables, and the specification of trajectory segments. Figure 2.17 shows the layout of this teach-box on the screen.

A completely specified command can be executed by the runtime system. It controls the execution carried out by various emulators. The programmer can check the results of the command and modify immediately the parameters, if necessary. The graphical-supported communications combined with syntax and semantic checks represent a comfortable interface for interactive generation of programs.

High level programming languages provided with user-defined data types, control structures, subprogram techniques, and other features are advantageous for complex applications. SRL (Structured Robot Language) is a Pascal-like language for programming of robots developed at the University of Karlsruhe [19]. SRL-written programs are translated by the SRL compiler to the standardized IRDATA code [20]. SRL programs may be executed by a translation into IRDATA statements of the internal command language of the simulation system. Along with the interactive method, the user is provided with a second alternative for program verification. Thus two powerful programming methods (combination of interactive teaching of frames and action sequences in SRL) are at the disposal of the user.

Experimental programming is done in the visualized world of the robot. All user instructions are executed immediately. Thus, the programmer can examine results in real time on the graphic screen. The interactive generation, modification, and testing of programs

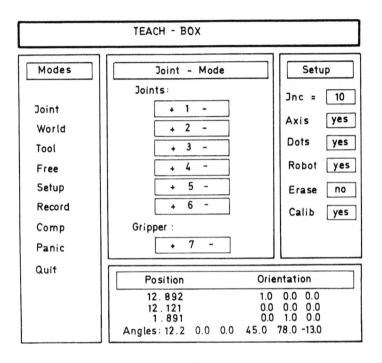

FIGURE 2.17 Layout of the teach-box on the screen.

for the workcell or of program parts are the essential advantages of experimental programming. This method allows programming in different coordinate systems (robot, world, tool coordinates). Also, complex statements like INSERT, GRASP, or MOVE AROUND OBSTACLE can be used. Alphanumeric information about joint angles, position, and orientation of cell objects, and so on, support the programmer to evaluate the performance of the program.

Debugging of SRL or IRDATA programs is another essential programming tool that is provided. Debug functions allow the modification of motion parameters, program variables, or the change of the program sequence.

Simulation as a Tool for Off-Line Programming

The simulation can assist the following essential programming functions [21]:

1. Support of the programmer in the testing and validation of the written workcell programs; graphical visualization and an intelligent knowledge-based analysis of the program performance are offered.

2. Evaluation and optimization methods allow the programmer to op-
    timize syntactically and semantically correct programs.
3. Integration of advanced planning methods for implicit programming
    (e.g., planning of coarse and fine motion for assembly, or testing
    for collision avoidance); the evaluation of the simulation assists
    the discovery of problems and bottlenecks.
4. To achieve the goals, the simulation module consists of a series of
    elementary functions. The user can compose the interpreter for
    simulation purposes from these elementary functions. The simu-
    lation performs the selected operations and records the operating
    results. Examples are records of joint variables, position and or-
    ientation of the gripper, and supervision of the collision space.

If the graphical simulation of a sensor is not possible, the user may
obtain information about the sensor, the time of measurement, and the
value of the signal via alphanumeric output. In addition, the simulation module allows real time operation, time compression, or time expansion. A debugging mode supports the step-by-step emulation,
definition of breakpoints, or definition of traces. Trace variables
are established by the programmer. For breakpoints, rule-based
knowledge to solve conflicts and to perform optimization can be made
available (e.g., knowledge about optimal gripping points or assembly
tasks).

For the realization of functions that require knowledge, artificial
intelligence methods must be implemented using the world model and
a knowledge base. The knowledge base provides rules for error diagnostics. For example, if an insert operation is defined, the restrictions about the operation must be described in the knowledge base.
During the conflict analysis the actual world model data are examined
to determine whether they fulfill the formulated rules in the knowledge
base. In case of inconsistencies, a diagnostic routine can be started.
Thereby, further rules and file entries of preceding emulated program
statements are evaluated. The analysis function identifies the error,
giving the user support for a correction.

For the testing of complex programs of sensor-guided robots, an
error simulator is made available which allows one to change the state
of the world with a defined error. Thus the reliability of an adaptive
sensor-guided operation and a sensor-integrated program can be examined during the execution, and corrections can be made. The incorporation of an error diagnostics system containing methods of automatic error corrections with the help of the simulation is a future objective.

Graphical Visualization

The graphic system displays the robot and workcell and the motions
and actions defined by the program. A detailed three-dimensional

graphical display allows the analysis of the state of critical situations and the detection of errors. The graphic module offers to the user functions that improve the graphical presentation. Hidden line and visible surface algorithms facilitate the presentation of static pictures. Hidden line and shading algorithms for dynamic pictures can also help to visualize the third dimension and can detect possible collisions in the manufacturing cell. However, they are time-consuming when used for realtime simulation. Variable definitions of viewpoints and the zoom function allow a detailed evaluation of the assembly operations.

The virtual graphic display manager mentioned earlier defines the interface between the simulation software and the graphic system. The simulation system calls on functions of the VGDM via pseudo-graphic statements to dynamically generate and manipulate the graphic data structure. Dynamic manipulations are changes of the picture structure done during the emulation and simulation. Examples are:

The display of robot joint motions, as well as transport motions (joint angles are computed by the robot emulator)
The display or the deletion of graphic and alphanumeric information to show emulation results
The visualization of emulated sensor operations
Highlighting of detected simulation errors (e.g., blinking or coloring of colliding objects)
Perspective display of objects
Zoom function
Change of viewing angles
Processing of inputs from the graphic peripherals (dials, light pen operation, etc.)
Retrieval of transformed data from the graphic data structure

The VGDM decodes the pseudographic statements and generates a pseudodisplay file. Graphic driver routines produce the data structure and control code of the connected graphic system. The driver routines know the local transformation facilities (perspective, zoom, etc.). Thus the VGDM is freed from these operations. For linking another graphic station to the simulation system, only the implementation of the new driver software is necessary. Complex changes of the simulation software need not be performed.

System Support Functions

To use the simulation system, several system support functions are necessary [21]. They include:

A runtime system to control program execution
A global sequence control to furnish timing functions to the simulation system

An internal command language to supply an interface between the programming modules and the runtime system

Runtime System to Control Program Execution: The runtime system controls the interpretive execution of one or more programs [21]. The number of active programs is related to the number of mechanical devices in the assembly cell. The simulation system executes the defined programs on a virtual cell. The manipulators, positioning devices, transport systems, sensors, and workpieces are stored in form of models in a database. For the robots and all other devices of the cell, functionality is modeled by software routines called emulator functions. Depending on the program instruction, the runtime system decides which emulator is to be activated next. To arrange the functionality of each device of the cell, the runtime system has access to the following emulator types:

Trajectory Planning: planning of robot-independent Cartesian trajectories defined in different reference coordinate systems of the cell.

Interpolation and Robot Coordinate Transformation: interpolation of the specified trajectory and execution of robot-dependent inverse coordinate transformation for each interpolation sample point. The joint variables are transferred to the runtime system.

End-effector Control: calculation of finger motions for the execution of grasp strategies. The results are control variables for the fingers.

Sensors: calculation of sensor values equivalent to those produced by real sensor systems. Continuous and discrete operating sensors can be emulated. The result is an equivalent value to the real sensor signal.

Peripherals: calculation of motions on transport systems. The result gives the position and orientation of workpieces on the transport system and its velocity.

Each emulator executes its specific functions. The results are transferred to the runtime system and utilized for the direct execution of the program on the virtual workcell that is visualized by the graphic system. Simultaneously a trace is generated, which protocols the internal states of the cell and each event. The trace functions can be activated with the help of the simulation dialogue before the start of the program; the trace generation is processed in parallel to the emulator functions.

Global Sequence Control: A global timer unit is used to control the execution of the programs by the runtime system [21]. The time step of the system timer is shorter than the time of the interpolation cycles for the trajectory calculation and the duration of sensor measurements.

The execution time for program statements in the real system is transformed into the system clock. It can be derived from trajectory planning, or it can be part of the world model (i.e., duration of measurement as an attribute of the sensor model). This allows the transformation of the real system execution cycles into the global system timer. Thus the runtime system can guarantee execution of the commands within the time parameters specified by the program. This is how the timing of interacting synchronized assembly operations can be tested through simulation.

The synchronization of robots and peripherals may follow implicitly through definite sequential action planning of individual robots and peripherals; relating to signals of the execution of the synchronized sequential definite assembly operations. The defined binary signals are related to the program and characterized by an identifier consisting of program number and signal number. Two operations on these signals allow to describe the coordination of different processes.

SET   < Signal ID >
            This operation sets a signal to the value ON. The SET
            operation allows the definition of an event.

RECEIVE   < Signal ID >
            This operation allows reading of defined signals. The
            control of the program execution is dependent on the
            value of the signals. After being read, the signal is
            reset to OFF.

These operations, applied to the defined signals, allow the definition of dependencies between manipulation programs. For example, a conveyor sets a signal to ON if a workpiece passes a light beam. The robot program delays its grasp operation until the signal read by RECEIVE obtains the value ON by the SET operation. The activated program is then synchronized with the conveyor and can grasp the workpiece from the transport system. Thus the execution of quasi-parallel, explicitly synchronized programs can be prepared in the off-line programming mode. The runtime system must interpret the synchronization signals and activate appropriate operations.

Structure of the Internal Command Language: The internal command language is the interface between the programming module and the runtime system [21]. The individual commands of the internal language are executed interpretively. The format of the commands is defined by an identifier and a list of operands. The semantic of the commands defines the following types of operations (Tables 2.2 and 2.3):

TABLE 2.2 Commands for the Programming of Robot Movements

Command	Semantics	Parameter
	Move to frame	
BFR	in robot coordinates	Frame, MTYPE, time
BFU	in world coordinates	Frame, MTYPE, time
BFE	in end-effector coordinates	Frame, MTYPE, time
BFW	in workpiece coordinates	Frame, MTYPE, time, name (workpiece)
BFK	along circular trajectory	Frame, ZWPKT, time
	Move to position	
BPR	in robot coordinates	Frame, MTYPE, time
BPU	in world coordinates	Frame, MTYPE, time
BPE	in end-effector coordinates	Frame, MTYPE, time
BPW	in workpiece coordinates	Frame, MTYPE, time, name (workpiece)
BDO	Move according to defined moving object	Name (moving object)
GIN	Joint movement, individual	Joint number, value, time
	Move	
BZA	to approach frame of object	Name (object), time
BZO	to object, linear movement	Name (object), time
BAO	via approach point to object	Name (object), time
BAB	back via departure point of object	Name (object), time
OPE	Open gripper	Value, time
OPB	Open gripper during next movement	Value
CLO	Close gripper	Value, time
CLP	Close gripper during next movement	Value
EON	Activate end-effector	
EOF	Deactivate end-effector	
NST	Initial position	Time
HAL	Timing stop	Time

1. *Move Statements*: definition of trajectory segments; the segment end points may be defined in terms of frames or positions that can be expressed in relation to different coordinate systems (world, robot base, end-effector, workpiece).

TABLE 2.3 Commands for the Generation and Specification of Robot Programs

Command	Semantics	Parameter
	Protocol of	
AEF	end-effector in robot coordinates	
AGW	joint angles	
AOF	object frame	Name (object)
PMX	maximum values (velocity accelerator)	
	Definition of	
DFR	frame variable	Name (frame variable), frame
DPO	position variable	Name (position variable), frame
DBO	moving object	Moving parameter
DBT	Define move of transport system	Name (transport system), velocity
BOT	Fix object to transport system	Name (object), name (transport system), time, frame
RST	Robot movement synchronously with transport system	Frame, time, name (transport system)
RAS	Robot asynchronous movement	Frame, time
WAI	Wait	Time
DHK	Define action sequence	Name (action sequence), sensor, name (program)
AHK	Execute action sequence	Name (action sequence)
GEN	Generate robot/transport program	Name (program), name (robot)
END	End of program	
RUN	Execute robot program	Name (program)
SYN	Synchronization of multiple programs	
PRO	Robot program	Name (program)
TPB	Transport program	Name (program)
ESY	End synchronization	

2. *Grasp Statements*: open and close of the gripper fingers.
3. *Sensor Statements*: activation of dedicated sensors and emulation of a discrete or continuous sensor measuring procedure.

4. *Synchronization Statements*: operations on signals and WAIT commands with defined absolute or relative time; definition of synchronized handling sequences.
5. *Program Flow Control Statements*: definition of program flow control CONDITION, START, STOP, and JUMP instructions.

Example: Programming of an Assembly Task for a Two-Arm Robot Configuration

The assembly task "pendulum assembly" of the Cranfield Institute of Technology is a benchmark for testing robot programming languages and manipulation devices (Fig. 2.18) [22]. It consists of typical task sequences and problems related to assembly procedures as follows:

Simultaneous handling of large and small workpieces
Picking, placing, and inserting of pins
Fixing of parts with hard constraints

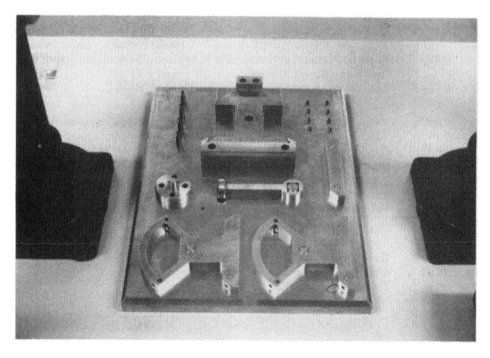

FIGURE 2.18 Pendulum assembly task of the Cranfield Institute of Technology.

Performance of different directions of compliant operations nonparallel
   to the axis
Handling of unstable position of parts during assembly

The pendulum device to be assembled consists of two sideplates, five
joining parts, and the pendulum itself. Fixing of the parts is performed through eight locking pins.

   Layout of the Assembly Cell: The design of the assembly workcell
is effected by various functions of the modeling module. The cell consists of two Puma 260 robots with different grippers and an assembly
table with the assembly base plate and the individual workpieces. Figure 2.19 shows two Puma 260 robots performing an assembly, and Figure 2.20 depicts a graphical scene with a Puma 260 robot and the
grasped workpiece "lever."
   The geometry of the objects is defined by the CAD modeler ROM-
ULUS. Figure 2.21 shows the geometry and additional description
attributes like approach/deproach frames and grip area of the benchmark workpiece "sideplate." The kinematic structure of the robot
models with the axis and the attached gripper are built up using the
robot modeling functions. The existing workpieces can be taken from
the object library and arranged in the workspace. All objects are
placed into the graphical world model by functions of the modeling
module. Graphical input functions support this process. For example,
the exact position of an object can be determined with six control dials
(x-, y-, z-translational and rotational values), which specify a frame
relation between the object and world reference coordinate system.
Relative frames between workpieces, workstations, and robots are
computed internally in the system. The result of the world modeling phase is the complete layout of the workcell with all functional
relations for the execution of the assembly.

   Interactive Generation of the Assembly Program: The program for
the assembly task can be developed interactively step by step. Constraints concerning the sequence of operations defined in the precedence graph for the assembly task must be considered. The approach
frame and the grasp position can be computed from the internally stored
position of the part relative to the robot and the additional geometric
attributes. The trajectory may be specified by the user through the
choice of an appropriate trajectory type. To arrive at a desired joining direction, the final orientation of the gripper must be specified.
Thus the whole assembly is generated by alternating motion and object handling commands. The explicit specification of position parameters is supported by the system to facilitate programming. The commands and parameter lists are interpreted and executed by the graphical simulation system. This enables the programmer to check the results and consequently to verify and optimize the program.

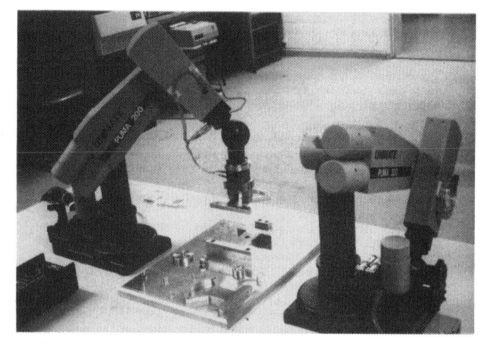

FIGURE 2.19 Configuration of the two Puma 260 robots and the assembly plate.

Assembly Programming with SRL: The high level robot programming language SRL has the following basic features (the reader will find more about SRL in Chapter 6):

Pascal conception of structured data
Integration of geometric data types for the robot
Arithmetic and geometric operations
Block structure
Program flow control
Trajectory specification commands
Command to handle sensor data
Effector/move statements

The programmer using SRL for a robot application has such advantages of a modern structure language as clarity and ease of modification and

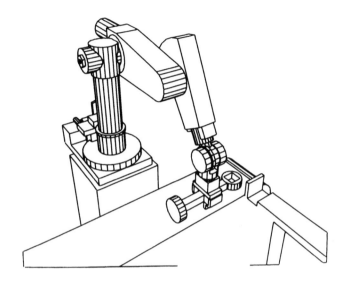

FIGURE 2.20 Graphical scene with Puma 260 robot and the grasped workpiece "lever."

maintenance. A subprogram concerning the assembly of four parts of the benchmark demonstrates the use of SRL:

```
Procedure fetch_spacer (pos:frame);
via approach_pos;frame; (* approach-frame *)
begin_procedure
        approach_pos:=spacer_approach *)
        SMOVE puma TO pos
            WITH velocity = 25
            WITH dep = spacer_depart_vector
            WITH VIAFRAMES (approach_pos)
end_procedure;
```

The procedure is called four times with new fetch_ positions for each of the spacers. In comparison to SRL, a program segment in the internal command language has the following form:

```
MOA{<Move to Object via Approachframe>}spacer1,traj_type_9,t = 2s}
GRO {<Grip Object>}
MPE {<Move to Position in Eff.Coord.System>}x=0,y=0,z=5,t = 1s}
MFW {<Move to Frame in Workpiece Coord.S.>}frame_C,sideplate1,
        traj_type_11,t = 4s}
```

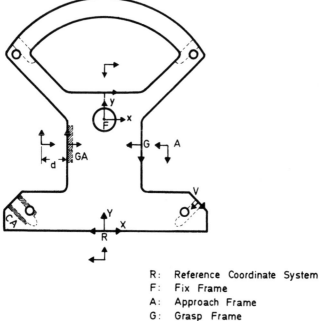

R: Reference Coordinate System
F: Fix Frame
A: Approach Frame
G: Grasp Frame
V: Joining Frame
GA: Grasp Area
d: Distance
CA: Contact Area

FIGURE 2.21 Geometry, frame and area description of the benchmark workpiece "sideplate."

Graphical Simulation: The runtime system executes the generated assembly program stepwise and displays the results on the graphic screen. The 3D geometrical models of all objects are transformed by viewing algorithms that allow the visualizing of each detail of the assembly cell. Single motions can be interrupted by a "stop" function key for the visual detection of collision. Details like the join operation of the locking pins can be checked using the zoom function. The gripping of parts can be considered from the view of the gripper by displacing the viewpoint into the wrist. Figure 2.22 shows a scene with a grasp operation from different viewpoints. The attainability of all assembly parts can be tested by graphical simulation. In case of error, the layout of the robot cell must be altered by functions of the modeling module so that each assembly part can be grasped by at least one robot.

During simulation, the individual robot and gripper emulators, which virtually imitate the functional behavior of the real devices, check all operations and physical constraints such as maximum acceleration, continuous velocity paths. Thus the program is executed realistically.

Interface to the Real System: The graphical simulation system allows the generation of almost completely verified programs. To run these programs on real robot systems, an interface is required between the simulation system and the target system. Computed output values must be adapted to the internal formats of the individual robot devices. This entails adaption of data formats, software drivers for communication between computer and robot joint controller, and interfaces for the integration of real sensors.

The Robotics Research Group of the University of Karlsruhe uses a low level interface between computer and Puma robot, which allows the transfer of joint angles computed by the simulation system to the robot. The integration of this interface into the simulation system demonstrated the advantages of graphical simulation to support the development of user programs for new applications.

2.4.2 Simulation of Cameras in Robot Applications

Programming of third-generation robot systems is very difficult, due to the necessity of programming sensor feedback data. A visualization of the sensor view of a scene, especially the view of a camera, can help the human planner to improve the programming and verification of a robot action plan.

Several simulation systems include sensor data in their simulation process, and they use very simple sensors [23–25]. Mostly, pure geometric modeling of distance and contact sensor is done because it is not easy to simulate physical effects with a model.

The work described here is an extension of the robot simulation system ROSI [26]. Various submodules for different sensors were developed, which are included in ROSI. This section describes the submodule to simulate a camera and the methods used.

In the first step the internal geometric model is presented [27,28]. The data structure chosen for the internal model is a boundary representation. To calculate the reflections of the object surfaces, the model of Torrance and Sparrow is used [29]. The modeling of the light source and camera with the setting of their parameters is described.

The algorithms used can be divided into two layers:

Geometric calculation of the visible surfaces within the scene, hidden surface projection, and shadowing
Physical interpretation of the calculated data

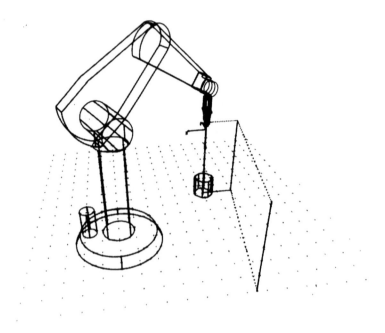

(a)

FIGURE 2.22 Grasp operation, regarded in the graphical simulation from different viewpoints.

The result of the physical part of the simulation is a gray level/color picture, based on the radiation flux for every pixel.

Modeling

The performance of a sensor simulation system depends primarily on the available information of the manufacturing cell and the sensors used. The data are stored in an internal computer model. The closer the data correspond to reality, the more accurate the results of the simulation will be. The database of the system is structured into the following four modules to increase adaptability:

A module for the geometric description of the cell objects
A module for the optical properties of the surfaces
A module for the physical and geometrical description of the light
  sources
A module for the physical and technological properties of the sensors

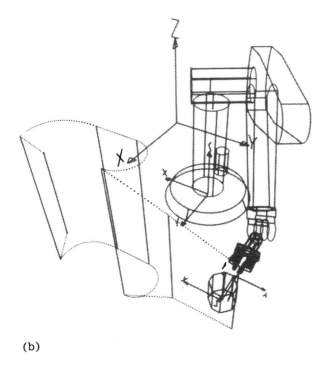

(b)

Object data are stored in external files and read into the internal models at runtime when needed. The internal models consist of lists of records. The records and lists are linked by pointers to allow fast access and to facilitate model extension (e.g., increasing the number of light sources at runtime). Figure 2.23 is an overview of the data structures of the model and the links between them. A detailed description of the different models follows.

The Geometric Model: The geometric description of an object is given by a boundary representation scheme based on CAD-generated data. The data are read at runtime into the internal geometric model by a transformation module. This allows the system to use a variety of CAD modules. For the use of a CAD module, only its transformer has to be specified.

The geometric model has a hierarchical structure consisting of three layers. The layers comprise nodes for the description of surfaces enclosing an object, contours bordering a surface or holes of a surface, and vertices defining a contour. To simplify the complex geometrical calculations, the method is presently restricted to planar, convex or concave polygons with holes.

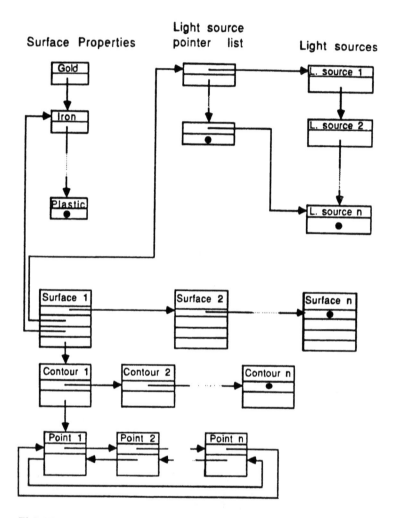

FIGURE 2.23 Data structure of objects for simulating a camera system.

A surface is represented by its normal vector and a point on the surface. A pointer to a list of contour nodes defines the border of the surface. Additional information includes the bounding box of the surface to speed up geometrical calculations, a pointer to a record in the list of optical properties to define the reflectance of the surface, and finally a pointer to a list of light sources, which is used for the calculation of the radiation flux. For every surface, a list of light

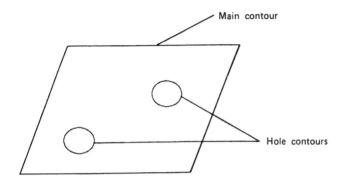

FIGURE 2.24 Exterior and interior bounderies describing a surface.

sources illuminating the surface is calculated during the shadowing process and accessed via this pointer.

There are two types of countour: a main countour describing the exterior boundaries of a surface and a hole contour defining the border of a hole cut into the surface (Fig. 2.24). Every surface is defined by one main contour and possibly several hole contours linked to a list of contour nodes. Every contour node has a pointer to a ring list of vertices defining the contour. The vertices of a main contour are linked in clockwise order and holes counterclockwise.

A vertex node contains the coordinates of the vertex in the three-dimensional space and the coordinates of the projection of the vertex onto the two-dimensional image plane. This projection needs to be calculated before activating the hidden surface algorithm.

Model of Optical Surface Properties: The algorithm for calculating the radiation flux that falls into the camera is based on the bidirectional reflection theory published by Torrance and Sparrow [29]. The parameters to describe the optical properties of a surface type are deduced from their theory. For every surface type that can be found in the simulated scene, a record containing the specific optical properties of that type must be entered into the list of surfaces before the radiation flux falling into the camera is calculated.

To understand the meaning of the parameters describing the optical properties, a brief summary of the theory is given. Torrance and Sparrow describe the reflectance of light from rough surfaces. Such a surface is assumed to be composed of a collection of mirrorlike microfacets, with random orientation. The reflectance of such a surface consists of a diffuse and a specular component. The diffuse component is generated by the multiple reflections from the various facets and from internal scattering. The specular component describes light directly reflected into the camera. This component can be computed by Fresnel's reflection law, the distribution function of the direction of the microfacets, and a geometrical attenuation factor that takes the masking and shadowing of adjacent facets into account, Equation (2.1) and Figure 2.25.

$$F = \frac{1}{2}\left(\frac{\sin^2(\phi - \psi)}{\sin^2(\phi + \psi)} + \frac{\tan^2(\phi - \psi)}{\tan^2(\phi + \psi)}\right) \tag{2.1}$$

F: Fresnel reflection
ϕ: angle of incidence
ψ: angle of refraction

The diffuse component is computed by Lambert's law:

$$I = L \cdot A \cdot \cos \psi \tag{2.2}$$

$$L = \frac{1}{\pi} \cdot r_d \cdot E \tag{2.3}$$

I = flux density of the reflectance surface
L = surface radiance
A = area of the surface
r_d = coefficient of the diffuse reflectance
E = incident flux density

In general the optical properties of a surface are the diffuse reflectance given by a constant, the index of refraction of the surface material, the ratio of diffuse to specular reflectance, and a factor describing the *rms slope* of the distribution function of the refractance. The diffuse reflectance and the index of refraction are wavelength-dependent properties and are represented in the computer by three-dimensional vectors. In this simulation the visible range of the electromagnetic spectrum (380–780 nm) was divided in three subranges, each 133 nm wide. Thus, every component of the three-dimensional vector gives the value of the spectral property according to one of the subranges.

Light Source Model: In the simulation system the light sources are approximated by a planar surface, evenly emitting light over the hemisphere of the surface. In physics such a radiator is called a Lambertian radiator. It is defined by the center point of the emitting surface, the normal vector, and the area of the surface and the

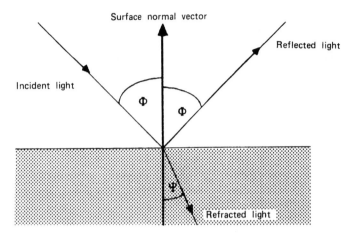

FIGURE 2.25 The effect of light falling on a surface.

spectral radiation density; the latter parameter is represented in the system described by a three-dimensional vector. The model is not restricted with respect to the number of light sources. The data of a light source are held in a record and the records are stored in a linked list.

In the modeling phase of the system three types of light sources can be chosen: incandescent light, neon light, and special light source. The spectral density of an incandescent light bulb is calculated by using Planck's law (Eq. 2.4) and taking into account that only 9% of the electrical power is transformed into visible radiation. Neon light is approximated to be wavelength independent; that is, it emits pure white light. It is assumed that one-quarter of the electric power is transformed into visible light. While these two types are characterized by their electric power, the spectral radiation density of special light sources must be given directly.

$$L_{\lambda_s} = \frac{2 \cdot h \cdot c^2}{\lambda^5} \cdot \frac{1}{e^{\frac{h c}{k \lambda T}}} \qquad (2.4)$$

L_{λ_s}: light source radiance
h: Planck's constant
c: speed of the light
λ: wavelength of the light
k: Boltzmann's constant
T: temperature

The Camera Model: A camera model consists of the lens system and the electronic hardware. The lens system projects the incoming radiation flux onto the sensor plane, and the electronic hardware converts the radiation flux to an electronic signal corresponding to the gray or color levels of the image generated. The camera model consists of three parts:

Parameters describing the lens system
Parameters describing the electronic hardware
Geometrical information about the position and the tilt of the camera

The lens system is approximated by a thin lens having an aperture diaphragm. It is characterized by the focal length of the lens, the aperture setting, and the distance setting of the lens.

The electronic hardware is described by the number of rows and columns of pixels, the dimensions of the rows and columns, and the sensitivity of the sensor.

A coordinate system is defined with its origin lying in the middle of the lens, whereby the z-axis is identical with the centerline of the lens, the x-axis is parallel to the pixel rows, and the y-axis is parallel to the pixel columns. The position and tilt of the camera are given by a 4×4 transformation matrix, which defines the position of the camera coordinate system with respect to the world coordinate system.

System Overview

According to the division of the camera model into the lens system and the electronic hardware, the succeeding stages of the simulation consist of the calculation of the radiation flux projected onto the individual pixels of the sensor plane and the generation of the corresponding gray/color levels of the pixels.

The algorithms used are divided into two parts. The first part comprises pure geometrical operations defined by the geometrical model and is thus applicable to sensors of all kinds integrated into the simulation system. The algorithms implemented range from calculating simple distance between two points to the hidden surface projection of a whole scene. The second part consists of sensor-specific physical calculations, which give the physical interpretation of the previously calculated geometrical results, for example. An overview of the stages of the camera simulation and the output of the stages is given in Figure 2.26; a more detailed description follows.

Recognition of Possible Objects: The simulation first determines those objects that can be seen by the camera, resulting in a rough reduction of the data [30]. Input of the algorithm is the measuring geometry of the sensor used. This geometry can be a point, straight

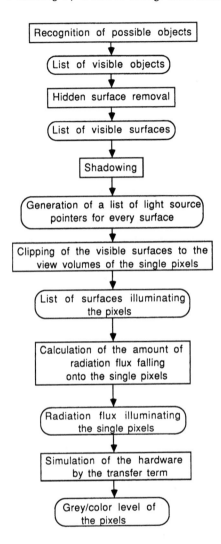

FIGURE 2.26 The various stages of simulating the operation of a camera system.

line, cylinder, cone, or sphere. The measuring geometry of a camera is approximated by a cone with its vertex coinciding with the origin of the camera coordinate system and the camera axis coinciding with the centerline of the cone. The angle of the cone 2ω is defined by the size of the sensor plane h and the distance of the sensor to the lens g:

$$2\omega = 2 \tan^{-1}\left(\frac{h}{2g}\right) \tag{2.5}$$

where

ω = angle of the view cone
h = size of the sensor plane
g = distance between lens and sensor

For recognition, the objects of a scene are approximated by spheres. The center of a sphere is assumed to be identical with the center of gravity of an object. The radius of a sphere is the distance from the center of gravity of the object to the vertex that is furthest away from it.

At first, all objects lying in front of the measuring plane are determined. Second, the objects whose spheres lie totally or partially inside the measuring geometry are entered into the list of recognized objects. They are sorted by the distance of the center of gravity of the sphere to the measuring plane.

Hidden Surface Projection: To determine the visible surfaces of the scene, it first is necessary to reduce the geometrical data in such a way that the input of the algorithm contains only surfaces oriented to the camera and belonging to the calculated list of recognized objects. The surfaces are copied from the original geometrical model into a second model, leaving the original data untouched for later calculations (e.g., shadow casting or error correction). In addition to the copying process, the vertex coordinates are transformed from the world coordinate system into the camera coordinate system. Next, the two-dimensional coordinates of the vertices are calculated by projecting the vertices onto a plane parallel to the sensor plane. This plane, called *image plane*, lies corresponding to the distance setting of the camera lens system in the object space and defines the plane of maximum sharpness of the camera (Figs. 2.27 and 2.28).

If there are several objects in the scene obstructing each other from view, hidden surface removal is necessary. This is achieved by projecting the three-dimensional surfaces onto the two-dimensional image plane, and clipping and removing the hidden surfaces so that the visible surfaces do not overlap in this plane. Then the resulting visible surfaces are projected back into the three-dimensional object space (Fig. 2.29).

The hidden surface algorithm used, which has been published in the literature [31], is based on a two-dimensional polygon clipper that is able to clip a concave polygon with holes to the borders of a concave polygon with holes. The polygon clipper is also needed for

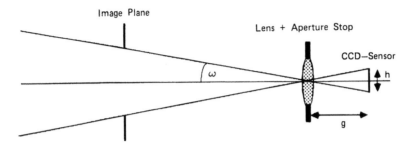

FIGURE 2.27 View volume of the CCD sensor as a cone.

shadow casting and clipping the surfaces to the view volumes of the individual pixels.

Input data of the polygon clipper are two polygons given in three-dimensional space and their two-dimensional projection onto the image plane. One of the polygons is called clip polygon; the other, the subject polygon, is clipped to the borders of the clip polygon. The output of the algorithm consists of two lists of polygons. The inside

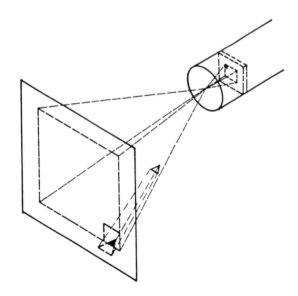

FIGURE 2.28 Clipping of an object projection with the view volume projection of a pixel.

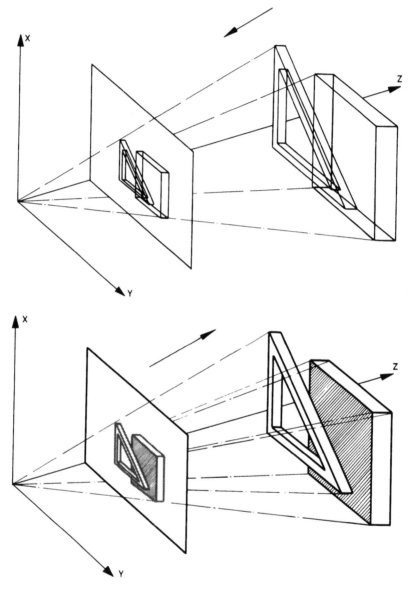

HIDDEN SURFACE PROJECTION

FIGURE 2.29 Removal of hidden lines.

list contains the parts of the subject polygon lying inside the area of
the clip polygon representing the obstructed parts of the view of the
object. The outside list contains the visible parts of the subject polygon lying outside the clip polygon (Fig. 2.30). The clip polygon is
not altered by the clipping process. The structure of the output data
of the polygon clipper is in the same form as that of the input data.
This allows the successive use of the algorithm for solving various
problems like hidden surface removal, shadow casting, and clipping
the visible surfaces to the view volume of a single pixel. The idea
of the algorithm is that any new borders created for the subject polygon are visible of the borders of the clip polygon. New polygons are
created by partial transversals of both input polygons.

The hidden surface removal is done in four steps. First, a rough
depth sort of the surfaces is performed. The next step is a two-dimensional polygon area sort to locate the area of the currently most
forward polygon. This is done by the polygon clipper. All polygons
are clipped to the borders of the polygon that is currently the most
forward. After that, the polygons of the inside list lying behind the
clip polygon are removed (Fig. 2.30). If some polygons remain in
the inside list, the last two steps are repeated with the now most forward polygon as clip polygon and the previous clip polygon as subject polygon. After that, the clip polygon is entered into the list of
visible polygons. The last three steps are repeated with the resulting outside list as input until the list is exhausted. Output of the
hidden surface removal is the list of visible surfaces given in the form
of the geometrical model.

Shadowing: To determine the light sources illuminating the individual visible surfaces, at present only a rudimentary approach is

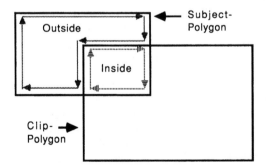

FIGURE 2.30 Clipping operation performed on the image.

implemented. Shadow casting will be added later. For every surface, a list of pointers to the nodes of the light source list is calculated. The light sources that are oriented in the direction of a surface are entered into its light source list.

For shadow casting we use an algorithm similar to the hidden surface removal with the light source as the viewpoint. The surfaces contained in the original geometrical model are taken as clip polygons and the visible surfaces oriented to the light source become the subject polygons. For every light source the visible surfaces are clipped to the other surfaces of the scene with the polygon clipper. After clipping, a pointer to the data of the light source is added to the light source list of the surfaces contained in the outside list of the polygon clipper. Surfaces contained in the inside list are not illuminated by the light source; they give the shadow casting.

Clipping to the View Volume of a Pixel: There are $n \times m$ pixels on the sensor surface of the camera. For each pixel, the view of the scene must be determined. This is done by a third clipping process in which the visible surfaces are clipped to the view volumes of the individual sensor pixels. For every pixel, all visible surfaces are clipped to the view volume of the pixel to get the proper boundaries of the surfaces illuminating the individual pixel. At present, the view volume of a pixel is approximated by a four-sided pyramid with the vertex lying in the origin of the camera coordinate system and axis along the principal ray from the center of the lens to the center of the image of the pixel in the object space. The boundary of a pixel is projected onto the image plane in object space and gives the clip polygon. All visible surfaces are clipped to that polygon, giving the list of inside polygons illuminating the pixel (Fig. 2.28).

This approximation of the view volume does not simulate the effect of blurred pictures, caused by a false distance setting of the lens system. To simulate such pictures, the view volume of a pixel is approximated by two four-sided pyramids (Fig. 2.31). The clipping process also must be changed, resulting in an increased program runtime. Every visible surface lying in the view of the camera must be projected on the image plane. This projection will be clipped by the view volume of the pixel, which is also projected on the image plane.

Calculation of the center points and area of the polygons of the inside list prepares us for the final calculations of the geometrical layer. With the calculation of the solid angles of the light sources, seen from the viewpoint of the surfaces, and the solid angles of the surfaces, seen from the viewpoint of the camera, all geometrical data needed for the calculation of the radiation flux of a pixel are provided.

Physical Calculation of the Reflectance: After the geometrical data have been prepared, the physical interpretation of the data represent

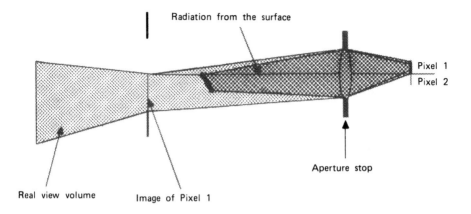

FIGURE 2.31 View volume of one pixel.

the amount of radiation flux falling onto a pixel. The physical calculations are based on the geometrical laws of optics and simulate the way light is emitted from the light sources, reflected from the surfaces into the camera lens system, and projected onto the CCD sensor pixel.

For every surface illuminating an individual pixel, the total radiation flux falling from the light sources onto the surface is calculated by Lambert's law. In addition to the direct illumination, the surface will be illuminated by ambient light. Its power per unit area E_a is estimated to be one-quarter of the radiation flux of all light sources emitted to a unit area at the center of the scene.

The radiation flux density of the light reflected from the surface in the direction of the camera is calculated by an algorithm based on the theory of Torrance and Sparrow [29], which describes the bidirectional reflectance of rough surfaces. The approach used resembles the algorithm published by Cook and Torrance [32]. It differs from Sparrow's method by the use of a slope distribution function D for a facet that describes the distribution of the normal vectors of the surface microfacets. Cook uses a distribution term with excessive computation time. It was published by Beckmann and Spizzichino [33]. The method used in the paper is a distribution function proposed by Trowbridge and Reitz, which is more easy to calculate but has the disadvantage of an arbitrary constant describing the rms slope [34]. The radiation flux density I_s of the light reflected from a surface in the direction of the camera is calculated by:

$$I_s = \frac{1}{\pi}\left(E_a r_d + \sum_1 I_1 (N\ L_1) d\Omega_1 (s\ r_s + d\ r_d)\right) \qquad (2.6a)$$

$$r_s = F \frac{D}{N L_1} \frac{G}{N V}$$ (2.6b)

where

r_s = specular reflectance of the surface
r_d = diffuse reflectance of the surface
s, d = fraction of specular s or diffuse d reflectance of a surface
I_1 = radiation flux density of the light source 1
$d\Omega_1$ = solid angle of the light source 1 when seen from the surface viewpoint
N = normal vector of the unit surface
L_1 = unit vector from the center of the surface to the light source 1
V = unit vector from the center of the surface to the center of the camera lens
F = Fresnel term; describes the reflectance of a perfectly smooth surface
D = facet slope distribution function; describes the distribution of the surface microfacets' normal vectors
G = geometrical attenuation factor; describes shadowing and masking of adjacent microfacets

All light reflected from the surface into the camera lens system is projected onto the pixel to whose view volume the surface is already clipped. Thus, the radiation flux F reflected from the surface onto that pixel is calculated by:

$$\Phi = I_s \left(-V_z\right) d\Omega_s \, B$$ (2.7)

V_z: z component of the vector V
$d\Omega_s$: solid angle of the surface, seen from the viewpoint of the camera
B : area of the lens system aperture—depends on the aperture setting of the camera

A pixel is illuminated by the parts of the surface projections that are visible to the pixel (Figs. 2.28 and 2.31). The three-dimensional object is projected on the two-dimensional image plane. This projection will be visible to some pixels and invisible to others. For this reason the projection will be partially clipped.

The camera hardware simulated consists of a CCD sensor with an 8-bit A/D converter. It is modeled with the aid of the minimum irradiance valued $B_{3(min)}$, a maximum irradiance valued B_{max}, and the relative spectral sensitivity $s_{rel}(l)$ of a pixel. To simulate the electronic hardware, a transfer term is applied to get the gray/color level Y of a pixel corresponding to the incoming radiation flux F. If the amount of flux ranges between B_{min} and B_{max}, the corresponding output is calculated by:

$$Y = ROUND\left(a \ s_{rel}(\lambda) \frac{\Phi}{A} + b\right) \tag{2.8}$$

$$a = \frac{255}{B_{max} - B_{min}} \tag{2.9}$$

$$b = -a \ B_{min} \tag{2.10}$$

where A is the area of a pixel.

Work in progress includes an additional module for simulating noise and blooming. Noise will be simulated by adding to the 8-bit integer a random value that will depend on the environment temperature. Blooming will be simulated by shifting the values of adjacent pixels. The physical calculation gives three color values for every pixel. Simulation of gray level cameras requires the calculation of the average of the color levels that give the gray level. The calculated gray/color levels are stored in an external file for later calculations like image processing or just viewing the simulated scene image on a monitor.

Results

The sensor simulation system was implemented in Pascal running on a Micro VAX 2 and a VAX 8700 under the VMS operating system. Execution code takes about 200 kb; program runtime varies from several minutes (<50 visible surfaces, resolution <200 × 200 pixels) to several hours (>100 visible surfaces, high resolution), depending on camera resolution and scene complexity. A special transmission program is implemented to display the pictures generated on an Evans & Sutherland PS 300 color raster display. The geometrical data of a scene are generated by the CAD module ROMULUS [28], running on the PS 300, which itself is linked to the Micro VAX. The geometric data of the scene are stored in XMT files, which are read and transformed into the internal geometrical model of the simulation at runtime. Additional data files contain information on the camera parameters, light sources, and a list of optical properties of metals and plastics. They can be edited, changed, and stored during program runtime via a comfortable user interface.

To show the capability of the camera simulation system, a scene was modeled containing workpieces of the "European benchmark" lying on

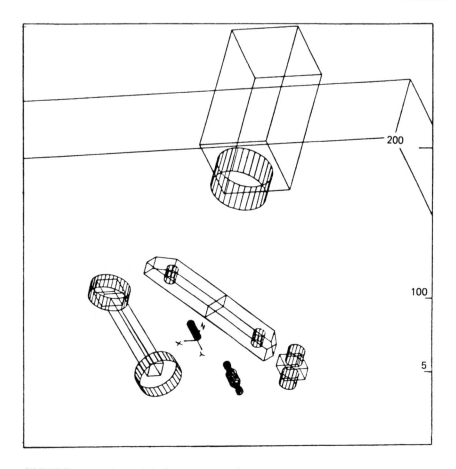

FIGURE 2.32 A modeled camera and scene.

a plastic desktop. Restricted by planar polygons, the round parts of
the objects were approximated by prisms. Figure 2.32 gives an overview of the modeled scene including the camera. The image was generated by ROMULUS and displayed on the PS 300 vector graphic monitor.

   The surface properties of the objects are diffuse (Fig. 2.33). Different metals were modeled (gold, silver, copper, iron, and aluminum)
in the other pictures. The camera is located 30 cm above the desk
in Figures 2.33 and 2.34, and 15 cm in Figure 2.35, always looking
straight down. The sensor size is 1 × 1 cm with a resolution of 200
× 200 pixels (Figs. 2.33 and 2.35) and 400 × 400 pixels in Figure 2.34.

FIGURE 2.33 Modeled workpieces of an industrial scene; all workpieces modeled with diffuse reflection. Resolution 200 × 200 pixels,
15 cm focal length.

Figures 2.33 and 2.34 were taken with a simulated lens system of 15
cm focal length and Figure 2.35 with a 7 cm focal length. The aperture setting of all pictures is 5.6. The scene is illuminated by a 100
W neon light located 100 cm above the objects. To display the different resolutions on the same monitor scale, one calculated pixed of a
200 × 200 picture is projected on four display pixels. The program
runtime is about 40 minutes for Figures 2.33 and 2.35 and 3 hours for
Figure 2.34. An example for a complete assembly situation is shown
in Figure 2.36.

2.4.3 An Expert System for Selecting Effectors and Sensors for Robots

The demand for high flexibility, especially with assembly tasks for
small and medium production runs, led to an increased number of applications of industrial robots. The gripper and sensors of a robot
determine to a large extent the flexibility of the manufacturing cell;
they are a robot's interface to the environment. The robot gripper

FIGURE 2.34 Modeled workpieces of an industrial scene; the reflection parameters for each workpiece are different.   Resolution 400 ×
400 pixels, 15 cm focal length.

determines which class of objects can be grasped and which tasks can
be performed.

Sensors are employed to handle uncertainties of the manufacturing
environment and to guarantee the operation of a computer-integrated
manufacturing installation.   The growing importance of sensors and
grippers has led to an increase in the market of grippers, sensors,
and gripper--sensor systems. Today, the complexity, cost, and prac-
ticability of an automation project may depend to a great extent on
how far the project engineer can be assisted in the selection of suitable grippers and sensors.

This section describes an expert system for the selection and configuration of gripper--sensor systems and discusses experience gained
in the design and impmentation process of this tool. The work was
based on a product line of 30 mechanical effectors and 8 sensors,
which may be configured in more than 200 combinations. The problem was to find the best gripper--sensor system for a given automation task. Usually, the selection is carried out by a highly qualified
engineer whose decisions are based on an unstructured collection of
information consisting of a project description, technical drawings,

FIGURE 2.35 Modeled workpieces of an industrial scene; the reflection parameters for each workpiece are different. Resolution 200 × 200 pixels, 7 cm focal length.

sample parts, and, in some cases, impressions gained from an inspection of the manufacturing cell. The human expert derives the required performance for the gripper—sensor system and uses data sheets and manufacturer's information to choose the components to be used. This procedure should be automated for two reasons. First, the effort and consequently the cost for gathering all the relevant information and for securing their evaluation by an expert is often out of proportion to the price of the gripper—sensor system. Second, experts in the field of assembly automation are very few and therefore often not available.

The goal of the above-mentioned endeavor was the design of an automated component selection system for robot effectors with which a less qualified user can describe an assembly task via a formatized input form. This information is entered into an expert system, and it proposes a suitable gripper—sensor configuration.

In the first phase of the project, literature on sensor selection and the proposed solutions were reviewed [35–39]. The literature was helpful mainly in identifying characteristic problems. Attempts to use already developed planning strategies for the problem had to be abandoned because too much information was required to describe the

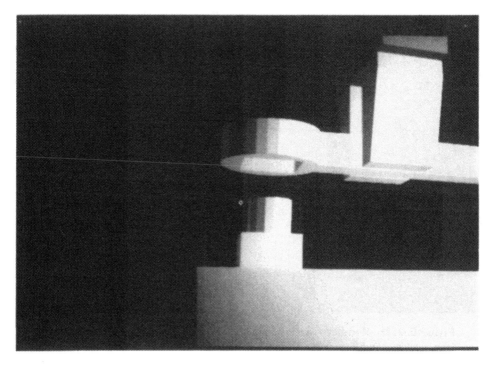

FIGURE 2.36 Modeling of an assembly situation; most parts modeled with diffuse reflection. Resolution 400 × 400 pixels, 15 cm focal length.

problem, which made the solution impractical. In addition, the information was often not available at the time of selection.

Good results were obtained only through close interaction with an expert. The input form served as the tool for analyzing the problem. This form had to be redesigned time and again until the expert could find a solution for the problem without further information. The information contained in the input form is shown in Table 2.4. From the way experts solved a number of practical cases, a set of approximately 100 rules was extracted.

The final design of the input form, the data sheet from the manufacturer, and the extracted expert knowledge were the basis of the first implementaion, called ROKON1. The prototype was implemented in the Franz Lisp version of the rule-based programming language OPS5 running on a VAX 750 under Unix [40].

In the following sections, the gripper and sensor databases are described and the architecture of ROKON1 is explained. A more detailed description of the implementation can be found in Reference 40.

The Gripper and Sensor Databases

A functional model of a sensor gripper is shown in Figure 2.37. It defines the general structure of the product gripper.

With a gripper, the energy is transferred to the object by the jaws through the gripping force. The reacting forces and moments are transmitted to the robot via the gripper housing. A sensory system in connection with a controller can guard and control all the actions of a gripper.

A total of about 3000 parameters describing the above-named components of a manufacturer's product line were placed in the database. The data structure was designed so that the database can be expanded and maintained easily. For these reasons it appeared appropriate to build a framelike structure. OPS5 was used to construct the expert system; however, only a very primitive database can be built with it. A compound data type is used, called an element class (e.g., the drive system is such a data type: Fig. 2.38). It consists of a list of a combined attributes and values (e.g., the drive system is an attribute and "pneumatic" a value). Each element class contains one vector attribute to which several values may be assigned (e.g., various external dimensions). Another restriction is that an element class may have 127 attributes at the most. Therefore, the database must consist of several element classes to describe completely a gripper or sensor.

In ROKON1 there are two attributes common to all element classes. They are used for interconnecting the entries stored in the database and for describing the current state of the database with respect to the element classes.

According to the structure of the effector, the gripper database consists of a module for the drive system and one for the transmission, together with a special data object combination that contains information on how the different drive and transmission systems can be combined. The features of available jaws are included in the transmission module because of the large number of interdependencies between the two elements.

Figure 2.38 shows a schematic diagram of the three top levels of the gripper database. The gripper and sensor databases contain 4530 attributes in 469 element classes. (The number 4530 is obtained from the sum of 3000 product parameters plus 1530 parameters for the database management.)

The expert system has a static component and a dynamic component.

Architecture of the Expert System ROKON1

The Static Part of the System: The expert system ROKON1 is composed of the following three main software components (Fig. 2.39).

TABLE 2.4 Questionnaire for Specifying Sensors and Grippers: Contents of Input Form

1. *General information*
   Project name
   Short project description
   Number of applications
   Type of industry
   Manufacturing operation
     manufacturing
     assembly
     quality control

2. *Robot-specific data*
   Number
   Type
   Payload
   Type of controller
   Computer interface

3. *Part-specific data*
   Part description
   Weight of part
   Part surface sensitive to
     impact
     pressure
     scratching

4. *Handling data*
   Temperature of part
   Temperature of manufacturing environment

5. *Gripping data*
   Specification number of manufacturing process
   Specification number of handling operation
   Specification number of gripping point
   Type of gripping operation
   Surface clearance condition
   Necessary clearance between open fingers
   Gripping distance of closed fingers
   Material of gripping surface
   Gripping surface sensitive to?

6. *Work-specific data*
   Work cycle number

(continued)

TABLE 2.4   (Cont.)

6.   [*Work-specific data*]
     Robot number
     Workpiece number
     Workpiece container
     Positioning accuracy of workpiece container
         coordinates x+/-Δx, y+/-Δy, z+/-Δz
     Number of picking locations
     Are all picking locations accessible to robot?
     Minimum distance between neighboring workpieces
     Accuracy of picking position:
         x+/-Δx, y+/-Δy, z+/-Δz
     Clearance of handling object:
         x+/-Δx, y+/-Δy
     Type of operation
         mating
         inserting
         storing
         processing
         others
     Auxiliary equipment
         bin
         box
         pallet
         tool
     Additional forces to be considered

7.   *Supervision of work cycle*
     Can workpiece be recognized?
     Can workpiece be identified?
     Characteristic features of workpiece
     Distance of workpiece surface from supporting surface
     Can workpiece container be recognized?
     Identification number of work cycle
     Any special observance needs
     Force in z direction
         measuring range
         measuring accuracy
         torque about x-, y-, and z-axes
         forces in x-, y-, and z-directions
     Minimum gripping force
     Preventive measures to protect gripper

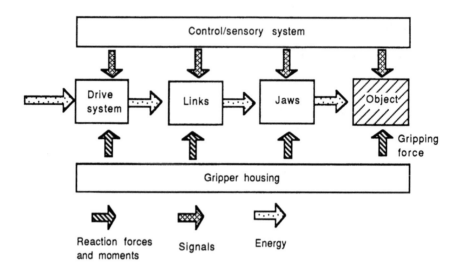

FIGURE 2.37 Functional model of a sensor gripper.

Definition module
Rule base
Database

The definition module is a component specific to OPS5. Before
the first application of OPS5, all element classes of the language must
be defined. Thus, attributes are assigned to certain element classes
and transformed into an internal cell representation with the Rete—
Match—Algorithm. For dividing the attributes into element classes, a
compromise must be found between readability and administrative ex-
penses. Dividing the attributes into a high number of element classes
keeps the program easy to read and facilitates its maintenance. In the
case of ROKON1, 98 element classes were defined for 546 attributes,
which means an average of approximately 6 attributes per element
class.

The rule base is divided into nine rule blocks, as shown in Figure
2.29. The rule block context has a special function; it is structured,
similar to that of a real expert. It decides which blocks should be executed. The knowledge necessary to solve the problem is contained in
eight other blocks, each of which runs in its own context. A jump to
or return from a rule block is realized by a change of context. For
this reason, the object context was defined containing the following
four attributes: block membership, processing type, processing detail, and status.

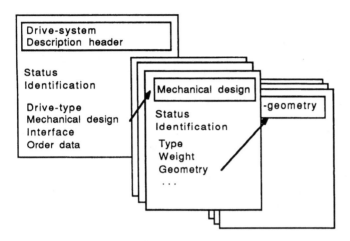

FIGURE 2.38 The gripper database.

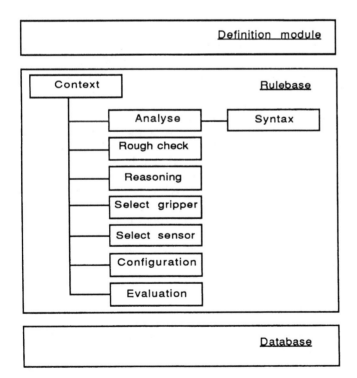

FIGURE 2.39 Rule block of ROKON1.

Block membership represents block affiliation at the top level and expresses membership to one of the rule blocks. Processing type describes the kind of processing (e.g., read), which can be described in more detail with the attribute processing detail. The status attribute is used to add some degree of relativity to certain goals with the help of such qualifiers as active, passive, or initial.

The database that contains all problem-relevant parameters consists of a fixed and a dynamic part. The fixed part contains the description of all available products, whereas the variable part of the database contains temporary data that are created, changed, or deleted in the process of solving the problem. The current state of the varying part of the database represents the state of the problem-solving process.

Strictly speaking, a program is processed by the OPS5-specific inference mechanism called Recognize—Act—Cycle, which is not described in this chapter.

The Dynamic Part of the System: This section describes the execution part of the expert system, its use, and its functions.

The input to ROKON1 is the contents of an input form, which contains a condensed description of the assembly to be automated.

The rule block Analyze controls the input dialogue, creates all necessary data elements, and divides the assembly description into subproblems that may be solved independently. Each subproblem consists of one or several assembly operations that can be carried out by one robot. The data elements created, which have been assigned to the subproblems, are the basis for the dynamic database and for the subsequent application of rules. If desired, a syntax check of the input can be executed by calling the Syntax module.

The rule block Rough check performs a rough semantic check to detect such errors as object weight exceeds maximum load capacity of robot and also tries to decide whether the task can be solved with the available product line of effectors by testing such parameters as maximum environment temperature. Also a check is made to locate erroneous data entered into the input form (e.g., the gripper clearance might be too small, the permissible payload of the robot insufficient to handle the workpiece, or the reach of the manipulator arm too short).

The rule block Reasoning contains the rules that were extracted from the human expert. Here, the necessity for (or recommendation of) obtaining certain components of the gripper or sensors is deduced from the data elements and performance data are specified. Three kinds of reasoning can be distinguished:

1. Reasoning that directly leads to a product with given specifications
2. Reasoning that leads to the recommendation of a component
3. Reasoning that leads to specifications (e.g., an estimation of the maximum grip width required)

The output of this module is a list of necessary and recommended components with a more or less complete specification of performance data. The number of performance data that can be derived depends directly on the completeness of the information on the input form. In the process of solving the selection problem, the set of recommended components can be reduced to a (smaller) set of necessary ones. This reduction is realized by the context data object, because conclusions can be related to a certain context frame.

When activating the modules Select gripper or Select sensor, the following actions are performed:

Load the static database
Compare required performance data with that of the data sheets of the
    product line
Choose components
Build configuration

A gripper may consist of numerous parts. It is the task of the Select gripper module to pick the parts from the database and to propose the final design of the gripper or to select an alternative solution if necessary.

One of the major problems in these modules is the handling of incompletely specified performance parameters resulting from missing entries in the input form. For this reason a default assignment procedure was implemented. It allows the efficient simulation of the way the human expert handles situations, falling back on solutions that have proved successful in the past.

The assignment of default values, however, leads to a preference of standard solutions and therefore must be carefully checked with the expert's experience. In the current version, each default assignment is reported to the user, who has the choice of either accepting it or assigning a different value.

Since a robot gripper consists of a drive and a transmission system, the selection procedure follows the steps shown in Figure 2.40. It is also here that a subconfiguration of the drive and transmission is built. The sensor selection is processed in a similar way.

The module Configuration combines drive/transmission submodules with sensors to build alternative solutions separately for each subtask. It contains all information necessary to configure a gripper—sensor system and derives necessary interfaces to connect the gripper—sensor system electronically and mechanically to the robot.

The module Evaluation was designed to include economic and organizational information. Currently, this module chooses the three least expensive solutions and suggests them to the user. If a similar design had been used previously, this fact will be pointed out to the user.

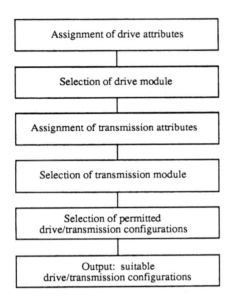

FIGURE 2.40 Functions of the module Select gripper.

The rule base in the current version contains 973 rules, 163 of which were derived more or less directly from the conclusions of the expert. There are 32 rules needed for the context control in the Context module. Most components of the rule base (778 rules altogether) are organizational rules needed to divide an assembly description into subproblems, to control I/O, and to perform a syntax check.

Conclusion

This section describes the design of an expert system for the selection of gripper--sensor systems. The most time-consuming part of such a system is the design and verification of the input form, which must contain all information needed by the expert to solve a given selection problem. In its final draft, the input form is a very good tool for revealing the human expert's approach to solving the selection problem.

Useful results are obtained only through very close cooperation with a human expert. Early prototype implementations based on selection procedures described in literature were impractical for the following reasons: an excessive amount of input data is required, and the requested data are often not available at the time the gripper and sensor are selected.

The expert system described here was implemented and proved to be successful for a number of examples. Future enhancements will include a user interface to keep the static database with its 469 element classes and its 4530 attributes up to date and consistent. Also, a more suitable language would be of advantage for I/O-intensive parts, which would result in a better balance between knowledge and administrative rules.

2.4.4 An Expert System to Determine the Layout of a Sheet Metal Manufacturing Cell

The layout of a manufacturing cell requires thorough manufacturing know-how, which normally is obtained through many years of production experience. Numerous parameters must be considered when components are selected for a manufacturing cell, including the workpiece spectrum, piece rate, manufacturing machines, tools, fixtures, feeders, material handling, and transportation equipment. This section describes an expert system for the layout of a manufacturing cell for making sheet metal parts. The methods used can also be applied to cells for the manufacture of rotational, cubical, and other parts. There are several requirements for an expert system. First, it must be able to use existing descriptions of the workpiece as they are contained in the CAD database of the designer. Second, the most up-to-date machine tools must be stored in a file. Third, the system must be capable of making return on investment calculations for different manufacturing runs and machine tool configurations. Fourth, a non-expert must be able to access the system in a user-friendly mode. Fifth, it must be easy to update the system to include new equipment and fabrication methods.

Analysis of the Workpiece

Planning starts with a description of the workpiece. A typical workpiece spectrum is shown in Figure 2.41. The data files containing the workpiece description are set up by the designer with the help of a CAD system. It will not be necessary in every case to work with the actual workpiece. Often, it suffices if typical representatives of the workpiece (e.g., workpiece variants) are considered. In this case, however, the user must be sure that the actual workpiece is very similar to its representative and that the same fabrication processes are used.

The data file for storing the workpiece description has a hierarchical structure. It contains the following information:

General Description of the Workpiece: name, CAD model, function, number of pieces to be produced, type, etc.

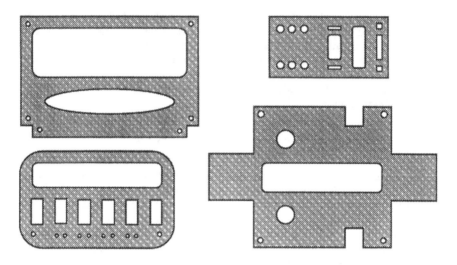

FIGURE 2.41 Typical sheet metal parts to be manufactured.

Physical Parameters of the Workpiece: raw material, blank size, dimensions of the finished part, gage size, outer and inner contours,
   bending edges, quality parameters, etc.
Bending Edges: number, length, angle, type of bending contour
Bending Contour: simple geometry, complex geometry, compound geometry
Type of Cutout: square, rectangular, circular, defined geometrical
   curve, etc.

The principal of the hierarchically structured data file is shown in
Figure 2.42. Easy access to the data file is provided to allow the user
to add or delete parts and to change the workpiece parameters. The
workpiece data file describes the part spectrum to the expert system,
and it can be considered to be the principal entry point to the planning system.

   Determination of the Machining Principles

   From the description of the workpiece, the machining principles are
selected (Fig. 2.43). The database knowledge contains information
about nibbling, simple and progressive punching operations, laser
beam cutting, and other sheet metal manufacturing methods. There
are numerous parameters to be considered, such as the dimensions of
the workpiece, the required number of workpieces, the thickness of

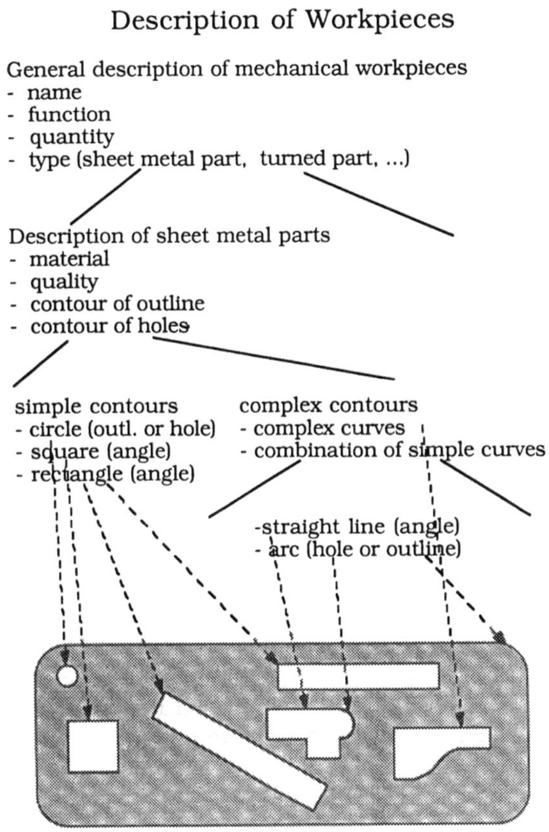

Description of Workpieces

General description of mechanical workpieces
- name
- function
- quantity
- type (sheet metal part, turned part, ...)

Description of sheet metal parts
- material
- quality
- contour of outline
- contour of holes

simple contours
- circle (outl. or hole)
- square (angle)
- rectangle (angle)

complex contours
- complex curves
- combination of simple curves

-straight line (angle)
- arc (hole or outline)

FIGURE 2.42 Principle of the hierarchically structured workpiece data file.

the sheet metal, the shapes of the contours to be produced, and the material. For example, a workpiece made from heavy gage sheet metal or one that has a complex contour may most economically be produced by laser cutting. This technology may also be a candidate when only a small number of parts is needed and it is too costly to build special punching tools. On the other hand, when standard shaped holes and contours are produced and the piece rate is high, punching dies may render the best solution. Via its problem-solving knowledge, the expert system suggests a solution to the user. The rules are those of procedural structure, and an action is proposed when a condition is met.

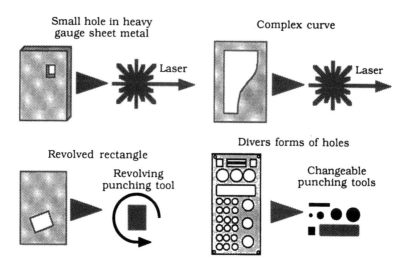

FIGURE 2.43 The manufacturing method depends on the part and piece rate to be produced.

Knowledge about the problem is presented to the expert system via frames. Figure 2.44 shows the typical structure of a frame, describing the database knowledge and problem-solving knowledge.

Selection of the Machine Tools

In the selection phase the machine tool to produce a part is chosen. The system will choose a machine that will render the exact output. If, however, other workpieces will require the use of the same type of machine tool, a more powerful unit may be selected. This can also be specified by naming the desired machine via manual input. If the production capacity of a machine tool is insufficient, several identical or different units may be suggested. Most workpieces will require different machining sequences. In such a case, the selection process may be repeated several times.

Selection of Other Cell Components and the Cell Layout

After the machine tools have been selected, the additional cell components will be determined. Figure 2.45 shows the typical life cycle of a workpiece in a manufacturing process. In this case the manufacturing cell consists of machine tool, manipulator, transportation unit, and material storage. For example, the manipulator must be able to reach every point of the work area (Fig. 2.46). With the

Frame (data base knowledge)

Definition: Example (an instance of a frame):

(object_name   (sheet_metal_part
    attribute_1   identification *housing_control_65*
    attribute_2   weight_kg *5,8*
    ...   gauge_mm *3,5*
)   )

Production (problem–solving knowledge)

Definition: Example:

(p **production_name**   (p **define_robot_requirements**
    condition_1   (sheet_metal_part
    condition_2   weight_kg *<pointer_to_value>*
    ...   )
-->   {(list_of_robot_requirements
    action_1   load < *<pointer_to_value>*
    action_2   ) *<pointer_to_frame>*}
    ...   -->
)   (modify *<pointer_to_frame>*
Conditions refer to   load *<pointer_to_value>*
instances of frames.   )
Actions make, modify or   )
delete instances of frames.

FIGURE 2.44 The frame concept used in the expert system.

description of the robot and machine tool, the placement of the robot is determined. A safety area also is considered to protect workers from possible injury. The placement of all cell components is calculated relative to a cell reference point. Thus, it is possible to arrange the setting of several cells in a more complex manufacturing system.

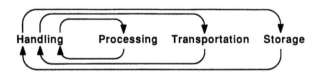

Handling   **Processing**   **Transportation**   **Storage**

FIGURE 2.45 Life cycles of a workpiece in the manufacturing system.

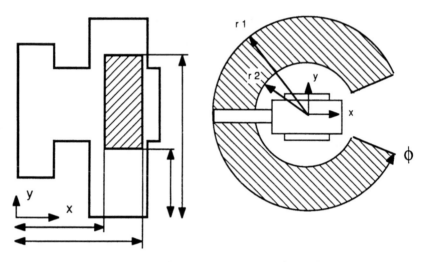

FIGURE 2.46 Placement of a robot in a manufacturing cell.

The Man—Machine Interface

The expert system for the manufacture of sheet metal parts described in this section was implemented with the help of the OPS 5
shell. At the present time, a KEE version is being designed. The
output of the system can be done in graphical form on a display terminal (Figure 2.47). The user enters the description of the workpiece

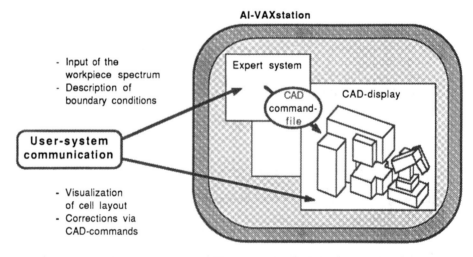

FIGURE 2.47 Interactive communication of the user with the cell layout
planning system.

spectrum via the interactive terminal and describes the boundary conditions. After all cell components have been computed, the layout of the cell will be displayed on the screen. This is done with the help of the CAD system ROMULUS. The user may now accept the solution as it is, or if desired, system components may be changed interactively.

REFERENCES

1. H. M. Bloom, C. M. Furlani, and C. R. McLean, "Emulation as a Tool in the Design of Factory Automation Systems," Internal Report, Factory Automation System Division, Center of Manufacturing Engineering, National Bureau of Sandards, Gaithersburg, MD, 1984.
2. B. F. Soroka, "Debugging Robot Programs with a Simulator," *Proceedings of the Eighth CAD/CAM Conference*, Anaheim, CA, November, 1984.
3. M. A. Wesley et al., "A Geometric Modeling System for Automated Mechanical Assembly," *IBM J. Res. Dev.* 24, 1–12 (1980).
4. T. Sata, F. Kimura, et al., "Robot Simulation System as Task Programming Tool," *Proceedings of the 11th ISIR*, Tokyo, October, 1981.
5. McDonnell Douglas Automation Company, Computer Graphics for Robot Off-Line Programming 1981," Product description, 1981.
6. McDonnell Douglas Automation Company, "PLACE," software product description manual.
7. Dassault Systems, "CATIA Robotic" User manual.
8. U. Schmidt-Streier and A. Altenheim, "Geometriedatenerfassung mit Computer Graphik und Sensoren zur Programmierung von Industrierobotern," *Proceedings of CAMP '89*, Berlin 1984, pp. 155–159.
9. R. Dillmann, "A Graphical Emulation System for Robot Design and Program Testing," *Proceedings of the 13th ISIR/ROBOTS*, 7, Chicago, April 1983, pp. 7.1–7.15.
10. B. Hornung and M. Huck, "Entwurf und Implementierung eines interaktiven 3-D Echtzeit-Roboter Emulationssystems," Thesis, University of Karlsruhe, 1984.
11. H. Maier, "Verwaltung von Methoden und ihre Kopplung an ein Werkstückmodell in einem integrierten CAD-System," *VDI-Fachberichte*, Ser. 10, No. 13, Düsseldorf, 1984.
12. E. Müller and R. Pods, "Entwurf einer Roboter-Datenbank," Thesis, University of Karlsruhe, 1982.
13. M. Rupp, "Tabellengesteuerte Entschlüsselung von Kommandosprachen mit geringem Speicheraufwand," *Elektron. Rech.* 17, 271–276 (1975).

14. A. Schmitt, *"Dialogsysteme,"* B. I.-Wissenschaftsverlag, 1983.
15. J. Denavit and R. S. Hartenberg, "A Kinematic Notation for Lower Pair Mechanisms Based on Matrices," *I. App. Mech.* 77, 215−221 (1955).
16. G. Andre and R. Boulic, "Système graphique et capteurs proximetriques pour la programmation de robots," Laboratoire d'Automatique, Campus de Beaulieu, 1984.
17. J. S. Albus et al., "Theory and Practice of Hierarchical Control," 23rd IEEE Conference, Sept. 13−17, 1981, Washington, D.C.
18. J. Wittenburg and U. Wolz, "MESA VERDE," Ein Computerprogramm zur Simulation der nichtlinearen Dynamik von Vielkörpersystemen," *Robotersysteme* 1, 7−17 (1985).
19. C. Blume and W. Jakob, "Design of the Structured Robot Language (SRL)," *Proceedings of the Advanced Software for Robotic Conference*, Lüttich, 1983.
20. C. Blume and B. Frommherz, "IRDATA−Eine Einführung," *Elektronik* 25, 60−66 (1984).
21. E. R. Argyle, "Romulus to Femgen Interface," *Shape Data*, April 1984.
22. K. Collins, A. J. Palmer, and K. Rathmill, "The Development of a European Benchmark for the Comparison of Assembly Robot Programming Systems," in *Robot Technology and Applications*, Springer-Verlag, Berlin, 1985.
23. J. Meyer, "An Emulation System for Programming Sensory Robot," *IBM J. Res. Dev.* 25:6 (November 1981).
24. R. Weinberg, "Computer Graphics in Support of Space-Shuttle Simulation," *[SIGGRAPH 1978 Proceedings]*, *Comput. Graphics* 12:3 (August 1978), 82−86.
25. T. C. Henderson, E. Weitz, C. Hansen, R. Grupen, C. C. Ho, and B. Bhanu, "CAD-Based Robotics," International Conference on Robotics and Automation, Raleigh, NC, 1987.
26. K. H. Mittenbühler and J. Raczkowsky, "Simulation of Cameras in Robot Applications," *IEEE GG&A*, Special issue on graphics for robotics and CAD/CAM/CIM planning, 1988.
27. K. Weller, "Sensorsimulation bei Roboteranwendungen," 1986, Institute for Real-Time Computer Control Systems and Robotics, Faculty for Informatics, University of Karlsruhe, 1986.
28. *ROMULUS Handbook*, Version 5.1, SHAPE DATA Ltd., Cambridge, United Kingdom.
29. K. E. Torrance and E. M. Sparrow, "Theory for Off-Specular Reflection from Roughened Surfaces," *J. Opt. Soc. Am.*, 57:9 1105−1114 (September 1967).
30. M. Huck, J. Raczkowsky, and K. Weller, "Sensor Simulation in Robot Applications, Advanced Robotics Program Workshop on Manipulators," Sensors and Steps Towards Mobility, Nuclear Research Center, Karlsruhe, May 11−13, 1987.

31. K. Weiler and P. Atherton, "Hidden Surface Removal Using Polygon Area Sorting, *SIGGRAPH 1977 Proceedings*, pp. 214–222.

32. R. L. Cook and K. E. Torrance, "A Reflectance Model for Computer Graphics," *SIGGRAPH 1981 Proceedings, Comput. Graphics* 15:3, 307–316 (August 1981).

33. P. Beckmann and A. Spizzichino, "*Scattering Electromagnetic Waves from Rough Surfaces*, Macmillan, London, 1963, pp. 33, 70–98.

34. T. S. Trowbridge and K. P. Reitz, "Average Irregularity Representation of a Roughened Surface for Ray Reflection," *J. Opt. Soc. Am.* 65:5, 531–536 (May 1985).

35. W. Eversheim and A. Hausmann, "Planung des Sensoreinsatzes für flexibel automatisierte Montagesysteme mit Industrierobotern" [Laboratorium für Werkzeugmaschinen und Betriebslehre" (WZL), Aachen], *VDI-Z* 127:1/2 (1985).

36. W. Eversheim and A. Hausmann, "Die richtigen Sensoren auswählen, Laboratorium für Werkzeugmaschinen und Betriebslehre" [WZL, Aachen], *VDI-Z* 127:10 (1985).

37. U. Cardau, "Systematische Auswahl von Greiferkonzepten für die Werkstückhandhabung," Thesis, University of Hanover, 1981.

38. Granow, "Stukturanalyse von Werkstückspektren, Planungshilfen beim Aufbau flexibel automatisierter Fertigungen," VDI-Verlag, 1984.

39. B. Nnaji, "*Computer-Aided Design, Selection and Evaluation of Robots, Manufacturing Research and Technology*, Elsevier, Amsterdam, 1986.

40. Huber, "Aufbau eines Expertensystems zur Unterstützung der Greifer- und Sensorauswahl bei der Einsatzplanung von Industrierobotern," Master's thesis, University of Karlsruhe, 1987.

3

Knowledge-Based Modeling for Computer-Integrated Manufacturing

PAUL LEVI*   *University of Karlsruhe, Karlsruhe, Federal Republic of Germany*

3.1 INTEGRATED MODELING

The notion of a model has multiple meanings in engineering, logic, and common speech [1]. A model can be conceived as a simulation schema (e.g., for engineering), as a realization in the logic sense, or as a paradigm of a prototype (e.g., in common speech).

We use the concept of a model for a simulation if we describe the organizational phase of the manufacturing process: such as product design, process design, resource scheduling, material handling, and manufacturing planning. If we speak about performing and monitoring a manufacturing process (e.g., by computer vision), we use the notion of a model as a prototype. This prototype serves as a pattern to analyze the situation or to supervise an operation (e.g., robot assembly).

With object-oriented programming, a model is an object that is composed of data (facts) and of standard operations on these data. For example, a geometrical model (CAD model) has standard operations such as extend, stretch, and rotate. Such a composed object is called an abstract data structure. This means that in the view of programming techniques a model is a data structure (independent of its implementation):

model = abstract data structure = data + structure + operations

Knowledge is more than a model. We need knowledge to generate a model and handle (control) it appropriately. The knowledge contained

---

Current affiliation: Technical University, Munich, Federal Republic of Germany

in the model should represent relevant attributes of the world as precisely as necessary and it should predict actions on the world. In
summary, the descriptions to generate and manipulate a model are
called knowledge. Therefore, the simple formula for knowledge is as
follows:

knowledge = model + control information

Control information not only defines how a model is to be built up and
managed, but also it specifies the a priori possibilities to recognize
and to modify reliably all individual features in every possible situation. A situation is defined by the constraints on the operations and
resources. For example, for the task in computer vision a so-called
visibility tree was created [2]. Such a tree defines how separate key
surfaces of workpieces can be extracted from camera images by different views (top, front, side, and additional fields of view). More details on the control component of the knowledge are given in Section
3.6, where the kinds of control operations that are mandatory in the
management of production planning, quality assurance, and maintenance of a manufacturing process are described.

In reality the complex concept of computer-integrated manufacturing (CIM) is the integration of subsolutions (islands) for a flexible
automation task. In the future the concept will be extended to the
entire life cycle of the manufacturing process. Thereby, artificial
intelligence (AI) will have a substantial impact on CIM. It will integrate and maintain the manufacturing knowledge throughout the whole
enterprise.

If the dominant role of AI is accepted for the installation of a CIM
concept, then the usual knowledge representation of AI (rules, frames,
semantic nets, graphs, etc.) should also be involved for generating
CIM models. This means that in addition to explicit geometrical and
topological attributes (classical models), symbolic representations must
be implemented.

Figure 3.1 shows the basic building blocks (islands) of CIM. They
are design (CAD), long- and short-range planning (CAL: computer-
aided logistics), and manufacturing (CAM).

A CIM system is characterized by the ability to integrate various
manufacturing stages to a functional entity whereby the semantics is
maintained from one manufacturing stage to the next. The decision
flow, information flow, and material flow should be passed down from
CAD to the CAM and finally to CAL [3]. For this reason, all models
of a CIM system must be integrated. An integrated model is bonded
together by inputs coming from various heterogeneous regions (e.g.,
from CAD and CAM regions). The parts of a model must be universal
in the sense of offering multiple usability. A local model (e.g., CAD)

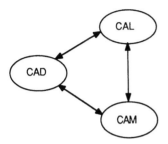

FIGURE 3.1 Basic components of a CIM system.

should also be accessible from any other CIM region (e.g., CAM). An example may illustrate these requirements.

A numerically controlled (NC) programming system needs technological data for the selection of tools, NC machines, and NC control parameters (e.g., cutting velocity, feed). To implement the final, executable program, the technological data must be merged with a CAD model describing the workpieces. To do this transfer, the original CAD model must be described on an abstract geometrical level. The high level geometrical representation of an object is then passed to the CAM region (NC programming system).

The same CAD model must also be adaptable as another abstraction model if it is to be used for object recognition and manufacturing monitoring by a camera. In this case, feature representation is not a pure geometrical task, but it also must contain all features needed to determine the identity of an object.

Two simple examples of the multiple use of a model demonstrate that for universality, a model must include the appropriate model representation required by the end user. The two basic demands that must be fulfilled by a CIM model are as follows.

1. Each component of a CIM system (CAD, CAM, CAL) is represented internally by a different hierarchical model level. Let us demonstrate this requirement with a typical CAM task. A robot may get the production order (workcell level) to perform a dedicated assembly (e.g., water pump). This global order must be refined by several stages (subgroups, joining operations, fixing, movements, etc.). Seven abstraction levels may be distinguished to perform this production order. The task is discussed in more detail in Section 3.3.1.

Figure 3.2 demonstrates that each CIM region can have an internal structure hierarchical, and the higher the level of a model, the higher is its abstraction.

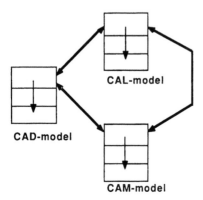

FIGURE 3.2 Universality of the models by high level of abstractions.

The upper levels of abstraction hide the detailed techniques and miscellaneous auxiliary operations used. For example, a high level CAD model hides its realization principle (e.g., wire model, constructive solid geometry model, or convolution areas). Only the topology, form, and size need to be represented at the highest level. A similar example is the operation sequence of robots as already mentioned. Sometimes the highest level of a model is called the conceptual schema (database technology).

2. To be integrated, a model must be separated into two parts. For the solution of each CIM problem, all local and isolated models (e.g., different turnkey systems) are converted into a neutral and universal representation (conceptional schema). The advantage of local integration is the ability of the various isolated models to communicate with one another. In principle, this local integration is possible at different higher levels. A good CIM application demands that the highest possible integration level be used.

Finally, all individual conceptual schemes can be combined to generate a global integrated CIM model. This model describes, for example, a product from the view of its geometry (CAD), manufacturing techniques (CAM), and planning functions (CAL). In addition, the model defines the interdependencies of all conceptual schemes that have been integrated (Section 3.6).

In summary, a CIM model has the following features: it is integrated on a global and local level, and it requires a high abstraction level to be useful for solving various manufacturing problems; the model must be universal, and it should be adaptable to specific areas.

A complete realization of this CIM modeling concept is still missing. In the next section we attempt to describe aspects of the modeling techniques that have been implemented, namely the implementation methodology (CIM shell) and robotics. Both fields are described in the view of AI techniques. Details of the database technology are discussed later.

3.2 CIM SHELL

A CIM shell is a conceptual framework that defines the factory model and the methodology of implementing knowledge-based techniques in a plant. The shell supports the knowledge acquisition and representation as well as CIM modeling, the architecture, and cooperation of expert systems.

3.2.1 Factory Reference Model

The factory reference model is a common model for the manufacturing knowledge used by the plant engineers. The three areas of information, material, and decision flow are discussed. It is assumed that numerous expert systems must be developed for the automatic operation of the manufacturing system that are capable of making decisions for part routing, machine assignment, inventory policies, make-or-buy strategies, and so on. It is unlikely, however, that an entire plant can be directed by one global expert system. A set of cooperating knowledge sources from subdomains such as monitoring, diagnostics, and prediction at different levels may fulfill the manufacturing requirements.

The reference model is based on the hierarchical factory control levels and maps these levels into the time and planning horizons of the factory (Fig. 3.3).

The special manufacturing functions to be controlled are production planning (P), quality assurance (Q), and maintenance (M). Each block of the PQM functions defines a controller of the manufacturing process. This controller is constructed from cooperating expert systems of the subdomains mentioned above. The internal architecture of the controllers is very similar (e.g., for a workstation or a workcell).

The main purpose of production planning (P) is the creation of a schedule for the manufacturing of machines and the product. Disturbances (unexpected events) of the manufacturing process necessitate a reactive management to be able to execute the factory schedule (rescheduling). Therefore, production planning needs a reactive revision of plans in response to various constraints (orders, resources, causal restrictions, temporal restrictions).

Quality assurance (Q) is concerned with the quality of the product, the consistent operation of the machines, and the dependence of the product on the uniformity and exactness of the production process.

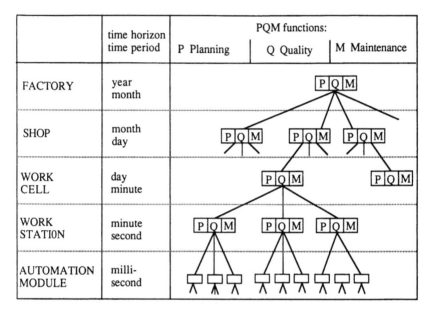

	time horizon time period	PQM functions:		
		P Planning	Q Quality	M Maintenance
FACTORY	year month			
SHOP	month day			
WORK CELL	day minute			
WORK STATION	minute second			
AUTOMATION MODULE	milli- second			

FIGURE 3.3 Factory reference model.

The maintenance requirements (M) can be characterized by three goals:

Prediction of the production rate for machines and products
Prediction of the breakdown of the machines
Intelligent assistance for the maintenance of the machines

It is through knowledge acquisition, a key topic in the development of expert systems, that an efficient and accurate formal representation is found. The problem can be divided into two phases. In the first step, an initial rough knowledge base is extracted from the human expert (first generation). In the second step, the initial knowledge base is progressively refined into a highly sophisticated knowledge base (second generation). This refinement can be thought of as an optimization problem in which one starts with a proposed general solution to a domain problem (P, Q, or M) and the goal is to improve it until the best solution is obtained. The second phase is characterized by a model-based approach. The model that is generated by the knowledge engineer defines the explicit formulation of detailed knowledge about the production (P), the product (Q), and the machines (M). For example, the acquisition of Q data of the multilevel diagnosis of a parametric test application must be done in parallel to the establishment of the models of the human operator, fabrication process, physical structure, and parametric measurements. A diagnostic reasoner has to generate the causal sequence among these four levels of knowledge representation [4].

The reference model is more suited for the generation of a model than the classical partition into CAD, CAM, and CAL. The three PQM functions are present at every level. The height of the level defines whether it is MRP (manufacturing resource planning) at the factory level (CAL) or planning of the explicit assembly operations of a robot (CAM, workstation level). For this reason, there exist P models for all three fundamental regions of CIM. This is also true for the M and Q models.

CAL is performed at three tiers: the factory, shop, and cell levels (Fig. 3.4). CAM starts with the automation module, then proceeds to the workstation, and finally to the cell level. On the cell level, the CAM and CAL regions are overlapping. This means that on this level,

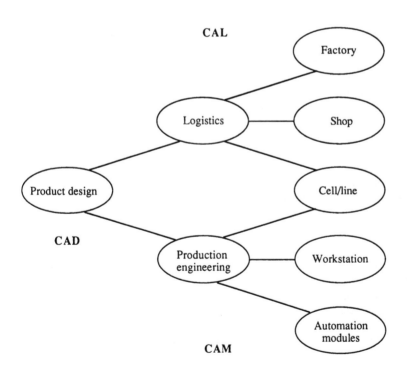

FIGURE 3.4 Conceptual view of a CIM architecture.

for example, the P models for CAM and CAL are merged. The CAD region is handled by the Q models.

3.2.2 Controller Architecture

Each PQM box in Figure 3.3 is represented as a controller. In analogy to the architecture of an autonomous robot, the controller performs the following three tasks: task decomposition from an upper to a lower level (planning), sensory data processing (interpretation, diagnostic), and world modeling. These tasks are performed for each of the PQM functions. As a consequence, the world model is divided into P, Q, and M models (Section 3.2.1).

Each controller, especially on the shop and workcell levels, has a structure identical to all the others (Fig. 3.5). A submodule has an expert system for the three above-mentioned functional blocks of a controller.

The modules (expert systems) of the controller are required to fulfill the following tasks:

Interpretation
Actualizing the performance of the devices (equipment, tools) and the flow of material and products (data acquisition of E_1, . . ., E_n; S_1, . . . S_m, and global task specification I).
Converting the data streams E_1, . . . S_m and I into appropriate representations for the same level (diagnosis, world model, planning) and the upper level (E).

Diagnosis
Explains the divergence between the actual flow of material, tools, and products and the planned (required) one.
Checks the correct operation of the lower level equipment.
Supports quick finding of faults.
Monitors product quality.

Planning
Decomposes the task C into the subtask C_1, . . ., C_p (planning module).
Initiates corrective action planning (replanning) if actions cannot be performed.
Generates the status report (S) about the execution of the task C.
Selects the strategy for the distribution of work orders (C_1, . . ., C_p).

Figure 3.6 describes the tasks of the controllers for each level of the factory reference model [5].

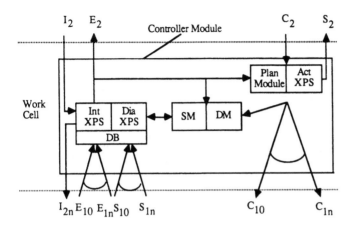

Expert Systems (XPS)
    Int   Interpretation XPS
    Dia   Diagnosis XPS
    Act   Action planning XPS

Data and Knowledge Bases
    DB   Real-time database
    SM   Statis world model (KEE)
    DM   Dynamic world model (Simkit)
    CB   Control knowledge base (metaknowledge)

Information and Decision Flow
    C_2 Plan from upper level (second level)
    $C_{10}-C_{1n}$ Commands to lower level
    S_2 Status report about the execution of the plan to upper level
    $S_{10}-S_{1n}$ Status signals about the execution of the command from the lower level
    E_2 Interpreted status reported by the resources to the upper level
    $E_{10}-E_{1n}$ Global signals from the lower level equipment
    I_2 Global information from the upper level
    I_{2n} Global information to the lower level

FIGURE 3.5 Architecture of a controller module.

In the following, the input/output functions of the workcell and workstation controllers are described in more detail.

A workcell controller receives the manufacturing plan (Fig. 3.7) from the shop level (CAL) and decomposes it into work orders to the workstation controllers. The controllers generate the control tasks for the different automation modules.

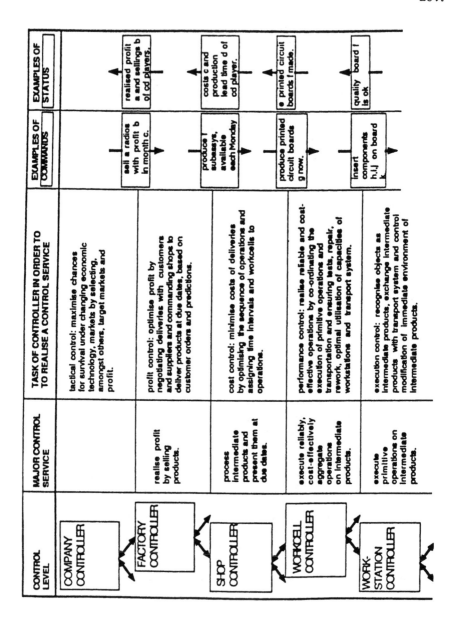

CONTROL LEVEL	MAJOR CONTROL SERVICE	TASK OF CONTROLLER IN ORDER TO TO REALISE A CONTROL SERVICE	EXAMPLES OF COMMANDS	EXAMPLES OF STATUS
COMPANY CONTROLLER		tactical control: maximise chances for survival under changing economic technology, markets by selecting, amongst others, target markets and profit.	sell a radios with profit b in month c.	realised profit a and sellings b of cd players.
FACTORY CONTROLLER	realise profit by selling products.	profit control: optimise profit by negotiating deliveries with customers and suppliers and commanding shops to deliver customer products at due dates, based on customer orders and predictions.	produce f subassys, available each Monday.	costs c and production lead time d of cd player.
SHOP CONTROLLER	process intermediate products and present them at due dates.	cost control: minimise costs of deliveries by optimising the sequence of operations and assigning time intervals and workcells to operations.	produce printed circuit boards g now.	e printed circuit boards f made.
WORKCELL CONTROLLER	execute reliably, cost-effectively aggregate operations on intermediate products.	performance control: realise reliable and cost-effective operations by co-ordinating the execution of primitive operations and transportation and ensuring tests, repair, rework, optimal utilisation of capacities of workstations and transport system.	insert components h,i,j on board k.	quality board f is ok
WORK-STATION CONTROLLER	execute primitive operations on intermediate products.	execution control: recognise objects as intermediate products, exchange intermediate products with transport system and control modification of immediate environment of intermediate products.		

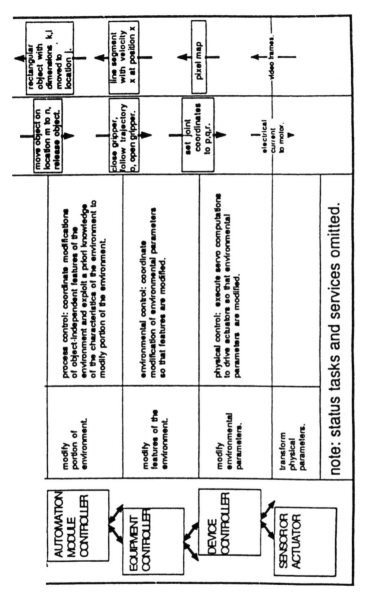

FIGURE 3.6 Controller tasks in the production control hierarchy.

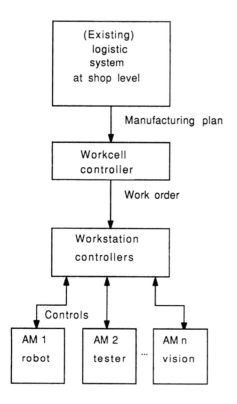

FIGURE 3.7 Order hierarchy of controllers at the workcell and work-station levels (AM = automation module).

Figure 3.8 demonstrates the input/output functions for the controllers at the workcell and workstation levels.

3.2.3 Realization of the Workcell Controller

Pilot-type controller modules for the workcell and workshop levels were installed in a Philips factory in West Germany for audio products (e.g., car radios). The production starts with the insertion of surface-mounted devices (SMD) into printed boards and ends with the automated assembly and test of the final product [6]. The operations are performed in a flexible flow line. One workcell of this line is the SMD workcell that supervises all operations (e.g., robots), that involve automatic insertion of about 20 chips.

The SMD workcell controller sets priorities for the production of each shift in a day. The planning frame is as follows:

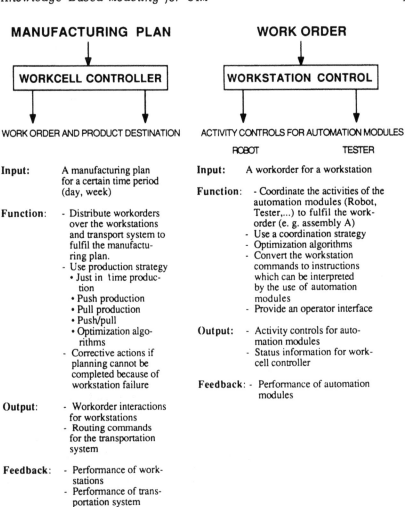

FIGURE 3.8 Input/output function of a workcell controller and a workstation controller.

10,000	orders of printed circuits per day
10	different orders per day
50	different products per day
20	physical machines

A controller defines a flexible plan based on heuristics and rules. It satisfies as many production constraints as possible. Standard MRP techniques, which optimize machine scheduling and maximize profitability, are not suitable for flexible short-term planning. A simplified basic planning cycle must solve the following steps:

1. Select an order from the higher level CAL plan (e.g., material release plan of COPICS for the planning horizon of about 1 week). The constraints are:
    orders to be processed for a day
    released material
    priority for next workcell
    batch size
2. Assign orders to the workstations. The constraints are:
    print type
    workload
3. Schedule the orders for all workstations. The constraints are:
    workstation capacity
    temporary buffer
    setup time
    maintenance time
4. Validate the derived plan by a simulation
5. Change the planning constraints by adjusting the batch size, maintenance time, and setup time. The planning horizon (1 week), due dates, and product quantity are fixed.

The development environment consist of the expert system shell KEE [7], the simulation tool Simkit [8], and a Symbolics machine (3670). Figure 3.9 shows an implemented workcell module modeled after the universal controller architecture. The generic terms of Figure 3.5 are replaced by terms specific to the SMD workcell controller.

There are also expert systems for capacity planning and production flow analysis implemented in this factory. CPLAN (capacity planning) represents the short-term production planning system. The foreman—user interface has the form of a semiautomatic planning module. It contains a set of functions, which interactively requests the status parameters of the workcell, and it is used as a decision aid for the foreman. CPLAN operates with heuristics and rules to plan the product routing in the workcell. The dispatching rule for the circuit layout is as follows:

IF

this print is of type Main-Print
and has axial-components inserted
or is the specific Main-Print-1
and the free-capacity of the SMD-workcell
is less than 90%
and the print is in the list of high priority
prints
and the currently processed print on the
SMD-workcell is setup compatible

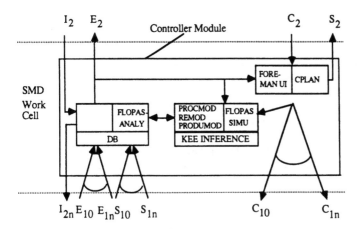

Expert Systems (XPS)
FLOPAS-ANALY Flow production analysis XPS (diagnosis)
CPLAN Capacity planning XPS (action planning)

Data and Knowledge Bases
PRODUMOD Product model
REMOD Resource model
PROMOD Process model
FLOPAS-SIMU Flow production simulation
KEE INFERENCE Inference engine, control

Information and Decision Flow
UI User interface
C_2 COPICS weekly plan: types, quantity, due dates
$C_{10}-C_{1n}$ Plan for individual workstations
Maintenance times spent by personnel
S_2 COPICS updating (continuously)
$S_{10}-S_{1n}$ Status of the workcell controllers
$E_{10}-E_{1n}$ Status of the workcell resources

FIGURE 3.9 Implementation of the workcell controller.

THEN
start immediately processing of this print
on the SMD-workcell

Flopas (production flow analysis) is a diagnosis expert system. It generates qualitative statements about the status of the SMD workcell. The expert system for the interpretation of the workcell status with the help of the sensors has not been implemented yet.

The implemented expert systems use mainly P models and some Q models. For the evaluation of P, Q, and M models, the entire production environment is considered as a set of operations, where products are manufactured by processes that are supported by resources (machines, material). Products, resources, and processes are modeled according to the P, Q, and M views of the CIM factory reference model. Figure 3.10 shows the basic P view of this reference model. Later this representation is extended to consider separate views of the Q and M models.

In the following we describe the product, resource, and process models PRODMOD, REMOD, and PROCMOD, which are P models for CAM.

Resource Model

The resource model contains the description of the workcells, workstations, and machines (Fig. 3.11), and the physical workcell layout (Fig. 3.12). The dashed lines in Figure 3.11 represent instantiations

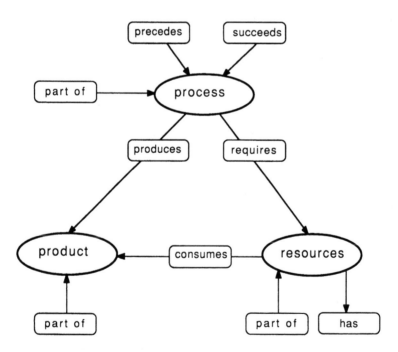

FIGURE 3.10 Structure of the basic P model of the factory reference model.

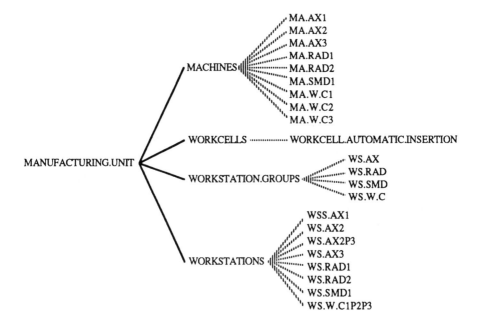

FIGURE 3.11 Resource model: type hierarchy (is-a relation).

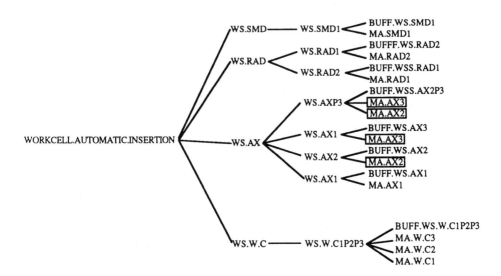

FIGURE 3.12 Resource model: workcell layout (part-of relation).

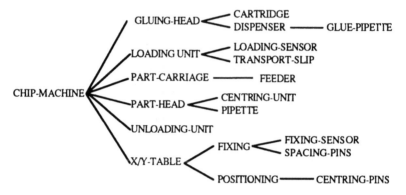

FIGURE 3.13 Resource model: chip insertion machine (part-of relation).

of the type classes. The hierarchical representation of a chip insertion machine and a resource model for the material used in such a machine are shown in Figures 3.13 and 3.14, respectively.

Each object shown in Figures 3.11–3.14 is further defined by a number of attributes. A frame of these attributes showing the detailed behavior of a manufacturing machine is as follows:

> Manufacturing Frame
> member:
> COMPONENTS (local)
> MEDIUM.AVAILABILITY (local)
> OPERATOR (local)
> PLANNED.AVAILABILITY (local)
> PLANNED.MAINTENANCE.INTERVALS (local)
> PROBABILITY.DISTRIBUTION.FOR.THE.DURATION.OF.A.
> BREAKDOWN.
> PROBABILITY.DISTRIBUTION.FOR.THE.OCCURRENCE.OF.A.
> BREAKDOWN
> VARIANCE.OF.AVAILABILITY

Product Model

Products are represented by hierarchical models (e.g., is-part-of, has-a-relation). The attributes of the product classes can be inherited through their instances. The inheritance mechanism is typical for knowledge-based modeling techniques. In Figure 3.15, which shows the product hierarchy of a printed circuit, the broken lines define the final product instantiation.

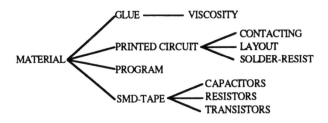

FIGURE 3.14 Resource model: material of the chip insertion machine (part-of relation).

Each member of this product model can be represented by a frame as shown above. It includes product attributes and relations (connections), material requirement information (e.g., printed wiring boards, transistors, resistors), and references to the processes that form this product (e.g., SMD chip handling processes: chip insertion, glue point selection, and gluing).

Process Model

The process model links the product to resources. Time aspects attached to this model are the relations before (precedes), after (succeeds), during, and so on. These relations are used to define the handling of time between the various ongoing processes.

The resource models are connected by the description of the causal relations (e.g., requires) to the process model. Therefore, by reasoning about the time relations [9], the resources and products in-

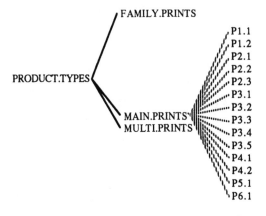

FIGURE 3.15 Product model: type hierarchy.

volved in the manufacturing process can be brought into context with
each other. In addition to the time aspects, the process model describes the orders, the scheduling strategies, and the products (Fig.
3.16). Time aspects are integrated into the attributes of these three
object units (e.g., breakdown time, setup time, earliest start time,
latest completion time).

3.2.4 Model Generation

The task of model generation (knowledge acquisition) is to provide
the ability to build P, Q, and M models. The following steps are
performed in generating a model:

Analysis of the production system (structure):
  decision flow
  information flow
  material flow

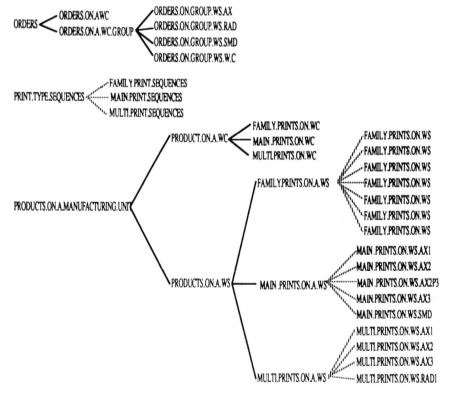

FIGURE 3.16 Process model: orders, scheduling, and products.

physical systems (machines, lines, etc.)

Specification of the critical hierarchical points for functions and decisions

Modeling of the decision activities

Architecture of the controllers

Connection of the controllers

Generation of the P, Q, and M models and specification of the interdependencies of these models

Figure 3.17 shows the partition of the knowledge acquisition phase into an analysis phase of the production environment and a generation phase for models and controllers (PQM blocks). The CIM shell must offer tools to be able to perform automatically the two phases of the knowledge acquisition.

The factory reference model performs hierarchical factory analysis mainly on the basis of the duration of the decisions (time horizon and time periods). The GRAI technique, which is well suited to provide the tools to perform this analysis [10], offers the basic tools to perform automatically the analysis, specification, and modeling phase.

The analysis of the production of the car radio mentioned earlier (Section 3.2.3) generates the hierarchical structure of Figure 3.18. On the left side of the grid the time horizon period (H/P) is shown, and on the right the functions (FCT) are indicated. The grid points represent the activities to be performed.

The SMD controller described in Section 3.2.3 supports the two lowest planning levels of Figure 3.18, which are highlighted by the heavy lines (orders availability, etc.).

The hierarchical GRAI grid can further be refined by a decision network (Fig. 3.19). The nodes of this network are the activity centers, and they have the graphical descriptors shown in Figure 3.20. The activity centers are supported by objects, decision rules, and decision variables.

Let us now return to the phases involving controller architecture and connections. All controllers are tied to the main decision activities in the factory. There are three types of links in the controller network [11]:

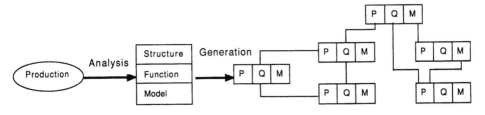

FIGURE 3.17 The knowledge acquisition phase: from analysis to implementation.

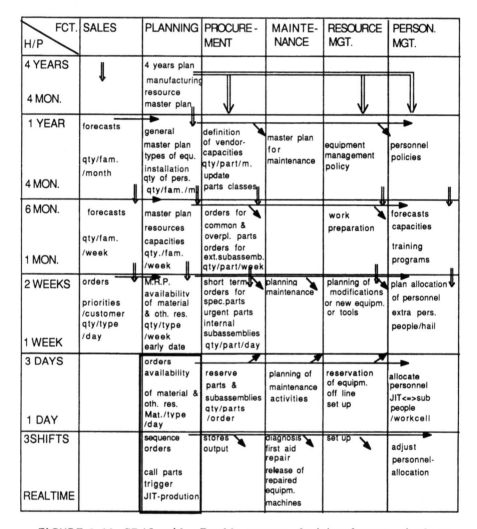

FCT. / H/P	SALES	PLANNING	PROCURE-MENT	MAINTE-NANCE	RESOURCE MGT.	PERSON. MGT.
4 YEARS / 4 MON.	⇓	4 years plan / manufacturing / resource / master plan				
1 YEAR / 4 MON.	forecasts / qty/fam. /month	general / master plan / types of equ. / installation / qty of pers. / qty/fam./m	definition / of vendor-capacities / qty/part/m. / update / parts classes	master plan / for / maintenance	equipment / management / policy	personnel / policies
6 MON. / 1 MON.	forecasts / qty/fam. /week	master plan / resources / capacities / qty./fam. /week	orders for / common & / overpl. parts / orders for / ext.subassemb. / qty/part/week		work / preparation	forecasts / capacities / training / programs
2 WEEKS / 1 WEEK	orders / priorities /customer / qty/type /day	M.R.P. / availability / of material / & oth. res. / qty/type /week / early date	short term / orders for / spec.parts / urgent parts / internal / subassemblies / qty/part/day	planning / maintenance	planning of / modifications / or new equipm. / or tools	plan allocation / of personnel / extra pers. / people/hall
3 DAYS / 1 DAY		orders / availability / of material & / oth. res. / Mat./type /day	reserve / parts & / subassemblies / qty/parts /order	planning of / maintenance / activities	reservation / of equipm. / off line / set up	allocate / personnel / JIT<=>sub / people /workcell
3SHIFTS / REALTIME		sequence / orders / call parts / trigger / JIT-prodution	stores / output	diagnosis / first aid / repair / release of / repaired / equipm. / machines	set up	adjust / personnel-/ allocation

FIGURE 3.18 GRAI grid. Double arrows, decision frames; single arrows, information flow. Qty. = quantity, fam. = family.

Hierarchical links (factory reference model)
Dependency links (information exchange)
Triggering links (synchronization of the controllers)

A controller is a supervision unit, and its task is event-driven. Events may be triggered from orders of upper levels, decisions,

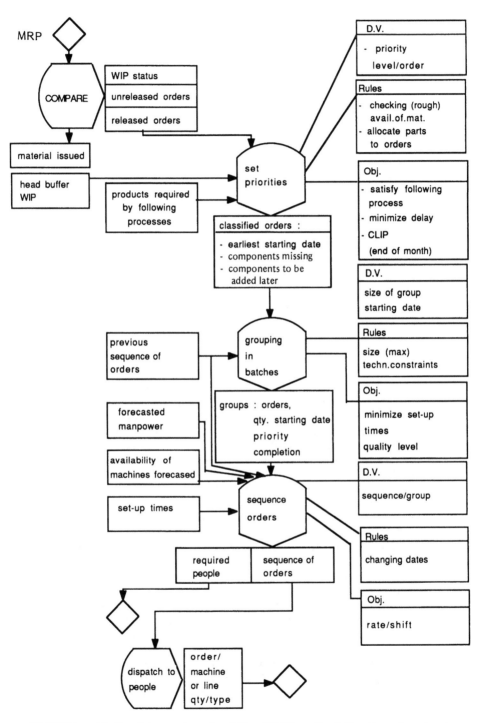

FIGURE 3.19 Decision network of the GRAI-method for the planning/
sequence of orders (D.V. = decision variable). WIP = Work in process.

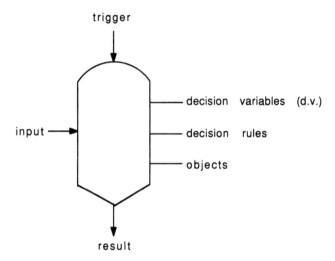

FIGURE 3.20 Schema of an activity center of the GRAI method.

execution of PQM functions, or information from the outside. An expert system architecture that is event-driven and operates in a flexible way is the blackboard concept of Hayes-Roth [12].

A blackboard is a structured and shared database that stores all
solution facts to solve a problem. It is the means of communicating
and cooperating among knowledge sources that get their input from a
dedicated blackboard level and deliver their output to another blackboard level. Each input and output of one blackboard to another is
handled as an event.

The separate modules of a controller (Fig. 3.5) can be implemented
as blackboards. For example, the planning module described in Section 3.2.3 was conceived as a blackboard. But, two additional requirements must be fulfilled if a controller is to be implemented as a
blackboard. The requirements extend the previously mentioned local
blackboard concept to a distributed one.

On one hand, each local blackboard of a controller must be extended
to an input/output blackboard (distributed blackboard), to synchronize
the external communication (dependency links). The original, local
blackboard is sometimes called a domain blackboard.

On the other hand, in addition to the structure of the controller
shown in Figure 3.5, a supervision module must be installed to monitor the cooperation (hierarchical and triggering links) between the
different controllers of the entire net and to synchronize all internal
PQM activities within the controller.

The decision network (Fig. 3.19) specifies the transformation of the decision activities into a controller (supervision unit). Usually, several decision activities are combined to a controller. In this way, all five decision activities of this figure are integrated into one controller (Section 3.2.3).

The supervision module itself can be implemented as a blackboard. This module is then structured internally by a decision and a control blackboard. The decision blackboard contains intermediate (abstracted results to permit the making of final decisions. The activities of the decision network are mapped into this blackboard. The control blackboard stores the control knowledge (goals, work plans, policies, and heuristics) for conceiving and monitoring of internal PQM functions of a controller.

Finally, the items generated by the GRAI grids and the decision network are transformed into P, Q, and M models by a controller (possibly as a blackboard). An automatic transformation does not exist yet. However, parts of such transformations were realized. The GRAI structures (grids, nets, objects, etc.) can be automatically transformed into a knowledge-based representation like the one discussed in Section 3.2.3 (e.g., the resource model of a workcell). The conversion can be rule-based. For each decision activity box, the attached arrows and the links between these boxes can be generated as separate classes of rules [13]. This topic is not discussed further because it involves many details and thorough understanding of the principles of expert system shells (e.g., KEE).

3.3 P MODELS FOR CAM

3.3.1 Planning Hierarchy for CAM

The requirements of an integrated and universal concept of CIM were discussed earlier (Section 3.1). Recall that the interfaces between the basic CIM regions, like the CAD/CAM interface (Fig. 3.3), must be implemented at a very high level. In addition, each level of the factory reference model has its own P model. The basis of each of these P models consists of the manufacturing process, resources, and products (Fig. 3.10). The notion of a process must be fitted into it, and the dedicated abstraction levels must be maintained. At the level of a robot, the assembly operations are the processes, the product is the part that should be mated, and the resources are specialized grippers (e.g., quick-change hand) and feeders, screwdrivers, and so on.

To produce an operable implementation, the level of the automation module (Fig. 3.6) must be further refined by three levels. This refinement ends with instructions to the actuators.

A hierarchy of planning models that starts at the workcell level (CAM region) is described in more detail later. Figure 3.21 illustrates

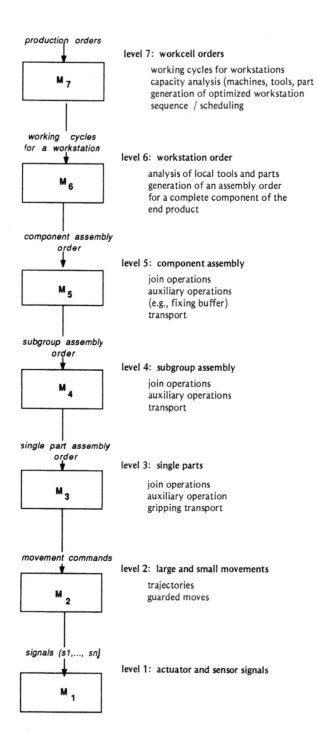

FIGURE 3.21 P model hierarchy for assembly planning.

this hierarchy ($M_1, \ldots, M_7$). An additional refinement of the generation of the models for levels 3–5 is described in Section 3.4.

Level 7 (Workcell)

The workcell planner receives from the shop level planner the production orders for the next shift. This means that the variants ($A_1, \ldots, A_n$) of a product A (e.g., chip, electric motor, water pump) must be produced by different dedicated workstations. The workcell planner analyzes its resources (capacity of machines, necessary tools, material supply) and decomposes a workcell order into workstation orders (working cycles).

Level 6 (Robot)

The robot planner receives the order to assemble an end-product. The planner's analysis function is similar to that of the preceding planning at the higher level. It locates tools (e.g., different grippers) and checks the availability of parts. Thereafter, it generates an order to assemble a component or product.

Levels 5–3 (Robot)

At level 5, the product (e.g., water pump) will be decomposed into its several assembly subgroups. At this abstraction level the joining operations for the product are defined by an assembly graph, and a subsequent precedence graph defines the optimized assembly sequences. Criteria for this optimization are tools, auxiliary operations (e.g., positioning, fixturing) execution time, handling efforts, transportation devices, and so on (Section 3.4).

At level 4, the assembly of the end product is planned from its subassemblies, and at level 3 the assembly of the subassemblies is planned from the single parts.

Note that contrary to the usual hierarchical decompositions, the model hierarchy is organized as a product hierarchy, not as an operation hierarchy (joining operations, transport operations).

Level 2 (Robot)

At level 2, the collision-free trajectories of the gross movements and the guarded movements for the approach and departure of the gripper are defined. The fine movements specify the final manipulation of the part. This tier is the classical level for which numerous planning techniques for gripping and movement were developed. Details on this subject can be found elsewhere in this book.

Level 1 (Robot)

The elementary robot movements are received as a sequence of signals. They are sent to the drives of the actuators and sensor controls (e.g., camera control).

It is characteristic of the P models of levels 2–7 that the internal models of the process, resource, and product are represented at each level. At lower levels, additional attributes of the preceding level (e.g., an accurate product description) or details of the robot (e.g., fine movement, gripping) are of importance.

3.3.2 Simulation: Support for Production Planning

Typical principles of simulation are event handling and process interaction. Classical operation research tools use only analytical methods. They require precise quantitative data and a known algorithm. These methods are often exact if they describe only a few—two to four—machines. They are unwieldy if the behavior of a complete sophisticated manufacturing line is modeled. This is the point where knowledge-based simulation (KBS techniques) should be applied. In addition, KBS can be used to investigate cause—effect relations [14].

For more precise discussions, we use the flexible flow line (workcell level) for the production of car radios, described in Section 3.2.3. The behavior of this line is represented as a sequence of process states. The state changes are the events that are entered as event notices into a calendar (queue). These notices are then ordered by execution time. Performing a simulation means the handling of the event queues in the calendar until all events are executed or there are some severe restrictions that stop the simulation.

The flexible flow line (ffl) is divided into one basic line and 11 sidelines. The product routing is as follows:

1. All parts of the input pool start with the first sideline.
2. All parts stay at sideline n ($n = 1, \ldots, 11$) until the capacity of n is exhausted.
3. If the capacity of n is exhausted, parts in sideline $n - 1$ must wait.
4. A part is finished if it is correctly completed by the last sideline. The finished part is transported to the output pool.

The physical objects of this line (workcell level) are [15]:

Switches
Buffers and assembly line (al)
Basic (bline) and sideline (slines) controllers
Several workstations (e.g., assembly stations)

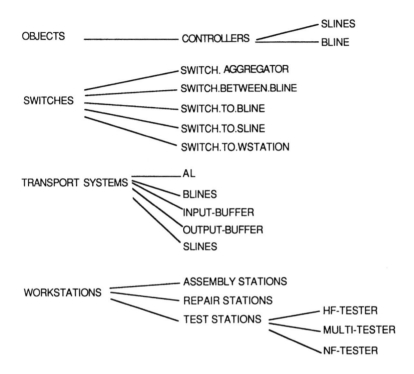

FIGURE 3.22 The four classes of objects of the flexible flow line.

A switch is a unit of the assembly line that accepts parts and transports them to a destination location (e.g., workstation). Figure 3.22 shows the possible objects of the flexible flow line.

Simkit is a knowledge-based simulation package. It is implemented in KEE. The predefined objects are shown in Figure 3.23. The items flow through the system (lines); thus simple items carry no information about the required processing and smart items carry their processing plan. Sources generate the items, servers operate on them, queues buffer them, and links delete them; finally, tasks are operations to be performed by servers on the items.

The physical objects of the flexible flow line must be transformed into simulation primitives. The following transformation is done:

items   → parts
source → entry pool of ffl
sink   → exit pool of ffl

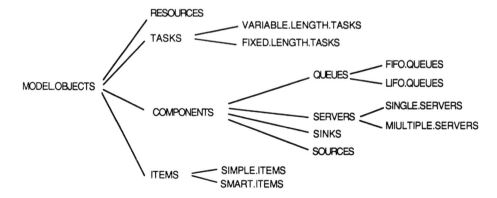

FIGURE 3.23 Predefined object primitives.

Queue   → buffer
single server   → workstation
(one part /time)
multiple server   → assembly line
single server   → switch

Every Simkit object gets a dedicated icon, and the layout of either the complete or manufacturing line parts of it is represented on a screen. Special values of different slots (variables) of these objects (e.g., queue length, buffer content of a workstation) can be shown in separate windows.

Transport and handling of parts are realized by the exchange of messages that contain the states of the items. In addition to the part flow, the decision flow of the flexible flow line is simulated by the exchange of messages (Fig. 3.24).

Switches in the basic line (bline) or sidelines (slines) are activated to route parts to the assembly or test stations in accordance with the routing plan, which is stored in the basic line controller and in the moving parts themselves. The switches are passive; they get all their instructions via message exchange with their controller.

A message is an event, and its handling is performed by the basic line and sideline controllers of the objects and workstations. Each object has a slot in which the sending and receiving of messages is defined by rules.

One purpose of a simulation is the optimization of the line parameters, including the transportation problems. The parameters (i.e., the speed of the moving parts, possible obstructions, part timeout, and size of the input and output buffers of the workstation) are strongly

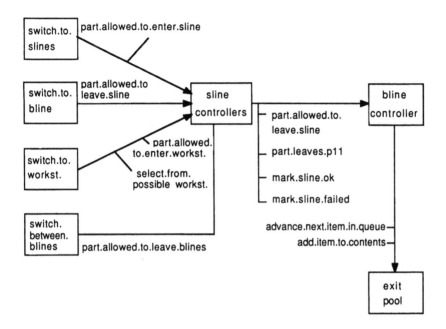

FIGURE 3.24 Message passing between switches, controllers, and the exit pool.

influenced by the control strategies of the workcell level. Therefore, a simulation at this level is an analysis tool to solve shop floor control problems in real time. An optimized product flow through the entire workcell means:

1. Throughput a maximum
2. Maximum number of parts in the workcell
3. Adaptation of the buffer sizes to maximum product flow
4. Minimization of the block and deblock limits of each sideline

The two scheduling methods described below are common to the workcell controller (Section 3.2.3).

1. Selection of the first free resource

    IF   a part (item) needs processing
                AND parallel workstations are available
    THEN   select the first workstation that is ready to work

2. Equal distribution of the workload

    IF   A part (item) needs processing
               AND parallel workstations are available
    THEN   select the workstation that has the lowest number of parts
               to be processed

A simulation of the flexible flow line has demonstrated that both scheduling strategies are not optimized in view of the phases (1 to 4). The following strategy tries to optimize this situation:

3. Equal work distribution with the use of tokens

    IF   a part (item) needs processing
               AND parallel workstations are available
    THEN   send a token through the available workstations and select the workstation that receives the token

Originally the evaluation of such a strategy was performed by the manufacturing engineer; the result of this analysis was to decide which product was to be produced on one shift. The decision steps can be automated with a knowledge-based simulation system. The system is event-driven, and the behavior of the events is documented by rules. In addition, this approach defines the causal chain between the events and postulates the way the events influence the object attributes. The causal chain handles the variables used by one or several rules.

    Let M and D be two object structures:

M: manufacturer		D: distributor	
production-rate:	attribute	*inventory*:	attribute
send-shipment:	event-rule	*stockouts*:	attribute
receive-manufac-		*receive-shipment*:	event-rule
turing order:	event-rule	*receive-order*:	event-rule

A rule that influences several attributes and events of these two objects is the following:

    IF   the goal is to minimize *D: inventory*
              AND   *D: inventory* is high
              AND   *M: production-rate* can be altered
    THEN   increase *M: production-rate*

If this rule is formulated, a knowledge-based system generates the causal chain of Figure 3.25. It demonstrates that the production rate is causally chained with the inventory.

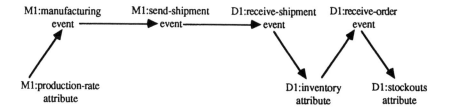

FIGURE 3.25 Causal chain for events and attributes.

The event M1: *manufacturing* uses the attribute M1: *production-rate* and causes a send-shipment event. The distributor D1: *receive-shipment* event accepts the shipment and increases the inventory of D1, thereby affecting D1: *inventory* attribute. The D1: *receive-order* event also accesses the D1: *inventory* attribute to check the material in stock. If the inventory is insufficient, D1: *inventory* records an inventory, which finally affects the D1: *inventory* attributes. Thus, the manufacturing order causes several effects by generating new events and by changing attributes.

If the causal chains generated for the user-defined rules are compared, the individual conflicting goals of these rules can be represented quantitatively. This kind of simulation is, therefore, suited to extract explicitly the interaction of the goals.

Another technique to detect the interaction of goals is the description of goals by constraints. The resolution of the conflicts will then be performed by constraint manipulation techniques. This topic is described in the next section.

3.3.3 Constraints Manipulation for Scheduling

The scheduling problem can be considered to be a constraint-driven search process. Examples for this approach are the ISIS system and its successor OPIS [16]. The modeling technique of ISIS and the opportunistic scheduling technique of OPIS are described now.

The P model is divided into the three blocks: process, product, and resources. Constraint-based modeling of the factory environment implies that the three blocks must be extended by suitable constraints (Table 3.1). The organizational restrictions (e.g., due dates and shop stability) are given to the process model.

These constraints can be attached to the P model. Figure 3.26 shows the P model (workstation level) in which the constraints are represented by rectangles and are attached by dashed lines to the relation attributes.

TABLE 3.1 Scheduling Constraints at the Workcell Level

Process model constraints	Resource model constraints

Organizational goals
  due date
  work in process
  shop stability
  shifts
  cost
  productivity
	Physical constraints
	  physical machine constraints
	  setup times

	Enable constraints
	  machine alternatives
	  tool requirement
	  material requirements
	  personnel requirements

Physical constraints
  processing time
  quality
	Availability constraints
	  reservation of resources
	  machine downtime
	  shifts

Enable constraints
  operation alternatives
  interoperation transfer
  time
	Preference constraints
	  machine preferences

Preference constraints
  operational preferences
  sequencing preferences	

The representation of constraints is included in the frame-based schema representation language (SRL). This language also provides different aspects of the constraints to perform an efficient scheduling operation. Aspects for quantifying constraints are as follows.

Relaxation: the specification of allowable alternatives and the preferences among them
Importance: the relative influence to be exerted by the constraints
Relevance: the conditions under which the constraint apply
Interaction: the constraint interdependency with regard to other constraints
Generation: a mechanism for creating and propagating constraints

ISIS (intelligent scheduling and information system) uses three levels of constraint estimations for performing scheduling. At the highest level there is order selection (organizational constraints). Thereafter, the capacity analysis is done to detect bottlenecks (e.g., enable and availability constraints). Finally, at the lowest level, resource analysis (e.g., physical constraints) is activated.

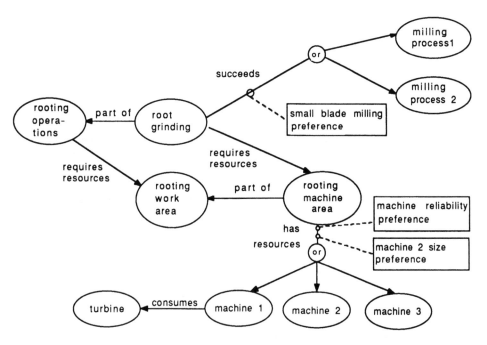

FIGURE 3.26 P Model to which constraints are added.

The need for reactive management to generate schedules in a factory is twofold. First, the system must respond to unanticipated events that occur during the execution of the plan (e.g., machine breakdown, failure of the raw material to arrive on time, failure of the manufactured part to meet quality control standards). Second, a plan can be revised if the synthesis of the partial scheduling operation necessitates generation of a completely new plan (e.g., if the original assumptions are not valid any more). Individual results of subproblems are often incompatible with previously generated solutions. Therefore, reactive scheduling should be used to recognize violation of a constraint of a specific solution and to resolve the components of this solution. With the results, an integrated solution may be obtained. The second case stresses the point that the activities of schedule generation and reactive management are identical. The generation of a schedule is an opportunistic process that merges the top-down and the bottom-up approaches. Therefore, all aspects of reactive management are identical with those of schedule generation.

The term "opportunistic" characterizes a problem-solving process, whereby the control activity is consistently directed toward the actions that appear most promising in terms of the current problem-solving state.

The order-oriented schedule (single-perspective scheduling) does not allow opportunistic scheduling. For this reason, in the OPIS (opportunistic intelligent scheduler) system, the order-oriented perspective is extended to the event- and resource-based perspectives (multiperspective scheduling). The combination of these perspectives is performed opportunistically. Opportunistic scheduling can be used to implement reactive management of schedules. The opportunistic schedule addresses issues relating to the revision of the plan as a response to violated constraints. A manufacturing plan has a large number of constraints. Plan generation and plan revision are triggered by the constraint propagation and consistency checking activities. Constraint propagation determines the interdependencies of one constraint with another. The activity of consistency checking characterizes conflicting situations like capacity limitations. Every inconsistency and unacceptable compromise on constraints is an event that may change a plan. The revision of a plan involves an analysis of the posted events to determine the most pertinent problems and a subsequent operation to generate a more appropriate plan.

3.3.4 Assembly Sequences for Robots

Assembly planning for the levels 2 to 5 of a CAM hierarchy (Section 3.3.1) that also considers the layout of the assembly system must fulfill the following requirements [17]:

Planning of the assembly flow
Optimization of the assembly flow
Configuration of the assembly system

The planning subtasks in the assembly flow can be refined as follows:

Generation of the product hierarchy (assembly, subassembly, single part, etc.)
Specification of the dedicated (static) joining operations (assembly graph) and definition of the auxiliary operations (e.g., feeding, positioning, connecting)

Figure 3.27 shows the components of a complete planning system for assembly robots, constructed according to the three global requirements mentioned before.

The task-oriented planning system for optimized assembly sequences (TOPAS) gives solutions for the planning and optimization points; the details of TOPAS can be found elsewhere [18]. The configuration of

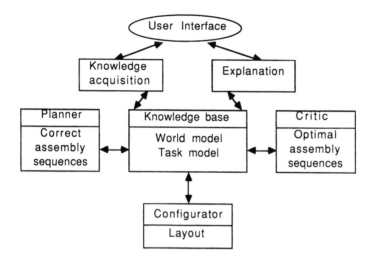

FIGURE 3.27 Components of a complete planning system for assembly by robots.

the assembly station (layout) is done with the assistance of a CAD system.

The main parts of the planning system (knowledge base) are the world model (product, process, resources) and the submodule for the specification of the assembly task. The plans for the task can be described by an attributed directed graph, which can be separately generated for every product level. The graphs are all connected, to generate a total assembly plan that includes all details. Three types of graphs are distinguished. They can be explained with the example of an assembly of a suds (lye) pump of a washing machine. Details about the entire product, its components, joining surfaces and joining surface matrix are available elsewhere [19]. Figure 3.28 shows the assembly parts of the component bearing cap of the suds pump.

Assembly graphs define the joining operations that must be performed to obtain the final product. The nodes of the graphs are defined by the parts to be mounted. They are attributed by the joining matrix, remaining degrees of freedom, assembly restrictions (e.g., precedence, requirements), and so on. The attributes of the directed edges are specified by the joining operation, additional joining parameters, and the auxiliary operations. The graphs are used for hierarchical planning. They can be generated hierarchically for each level

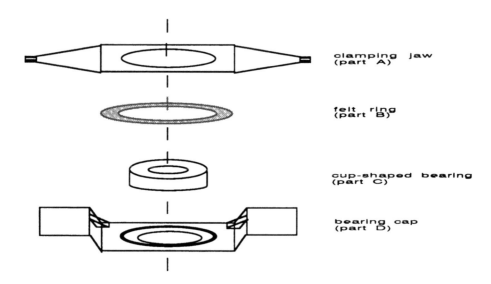

FIGURE 3.28 Parts of a bearing cap assembly.

of the product (e.g., subgroups, single parts) together with the desired set of attributes. Figure 3.29 shows the assembly graph at the part level for the assembly of bearing caps (Fig. 3.28).

The restrictions of the assembly sequence are: "begin with the single part *bearing cap*" and "join the *clamping jaw*" as the last part. The first restriction is needed because the part *bearing cap* is the base part of the assembly group. The second restriction is produced by the accessibility of the part. If the felt ring is put on first, the joining surface of the bearing cap is not accessible for the final joining operation.

AND/OR graphs are well suited for generating all possible assembly sequences and for choosing the optimal solution [20]. The advantage of these graphs is that they describe the complete problem space (all assembly sequences) for restrictions of all kinds (accessibility, auxiliary operations, tools, etc.). They also include temporary subassemblies needed for the assembly operation which are not a part of the product hierarchy and the assembly graph. One such A/O graph ha been shown (Fig. 3.26). Figure 3.30 depicts the A/O graph of the subgroup represented in Figure 3.28. The attribution of a hyperedge (several edges combined by an AND operation) for the joining operation is given on the lower part of this figure. The active part is referred to in the first position of the joining operation. For example, the expression *place on* (clamping jaw, cup-shaped bearing) defines

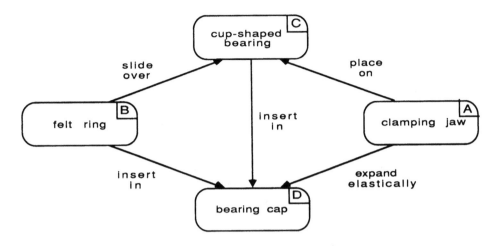

FIGURE 3.29 Assembly graph of the assembly of the bearing cap.

that the active part clamping jaw is placed on the passive part cup-shaped bearing. All operations defined after the first comma are automatically generated by the first operation. The complex graph is interpreted from the bottom up. For example, the sequence of hyperedges 12, 9, and 3 represents correct assembly operations.

This graph can be expanded as follows. We begin with the assembly graph and extract in the first step the joining operations and the mutual predecessor/successor relations. Thereafter, we eliminate the joining operations that are automatically performed by the preceding operation. For example, the joining operation *insert in* (felt ring, bearing cap) of Figure 3.29 is generated in parallel with the joining operation *insert in* (cup-shaped bearing, bearing cap). After this reduction only four out of the five joining operations of the assembly graph remain. In the next step the restriction "begin with the base part (bearing cap)" will be evaluated. As a result of this restriction, all nonterminal nodes of the A/O graph that do not have "bc" as a content will be invalidated. By this procedure only 11 nodes remain out of the 15 original nodes of the A/O graph. These 15 nodes are possible because four single parts exist (2^4 - 12). The remaining 11 nodes define the three alternative basic classes of the assembly sequences. They differ from one another in that the first part is maintained as the base part and the assembly of the three remaining parts is altered.

In the next step, we add the second previously mentioned restriction ("join clamping jaw as last part"). After this step all assembly sequences are eliminated which do not end with the hyperarc 3.

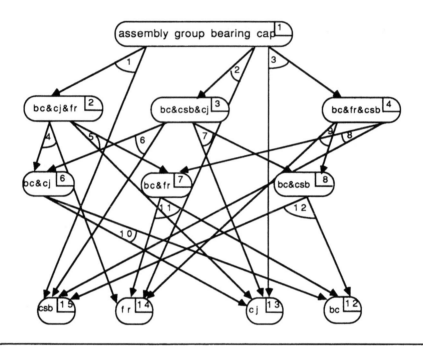

FIGURE 3.30 Correct A/O graph of the assembly group bearing cap: bc = bearing cap, csb = cup-shaped bearing, fr = felt ring, cj = clamping jaw.

These are the hyperedge sequences: 12, 9, 3, and 11, 8, 3. All other assembly sequences are incorrect. The remaining, correct assembly sequences can be represented as precedence graphs.

Precedence graphs show different correct assembly sequences. They order the assembly sequences in time and include the subsequent joining operations, which are automatically generated by a preceding joining operation. The nodes of these graphs can be derived from the assembly graph. The edges of the graphs are nonattributed.

Figure 3.31 demonstrates the precedence graphs for the two abovementioned assembly sequences. The graph on the right is defined by the hyperarc sequence 11, 8, 3. The sequence 12, 9, 3 correspond to the part sequence (C,D), (B,C), (A,C), (A,D). But (C,D) and

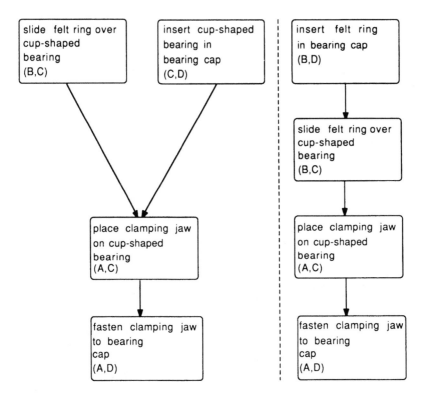

FIGURE 3.31 Two precedence graphs of the assembly of the bearing cap.

(B,C) can be performed in parallel. Therefore, the part sequence (B,C), (C,D), (A,C), AND (A,D) is also a correct assembly sequence. The last sequence is not drawn in Figure 3.31.

In the next step, the assembly sequences of the precedence graphs are optimized. Criteria for the optimal choice of an assembly sequence are tools, change of tools, execution time, handling effort, positioning effort, and difficulty of the joining operation. The assembly sequence shown on the right of Figure 3.31 is optimal for the number of tools, execution time, handling effort, and difficulty of the joining operation. If more criteria were to be used (e.g., joining direction), this sequence might change. For this case, for every set of criteria a local optimum is defined. A global optimum is difficult to obtain.

3.4 Q MODELS FOR CAM

The elementary form of a P model (Fig. 3.10) must be extended in two directions to make it suitable for describing the states of a product. First, the resource model must be subdivided into machine and material descriptions. This means that each of the two relations "consumes" and "requires" of Figure 3.10 must be divided also. In addition, a new relation must be created between the new models (nodes) "machine" and "material." This relation is called "handles."

Second, the new model "defect cause" has to be introduced. Figure 3.32 shows the resulting basic Q model [21].

A defect cause is "produced by" a process, it is "located at" a machine, it is "characterized by" material attributes, and it is a dedicated product that "has" this defect. For this reason, the four edges (relations) that are connected to the defect cause model are causal relations. In the following we give more details on this Q model by using the known example of a chip insertion process (Section 3.2.3).

The resource model describing the chip insertion machine appeared in Figure 3.13. Figure 3.33 depicts the chip insertion process model that is part of Figure 3.32.

The chip insertion process model contains a setup phase and the operating cycles, both of which are further refined. In addition, two time relations (precedes, succeeds) are defined. These two relations describe the temporal sequences of the individual subprocesses of the chip insertion process. A time axis can be used for the visualization of the temporal partial ordering of the subprocesses. It is shown as a vertical time axis in Figure 3.33 and points downward. The setup process is done in the sequence glue—cartridge setup, part—carriage setup, and so on.

Machine cycles are executed in the sequence: chip setting, glue dot setting, printed circuit load, printed circuit unloading.

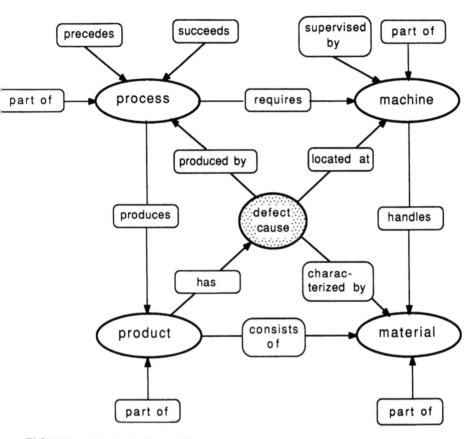

FIGURE 3.32 Basic Q model.

Causes for defective products (e.g., glue point missing), are usually incorrect states of the manufacturing machines (e.g., valve defect) and material problems. But, these "defect" states of machines and materials are also very important for the M model (Section 3.5). They are the intersection between a Q and an M model. Therefore, defect causes were included in Figure 3.32, not their symptoms.

The direct links between the defect causes and the effects of the causes (symptoms) are entered in the product defect model (diagnosis model). This model, a refinement of a basic Q model, represents not only the defect causes but also their effects. Therefore, the diagnosis model includes, in addition to the four causal relations of Figure 3.32, an effect relation that characterizes the product defects (Fig. 3.34).

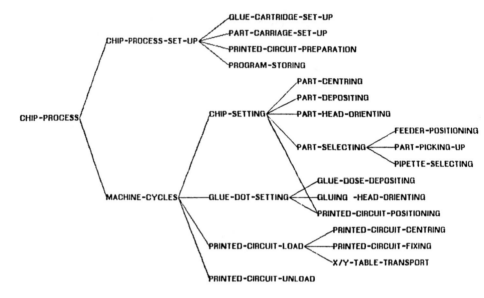

FIGURE 3.33 Model of the chip insertion process.

Figure 3.35 shows the diagnosis model of the product "SMD chip" (workstation level). For clarity, the product model is not shown. Let us examine the network from right to left. One possible defect is a missing glue dot. This symptom may be affected by 10 different kinds of defect (bubble in cartridge, dispenser choked, wrong program, etc.). The defect "bubble-in-cartridge" may be caused by the incorrect operating gluing head of the chip insertion machine (located at-relation), an incorrect glue cartridge, or missing glue. Similar to the representation shown in Figure 3.26, there may be constraints attached to the relations (edges) of the diagnosis model. These are primarily the resource model constraints of Table 3.1, which must be divided into machine and material constraints.

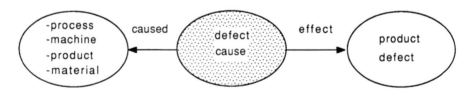

FIGURE 3.34 Cause—effect relations in the product defect model.

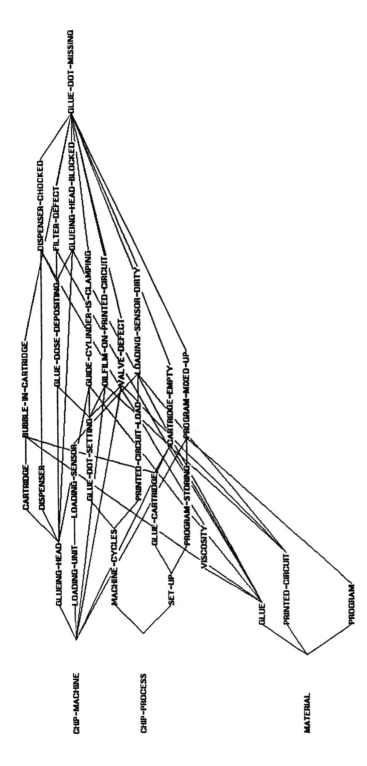

FIGURE 3.35 Product defect model for the chip manufacturing by SMD technique (workstation level).

In connection with the time relations of the process model, the product defect model can be used to localize the process that has caused a defect. For this case, the moment at which the product defect was detected must be registered. From the process model, the machine operation that is responsible for a defect and the material used by the operations can be inferred.

3.5 M MODELS FOR CAM

The task of maintenance is to minimize the production losses and if necessary to predict (planning) the reduction in process quality. The causes of production losses and poor process quality are bottlenecks in the production flow, faulty machines, defective raw materials, and defective products. This short enumeration demonstrates the close connection of the M model with the P and Q models. The basic M model is, therefore, conceived in analogy to the basic P model (Fig. 3.10) and the basic Q model (Fig. 3.32). Figure 3.36 defines the basic maintenance model [22].

If we compare this model with the basic Q model of Figure 3.32, it is obvious that the following analogy exists between the Q model and the M model.

Q Model		M Model
production process	→	maintenance process
machine	→	workers, tools
product	→	machines
material	→	spare parts
defect cause	→	machine fault

Maintenance is considered to be a separate process, which tries to remove or avoid (predicted maintenance) defects in machines. Workers perform these operations with the help of maintenance and spare parts. Machine faults are described by the four relations: changes, operates, needs, and has.

Similar to the product defect model, this basic M model is augmented by a malfunction model. An effect relation is attached to the machine fault that points to the malfunction node (Fig. 3.37). This node comprises the machine breakdowns and product defects. For demonstration purposes, we turn again to the chip insertion machine. Figure 3.38 demonstrates the symptom "x-y table badly positioned" as an example of the breakdown of the chip insertion machine.

The four external nodes of Figure 3.36 are included at the extreme right in Figure 3.38. They are further refined and point to four other machine faults that are possible, namely disconnection, defect of the

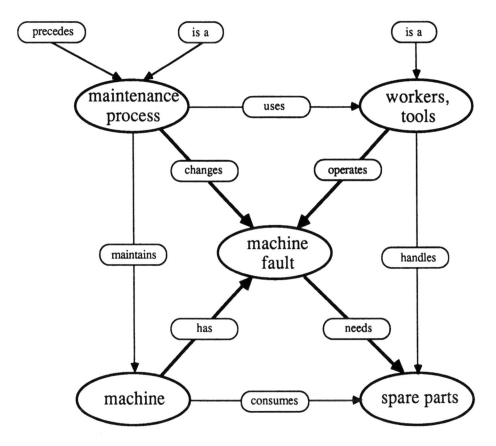

FIGURE 3.36 Basic M model.

x-y card, defect of a loose motor of the x-y table, and problems with the clutch. Each of the four machine faults is connected on the right to four external nodes. For example, the "changes" relation of Figure 3.36 pointing to the machine fault "x-y card defect" is refined

FIGURE 3.37 Cause–effect relations in the malfunction model.

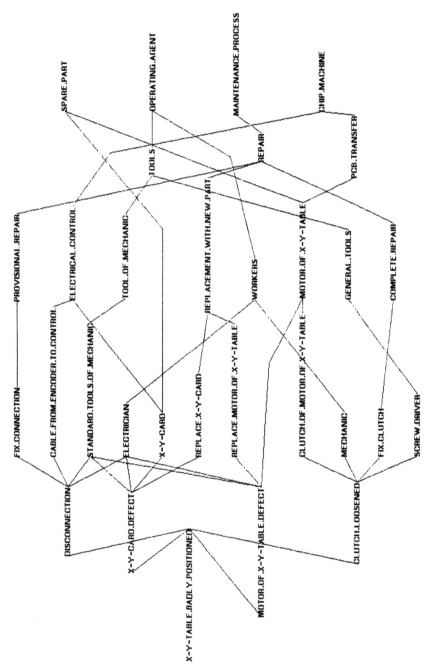

FIGURE3.38 "x-y table badly positioned" as an example of the malfunction "machine breakdown."

into the edges (to be read from right to left): repair, replace with a new card, and replace the x-y card. The four machine faults already mentioned (disconnection, etc.) are connected at the left, near the symptom "x-y table badly positioned" by the effect relation.

Product defects are usually considered by means of the Q view. But, in the malfunction model (refined M model), these defects must be considered. It is now interesting to compare the two different views as one defect. As an example, the product fault "glue dot missing" is shown in the Q view in Figure 3.35 and in the M view in Figure 3.39.

A comparison of these two different views shows that the M model (malfunction model) is essentially different from the Q model (product defect model, in the following respects.

1. The M model describes the chip insertion machine in more detail, as is done in the Q model.
2. The maintenance process replaces the chip insertion process.
3. The M model neglects material descriptions.
4. The M model does not have the node "product," whereas the Q model does not handle spare parts.

3.6 PQM INTERDEPENDENCIES

In this section we return to the definition of a CIM model. We have demonstrated that all P, Q, and M models have enough abstraction levels to be universal (at least for the CAM region). Therefore, these models can also be used to integrate different local and isolated models into one universal local model. The final task is, therefore, to demonstrate how these local universal models can be integrated into a global CIM model.

As the first part of this integration problem, the intersections of the P, Q, and M models are considered. For this comparison, the basic form of the models is used. This is the portion in four parts (process, product, material, and machine) shown in Figure 3.10. Figure 3.40 depicts the P and Q models as having one common base, such that the M model must be considered separately. The intersection between all three models is the machine model.

To generate a CIM model, all interdependencies between the P, Q, and M models must be specified. This means that the control components for the knowledge representations must be defined. A control component states when and how separate models of a life cycle of a production process are to be generated, modified, and used. We focus our attention on the dynamic manufacturing process and connect the control components for the P, Q, and M models to the P, Q, and

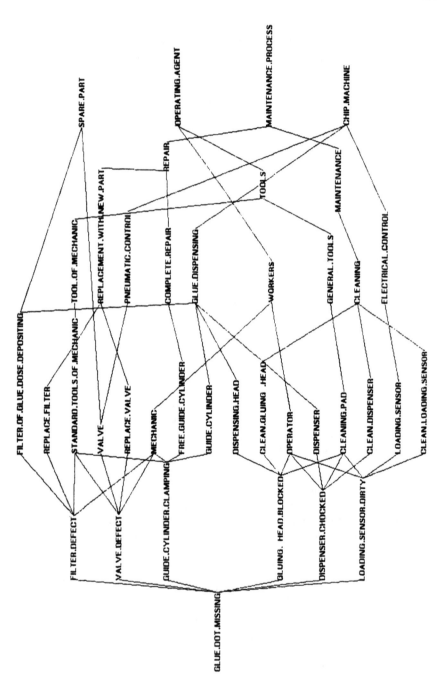

FIGURE 3.39 "Glue dot missing" is an example of the malfunction product defect.

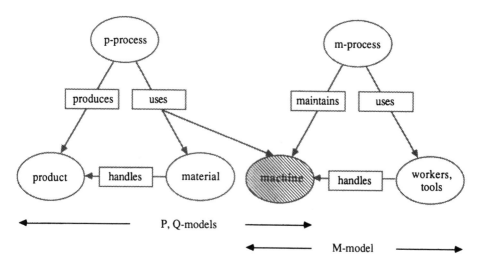

FIGURE 3.40 Communality between the P, Q, and M models.

M states [23]. Figure 3.41 shows that the P state is to be connected to the (manufacturing) process, the Q state to the product and the M state to resources.

The nodes process, product, and resources represent the static model frames. The states define the dynamic generation and handling of the P, Q, and M models in the factory.

The P state determines the manufacturing process (model) and knows how to extract all runtime information from the production line. For this reason, it is informed by the Q and M states, and vice versa.

The M state is concerned with the maintenance of the machines—that is, the actual state of the parts on these machines, the breakdown, and the repair strategies.

The Q state determines the product quality, keeps a record of the defects, and collects fault histories.

The task of specifying the information and the decision flow in the plant remains unsolved by this presentation. The GRAI grids or decision networks described in Section 3.2.2, are tools to monitor these two activities. The information flow is not a mere question of transmission protocols, but an activity that actualizes the data (slots) of the appropriate models. For example, if an attribute is connected with a machine, it will be used by the M state and P state during the production planning process. The operations performed by the P, Q, and M states are control operations that interpret the diagnosis and planning of a controller, defined by the factory reference model (Fig. 3.5).

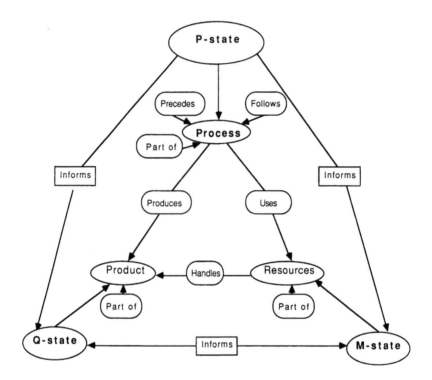

FIGURE 3.41 Definition of the P, Q, and M states.

In the next step, we discuss the interdependencies between all three states, which are caused by the information exchange.

3.6.1 Information Flow Between P and Q States

While the P state plans the products and the production time, the Q state plans the quality test strategies. Similarly, the Q state must predict (plan) probable production losses and quality requirements (time, test bed) to determine the required product quality features. The P states need the estimated quality problems to calculate possible product losses. The quality requirements serve the P state to plan the production times, which include test times.

3.6.2 Information Flow Between P and M States

A P state must diagnose bottlenecks in production (scheduling). It plans the production time, the production rate, the required resources, and the long-term maintenance strategies (maintenance plan). This information is used by the M state to plan the short-term maintenance of the operating machines and the appropriate starting time of the maintenance activities.

An M state diagnoses the availability of machines and plans the maintenance requirements (scope of the maintenance, estimated duration, affected resources, etc.). The P state needs the machine availability to plan or replan the production. The maintenance requirements are needed to plan a realistic production.

3.6.3 Information Flow Between M and Q States

A Q state must perform the diagnosis and interpretation tasks to detect product defects. Typical defect parameters are number of defective products, time when the defect was detected, machine part that might have caused the defect, manufacturing process, and symptoms of the defect. These information units serve the M state to detect failures that lead to a machine breakdown.

A M state must define (diagnose) the condition of a machine. Conditions are the machine part that was maintained, time when a machine part was maintained, and maintenance process (repair, replace, adjust, etc.). The M state needs this information to detect interdependencies between product quality and maintenance operations.

REFERENCES

1. J. F. Sowa, *"Conceptual Structures*," Addison-Wesley, Reading, MA, 1984.
2. J. Majumdar, P. Levi, and K. Su, "A Knowledge-Based Method to Determine the Orientation of the Operation Surface of an Object in Robot Manipulation," IFAC *Proceedings, SYROCO - 88*, Karlsruhe, October 1988, pp. 34.1–34.6.
3. H. Wedekind, "Problems of Computer-Integrated Manufacturing," *Informatik Spektrum*, 11:1, 29–39 (1988).
4. J. Pan and J. Tenenbaum, "P.I.E.S.: An Engineer's Do-It-Yourself Knowledge System for Interpretation of Parametric Test Data," *Proceedings of AAAI-86*, Philadelphia, 1986, pp. 836–843.
5. F. Biemans, "An Update: Reference Model of Production Control Systems," Philips Laboratories, New York, Report TR-87-031, 1987.
6. W. Meyer, R. Isenberg, and M. Hübner, "Knowledge-Based Factory supervision -- The CIM Shell," *Int. J. CIM 1*:1, 31–43 (1988).
7. KEE Software Development System, User's Manual, Version 3.0, Intellicorp, Mountain View, CA, 1986.

8.   Simkit Manual, Intellicorp, Mountain View, CA, 1986.
9.   J. F. Allen and P. J. Hayes, "A Common Sense Theory of Time,"
     Int. J. CAI 85, 528–531 (1985).
10. G. Doumeingts et al.; "Design Methodology for CIM and Control
     of Manufacturing Units," in *Computer-Aided Design and Manufac-
     turing*, U. Rembold and R. Dillmann, Eds., Springer-Verlag,
     Berlin, 1986, pp. 137–182.
11. S. Appel, "Implementation of a Blackboard Mockup, ESPRIT 932
     (knowledge-based real-time supervision in CIM)," Internal Re-
     port 4/1, SGN-Graphael, Paris, 1988.
12. B. Hayes-Roth, "A Blackboard Architecture for Control," *Artif.
     Intell.* 26, 251–321 (1985).
13. J. Foldenauer, "Representation of the SADT structure in KEE,"
     ESPRIT 932 (knowledge-based real-time supervision in CIM),
     Internal Report 3/1, FZI, June 1987.
14. Y. Reddy, M. Fox, and N. Husian," The Knowledge-Based Sim-
     ulation System," *IEEE Software*, pp. 26–37, March 1986.
15. G. Bünz, "Object-Oriented Programming for Simulation," ESPRIT
     932 (knowledge-based real-time supervision in CIM), Internal Re-
     port 1/1, Philips Laboratories, Hamburg, 1986.
16. S. Smith, M. Fox, and P. Ow, "Constructing and Maintaining
     Detailed Production Plans: Investigations into the Development
     of Knowledge-Based Factory Scheduling Systems," *AI-Mag.*, pp.
     45–61, Fall 1986.
17. K. Feldmann, "Von der Insellösung zur rechnerintegrierten
     Montageautomatisierung," *Proceedings of the third VDI-Fach-
     tagung Montageautomatisierung*," Würzburg, 1986, pp. 261–269.
18. P. Levi, "TOPAS: A Task-Oriented Planner for Optimized As-
     sembly Sequences," *Proceedings of the ECAI-88 Conference*,
     Munich, 1988, pp. 638–643.
19. P. Levi and T. Löffler, "The Use of Assembly Graphs to Pro-
     gram Robots," NATO ASI Series, No. 29, Springer-Verlag,
     Berlin, 1987, pp. 233–259.
20. P. Levi, "Planen für autonome Montageroboter," Informatik-Fach-
     berichte, 191, Springer-Verlag, Berlin, 1988.
21. P. Berg, "Real-Time Knowledge Acquisition," ESPRIT 932
     (knowledge-based real-time supervision in CIM), Internal Re-
     port 5/1, Forschungzentrum Informatik, University of Karlsruhe,
     West Germany (FZI), June 1988.
22. H. W. Dohr, "Knowledge-Based Supervision of Machines CIM,"
     ESPRIT 932 (knowledge-based real-time supervision in CIM),
     Internal Report 5/1, FZI, June 1988.
23. J. Foldenauer, "P-Q-M Planning: Interdependencies," ESPRIT
     932 (knowledge-based real-time supervision in CIM), Internal
     Report 5/1, FZI, June 1988.

4

Simulation: Layout of a Robot Cell

JIM BROWNE and SUBHASH WADHWA   *University College Galway, Republic of Ireland*

4.1 INTRODUCTION

Digital computer simulation is widely used as a design tool to ensure the desired material flow performance from manufacturing systems. A growing interest in the simulation technique is well reflected by case studies reported in the literature and the growth in available simulation software [1,2]. However no comprehensive work has been reported for material flow simulation of robot-based flexible assembly systems from an integration perspective. In this chapter we analyze the design of robotic assembly cells viewed as a part of flexible assembly systems (FAS) and focus our attention on the issue of material flow performance. The reason for choosing the FAS area for our presentation is twofold. First this area has vast potential for robotic applications, and second very limited research effort has been focused on the material flow performance aspect of the robot assembly cells and their impact on the design of FAS at large. We found it expedient to present our own research work in this area. In doing so we present to the reader six areas.

1. An appreciation for the scope, need, and potential of material flow simulation in robotic assembly cells, multirobot cells, and the flexible assembly systems (FAS).
2. The operationally significant elements of robotic assembly cells and flexible assembly systems that affect the material flow performance of the designed systems.

3. A framework to view the design of the assembly systems from a material flow control perspective, hence to aid specification and subsequent analysis through simulation.
4. Concepts underlying the need, development, and use of generalized data-driven models for evaluating, prioritizing, and modifying the designed systems.
5. The role of expert systems in providing simulation output analysis and FAS design modification suggestions for a given material flow performance goal.
6. The role of Petri nets in the representation, specification, and modeling of multirobot assembly cells for the material flow simulation purposes.

The presentation of the material in this chapter is highly informal (nonmathematical), to retain the interest of a wide spectrum of readers. The emphasis is on providing the necessary concepts to enable the reader to appreciate the need for simulation and to be able to develop his or her own simulation solutions. We believe that by presenting and discussing our own research work we are able to maintain an integrity in the development of the concepts we wish to present.

The organization of the chapter is as follows:

Section 4.2: We review the fundamentals of material flow simulation; by using three simple examples, a single robot workcell, a multirobot cell, and an FAS composed of a number of robot-based cells, we illustrate the modeling concepts in material flow simulation.

Section 4.3: We consider the role of material flow simulation in the design of robot cells and robot-rich assembly systems and point to the design features of operational significance that can be analyzed using simulation.

Section 4.4: We look at the difficulties of building simulation models— the expertise required, the long lead times to develop models, and so on—and argue for the case for generalized material flow simulators.

Section 4.5: We present a generalized data-driven simulation tool, based on the simulation language/package SLAM II, which models a single robot cell. This tool is called FACET (flexible assembly cell evaluation tool) and was developed by us.

Section 4.6: We present a generalized data-driven simulator for multirobot cells. This simulator, again developed by us, is labeled ROBSIM (ROBot SIMulator) and is based on Petri nets. As with FACET, ROBSIM is implemented using the SLAM II simulation language/package.

Section 4.7: We present FASIM (Flexible Assembly system SIMulator), which allows us to model flexible assembly systems composed of groups of robot-based cells and transport devices. FASIM takes

FACET and ROBSIM output data as a part of its input. Again
FASIM was developed by us and is implemented in SLAM II.

Section 4.8: Since expert systems may be used to help analyze the
results of simulation model experiments, we describe a tool called
FASDT (Flexible Assembly System Design Tool), a prototype expert system that analyzes the output of an experiment using FASIM;
based on a user-defined throughput goal for the FAS, the design
tool indicates whether the goal has been met. If the goal has not
been reached, FASDT offers suggestions to the designer on how
to modify the FAS design. If the designer accepts the suggestions
made, FASDT initiates the changes in the FASIM model and reruns
a simulation of the "redesigned" FAS. FASDT has been developed
using OPS5.

Section 4.9: We offer some general remarks on simulation modeling
and draw some conclusions.

4.2 THE FUNDAMENTAL CONCEPTS OF MATERIAL FLOW SIMULATION

In this section we introduce the concept of material flow simulation and
establish its importance as a design tool for robot-based systems. We
define material flow and its elements, then illustrate these concepts
through simple examples.

4.2.1 Material Flow

Material flow refers to the *flow of entities* such as jobs, resources, or
information through a system. For instance, in flexible assembly systems the jobs may be assemblies, subassemblies, or components; the
resources may be robots, automated stations, or transporters; the information may be the status of jobs and resources. Each *entity flow*
requires a certain time.

The design and control of a system determines the nature of this
flow. In general, the material flow through a system involves a number of interacting processes, where each process involves the flow of
one or more entities through the system. Each process for an entity
may consist of one or more activities. The flow in each process is governed by certain logical relationships imposed by the design and control of the system. The interaction of the processes also entails some
logical relationships. In the actual operation of the system, the interactions may cause delays or alterations in the interacting processes if
the latter are not synchronized over the operating time. The objective
of the design and control of the system is to synchronize the processes
so that the interactions cause minimum delay or the alterations take a
desired change in flow. The interaction elements in the processes are

the start and end of the activities and are called *events*. These elements may be delayed or altered according to the designated control structure.

From an operational perspective, a good system design is one that ensures good synchronization over the interacting processes. A good system design tool is one that can analyze, evaluate, and aid in modifying the system design in achieving acceptable synchronization over the processes. Such an analysis is difficult for several reasons:

The processes are dynamic; that is, the activities comprising a process take place over a time domain.

There may be a large number of processes in the system.

The logical relationships within each process and their interactions may be complex.

Some processes may have entity flow activities with stochastic times.

Real-time control on the interacting processes may be a crucial part of the system design. Real-time control requires among other things a real-time status of the processes and entities in the system.

Robot-based systems involve high investment, and there must be a high level of confidence in their design before they are implemented. The motivation for investing in such systems is to achieve high productivity, flexibility in production, and reliability in meeting the production requirements. As much as 95% of the total assembly and manufacturing time of a product may be wasted in transport or in delays. This reflects poor synchronization and control over the processes in the system. Thus to justify the high investments in robot-based systems, the greatest potential lies in the area of efficient material flow designs.

4.2.2 Material Flow Simulation

A simulation model is a mathematical logical representation of a system that can be exercised in an experimental fashion on a digital computer [4]. These experiments, or simulations, permit inferences to be drawn about the systems:

Without building them, if they are only proposed systems.

Without disturbing them, if they are operating systems that are costly or unsafe to experiment with.

In this way the simulation models can be used for design, analysis, and performance assessment. A material flow simulation model mimics the logical relationships of the processes and their interactions and allows us to envisage the material flow behavior of the designed system, over a time domain, on a digital computer. A high level of confidence is achieved from digital computer simulation simply because it portrays the actual operation over a defined time frame.

4.2.3 Material Flow Simulation in Robot Cells

In this section we present three simple examples: a single robot cell, a multirobot cell, and an FAS composed of individual cells. We then indicate how they are modeled in a material flow simulator.

A Robot Assembly Cell

A conveyor supplies parts to a presentation table from which a robot picks up a part and assembles it on a base assembly. The supply of parts to the presentation table is exponentially distributed with a mean time between arrivals of 5 seconds. The presentation table can hold at most one part and until it is empty, all part arrivals are thrown into a tote bin. The robot assembly cycle consists of picking a part, traveling to the base assembly, finishing assembly, and traveling back to the presentation table. The robot cycle time is uniformly distributed between 4 and 6 seconds. If the robot cannot find a part, it waits at the table. It is required to analyze the operation of the system and assess the number of parts that will be assembled over an hour.

Material Flow Elements in a Single Robot Assembly Cell

There are two material flow processes in a single robot assembly cell, namely, the conveyorized part flow and the robot flow. The latter is a resource flow. The two processes interact at the presentation table. The logical relationship in the part flow process is to supply a part to the presentation table only if it is empty; otherwise the part is discarded. The logical relationship in the robot flow process is to arrive at the presentation table uniformly at intervals of 4–6 seconds and, if there is no part to be picked up, wait until one is available. The robot flow process is delayed if the interacting part flow process is out of synchronization (i.e., a part is not present at the table when it is needed). The part flow process is not delayed but is only altered by the state of the interaction area; that is, the arriving part goes into the tote bin if the table is not empty. The material flow performance parameter is the number of assemblies per hour. Obviously it depends on the robot cycle time and the interaction delay on the robot process. One cell equipment design feature that influences this delay is the inability of the presentation table to hold more than one part.

A Multirobot Assembly Cell

Now consider two robots in the cell just described. Both robots follow the same process. But their trajectories of travel are different.

However the cell operation control, to prevent collision, cannot permit both robots simultaneously to access the presentation table or the assembly area. The presentation table design now allows two parts to be held at the table. The cycle time for the conveying of assembly components is exponentially distributed, with a mean of 2 seconds.

Material Flow Elements in a Multirobot Assembly Cell

In the multirobot setup there are three interacting material flow processes. The first process is the part flow process and the other two are the robot flow processes. The interaction area between the part flow process and each robot flow process is the presentation table. The presentation table is also an interaction area for the two robot processes. The logical relationship at this area is that only one robot can be present at a time. The other interaction area is the assembly area where the two robot flow processes interact. The logical relationship here is again that only one robot can be present in this area at a time.

Again the two robot processes can cause delay to each other at the interaction areas. Furthermore, each robot may have a delay at the presentation table. The part flow process can have an alteration in its normal flow if two parts are already present on the table.

The material flow performance measure of interest is the number of assembled parts over one hour. However from the designer's point of view performance measures such as interaction delays caused to various process entities are equally important. They indicate to the designer the causes for poor performance and thus aid in the improvement of the system design.

A Flexible Assembly System

Consider an FAS consisting of five cells connected on spurs around a track served by two automated guided vehicles (AGVs) as transporters. Let one of the cells be that described in our first example above and the second as described in the second example. In addition let there be one inspection cell and one repair cell. Finally let there be a load/unload station. The base assemblies enter the system through the load/unload station and are transported to the robot assembly cell or the multirobot cell, depending on which one is free. Each assembly needs each of the assembly cells and the inspection cell. At the inspection cell a proportion of the assemblies are sent for the repair station and the others are sent for load/unload station to be removed as finished assemblies.

Material Flow Elements in a Flexible Assembly System

The material flow processes in this system can be divided into two groups: cell level and system level.

The cell level processes are as described above. The system level processes are as follows:

Base assembly flow process
AGV flow process
Cell flow process
Information flow process

The base assembly flow and the AGV flow are physical flow processes, with the former involving the job flow and the latter involving the resource (AGV) flow. The cell flow process is not a physical flow but is a state flow process. These states are the busy and free states, respectively. The logical relationship for each cell in the system is that if an assembly arrives at a cell and the cell state is free, then the assembly is loaded into the cell and the cell status is reset to "busy." If the cell is a single robot or a multirobot cell as described above, the cell busy state corresponds to the interaction processes defined above. The cell busy time is the cycle time for the cell. If the cell is an inspection or a repair cell, the busy status will correspond to the interactive processes within those cells.

The base assembly flow process interacts with the AGV flow process and the cell flow process. The logical relationship with the AGV flow process is that if the assembly needs to be transported from one cell to another, it is transported if the AGV is available; otherwise it waits. If it waits, it sets the information flow process to request an empty AGV to arrive at this waiting point. The base assembly flow process also interacts with the cell flow process. If an assembly arrives at a cell whose status is busy then the logical relationship is to allow the assembly to queue until the cell is free.

The AGV flow process interacts with the assembly flow process and the information flow process. If a request for an AGV is made, the AGV is dispatched to the request point.

The example above only highlights some of the material flow interactions and the logical relationships in the operation of the system. We can see that the complexity of the material flow increases as we move through the examples. One important observation though is that the nature of material flow interaction remains the same.

If when two or more processes interact, the status of one process is not as the other process requires, a flow is either delayed or altered. That is, the flow is controlled at each interaction area. Some of these controls may be well defined for the type of equipment in use; the others may be defined as part of the designer's control strategy for achieving certain material flow performance goals. As we shall see later, this observation is useful in viewing complex material flow systems in a way that facilitates their design and analysis. This is especially true for flexible systems, which inherently provide a control on material flow to minimize delays and provide alterations in a desired direction (priorities). Another important observation is that some process interactions can be integrated into modules and these modules can be viewed as a single process. For instance, the interactions within a single robot cell are seen as a cell flow process. Such a view can reduce the modeling and analysis effort considerably with an appropriate level of approximation to reality.

4.3 THE ROLE OF MATERIAL FLOW SIMULATION IN THE DESIGN OF ROBOT CELLS/SYSTEMS

The design of a robot cell can be broadly divided into two activities:

1. Technological design, dealing with the design of the equipment, accessories, programming, and control of the robot in the cell, with the objective of optimally fulfilling the functional requirements of the cell.
2. Operational design, dealing with the design issues such as layout, operation sequence, and control strategies to obtain high productivity from the cell.

Material flow simulation is used as a design tool for the operational design activity.

The role of material flow simulation in the design of robot cells is that of a design evaluation tool. First of all it is necessary to point out that digital simulation is only an experimentation tool in that it allows us to study the behavior of a designed system. It does not yield a satisfactory design on its own but allows the designer to evaluate and analyze a proposed design.

The procedure for using simulation involves three steps:

1. Develop an initial design.
2. Simulate the initial/modified design.
3. Evaluate and analyze the simulation results to modify the design.

Repeat steps 2 and 3 until a satisfactory design is reached.

Material flow simulation can be used in two roles for the robot cells and robot-based systems: to design a new cell/system or for the operation control design of an existing cell/system.

The first role involves the design of a new system. Here the designer's task is to decide the equipment, layout, configuration, operation sequence, and the control strategy to fulfill a given material flow requirement (e.g., cycle time). In the second role the design involves changes in the operation sequence and the control strategies as the cell/system already exists.

It is necessary to point out that the available graphical three-dimensional packages for the design of robot cells are weak on the material flow aspect. These simulators focus primarily on the graphical illustration of the operation of robotic cells and on the detailed evaluation of a particular robot for a certain application. Emphasis is not placed on the evaluation of material flow using robots, including their impact on the overall system performance, but rather on the local tasks that the robot performs and its interaction with other components of the cell [5]. Material flow simulation should not be seen as an alternative to graphical simulation packages but as an extension of it.

Material flow simulation does not attempt to see the different aspects of a cell design in isolation but views them as part of an integrated operational framework. Thus the layout design is evaluated not in a static and deterministic environment but under the actual stochastic environment with assigned operation sequences and the designed control strategies. The result of this analysis is an estimation of realistic cycle times and their variability with regard to the flow of the assemblies through the cell. These material flow parameters are of interest to the designer of the FAS of which this cell may eventually be a part. The designer may also understand the nature and magnitude of various parameters that are affecting the flow of assemblies.

Various cell/system design features of operational significance that can be analyzed by simulation are as follows:

Layout
Operation sequence
Equipment capacity
Use of sensors
Control structure

All the features above define the material flow processes, the interaction, and the logical relationsips, and hence influence the material flow performance.

4.4 THE NEED FOR GENERALIZED MATERIAL FLOW SIMULATORS

Up to now we have described material flow simulation and outlined its need and scope for the design of robotic cells in an FAS. Now we focus on the designer's requirements from the material flow simulation software. To understand this we must appreciate the environment within which FAS design takes place. Usually there are many possible design alternatives, but there are limits on the time and expertise that can be justifiably devoted to their evaluation. If material flow simulation is performed on each design alternative using a conventional

simulation language, the time and effort involved become extremely
large. Furthermore, the designer must have simulation expertise at
his disposal. The simulation expert must understand the material flow
interactions in the designed alternative, build a simulation model, verify the model, and experiment with the model. The latter implies that
the expert should be able to analyze the simulation results to understand the behavior, locate the flow bottlenecks in the design, and ad-
vise the designer on possible modifications.

Thus there are three constraints on the widespread use of material
flow simulators:

Availability of simulation model building expertise.
The long lead times to develop detailed simulation models.
Simulation output analysis to suggest design modifications. This re-
    quires simulation expertise as well as FAS design expertise.

To overcome these constraints, a design procedure involving the
use of *generalized models* may be used. This procedure is shown in
Figure 4.1. In this procedure the aggregate design solutions are simulated using a generalized simulation model. The designs are modified
to reach an acceptable performance level. Usually these modifications
do not involve major equipment or configuration changes. Finally the
highest priority design is selected from the acceptable set of designs.
This design is then detailed, and to verify the material flow performance, a tailor-made emulation model is developed. This procedure allows us to minimize the time devoted to the early planning stages before a final design is selected for detailed planning.

A generalized model is built by a simulation expert for a particular domain and is data driven. Such a model involves an identification
of a finite set of operationally significant parameters in the application
domain. The possible interactions between these parameters are then
analyzed for the possible and feasible design alternatives. Next it is
necessary to identify a generalized set of basic interaction elements in
a way that permits all alternatives to be expressed as a combination of
one or more of these basic interactions. The modeler can then define
the logic behind each of the basic interactions and provide a special
nomenclature in the data input formats of the user interface. This
nomenclature is usually application specific, and the user can readily
define the operationally significant elements and interactions of the
designed alternative in application specific terminology. Thus a generalized data-driven simulation model can be used for material flow
simulation by a designer who needs to know very little about the simulation technology. The model development lead time is drastically reduced and is equal to the data input time to the model.

The third constraint can be alleviated by providing a decision support environment for the designer who is using a generalized material

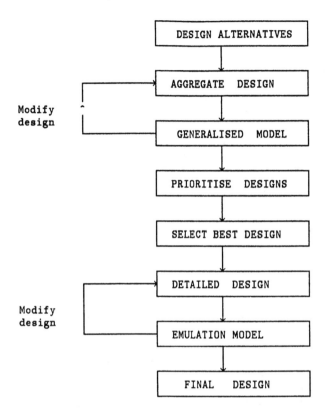

FIGURE 4.1 A design evaluation procedure showing the scope of generalized models.

flow simulator. This decision support system has an expert system as its key element. The expert system holds the simulation analysis expertise and can suggest the necessary design modifications based on an expert's knowledge. Needless to say, both the expert system and the generalized model are domain specific.

In some applications it is difficult to select a set of basic interactions from which all of the design alternatives can be unambiguously defined by the application-specific terminology. In this case it is useful to use some simple representation and modeling language. For instance, the multirobot cell operation may have a large variety of alternative designs, and it may be difficult to unambiguously express the operation logic of the design by mere data input. We suggest the use of Petri nets as a simple design representation, specification, and

modeling tool. Alternatively, one can build simulation models using a
high level simulation language (e.g., SLAM II, SIMAN) for such applications.

The discussion above outlines a solution for using material flow simulation effectively to evaluate the alternative designs during the early
planning stages. Now we define the domains for which such models
can be built. The effectiveness of material flow simulation lies in an
integrated view of the FAS rather than its constituent parts. Therefore we need to select the domains that can be naturally integrated to
specify an FAS design. These domains are as follows:

1. A single robot assembly cell
2. A multirobot assembly cell
3. The flexible assembly system, consisting of the cells above and
    a transport system, auxiliary cells (e.g., inspection, repair,
    carousel) and system control strategies.

We have developed a data-driven generalized simulator for each of
these domains. In the following sections we describe the basis for
these and an application example to illustrate their use. The simulators are:

FACET (Flexible Assembly Cell Evaluation Tool): Allows the user to
   define the design of a single robot assembly cell through specially
   designed user input menus. The simulator carries out the material
   flow simulation to yield various cell performance measures of interest. FACET is presented in Section 4.5.
ROBSIM (ROBot SIMulator): Allows the user to define the operation
   of multirobot assembly cells using Petri nets through specially designed user input menus. The simulator carries out the material
   flow simulation to yield various cell performance measures of interest. ROBSIM is presented in Section 4.6.
FASIM (Flexible Assembly SIMulator): Allows the user to define the
   design of an FAS through specially designed user input menus.
   The simulator carries out the material flow simulation to yield various system performance measures of interest. FASIM is presented
   in Section 4.7.

Figure 4.2 summarizes our view of the FAS with its constituent domains and corresponding data-driven simulator for each domain. The
material flow parameters evaluated by the cell modules are the cycle
time and its variability. At the system level these two parameters
along with the setup times and the reliability of the cell are important.
Thus the integration of the material flow simulator cell modules FACET
and ROBSIM [9] to the system module FASIM is achieved through the

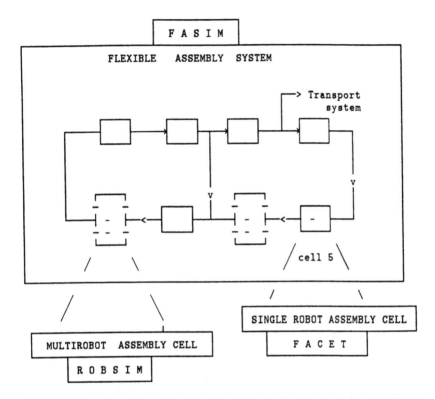

FIGURE 4.2 A schematic of a Flexible Assembly System (FAS) showing domains of the generalized simulators FASIM, FACET, and ROB-
SIM.

cycle time distribution for each robotic cell. Although it is possible to
build an expert system attached to each module, we believe it is necessary only for the systems module (i.e., FASIM). One such expert system developed by us is described in Section 4.8. We refer to this as
FASDT (Flexible Assembly System Design Tool). Figure 4.3 illustrates
the integration of various simulators and the expert system module.

4.5 A GENERALIZED SIMULATOR FOR A SINGLE
ROBOT ASSEMBLY CELL: FACET

FACET is a generalized data-driven simulator for evaluating the material flow performance of a single robot assembly cell. It is coded in

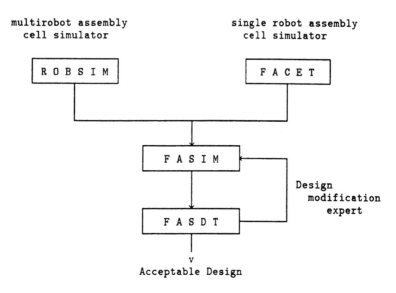

FIGURE 4.3 The integration of ROBSIM, FACET, FASIM, and FASDT.

the simulation language SLAM II, and its user interface is supported
by FORTRAN and SMG routines on a VAX 780, under VMS.

FACET allows the user to define the static configurations (layout,
equipment, sensors), the assembly requirements (precedence relationships), gripper requirements, special processes (inspection, soldering, etc.), the interaction control (control strategies), the reliability
parameters (feeder jams, robot breakdowns, etc.), and the operation
control for various processes (component feeding, robot flow as operation sequence, workpiece flow, and fixture flow). This is accomplished
by guiding the user through an interface built with a number of appropriate menus. An analysis of the operation of a number of assembly
cells indicates that with these definitions, there exists a one-to-one
correspondence in the logical relationships for various processes in
the cell. Thus FACET simulation code is a mapping of these logical
relationships from the definitions of the cell design and operation.

The material flow performance statistics generated by FACET consists of the cycle time characteristics, utilization level of equipment
(grippers, tools, feeders, robot, etc.), and the interaction delays
due to the lack of synchronization on the interacting processes. The
cycle time evaluation is the primary objective of FACET. If the cell
is a stand-alone unit, the cycle times are a measure for prioritizing
the design alternatives. The utilization and delay statistics are

normally used for the design modifications. At the system level (i.e., when the designed cell is a part of a system), the cycle time characteristics may be important depending on the type of the system configuration.

4.5.1 Flexible Assembly Cell System Analysis

Figure 4.4 shows a schematic of the configuration of a typical robot-based assembly cell. The cell typically consists of a number of component feeders which present the components to be assembled on a base assembly or workpiece. The workpieces on which the components need to be assembled are positioned on a work table by means of a fixture. The job of the robot consists of traveling to the feeder, picking up the component, traveling to the work table, and assembling the component on to the workpiece. The layout of the cell determines the relative positions of the feeders and the work table.

Thus the robot travel times are layout dependent. The assembly of a workpiece may call for components of one or more types. The required components may be fed and presented at different locations in the cell depending on the design of the cell. The handling of a component by the robot requires a gripper. Different component types may require different grippers. These grippers are typically located on a tool-changing mechanism in the cell. Again the layout of the cell determines the relative location of the grippers with respect to the work table and the feeders. The robot needs to change its present gripper with the gripper needed for handling a different component type. The robot movement in this case consists of traveling to the

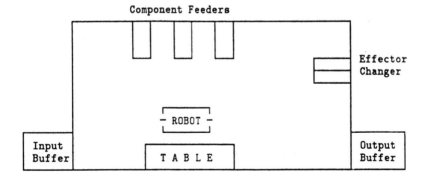

FIGURE 4.4 Schematic of a single robot assembly cell.

tool changer, changing the gripper, traveling to the required component, picking up the required component, and finally returning to the
work table and assembling the component.

The description above outlines the typical operation logic of a robot-based assembly cell for discrete part assembly. A number of variations can exist for the component feeding, tool changing, workpiece
flow (in and out of the cell), and so on. For instance, components
may be fed by means of pallets holding a number of components in a
known position and orientation, or the components may be fed one at
a time by a bowl feeder at a known rate. Similarly a tool changer
may exist on the robot itself as a multigripper. In this case the robot simply indexes the necessary gripper on a turret-like mechanism
that holds a number of grippers. Thus the robot saves the travel
time to a tool changer but at the cost of carrying extra weight (multigripper vs. single gripper) and thus slowing it down for other activities in the cell. Another option on the tool changer is to have a multipurpose gripper called a universal gripper. This saves gripper indexing time but may be more expensive. The workpiece may arrive at the
work table by conveyor or by means of the robot. In the former case
the robot can carry out tasks that do not involve the workpiece, such
as component pickup or tool change, while the workpiece is being
transported to the work table by the conveyor.

Some cell operations may require processes such as inspection and
soldering, which may or may not involve the use of the assembly robot. Some designs present components preoriented while others present components in a random orientation. With respect to the sensors,
some sensors send information about the status of the components to
the robot controller in advance of the robot travel to the corresponding feeder. For instance, a vision system at each feeder will send
such information to the robot controller. In such a case the robot
may orient itself while traveling to the feeder; that is, the orientation operation is done in parallel with the travel operation. On the
other hand, a vision system provided on the robot but not at the
feeders will cause the robot to travel to the feeder and exercise the
orientation operation after the travel operation; that is, the former
operation is in series with the latter.

Another important feature of the robotic assembly is the probability of accomplishing a task successfully. For instance, the component
pickup and assembly tasks may require repeated attempts by the robot
until success is achieved. Furthermore, these operations by the robot may involve stochastic times. A robot with compliant devices for
the assembly of a component is one such case. The control strategies
designed for robotic assembly specify the maximum number of attempts
to accomplish a task. If the task is not successful, a remedial control
action to alter the planned process is required. For the component

assembly task this control action may involve discarding the component and/or discarding the workpiece.

One last feature of the robotic assembly is the interaction of various processes in the cell. The typical processes in the cell are as
follows.

The component flow process: a controlled flow of components into the
cell presentation table. The feeding equipment along with its operational settings define the logical relationships in this flow.
Feeder jams and the reliability of the feeders affect the controlled
flow of components in a stochastic fashion.

The robot flow process: the controlled flow of the robot to accomplish
travel to and from the various task areas to accomplish the tasks
such as component pickup, assembly, and tool change defines this
process.

The information flow process: the information regarding the status of
a task or a flow process in the cell by means of various sensors defines this process. The times taken for sensing and information
processing define the time relationships for the activities in this
process. These times may be significant in robotic assembly.

The tool flow process: tool changing by the robot at the tool changer.
The tools may be statically situated in the tool changer or may flow
to the changer if they are shared by other cells adjoining this cell.
While the grippers may flow as a result of sharing, other tools may
flow for replacement as they may have limited tool life.

The base assembly flow process: the movement of a new base assembly on to the work table to replace a finished assembly, which moves
out of the cell. The base assemblies themselves can flow into the
input buffer at specified rates.

The fixture flow process: the fixture opening and closing at the time
of holding a new workpiece and releasing a finished one.

The coordination and synchronization of these processes is the task
of planning, design, and control of the cell. The robot flow process
interacts with all other processes, and it is this interaction that is
most critical for the coordination of the cell activities in the cell. The
cycle time of the cell depends on the nature of this interaction, and
the evaluation of the same is one major objective of the FACET simulator. The evaluation needs simulation because some components of the
interacting processes are stochastic, and the control of some processes
and the interactions involves real-time control strategies.

Figure 4.5 illustrates a generalized framework to describe the operation of robot-based assembly cells. The robot flow process consists of the tasks and the movement between the task areas. The tasks
are defined by the functional requirements (i.e., tool change task,

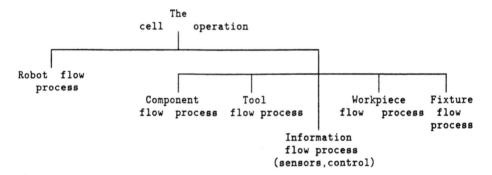

FIGURE 4.5 A generalized framework to describe the operation of a single robot assembly cell.

orientation task, assembly task, and component pickup task). It is the robot tasks that interact with other processes in the cell. For instance, the robot's component pickup task interacts with the component feeding process, which determines whether a component is present for the pickup. If the component is present, the interaction control allows the robot to proceed with the task. If it is not, the robot process is delayed—that is, the robot waits for the task to be accomplished. The task is started as soon as the feeding process presents a component. An alternative control strategy could be to alter the planned robot process; that is, the robot moves to a feeder where a component is present. This is done only if there exists sufficient flexibility in the assembly precedence relationships to allow the alternative component to be assembled before the planned one.

4.5.2 FACET Modeling Framework

The foregoing description shows that the robot flow process and its interaction with other processes are basic features of the operation of any single robot assembly cell. At each interaction a robot task is exercised. For cell design purposes, the robot motions are actually divided into two components as shown in Figure 4.6.

1. *A path motion*: the gross motion period of the robot. Characterized by high velocity and low accuracy motion, it is used for traveling between the task areas. The times taken for path motions are considered to be deterministic.
2. *A task motion*: an interface motion and a fine motion. The interface motion corresponds to the distance H. The fine motion

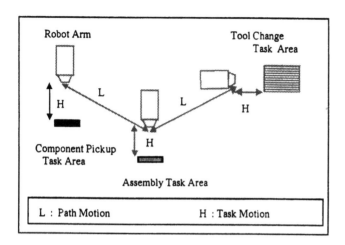

FIGURE 4.6 Illustration of the robot motion times into path times and task times (pickup, assembly, tool change times).

starts when the robot nearly touches the task surface. After the task is accomplished, the robot moves back a distance H. These motions are relatively slow and accurate. The times associated with task motions in assembly are usually stochastic, especially when compliant devices are in use [10].

From a modeling perspective, the cell is viewed as a framework of task areas. The robot visits these task areas according to a planned operation sequence. This framework reflects the layout of the cell in terms of the relative locations of the equipment (feeders, tool changer, fixture) associated with each task area. The layout is quantified by the times taken to travel between the task areas. Each task area is quantified by the task time depending on the task to be accomplished there. Each task area is associated with an interacting process, for example, the component C1 pickup task area is associated with the feeder process corresponding to component C1. We define *a control or decision point* at each task area. This point dictates the next action to be taken by the robot. If the task has been accomplished, this decision point determines the next destination of the robot according to the operation sequence. If the robot has just arrived to accomplish a task, the decision point determines whether robot needs to be delayed, oriented, or allowed to proceed with the task. The nature of decision making at each decision point is inherited from the design and control of the cell.

From the cell simulation perspective, the robot flow in the cell is modeled as the time relationships of the robot travel and the robot tasks in the above-described framework. The logical relationships of the robot flow process are inherited from the decision points for each task area. Similarly, the logical relationships for other processes are viewed in relation to the task areas.

In FACET a material flow logic relationship is mapped to every FAC design element. Such relationships are summarized in Tables 4.1 to 4.4.

4.5.3 FACET Data Input

The FACET input interface guides the designer to input all operationally significant features of an assembly cell. It takes into account the compatibilities of the equipment and the corresponding operation and control strategies. For example, if the user specifies that the components must arrive unoriented at the feeders, FACET prompts the user to input the necessary sensor to feed the orientation information to the robot and the necessary orientation times.

Figure 4.7 shows the sample data input screens from FACET. The top screen presents a summary of the current setup of the cell. The bottom screen illustrates how the designer is aided in inputting the operation sequence with respect to the order of assembly of the components. Based on the gripper requirements for each component, FACET prompts with an effector change indication between the components. Such a screen is presented for each product type to be assembled at this cell.

4.5.4 FACET Output

FACET output consists of an on-line simulation run status, simulation summary reports, and an optional simulation trace. In this section some of the output reports are presented and discussed.

FACET generates the cycle time characteristics over each product type and overall product types. Figure 4.8, a typical result from FACET, illustrates that a mean cycle time of 133 time units can be expected for the product. It also shows that 61% of the products will have a cycle time below 125 time units and 87% will have cycle times below 176 time units. The variation is caused by various breakdowns and the effect of various process interactions in the cell. Similarly, Figure 4.9 presents utilization statistics on the grippers used in the cell. Such statistics indicate to the designer the possible advantages of selecting a multigripper or a general-purpose gripper instead of a special-purpose gripper as used in this cell. The feeder performance statistics (not shown) similarly allow the designer to alter such variables as the feed rates to synchronize them with the robot flow

TABLE 4.1 The FACET Input for Sensors and Hardware Equipment and Their Implied Material Flow Logic

Cell description	Material flow logic
Sensors	
Compliant devices	Stochastic activity times.
Robot vision	Process interaction control decision at robot's present task only.
System vision	Process interaction control strategies possible in a look-ahead mode when the status of next planned task is known.
Presence detectors at feeders	Look-ahead control is used only in connection with the availability status of a component in the feeder.
Equipment	
Unigripper	Robot needs to travel to the tool changer for the necessary gripper change.
Multigripper	Robot indexes to change to the required gripper.
Universal gripper	No gripper change required.
Soldering tool	Robot performs a process on the assembly after the tool change.
Orientation devices	Robot does not require orientation time at the feeder because components are preoriented
Single component feeders with re-circulation	Specify: time between arrival (feed rate) 　　　　　starting buffer level 　　　　　maximum buffer level 　　　　　feeder jam characteristics Components may or may not be oriented. Component unavailability can occur.
Pallet feeding	Pallets with kitted (oriented) components arrive at a certain rate. Component unavailability is rare.

TABLE 4.2 The FACET Input for the Operation Sequence in a Robot Cell and Its Implied Material Flow Logic

Cell description	Material flow logic

Operation sequence

Symbols
  Tn   gripper no. n
  Cn   component no. n
  STn solder tool no. n

Assembly sequence

  ⌐T3-C1-C2-T1-C3-ST2⌐
  └─────────←─────────┘

First the robot moves from its start position to the tool changer to access tool T3. After acquiring T3 it moves to the feeder that supplies component C1. After component C1 pickup the robot travels to the assembly area and assembles component C1. Then the robot travels to the feeder supplying component C2. C2 is picked up and the robot travels back to assembly area and assembles the component. The next component C3 requires gripper T1; hence the robot travels (or indexes) to the tool changer to acquire gripper T1. With T1 the robot moves to the feeder for component C3, picks up C3, and travels back to the assembly area to assemble it. Finally the robot travels to the tool changer (or indexes) to acquire the soldering tool ST2 and returns to do a solder process on the assembly.

The equipment, feeder process control, sensors, and control strategies affect the delays or the robot process alterations. The operation sequence is a plan and the delay and alterations are real time process controls. The combination of the planned activities and real time control yields realistic cycle times.

The order of C, T, ST defines implicitly the next action of the robot.

The type of gripper used determines whether the robot travels to the tool changer, indexes, or takes no action corresponding to external tool changer, multigripper, or universal gripper, respectively.

TABLE 4.3 The FACET Input for the Cell Layout/Configuration, Assembly Handling, and Reliability Data, and the Implied Material Flow Logic

Cell description	Material flow logic
Layout/configuration	
Number and type of feeders, component–feeder pairs	Number of feeders determines the number of task areas. The type determines logical relationships.
Number and type of grippers, component–gripper pairs	Is tool change travel needed?
Location of feeders, tool changer, assembly area	Travel times are specified between: assembly area and tool changer assembly area and feeders tool changer and feeders
Base product handling	
Robot	The robot process for each assembly starts by picking the base product from the cell input buffer and then placing it on to the fixture. Similarly, the finished assembly is transferred to the output buffer by the robot.
Conveyor	The conveyor transfers the assemblies in and out of the cell. The robot can simultaneously proceed for its new cycle.
Reliability	
Robot breakdown, feeder breakdown	Specify: time of first breakdown time between breakdowns time of repair
Tool changing	Probability of tool change; if failed, then manual intervention delay is caused.
Fixture operation	Probability of correct fixture operation; if failed, then manual intervention delay is caused.

TABLE 4.4 The FACET Input for Various Control Strategies and Its Implied Material Flow Logic

Cell description	Material flow logic
Task attempt strategies	
Number of assembly attempts, "Na"	The probability of successful assembly determines in real time whether an attempted assembly operation is successful. If not, the robot makes up to "Na" attempts. If it still fails, the failure strategy is used.
Number of pickup attempts, "Np"	The probability of successful pickup determines in real time if an attempted pickup is successful. "Np" attempts are made if needed.
Assembly failure strategy	
Reject component	If assembly is not successful in "Na" attempts, then the robot rejects the component.
Reject assembly	If assembly is not successful in "Na" attempts, then reject assembly and start afresh.
Inspection strategies	
No inspection	The robot is not delayed due to any inspection.
Inspect each component assembly	The robot is delayed for the inspection time for each component assembly.
Inspect randomly a percentage of component assemblies	The robot is delayed randomly for the proportion of components needing assembly.
Inspect each assembly	The robot is delayed at the end of each assembly for the inspection time.

(continued)

TABLE 4.4 (Cont.)

Cell description	Material flow logic
Interaction controls	
The robot process	
waits	If processes are not synchronized at a task area, then the robot waits.
alters	The robot process is altered to try other tasks.
The feeder process alters	If the feeder process is constrained by maximum buffer size, then the new fed component is directed into a bin.

process. If feed rates are too high, a large proportion of the components are recirculated or rejected to a tote bin. If the feed rate is too low, the robot may need to wait for the required component at a feeder. Usually the feed rates are set marginally higher than the rate at which they are expected to be needed by the robot. Simulation modeling helps the designer to determine these rates.

4.5.5 Role of FACET in Cell Design

At the cell design level there usually exist a number of alternatives for the layout, configuration, equipment, sensors, and the control of the cell activities to fulfill the functional requirements of the cell. However, different alternatives incur different levels of cost. Here "cost" refers to the effort in planning, control, and implementation along with the cost of equipment and software. It is at this stage that the designer may wish to analyze the operational performance of various alternatives and quantify them. The cycle time characteristic is one such parameter on which the cell operational performance can be prioritized and quantified.

It is important to understand the role of material flow simulation in the context of three-dimensional graphical simulators and analytical methods. The use of graphical 3D models allows the designer to plan the layout and the robot travel and task approach paths and trajectories for optimizing the time requirements for individual activities in the robot flow process. This optimization is constrained by space limitations and potential collision problems, which can be addressed only by interactive 3D models. During this stage the designer may need

FACET - THE CURRENT SETUP

Fee der	Part name	Part no.	Presence detector	Pre- orient	Vision	Pallet	start bfr	Max bfr
1	Cmp1	CP/1	yes	no	yes	no	0	5
2	Cmp2	CP/2	no	yes	no	no	0	5
3	Cmp3	CP/3	yes	no	yes	no	1	5
4	Cmp4	CP/4	no	yes	no	no	0	5
5	Cmp5	CP/5	yes	yes	no	no	0	5

Grippers/Tools

Suction gripper	Magnetic gripper	Vacumm gripper	3-finger gripper	Solder tool	Inspect tool
External	External	External	External	no	no

COMPONENTS		Assembly Precedence	SEQUENCE
Part name	**Required**		Cmp1
			Gripper change
Cmp5	1	Product : PROD-1	Cmp2
Cmp4	1	Part no.: MP/1	
		Use select key after moving arrow to the component row.	

FIGURE 4.7 A sample input data screen from FACET.

analytical tools to lay out the cell equipment to minimize the travel times for the overall assembly requirements. This problem, termed the assembly plan problem, may be solved by algorithms such as the traveling salesman's algorithm. At this level the process interactions and control issues are not addressed and the planning assumes a deterministic environment, conforming to the requirements of the robot

TIME IN ASSEMBLY

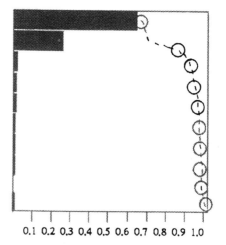

Total No of Obs		175		
Average Value:		133		
Lower	Upper	Obs	Freq	Cum
75	125	108	0.61	0.61
126	176	45	0.25	0.87
177	227	6	0.03	0.90
228	278	5	0.02	0.93
279	328	4	0.02	0.96
329	379	3	0.01	0.97
380	430	2	0.01	0.98
431	481	1	0.00	0.99
482	532	0	0.00	0.99
533	Inf	1	0.00	1.00

0.1 0.2 0.3 0.4 0.5 0.6 0.7 0.8 0.9 1.0

FIGURE 4.8 Cycle time characteristic for a robot assembly cell generated by FACET.

flow process. Estimates of the cycle times can be obtained using simple calculations and/or analytical methods.

The final stage in planning is to define the operational control of
the various processes to synchronize them with the robot flow process.
The reliability of the equipment and the stochastic nature of various
activities in the cell are defined and quantified. Control strategies
are designed to cope with these situations. At this level material flow
simulation comes into play for the evaluation of realistic cycle times for
various designed alternatives.

Table 4.5 illustrates the respective material flow issues addressed
by graphical 3D models, analytical methods [11], and material flow simulation (FACET). The cell off-line design and evaluation requires a
hierarchy of tools to analyze various levels of material flow details.

4.5.6 FACET and FASIM

As noted earlier, the integration of FACET and FASIM is accomplished
through the cycle time characteristics for each product through the
designed cell operating as part of an FAS. At the FASIM level each
such cycle time is treated as an operation on the product, which flows
through a number of cells before getting finished. Thus at each cell

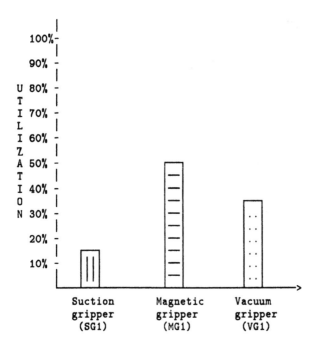

FIGURE 4.9 The utilizations for the grippers used in the cell.

the operation time is sampled from the cycle time distributions evaluated by FACET for the product flowing through each cell (assuming
each cell is evaluated by FACET). We review FASIM in detail in Section 4.7. First we shall look at a Petri net based simulation for multirobot cells (ROBSIM).

4.6 A GENERALIZED SIMULATOR FOR
MULTIROBOT CELLS: ROBSIM

ROBSIM is designed as a Petri net based simulator for the material
flow analysis of multirobot cells. It is developed in the simulation
language SLAM II and is supported by a user interface written in
FORTRAN and SMG routines in VAX/780 under VMS. Since ROBSIM
is based on Petri net data, its application domain is not restricted to
a particular configuration of the cell/system.

  In this section ROBSIM [6,9] illustrates the analysis of design issues in a multirobot assembly cell configuration. The operation of
the cell is constrained by the assembly precedence requirements,

TABLE 4.5 The Purpose and Scope of 3D Graphic Models, Analytical Tools, and Material Flow Simulation (FACET) in the Cell Design Process

Material flow process analysis	Tool	Operational analysis and design issues
Robot flow activities (e.g., tasks and travels to minimize robot activity times)	Graphical simulation (3D modeling)	Layout of cell hardware Robot path planning to avoid collisions Generate travel and task times
Robot flow process (e.g., order the tasks and travels to minimize robot flow process cycle time)	Analytical methods (assembly plan problem)	Operation sequence planning Modification in cell layout Estimate cycle time
Integrate all flow processes in the cell (e.g., robot flow, component flow, tool flow, workpiece flow, and fixture flow)	Simulation (FACET)	Evaluate realistic cycle times Synchronize various cell processes Evaluate stochastic flow, reliability, and control strategies

work coordination by robots, equipment constraints and collision constraints. Activity or branch network simulation packages, such as Q-GERT, can also be used to simulate multiple branch activities of a multiarm robot station [16]. Other Petri net simulators [12–15] exist and are oriented toward the simulation of FMS.

4.6.1 System Analysis of a Multirobot FAC

A multirobot flexible assembly cell typically involves the synchronization of two or more robots to carry out various assembly tasks. An assembly task in discrete parts assembly consists of robot movement to pick up the required component, return to the workstation, and assemble the component. The components may be fed from different feeders or presented at a shared buffer. In the latter case the two robots face a constraint that during the cell operation they cannot simultaneously access the required components. The availability of a given component at the time of need may impose another constraint on the pickup activity of the robot. The assembly requirements impose precedence constraints on the order of assembly operations. In a cell where a stochastic environment exists, it may be difficult to plan and assign assembly tasks to each robot so that perfect synchronization is obtained. A robot may have to wait at the assembly station for the other robot to finish the preceding task.

The robots may require different grippers or tools to accomplish different assembly tasks. The gripper storage unit may have a limited number of grippers, and one robot may need a gripper that is currently being used by the other robot. The gripper requirement thus becomes another constraint on the activities of each robot. The grippers may be stored in a common buffer, so that only one robot can access the common buffer for the gripper change at a time. The other robot may have to wait until the common buffer resource is free. The common gripper change location may then be a constraint on the activities of each robot.

Furthermore, there may be spaces in the cell where the two robots may not be allowed simultaneously to avoid possible collisions. These spaces or areas then act as shared resources also and impose a constraint for synchronization.

The simplest control strategy to deal with the constraints is to wait until the constraint is satisfied. The relative location of the control area (i.e., the area of checking the constraint and its consequent wait) has an influence on the performance of the cell. The closer this area is to the constraint, the better the cycle times that are likely to be obtained.

Since the operational issues are related to the cycle times, the analysis of the system operation needs to focus on the time aspect. The cycle time of a multirobot cell has the following components:

1. Gross motion times
2. Task times
3. Interaction delays

The gross motion times refer to the travel times for the robots in the cell. The task times refer to the tasks accomplished by the robots (component pickup, assembly, tool change, etc.). The interaction delays refer to the waits caused by the lack of synchronization of the constraints on the planned flow of the processes.

The gross motion times and task times can be evaluated in the same manner as for a single robot station. For the interaction delays, however, the analysis cannot overlook the presence of other concurrent and asynchronous activities in the cell. In fact it is how well these activities are synchronized that will influence the cycle times for a

multirobot cell. Thus the critical analysis area for the multirobot cell
is that of interaction delays.

Some operational issues of interest in multirobot applications are:

Cycle time evaluations
Assignment of tasks to robots
Synchronization of the operation sequence assigned to each robot
Evaluation of the cell control strategies
Determination of equipment type and capacity

All these issues should be focused toward the analysis of interaction delays.

Since multirobot cell operation involves both stochastic and real-
time control elements, it is unlikely that an analytical solution can
evaluate precisely the interaction delays, hence the cycle times. Material flow simulation is an expedient tool for these situations.

The factors that influence the interaction delays in a multirobot cell
can be summarized as follows:

1. Number of robots
2. Gross motion times and task times for each robot
3. Process of each robot and other processes
4. Number and type of the constraints
5. Location of the constraints in the process
6. Control strategy at each location

4.6.2 Cell Design Specification

A framework for the representation and specification of the cell activities has two requirements: it should unambiguously represent the operationally significant parameters of the cell activities, and it should
provide a framework for easy development of the model.

The cell activities have two components:

FLOW ELEMENT
    The assigned movements of the robots to accomplish the assembly
    tasks.

CONTROL ELEMENT
    The constraint controlled movements of the robots.

We assume the control action to be as follows:

"Allow the flow element to continue on a process until an unsatisfied
condition is encountered. The robot associated with the flow waits
until the condition is satisfied. This means the control action is

either to allow a robot process to continue or to delay it until the condition is satisfied."

The delay caused in a cycle in this manner is called the constraint-controlled interaction delay. The representation and specification of the cell activities then have two corresponding components:

Representation for the flow element
Representation for the control element

We represent the flow element by viewing the cell as a framework of interaction areas and the control element by the constraint controlled framework. The next two subsections describe the interaction area framework and the constraint-controlled framework.

Interaction Area Framework

First we divide the cell space into areas that both robots visit during their assigned activities in the cell. Then we superimpose the location of control areas or decision points on to these areas. We refer to all these areas as the *interaction areas*. We view the robots as if moving from one area to the next. This division provides us with an elegant way to view the continuous motion of the robots in discrete steps. Later this will help us in modeling by Petri nets and in evaluating the nets by discrete event simulation. The size of the interaction areas will largely determine the accuracy of our model.

An analysis of the cell activities suggests the division of the cell space into the following logical areas:

1. Task areas
2. Control areas
3. Collision areas
4. Path areas

To each area an associated operation time is assigned. This is the time between entering and leaving the interaction area.

A *task area* is an area in the cell where the robot is required to carry out some task (component pickup, assembly, tool change, etc.). The task time consists of the time between entering this area and leaving it after completion of the task. The *control area* defines the places in the cell where each robot must make a decision on its next action. The *collision areas* are the areas in the cell where two robots can possibly collide. The collision area time for each robot is the time between the entry into and the exit from this area. The *path area* of the robots is defined as the gross motion between the other areas. The path time involves the time to move from one area to another.

The interaction area framework is then a representation of the cell
space into areas of activities by one or more processes. The planned
operation sequence for each robot can be represented by a sequence
of visits to these areas. Each area associates a time element to the
activity taking place in that area. For different processes the activity times can be different for the same area. Furthermore, an area
can be visited by the same robot more than once with a different activity time on each occasion. Thus the interaction area framework allows us to describe the robot processes as a sequence of area activities with associated times. Each activity time refers to an area and
is the time between entering and leaving of the area after the accom-
plishment of the assigned task. We refer to the start or end of an activity at an area as an event.

Constraint Controlled Framework

Having outlined constraints of various types and discussed how
they control the operation of the cell, we now classify these constraints
and develop a framework to represent the complete cell operation.
Two types of constraint control the operation of a cell:

1. Process conditions:
     assembly precedence conditions
     work sharing conditions
2. Entity conditions:
     component condition: availability of required component when
     needed
     tool condition: availability of required tool when needed
     collision condition: availability of a shared interaction area

For the process conditions, the timing of the individual events in
each process must be synchronized. The conditions are generated
and changed by the events of the process. For instance, suppose
that robot 1 is assigned to assemble components C1 and C2, robot 2
is assigned to assemble components C3 and C4, and the precedence
sequence for the assembly is C1, C3, C4, C2. Then the operation
sequence and the timing of each event of the process must be so
planned that each robot proceeds for the assembly of the components
to satisfy the precedence condition. If either robot arrives for assembly before the preceding assembly is finished by the other robot,
it must wait. This wait is the interaction delay resulting from the
lack of synchronization of the processes of robot 1 and robot 2. One
may try alternative work assignments to minimize this. The cell layout may be altered to effect the timing of the events. For instance,
if the assembly of components C3 and C4 by robot 2 require much more
time than assembly of components C1 and C2 by robot 1, the cell layout

should be modified to make the paths traveled by robot 2 much shorter, even at the cost of much longer paths for robot 1. In general the lay-out should be such that all factors that influence the travel times for robot 2 are given priority.

The latter (entity) condition is independent of the order of the events in the process. It is equipment related. By "equipment" we imply the number and type of tools in the tool changer, the selected component feeders (feed rates), the choice of whether the tool changer and component feeders are shared resources, and the cell space. "Shared resources" clearly implies that only one robot can access the equipment at a time. Thus a shared resource implies an extra collision condition, hence a possible extra interaction delay. A shared resource will normally imply less space and less equipment cost. The cell space may influence the number of collision areas (i.e., the locations where the paths of the robots interfere). The smaller the cell space, the greater the number of possible collision areas and consequently the num-ber of conditions. The greater the number of conditions, the greater the likely interaction delay. On the other hand, the greater the cell space, the higher the travel times and perhaps the greater the size of the robots (greater work envelope). This normally implies greater cost. From the cell design point of view here, we are compromising between the interaction delays (wasted cell time) and the cost of having extra equipment. In terms of cell performance, the compromise is be-tween acceptance of high interaction delays with high utilization of equipment and securing low interaction delays at the cost of low equip-ment utilization. The common point about the two types of condition is that each affects the interaction delay in the same manner.

To summarize, from the cell design point of view the process con-dition is related to the planning of operation sequence, the work as-signment of the robots, and the cell layout design. The entity con-dition is related to equipment design. The goal of planning and de-sign is to synchronize the activities of the robots and other processes (e.g., component feeding) to minimize the interaction delays caused by the various conditions at minimum cost.

So far we have discussed various types of condition and how they effect the interaction delay. We have also outlined the planning and design implications of the cell with regard to these conditions. Now we focus on the control of the process with respect to these conditions and its influence on the interaction delays.

The control of a process with regard to a condition has two ele-ments:

1. How far ahead in time we need to take the control action.
2. What control action is taken.

Both these elements affect the performance of the cell.

4.6.3 Complete Cell Specification

Our representation and specification for the cell activities should be completely expressed in relation to the interaction areas and the conditions. We express all the robot activities as movements between these areas. A complete multirobot cell specification for cycle time evaluation can be summarized as follows:

1. The definition of the interaction areas.
2. The identification of the conditions.
3. The number of robots visiting all or a subset of the interaction areas.
4. Task assignments to each robot in terms of task interaction areas.
5. The operation sequence for each robot in terms of the sequence of visits to the interaction areas in order to carry out the assigned tasks.
6. The process times for each interaction area for each robot.
7. The location of control areas with respect to the conditions.
8. The relationship of the conditions with each control area.
9. The control action.

4.6.4 Modeling Framework

We can view a condition as a "resource." Each robot process can be viewed as an entity flow through the interaction areas. The flow is controlled at the control areas based on the status of the resources required for the movement to the next interaction area. If the resource is not available--that is, the condition is not met--the entity waits at the control area, causing an interaction delay. Thus the cell has a process related to each robot. The entity in the process is the robot. There can be other processes in the cell, too. The component feeding process, for instance, replenishes the entity resource (i.e., the condition that components are in the shared component buffer).

The process conditions are viewed as resources. They are created by one process and are consumed by the other process. The entity conditions can be viewed as fixed resources. They are used by each process for a certain time and then relieved or replenished. For instance, the shared tool changer resource is used for the tool change process time and then released. The interaction delay occurs because a process needs a resource that either has not yet been created by the other process or is currently being used by the other process. This means that if the cell operation is not locked, at least one process is always taking place. Thus the status of the resource causing a delay to a waiting process will eventually be changed by the active process. The magnitude of the interaction delay caused by a condition not being met is then governed by the status of each process with respect to a

condition. For instance, if the robot 1 process is in the occupation of
the tool change resource when the robot 2 process requires the same
resource, the interaction delay on the robot 2 process is equivalent
to the balance of the tool change time for the robot 1 process. The
resulting simulation on the flow of these entities through these resources is the classical material flow simulation. The control points
give a decision point framework [17] for control on the flow of these
entities.

The modeling framework for the cell operation can be summarized
as follows: The two robots are competing for the "resources." If
one robot is undergoing a process at a resource, the other robot must
wait if it requires the same resource. The competition for the resources arises because each resource has a unit capacity and the two
robots require some common resource (e.g., shared tools and the
shared interaction areas). We model the system as two entities (robots)
flowing under the constraint of their operation sequence to carry out
assigned operations at the free resources.

4.6.5 Petri Net Based Model

Referring to the modeling framework just described, we can convert
the cell specification into a Petri net as follows: *Each condition is
viewed as a resource and is modeled as a place. If a token is present in a place, it implies that the condition is satisfied or, equiva-
lently, that the resource is free.*

Each process is modeled as a disjointed Petri net with some common
places among each such net. The events in each process signify the
entering and leaving of the interaction areas. The operation sequence
determines the order of these events. Each interaction area then has
two events associated with it. Each event is modeled as a transition.
Entering into a new interaction area may require certain conditions to
be satisfied. Some interaction areas themselves are the shared entity
conditions, and their status at any time is modeled by the presence
or absence of tokens in the places representing them. As each process proceeds, the tokens are created or consumed. For the Petri
net, this implies that tokens are created as output tokens by certain
transitions, consumed permanently by some transitions, and tempor-
arily consumed by some and released back by others. Each transition, when it fires, schedules the next transition as in the process
where the end of each event schedules the next. The process time
associated with each interaction area is portrayed by the Petri net
as the time between scheduling of the next transitions. A transition
fires only if it is scheduled, and its input places, if any, have the
required number of tokens. The only interaction areas associated
with conditions are the control areas. Hence only transitions that
have input places correspond to the events for the control areas.

Each disjointed net is a sequence of transitions with a common sub-set of input places. The status of the places over different nets con-trols the firing of a scheduled transition at the control area. The marking of the Petri net is the union of the marking of the disjointed nets. Thus if a place on the net number 1 has a token deposited through a particular transition, the marking for that place on the other nets is a place with a token. The idea of the disjointed nets is that they model a number of processes competing for some common resources modeled as places. The transitions and activities on each net are completely different and independent.

4.6.6 Application Example

A multirobot assembly cell layout (Fig. 4.10) features a magazine M1, which acts as a shared resource for carrying the required components for the assembly. Five types of components (C1, C2, C3, C4, C5)

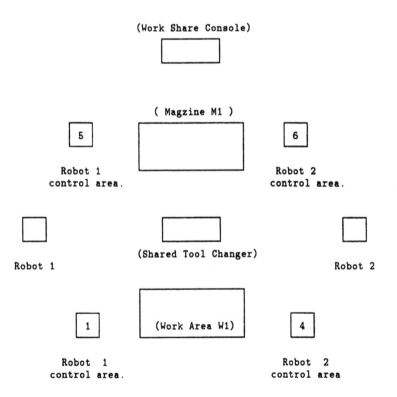

FIGURE 4.10 A schematic of a multirobot flexible assembly cell (FAC).

are to be assembled on a base assembly, which is situated on a fixture
in work area W1 as shown in the cell. An input conveyor is used to
bring the base assembly to the work area after every cycle. An output conveyor carries the finished assembly out of the cell. Two robots fulfill the assembly requirements, which consist of a series of
assembly tasks. An assembly task involves a robot moving to the
magazine, picking of a component, carrying it over to the work area,
and assembling it at the work area. One assembly task needs the coordination of the two robots; it involves the assembly of component
C2 to component C4. It is accomplished in a work share task area as
shown. The resulting assembly is denoted by C24. The assembly of
each component needs to be done in strict order: C1-C24-C3-C5
(i.e., component C1, followed by assembly C24, followed by components C3 and C5).

The work assignments to each robot are as follows:

Robot 1: C1, C2, C3
Robot 2: C4, C5

Two types of grippers or tools are needed to handle various component types. There is a shared tool changer, where the grippers
can be exchanged by each robot. The tooling requirements for handling these component types are as follows:

Tool T1: C1, C3, C4, C24
Tool T2: C2, C4

There are two tools of type T1 and one of type T2 allocated to the
cell. The tool not in use is stored in the shared tool changer area.
The components are fed by feeders, located in the shared component
area. Since one component type is needed in every cycle, the feeding rates are adjusted to be just below the estimated mean cycle times.

All shared resources imply potential collision areas. To control the
collision-free movement of the robots, control areas as illustrated in
Figure 4.10 are defined. For instance, the control area 5 for robot 1
controls the movement of the robot 1 to magazine M1. Such a move is
assigned only when magazine M1 is not in use by robot 2 and the component required by robot 1 is currently available at the magazine.

To help plan and design the cell more effectively, it is necessary to be able to estimate the cycle time and other performance
data.

Cell Specification

Figure 4.11 shows a schematic of the cell with various interaction
areas marked on it. The path areas are shown as lines joining other
areas in the cell. Figure 4.12 presents various process constraints

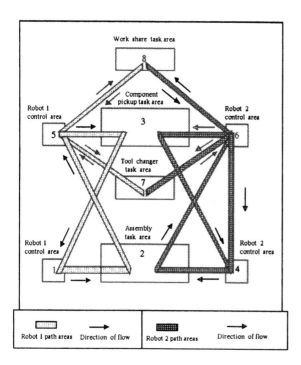

Robot 1 path areas   Direction of flow   Robot 2 path areas   Direction of flow

FIGURE 4.11 A schematic of a multirobot flexible assembly cell (FAC) showing various interaction areas. All possible paths for each robot are shown.

(e.g., work assignment, precedence constraint, tool assignment constraints, and robot coordination constraints). These constraints determine the operation sequence for each robot, as shown in the figure. Note that the precedence constraints can influence only the operation sequence for each robot. The actual control on the constraint will be determined in real time when the cell operates.

Figure 4.13 illustrates the entity conditions with their maximum capacities. The entity conditions are always related to task areas, and the process is controlled at the control areas adjoining the task areas. Table 4.6 summarizes the location, action, and relationship of the conditions with each control area.

Petri Net Model

The Petri net model development involves two steps:

1. Identify and define transitions, places, and activities.

Definition of the Constraints

WORK ASSIGNMENT

Robot 1 :   C1,C2,C3.
Robot 2 :   C4,C5.

WORK SHARING REQUIREMENT

C2 on Robot 1 is assembled to C4 on Robot 2
at work share area 8.

The resulting assembly is : C24

ASSEMBLY PRECEDENCE REQUIREMENT

C1—>C24—>C3—>C5

TOOL ASSIGNMENT

Tool 1 :   C1,C3,C4.
Tool 2 :   C2,C5.

OPERATION SEQUENCE ASSIGNMENT

Robot 1 : ┌─>5->3->1->2->5->7->5->3->5->8->5->7->5->3->1->2->5->7->─┐

Robot 2 : ┌─>6->3->6->8->6->4->2->6->7->6->3->4->2->6->7->6->─────────┐

FIGURE 4.12 The specification of a multirobot cell illustrating various process conditions and the operation sequence that is generated.

2. Develop a disjointed Petri net for the operation of each robot by linking transitions, places, and activities.

Table 4.7 gives a sample of identified transitions representing the events in the process for robot 1. Each transition is clearly defined.

Definition of the Constraints
_____

CAPACITY CONSTRAINTS
_____

Task Area	Description	Capacity
2	Assembly Workstation	1
3	Shared Component Buffer	1
7	Shared Tool Changer	1

Tools	Description	Capacity
T1	Located in area 7	2
T2	Located in area 7	1

Components	Description	Feed rate
C1	Located in area 3	10
C2	Located in area 3	10
C3	Located in area 3	10
C4	Located in area 3	10
C5	Located in area 3	10

FIGURE 4.13   The specification of a multirobot cell illustrating entity conditions of various types.

Table 4.8 shows the definitions and identifiers for various places representing the conditions in the multirobot cell. Table 4.9 illustrates the identifiers and the definition for the various activities. The activities on a Petri net graph indicate the scheduling of the next transition from a previous one.

Before describing the actual Petri net model, it is important to point out that the identifier symbols (pw1, pws1, t5c1, a5c1, etc.) are arbitrary. The symbols do not have any logical implications on the Petri net model. For convenience and easy interpretation of the Petri net graphs, they are chosen to be easily interpreted by the user. The identifier symbols chosen for our example bear some informal resemblance to the definitions. However we have maintained no rigorous formal structure. In some modeling approaches, such formal structures are well designed to enhance the translatability of the graphic model into a simulation code. The network modeling approach in SLAM II [4] uses graphical

TABLE 4.6 The Control Strategy for the Multirobot Cell

Control area	Control action
1	Robot 1 waits at area 1 if the required constraints at area 2 are not met.
5	Robot 1 waits at area 5 if the required constraints at areas 3, 7, 8 are not met.
4	Robot 2 waits at area 4 if the required constraints at area 2 are not met.
6	Robot 2 waits at area 6 if the required constraints at areas 3, 7, 8 are not met.

identifiers such as QUEUES and ACTIVITY, and these are translated directly into SLAM II simulation code. Since ROBSIM is designed to be data-driven (i.e., no simulation code translation is required), such formal structures are of little significance. However, we must reiterate that during a Petri net model development the use of meaningful symbols eases the task of the developer.

TABLE 4.7 Example Definitions for Transitions Describing the Petri Net for the Activities of Robot 1 in the Cell

Identifier	Definition
t5C1	Robot 1 leaves path area 7-5 and takes control area 5 to check constraints for component C1 pickup.
t53C1	Robot 1 leaves area 5 and takes path area 5-3 to approach component C1 pickup.
t3C1	Robot 1 leaves path area 5-3 and takes area 3 to pickup component C1.
t31C1	Robot 1 leaves area 3 and takes path area 3-1 with component C1.
t1C1	Robot 1 leaves path area 3-1 and takes control area 1 to check constraints for component C1 assembly.

TABLE 4.8 Example Definitions for Places Describing the Petri Net for the Activities of Robot 1 in the Cell

Constraint	Identifiers	Definitions	
Collision	p3	Shared component pickup	area 3 is free
	p2	Component assembly task	area 2 is free
	p7	Shared tool change	area 7 is free
Precedence	pc0	No component	assembled
	pc1	Component C1	assembled
	pc24	Subassembly C1, C24	assembled
	pc4	Subassembly C1, C24, C3	assembled
	pc5	Subassembly C1, C24, C3, C5	assembled
Work share	pws1	Robot 1 with component C2 waits in work share area 8.	
	pws2	Robot 2 with subassembly C24 leaves work share area 8.	
Component availability	pw1	Component C1 available in shared component buffer area 3.	
	pw2	Component C2 available in shared component buffer area 3.	
	pw3	Component C3 available in shared component buffer area 3.	
	pw4	Component C4 available in shared component buffer area 3.	
	pw5	Component C5 available in shared component buffer area 3.	
Gripper capacity	pT1	Tool T1 is available in shared tool change area 7.	
	pT2	Tool T2 is available in shared tool change area 7.	

Figure 4.14 illustrates a timed Petri net developed to model the process followed by robot 1. We now explain the execution of this Petri net to show how it simulates the process followed by robot 1. At the start time robot 1 is ready to enter conrol area 5. At this area the robot is required to check the status of task area 3, where it should proceed for the pickup of component C1 as the first task in its planned operation sequence. Two conditions must be satisfied before the robot can enter and leave the control area: component C1 must be present at

TABLE 4.9 Example Activities Associated with the Scheduling of the Transitions on Petri Nets for Robot 1 in the Cell

Identifier	Definition
a5C1	Time spent by robot 1 in control area 5.
a53C1	Time spent by robot in path area 5-3.
a3C1	Time spent by robot 1 in component C1 pickup at area 3.
a31C1	Time spent by robot 1 in path area 3-1.
a1C1	Time spent by robot 1 in control area 1.

shared component buffer area 3, and area 3, as a space in the cell, must be available (i.e., robot 2 is not currently using it). A similar strategy is designed for robot 2, which needs to start from control area 6 for the collection of component C4 from the shared component buffer area 3. At the start robot 1 holds a tool T1 and robot 2 also holds a tool T1 to enable them to pick C1 and C4, respectively. The component feeding processes for each component type also start at this time. An infinite input buffer is assumed to supply each component at a defined feed rate to the area 3.

Corresponding to this starting status, the initial Petri net marking contains a token for every place, except pT1, where there is no token because both of the tools T1 are being used by the robots. To start the timed Petri net execution (i.e., the simulation of the cell operation by the Petri net model), we invoke various start transitions to fire at time 0. For robot 1 the start transition is t5c1, for robot 2 it is t6c4, and for the component C1 feeding process it is tW1.

Now we briefly outline how the Petri net in Figure 4.14 models the process followed by the robot 1. The start transition t5c1 is invoked to fire at time 0. The transition can fire only if its input places p3 and pw1 have an enabling token each. This represents the action of robot 1 at control area 5, where the condition on the movement of the robot into area 3 is the availability of the area 3 (place p3 on the Petri net) and the availability of component C1 (place pw1). If any of the conditions is unsatisfied (i.e., if a required token is unavailable), this transition waits until a token becomes available. That is, the robot waits at control area 5. If the required tokens are available, the transition t5C1 is enabled to fire. It fires instantaneously and removes the enabling tokens from places p3 and pw1. Since there

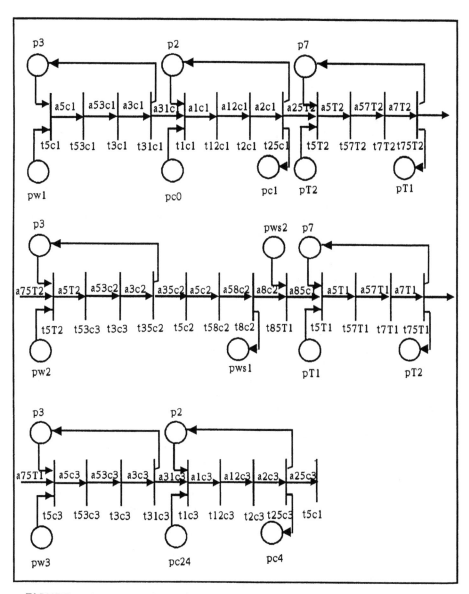

FIGURE 4.14 The Petri net for representing the process followed by robot 1. NOTE: In the strict Petri net graph, each transition is followed by a place. For ease of representation we have used activity arcs between some transitions to represent time durations.

are no output places, there is no deposition of the tokens. However, there is a next transition t53c1, which is scheduled to be invoked immediately after the activity period a5c1 as marked on the net.
This implies that the space area 3 and component C1 are assigned for
the use of robot 1 and are unavailable for use by robot 2. The activity a5c1 represents the time spent in control area 5 in preparing for
a move to path area 5-3. This time can be viewed as sensor information or program loading time in control area 5 for robot 1.

When time corresponding to a5c1 has elapsed, the transition t53c1
is invoked to fire. Again the conditions for enabling it to fire are
checked. Since there are no input places to this transition, it fires
as soon as it is invoked. There are no output places to it, so there
is no deposition of tokens as it fires. However there is a next transition t3c1, which is scheduled to fire after activity time a53c1 (representing the path 5-3 travel time). The robot 1 activity in progress
is now the travel from area 5 to area 3. After the travel time has
elapsed, transition t3c1 is invoked. Since there are no input places,
this transition also fires instantaneously as it is invoked. It schedules the next transition t31c1 to be invoked after activity time a3c1.
Note that transition t3c1 represents the event of robot 1 entering
area 3 and leaving path area 5-3. The robot activity in area 3 is the
component pickup activity. The activity a3c1 corresponds to this.
After the pickup time has elapsed, the robot is ready to move to the
control area 1 through path 3-1. The transition t31c1 represents this
event. When this transition is invoked it fires immediately, since there
are no input places to it. However as the robot is about to leave area
3, this area is to be set free. This is represented by the output place
p3, to which a token is deposited as t31c1 fires. The transition t31c1
is scheduled to invoke the transition t1c1 after activity time a31c1.
This activity corresponds to the travel time for path 3-1 for robot 1.
When robot 1 travels this path (i.e., the a31c1 activity time has
elapsed), the transition t1c1 is invoked.

The robot needs to check the status of assembly area 2 before the
next move can take place. This is shown by the input places p2 and
pc0, which represent the necessary conditions for the next robot
1 activity. The invoked transition t1c1 can fire only if these input
places have a token each representing that the necessary conditions
are satisfied. The place p2 represents the condition regarding the
availability of area 2 (i.e., it is not being used by robot 2 at the
time). The place pc0 represents the condition that an assembly with
no components deposited on it is present in the assembly area. If
these conditions are met (i.e., the tokens are present in the places
p2 and pc0), the invoked transition t1c1 is enabled to fire. In this
way the Petri net execution simulates the complete process followed
by robot 1.

The process for robot 2 shown in Figure 4.15 can be similarly interpreted.

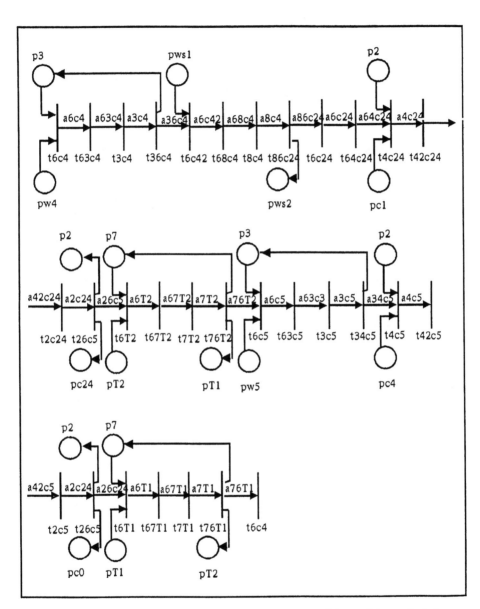

FIGURE 4.15 The Petri net operation sequence followed by robot 2.

The interaction delays are represented by delays in the firing of an invoked transition which occur because the input places to it may not have an enabling token. These delays are caused because the necessary conditions are not met at a time when they are needed by a process. Such waits can be evaluated only when the Petri nets representing each process are executed simultaneously to mimic the operation of the cell. ROBSIM is designed to execute these Petri nets in this manner. Note that the interactions between processes are represented by some common places among the Petri nets representing the processes.

We refer to these Petri nets as disjointed Petri nets. In this form of disjointed Petri net representation, the graphs are more readable and their development is eased. From this perspective the Petri nets shown here differ from the conventional Petri nets. In the latter each place is mentioned once and only once on a Petri net graph, and all the processes are represented in a single Petri net graph. ROBSIM is designed to collect the interaction delay statistics for each place along with the utilization statistics that are collected for each activity. Similarly other statistics such as the cycle time and slack time are collected. The cycle time refers to the time elapsed between the firing of the start transitions for each robot. Note that on each disjointed Petri net for a robot process, the start transition is invoked after the last activity in the robot cycle. A Petri net for the component arrival process is illustrated for component C1 in Figure 4.16. The initial marking allows a large number of tokens in the place FB1.

4.6.7 ROBSIM Input Data

Table 4.10 is a sample logic table of the type that summarizes the Petri net for robot 1 and is used in ROBSIM. For each disjointed Petri

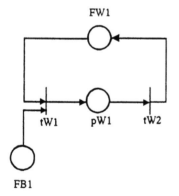

FIGURE 4.16 A Petri net for the component arrival process.

TABLE 4.10 Sample of the ROBSIM Logic Table to Input Data About the Petri Net for the Activities of Robot 1

Source transition	Input places	Output places	Scheduled transitions	Activity identifiers
t5c1	pw1, p3		t53c1	a5c1
t53c1			t3c1	a53c1
t3c1			t31c1	a3c1
t31c1		p3	t1c1	a31c1
t1c1	pc0, p2		t12c1	a12c1

net, one such table is used. Each transition in the net is specified in the first column. For each of these source transitions, the input and output places and the scheduled next transitions are input. For each pair of "source to next" transitions, an activity number is assigned. Later, through the activity times table, the activity times can be specified. Table 4.11 illustrates the activity times table, showing some sample activities for robot 1 and robot 2 on their corresponding Petri nets. These times can be input as any standard time distribution by specifying the parameters defining that distribution. For example, the time for activity a5c1 is input as a normal distribution (code 2) with mean of 1 second, minimum of 1 second, maximum of 4 seconds, and a standard deviation of 0.5 second.

4.6.8 ROBSIM Output

Figure 4.17 shows a sample from the ROBSIM output report for the present example.

An average cycle time of 66.78 seconds is expected from the cell. This time can be viewed as consisting of two components: (a) actual time required to assemble, assuming perfect synchronization of the two robots, and (b) time wasted because of lack of synchronization. The second component is reflected in the following output data.

1. The mean cycle slack time of 7.22 seconds, which represents the time difference between the finishing of a cycle by robot 1 and

TABLE 4.11 Sample from ROBSIM Activity Times Table

Activity numbers	Distribution code	Parameters				Random number code
		1	2	3	4	
Robot 1						
a5C1	2	1.	1.	4.	0.5	1
a53C1	2	1.	0	0	0	1
a3C1	2	1.	1.	3.	0.5	1
a31C1	5	1.	0	0	0	1
a1C1	2	1.	1.	3.	0.5	1
a12C1	5	1.	0	0	0	1
a2C1	2	1.	1.	3.	1.	1
Robot 2						
a6C4	2	2.	2.	5.	2.	1
a63C4	5	3.	0	0	0	1
a3C4	2	8.	2.	12.	3.	1
A36C4	5	3.	0	0	0	1
a6C42	2	4.	2.	7.	2.	1

robot 2. This indicates that one robot is slower in finishing the cycle than the other.

The result suggests that it may be worthwhile to increase the speed of the slower robot. Alternatively, the layout of the cell may be modified to reduce the paths traveled by the slower robot. In general, it reflects the imbalance in the assignment of operations to the two robots and indicates that design changes should be made to aid the slower robot.

2. The mean interaction delay times at assembly area 2 and magazine area 3 indicate that there is a potential for improvement in the coordination of robot operations. Such operation policy changes as different cycle start times for each robot may be analyzed. The control strategies may be changed to look sufficiently forward in real time to do an alternative operation, if the interaction delay is likely to occur with the current sequence of operations.

A further result of interest is the utilization of the task areas. The utilization of area 3 indicates that 30.0% of the time in a cycle the robots are doing useful assembly work. The higher this percentage, the higher the assembly efficiency of the cell.

```
┌─────────────────── SIMULATION  OUTPUT ───────────────────┐
│                                                          │
├────────────────────── Cycle Time ──────────────────────┤
│            Mean  Cycle  Time      :   66.78              │
│            Standard Deviation     :    3.6               │
│            Mean Cycle Slack Time  :    7.22              │
├──────────────────── Interaction    Delay ──────────────┤
│ Mean Interaction Delay/cycle  at assembly area 2        :   8.32 │
│ Mean Interaction Delay/cycle  at shared component area 3 :   1.35 │
├─────────────────────── Utilizations ───────────────────┤
│      % of time spent at assembly area  2        :   30.0 │
│      % of time spent at shared component area 3 :   47.5 │
│      % of time spent at tool changer area 7     :   28.5 │
│      % utilization for tool T1                  :   48.0 │
│      % utilization for tool T2                  :   80.0 │
└──────────────────────────────────────────────────────────┘
```

FIGURE 4.17 A sample from ROBSIM output report.

The utilization of tool T2 is high. However that of tool T1 is low, which suggests that we may reduce the number of available tools T1. A review of this suggestion indicates that such a change would raise the interaction delay caused by unavailability of tool T1.

4.6.9 ROBSIM Integration with FASIM

ROBSIM can be used in stand-alone mode to analyze the material flow performance of a multirobot cell. If this cell forms a part of an FAS, then like FACET, ROBSIM will be required to furnish the cycle time data to FASIM.

4.7 A GENERALIZED SIMULATOR FOR
FAS: FASIM

The material flow simulator called FASIM has been designed to analyze the integration of robot assembly cells, auxiliary cells (inspection, repair, carousel, etc.), the transport system, and the control strategies from the perspective of an efficient material flow in the system. The simulator is generalized, and a wide spectrum of FAS configurations can be modeled. It is data-driven, as the FAS configuration can be described in specially designed user input formats.

Before describing the various features of FASIM, it is necessary to discuss the integration of ROBSIM and FACET with FASIM. As noted previously, ROBSIM and FACET are used for the detailed modlling of cell designs to yield the cycle time distribution for the subassemblies to be assembled at each cell. At the systems level, other cell characteristics, such as setup times and the frequency of major breakdowns of each cell along with its repair times, are as important. Furthermore, the level of flexibility in each cell with respect to the subassemblies that can be processed at that cell is important. Knowing these characteristics, one needs to understand the material flow performance of the system over a given mix of assemblies. The effect of cycle time variability needs to be accommodated by proper design of the intercell buffers.

In the following section the modes of application of FASIM to simulate the material flow performance of the system involving robot cells is described. It is important to reiterate that at the systems level a robotic cell is simply a black box with certain material flow characteristics (evaluated and supplied by ROBSIM and FACET). Certain utilization levels are desired of them, and the system design and the control policies should attempt to achieve these levels.

4.7.1 FAS and Simulation

The role of simulation in FAS area can be categoried in two areas: FAS design and FAS operation.

4.7.2 FAS Design

An FAS design must identify the system equipment, layout, and configuration to meet the desired production requirements for finished assemblies. Using simulation for FAS design implies performance evaluation of alternative equipment configurations, then prioritizing the best solution. The equipment is varied intuitively to find the minimum equipment needed for production. Simulation provides a high level of confidence that a particular solution will meet the production objectives.

The operationally significant parameters in system design include the level of flexibility in assembly cells, number of assembly cells, number of inspection and repair stations, relative location or layout of cells, transport configurations, buffer levels allowed at various cells, and so on. The role of simulation lies in studying the integration of these elements and their compatibility with software control. Flexibility in equipment requires the designer to study the compatability between transporter type and amount of in-process storage. For example, an AGV-based transport system may require more in-process storage than a conveyor. The transporter and the in-process storage must serve the assembly cells without constraining the system performance. The design of a transport system and its compatibility with processing stations are more critical in FAS design than in FMS design. Some comparisons between FMS design and FAS design are presented in Ref. 18. Within the application domain of FASIM and its integration with FACET and ROBSIM, these factors have been appropriately taken into consideration.

Flexibility in cells and in the system is expensive. The designer has the complex task of identifying the level of flexibility that will meet the production requirements. The designer also tries to identify the number of pallets, fixtures, the mix of parts, and so on. The maximum number of pallets simultaneously in the system is a crucial design parameter for capacity-constrained systems. This must be compatible with the in-process storage capacity or designed buffer levels at each cell. Furthermore, the work should be well distributed over the cells. Since the cells have a degree of flexibility, proper material flow controls should be used to take advantage of this flexibility. With too few routing assignments, the cells are underutilized. With too many, congestion keeps the parts from moving freely. An imbalance in part mix causes a surge that overutilizes one cell and underutilizes another.

4.7.3 FAS Operation

The use of simulation in designing FAS operation control refers to the situation of an existing FAS system. However the production requirements for which the system was initially designed may change over time. New part families may appear, or parts mixes may change. Now the designer needs to vary the control strategies to meet the new production requirements.

The control strategies for an FAS are similar to the ones described by Lenz [19] with reference to an FMS operation control. These are as follows:

Part balance control: determines what assemblies are scheduled into the system at what time. Usually part mix control and the maximum number of pallets in the system are controlled. Proper control assures uniform utilization of various resources in the system without overloading the buffers.

Operation sequence control: determines the flexibility in the precedence sequence for assembly operations for each assembly. This
could be an effective way of balancing the utilizations of various
resources.

Loading control: provides a selection criterion for loading the next
assembly onto the cell from a queue of waiting assemblies in the input buffer of the cell. Some of the popular criteria are SPT (Shortest Processing Time), FCFS (First Come, First Served), and EDD
(Earliest Due Date).

Dispatching control: determines the next assembly station for an assembly that has finished one of its assembly operations at a station.
The selection criteria for this control may be the availability and
backlog at the next station.

WIP control: allocates semiprocessed items/assemblies to a centralized
storage area, if it exists in the system.

Transport system based controls: if the transport system is a conveyor system, this control is trivial. With conveyor systems, usually
the capacity is essentially unlimited and is treated as always available.

However with AGV-based system, the following controls will be significant:

AGV-initiated assignment control: exercised when an AGV becomes
free after finishing an assembly delivery. There are usually a number of requests pending in the system for assemblies waiting at the
output buffers for a pickup and transportation to next station. The
selection criteria seek to ensure high utilization of the transport resource and proper attention to priority assemblies.

Part-initiated assignment control: used when an assembly has a request for movement to next station and there are a number of idle
AGVs available. The control determines which AGV should be assigned for the job. The control offers uniform utilization of the
AGVs and often higher "loaded AGV" utilizations.

The designer chooses the appropriate control policy for each of
these controls. Only simulation can aid in determining the influence
of these control strategies on the overall system performance.

The foregoing discussion illustrates the breadth of FAS design and
control studies. For the complexities involved and high investments
associated with FAS, only digital simulation can provide high level of
confidence in satisfactory design or operation of such systems.

4.7.4 FASIM

FASIM: An Overview

FASIM is a data-driven generalized model for FAS simulation. It
is based on the discrete event framework of the simulation language/

package SLAM II and is implemented on a VAX 11/780 under VMS. The methodological development features of FASIM are described in Ref. 17.

FASIM views the FAS system as follows.

1. The FAS physical structure (type of workstations and their relative location, transportation system, etc.).
2. The FAS control structure (the production control procedures that manage the flow of work through the FAS physical structure).
3. The assembly requirements for the products to be processed through the system.

FASIM defines the FAS physical structure in terms of three areas:

1. Robot-based assembly cells or automated assembly cells for the discrete parts assembly.
2. Auxiliary cells such as inspection, repair, pallet loading/unloading stations, carousels, special processing stations, and the pallet/fixture types for moving the assemblies through the system.
3. A transport system interlinking these cells.

FASIM views the FAS operation control in terms of two factors:

1. A decision point framework [17].
2. A control decision associated with each decision point in this framework.

FASIM provides a set of selection criteria and algorithms as input data menus for each class of control options. The designer selects from this menu and assigns a decision criterion for each decision point in the framework. Therefore FASIM models the flow of material, resources, and information through a designated FAS. The decision points exercise "control" on this flow. We now review in more detail how FASIM views the physical components of an FAS.

FASIM: Details

The Assembly Cells: The first important elements in FASIM are the assembly cells. FASIM allows the designer to input the processing capabilities of each of these cells in one of three modes:

1. Deterministic estimates of the assembly operation times.
2. The assembly operation times modeled as standard distributions.
3. Previously evaluated cumulative probability functions of the processing times. To support the evaluation of these functions, two special-purpose submodules, namely FACET and ROBSIM, were described earlier (Sections 4.5 and 4.6).

FACET and ROBSIM are relevant only in the context of robot-based
assembly cells. FACET (flexible assembly cell evaluation tool), as we
have seen, is in itself a generalized data-driven simulator for modeling
a single robot-based assembly workstation. ROBSIM is a Petri net
data-driven simulator for modeling individual assembly cells that contain multiple robots.

The Auxiliary Cells: As noted previously, auxiliary cells include
inspection stations, repair stations, pallet loading and unloading stations, carousels, special processing stations, and the pallet/fixture
types for moving the assemblies through the system. FASIM models
the operating logic of each of these workcell types. For instance, a
percentage of assemblies are sent for inspection. A proportion go for
repair. In repair cells a proportion of assemblies are scrapped. The
remainder either are sent back to inspection or proceed to further operations. The parameters are dictated by the designer. Similarly
FASIM considers a carousel as a generalized storage facility for work
in progress.

FASIM provides an input format to capture operationally significant
features of these cells. Also FASIM allows the designer to specify how
many types of pallet are being used, which pallet is used for which assemblies, and where the pallets are located. The designer also specifies the maximum number of pallets that can simultaneously exist in
the system.

For each station FASIM allows the designer to specify the maximum
buffer capacities at input and output. Furthermore, for each cell the
loading, unloading, and setup times are specified. FASIM provides
the facility to pool cells by allowing the designer to input the number
of servers (i.e., identical cells in parallel) for each station.

Transport System: The third component in FASIM is the transport
system. FASIM allows the designer to simulate the following types of
transport systems:

1. An AGV system where the AGVs act as a transport device only.
2. An AGV system where the AGVs act as both workstation and transport.
3. Conveyor system.

FASIM allows the designer to detail the interface of the cells with
the transport system (i.e., are the cells located on spurs or on a line
on the track?). For the latter case, the designer is prompted to specify the collision avoidance time for the approaching AGV if another
AGV is serving a station while staying on the track line.

Finally the designer inputs the transport times for movement between the stations in the system.

With respect to all equipment, FASIM allows the designer to specify the reliability data and the repair data in terms of breakdown and repair time distributions.

Finally FASIM input requires the routing data about all the assemblies to be produced by the system. For each assembly it allows the designer to specify priorities and the maximum proportion of each assembly type that can reside in the system. This keeps a control on the part mix in the system. The precedence constraints of each assembly are input as operation sequence.

From the simulation control perspective, FASIM has designer-requested statistics clearing, trace generation, and simulation termination condition options.

The designer can also specify the starting conditions of the FAS to be simulated. It can be started as empty or with specific assemblies at specific stations at various stages of their operation requirements.

FASIM has built-in data entry error check routines, which provide appropriate message on the screen and stop simulation if certain discrepancies are found while the model is running.

4.7.5 FASIM Output

FASIM provides an on-line intermediate report at regular user-defined intervals. These reports are put on the screen while the simulation runs in the background. Besides the intermediate reports, FASIM presents a detailed summary report about various significant statistics. The utilization and part statistics reports are also presented graphically for review by the designer. Some common statistics generated by FASIM are:

Part statistics: the average time in the system for each assembly type, the average cycle times, the minimum and maximum times for each, the standard deviations, the proportion of time spent in storage and transport, etc.

Resource utilizations: the utilization of cells, pallets, and the AGVs. The proportion of time each station is busy, idle, down, or blocked is presented. The blocking statistics are a measure of congestion due to inappropriate control and/or insufficient buffer capacities. The cell utilizations indicate underutilized and bottleneck cells.

Queue statistics: an effective measure of FAS congestion. FASIM provides maximum, minimum, and average queue lengths for each station input and output buffer.

An application example to illustrate the use of FASIM is presented later (Section 4.8.4).

4.8 FLEXIBLE ASSEMBLY SYSTEM DESIGN
TOOL: FASDT

FASDT [8], the flexible assembly system design tool, contains an *expert system* and a user-friendly interface for the FAS designer. The former is written in OPS5 and the latter is implemented with Pascal and SMG run-time routines available on VAX/VMS 780.

This section outlines a system that is a combination of a generalized data-driven simulator (FASIM) for the modeling of an FAS and a prototype expert system (FASDT), which analyzes the model output and suggests modifications to the FAS design, as captured in the model, to meet the previously defined design goals. This approach reduces the need for the simulation expertise and FAS design expertise, as well as the lead time to reach a satisfactory FAS design.

Figure 4.18 illustrates the cooperation of FASIM and FASDT. The system starts with an initial design input to FASIM. This FASIM input is simulated, and output performance is generated and stored in output data files. FASDT invokes the designer to define his objective as a target or goal. Then FASDT accesses the output files of FASIM and examines it to see whether the design objective is fulfilled by the initial design. If the objective is fulfilled (i.e., goal is reached), FASDT indicates this to the designer as "Target Fulfilled." If the goal is not reached, it uses its expert knowledge to locate the symptoms through relevant analysis of the output. For the identified symptom it suggests a number of design changes. The designer can accept, reject, or suggest his own changes. FASDT acts as a decision support

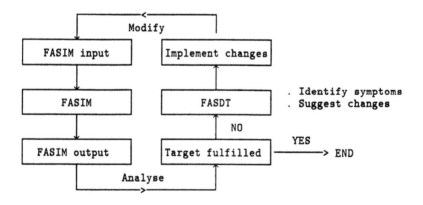

FIGURE 4.18 Association of FASIM and FASDT.

tool. The accepted design change is automatically implemented in the FASIM input files, and FASIM is rerun on this new input data. The cycle is repeated until the goal is reached.

4.8.1 Knowledge Representation in FASDT

There are three principal elements in FASDT: targets, symptoms, and suggestions. The target is input by the designer as the objective to be met by the final FAS design. Symptoms result from simulation output analysis. Suggestions are originated from FAS design expertise built into the expert system.

4.8.2 FASDT Knowledge Base

Figure 4.19 summarizes some of the rules in the FASDT knowledge base. For example, if the objective for design is to attain a certain throughput in a certain time, we say that the throughput is the target. If a system design fails to meet it, possible reasons are low WIP, station blockings, and AGV blockings. Each such reason is called a symptom. The particular symptom that applies to the present state of the design is determined from an analysis of its performance measures using some expert rules. After a symptom has been identified, a set of possible

TARGET	SYMPTOMS	SUGGESTION
Throughput	Low_Work_In_Progress	Increase_Work_In_Progress
		Increase_Buffer_Levels
		Increase_FAS_Loading_Rate
	Station_Blockings	Increase_Station_Output_Buffers
	AGV_Blockings	Increase_Station_Input_Buffers
	Station_Bottleneck	Redistribute_Work_Assignment
		Reduce_Processing_Time
		Add_Station
	Transport_Bottleneck	Add_AGV
	Wrong_Product_Mix	Change_Product_Mix

FIGURE 4.19 A sample of FASDT knowledge.

design modifications is sought. These possibilities are called suggestions; for example, if the symptom is station blockings, the suggestion is to increase the size of the station output buffers.

4.8.3 User Interface of FASDT

The FASDT user interface is built with Pascal routines calling the VAX runtime library routines for screen management (SMG). These routines allow the designer to have maximum information on the screen with minimum input effort. A FASDT user interface screen is shown later for an application example. The screen is divided into special areas or windows. An appropriate window is displayed as actions are required during an interactive session with the designer.

4.8.4 An FAS Example and Use of FASDT

This section describes a simple FAS configuration. FASIM is used to simulate this configuration. An objective is laid out for the design. FASDT in conjunction with FASIM guides the designer to design modifications necessary to achieve the stated objective. This is illustrated by sample screens from the FASIM results and FASDT suggestions. The schematic in Figure 4.20 shows 4 cells (LS, SR1, MR1, SR2) linked together by a track forming a loop. Three AGVs (AGV-1, AGV-2, AGV-3) serve as a transport system and move on this track. The track is unidirectional; that is, the AGV can move in only one direction. The system is required to process three types of assembly— say A1, A2, A3. Each cell has an input buffer where the assemblies wait in a queue if the cell is busy. Similarly each cell has an output buffer where the assemblies are placed after the completion of the required operation in the cell. The number of assemblies each cell buffer can hold is limited. The AGVs pick the assemblies from the output buffer of a cell and drop them to the input buffer of another cell at which its next operation is desired. The pickup and drop points for the AGVs are called decision points and are denoted as DP1, DP2, etc. Each DP is physically associated with an area called the checkzone area (as marked with DP8). The checkzone area controls the collision avoidance between AGVs. If one AGV is busy (picking up or dropping operation) at a DP, the other AGV cannot pass through this DP and is forced to wait in checkzone area. The AGVs travel from one DP to another at constant speeds. Thus the layout of the system can be defined in terms of the relative location of the DPs and the travel times between one DP and its next DP in the AGV travel direction.

The assemblies enter the system as base assemblies from the loading station LS, which loads them on to pallets (one pallet carries one type of assembly). The loaded assembly is placed on the output buffer

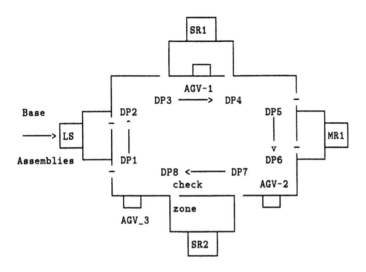

FIGURE 4.20 A sample FAS configuration.

of the load station. An AGV is required to pick up this assembly
and transport it to the cell at which its first assembly operation is
designed to take place. For each assembly, the desired sequence
of operations is defined by the operation routing. The assembly
cells SR1 (single robot cell), SR2 (single robot-based cell) and MR1
(multirobot cell) allow a certain degree of flexibility. This flexibility is defined by operation routings as shown in Figure 4.21. One
assembly at one cell undergoes one assembly operation. Operation
routing describes the precedence sequence of operations from start
to finish for each assembly type. Consider the routing for assembly
A1. The first operation, the loading operation at station LS, requires
operation time alf0. The next operation OP2 can take place either at
cell SR1 with operation time alf1 or at cell SR2 with operation time
alf2. The third operation OP3 can take place on MR1, with the operation time alf3.

The operation times at assembly cells may be constants, samples
from standard distributions, or expressed as cumulative distribution
functions (cdf). If tools like FACET are used to evaluate the assembly times for each of the operations, they can be obtained as a probability distribution function.

The foregoing description covers the physical structure of the system. The next part of the system description is the real-time system
control that controls the flow of assemblies, cells, and AGVs based on
the cell status.

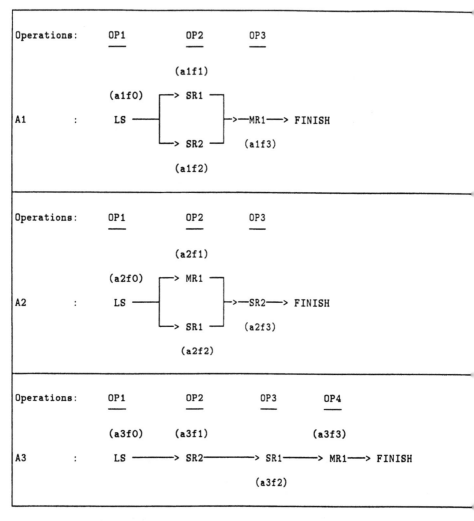

FIGURE 4.21 The operation routing for each assembly type (i.e., A1, A2, A3).

1. *Part entry control*: Parts (base assemblies) enter the system through the load station sequentially (i.e., A1-A2-A3). A limited number of assemblies are allowed in the system at any one time. This is a critical design parameter, and part entry control seeks to maintain a level of work in the FAS by allowing fresh parts to enter the system.

2. *Loading control*: Assemblies waiting in the input buffer need a priority control mechanism to determine which assembly is selected to be processed when the cell is free. This control is exercised by loading control through heuristics such as FCFS and SPT. In this system loading control is by the FCFS loading rule at each assembly cell.

3. *Dispatching control*: After an assembly has finished its current operation, it is placed in the output buffer. A request for an AGV to pick this part for its next operation is made. When an AGV arrives to pick this assembly for its next operation, a dispatching decision needs to be made to select the next station where the operation can be performed. The flexibility in operation routing allows this choice. This system dispatches the assemblies to the cell with the minimum number of assemblies in its queue.

4. *AGV-initiated assignment control*: When an AGV finishes its assigned job, its next job assignment must be determined. Usually there are a number of assemblies waiting to be picked up for dispatching. Each such assembly is a request job for the AGV. The AGV can be assigned based on rules such as assign to the nearest job or assign to the job that has been waiting the longest. Our example uses the nearest job assignment rule.

5. *Part-initiated AGV assignment control*: Sometimes, though rarely, more than one AGV is free or idle. It is necessary then to decide which AGV should be assigned for a job. Heuristics such as assign the nearest AGV or assign the AGV with minimum utilization can be used. Our example uses the latter rule (i.e., the idle AGV with minimum utilization is assigned).

6. *AGV transport control*: This control is used to avoid collisions in the system. As noted earlier, in our example this is achieved by forcing the traveling AGV to wait in the checkzone area until the serving AGV clears the interfering decision point.

7. *AGV blocking control*: The AGV is blocked at a delivery point if it cannot unload the assigned load. This happens if the input buffer at the assigned delivery point is full. The AGV blocking control is exercised under such a situation. The options are to allow the AGV to return to this buffer after a complete loop and try again, let the AGV be blocked along with the track (i.e., no other AGV can pass through the blocked track), let the AGV be blocked but the track cleared (by shifting the AGV to a side track), or send the AGV to unload at a temporary storage. In our example we use the third option—block the AGV but keep the track free.

8. *Free AGV control*: An AGV is freed at a delivery point after depositing its load. At this point, if no request is present for the AGV, it is free. However free AGV control is needed to decide whether this AGV stays on the track (blocking the passage of other AGVs), leaves the track and moves to a side track, or keeps

moving on the track until a request is received. In our example
the second option is used—the track is freed and the AGV stands
idle.

9. *Blocked cell control*: A cell is blocked when its output buffer is
   full and it cannot unload the assembly that has currently finished
   its operation on this station. A blocked cell is relieved only when
   the output buffer has the necessary capacity to relieve the cell.
   This is the default control for the described FAS.

FASIM allows the designer to input all the data regarding the foregoing FAS description by means of special input data menus. If
FASDT is used in association with FASIM, the design objectives are
also requested from the designer. For our example an aggregate objective of finishing 100 assemblies over one shift is chosen. In the
following section we present the output results obtained from a sample
of data input for our example FAS. We illustrate the working of the
FASDT using this example.

A set of data values were input to FASIM with the goal of producing 100 assemblies per 8-hour shift. The following is a summary
of an interactive session with FASDT and FASIM.

Figure 4.22 shows two screens. The top screen is extracted from
the FASIM simulation output run for the FAS configuration described
above. The top two rows summarize the information about simulation
parameters. The simulation is run for 600 time units and the statistics on material flow performance are cleared at 120 time units. The
latter action ensures that the system is in steady state before we collect the performance data. For each of the four stations, the blocking
and utilization statistics are collected. For each of the three AGVs,
the proportion of idle time, blocked time, and load utilization is presented. For each of the three assemblies (indicated under the column
"part number"), the current status of the finished assemblies, the
time of finishing of the last assembly, the number of assemblies in the
system, the average time in system, and the cycle time are shown. Finally the last two rows indicate the total number of assembly pallets
(parts) in the system, the maximum number of parts allowed, the total
number of finished assemblies, and the time at which the last assembly
was finished.

The bottom screen on Figure 4.22 is taken from FASDT. The simulation results in the FASIM screen indicate a high blocking at stations
1 and 2. This indicates that the assemblies are stacking up in the output buffer of these stations. Thus either the designed buffer capacity
is inadequate or the transport system is inefficient in moving the material from these stations. FASDT suggests that a higher output buffer should be provided at these stations. The designer can accept,
modify, or reject this suggestion and look for new suggestions from
FASDT by choosing an appropriate option from the "please choose"

```
───────────────────── F A S I M ─────────────────────
Simulation time : 600.0              Statistics clear time : 120.0
```

Station #	Block %	Utilization %	AGV #	Idle %	Block %	Load utilization %
1	45.0	16.0	1	0	2.8	59.6
2	53.0	30.0	2	0	1.5	63.0
3	4.4	52.3	3	.2	3.5	64.7
4	0.1	37.6				

Part #	Finished number	At time	Number in system	Average time in system	Cycle time
1	27	593	3	55.3	17
2	26	578	2	60.0	17
3	27	590	3	92.0	17

```
Total number of parts in system now : 8      Maximum Allowed : 9
Total number of finished parts      : 80     Last time       : 593
```

```
───────────────────── F A S D T ─────────────────────
```

User Input	Actual Target
Target : THROUGHPUT Not fulfilled !	THROUGHPUT ────────────
	──── Actual Symptom ──── STATION_BLOCKINGS
Actual value : 80 Desired value : 100 Reason : STATION_BLOCKINGS	── Possible Symptoms ── Low work in progress. Stations blocked. AGVs blocked.
FASDT suggestion ────────────────	Bottleneck station. No. of AGVs too low. Wrong product mix.
Higher output buffer levels from 1 to 2	
Station: 2 ┌─ PLEASE CHOOSE─ Accept Own suggestion Next item Next symptom See last change INFO HELP QUIT	── Possible Suggestions ── HIGHER_OUTPUT_BUFFER_LEVELS

FIGURE 4.22 Screens from FASIM and FASDT to illustrate the application of the goal-directed, data-driven simulator.

window. We accepted the suggestions, and the FASDT software implemented them automatically into the FAS design files.

FASIM was run again with this modified design. Figure 4.23 shows the performance results obtained. The station blockings dropped and

```
─────────────────────── F A S I M ───────────────────────
   Simulation time : 600.0            Statistics clear time : 120.0

   Stn #   %Block   %Utilization  │  AGV   Idle   %Block  %Load utilization

    1       7.0       17.7        │   1     0     2.8      65.3
    2       4.3       36.0        │   2     0      0       66.5
    3       5.2       50.5        │   3     0     3.8      60.8
    4        .3       37.5        │

   Part    Finished      At       Number in      Average       Cycle
    #       number      time       system      time in system   time

    1         28         586          1           52.5           17
    2         27         595          4           56.5           17
    3         31         596          4           97.2           16

   Total number of parts in system now : 9       Maximum Allowed : 9
   Total number of finished parts      : 86           Last time    : 596
─────────────────────── F A S D T ───────────────────────
       User Input                         Actual   Target

   Target : THROUGHPUT                      THROUGHPUT
   Not fulfilled !                          ──────────
                                   ──────── Actual Symptom ────────
   Actual   value : 86             LOW_WORK_IN_PROGRESS
   Desired value  : 100
   Reason       : STATION_BLOCKINGS  ─────── Possible Symptoms ───────
                                     Low work in progress.
   FASDT suggestion                  Stations blocked.
   ────────────────                  AGVs blocked.
   Increase W.I.P.                   Bottleneck station.
    from 9 to 10                     No. of AGVs too low.
   Maximum: 19.                      Wrong product mix.
                     ─ PLEASE CHOOSE─ ──── Possible Suggestions ────
                     Accept           INCREASE W.I.P.
                     Own suggestion   HIGHER_OUTPUT_BUFFER_LEVELS.
                     Next item        INCREASE_RATE_OF_ENTRY.
                     Next symptom
                     See last change
   ── Messages ──    INFO
                     HELP
                     QUIT
```

FIGURE 4.23 Screens from FASIM and FASDT to illustrate the application of the goal-directed, data-driven simulator.

the number of assemblies produced rose from 80 to 86. The station utilizations were still low, indicating that there is potential to improve the output if these utilizations can be improved. To accomplish this, it seemed that we should increase either the number of pallets in the system or the transport capacity. The former would increase the work in progress (WIP) and should increase the utilizations. The latter should result in the increased movement of the assemblies to the idle stations. Since the average time in system for the assemblies is high compared with the cycle time, it appears that assemblies are spending too much time waiting in the output buffers. However FASDT suggested that we increase the WIP by one unit. We accepted the suggestion and it was implemented. FASIM was run again. This modified design produced only 82 assemblies. This drop may be attributed to the dynamic congestion in the system.

Figure 4.24 shows the FASDT screen on top, with a suggestion to increase the transport capacity by adding another AGV. (FASDT assumes that the transporter speed cannot be increased.) The suggestion was implemented and FASIM was run again. The output results on the bottom screen show that the number of finished assemblies rose sharply from 82 to 105. Thus the goal was achieved: an overall improvement in station utilizations was obtained. Moreover, the average time in system for each assembly went down and the cycle times improved for each assembly. The performance analysis, however, indicates that there is still a large scope for improvement.

This example shows the application and necessity of a simulation model. A design that has ample static capacity to produce a certain throughput may not do so under the dynamics of the system operation. Simulation can help us evaluate the effect of such dynamics, and the performance statistics on material flow can iteratively guide us towards a design that fulfills a design goal. Furthermore, since the knowledge of simulation analysis and subsequent FAS design modifications is heuristic, qualitative, and rule-based, an expert system like FASDT has vast potential as a design aid. The implementation of FASDT-type tools should be done in a decision support mode because the designer may have certain priorities in design changes. Thus the designer can alter, modify, or differ with the suggestions given by FASDT.

4.9 CONCLUDING REMARKS

A number of simulation software tools are currently available in the market [2]. New features are constantly under development to make simulation more widely available to designers of manufacturing systems. Data-driven generalized models represent one such approach. There are commercially available packages such as MAST [19] for applications in the flexible manufacturing system domain. The data-driven simulators described in this chapter are not commercial packages and are at best research prototypes that address the area of robot-based assembly cells and systems. The primary reason for presenting them here is to pass on to the reader the firsthand experience of the authors in developing such tools. In so doing, we hope we have

```
┌─────────────────────────── F A S D T ──────────────────────────────┐
│        User Input                    │         Actual   Target      │
├──────────────────────────────────────┼──────────────────────────────┤
│  Target : THROUGHPUT                 │         THROUGHPUT           │
│  Not fulfilled !                     │         ──────────          │
│                                      │                              │
│                                      │  ──────── Actual Symptom ─────│
│  Actual  value : 82                  │      NUM_OF_AGV_TOO_LOW      │
│  Desired value : 100                 │                              │
│  Reason : NUM_OF_AGV_TOO_LOW         │  ─────── Possible Symptoms ───│
│                                      │  Low work in progress.       │
│   FASDT suggestion                   │  Stations blocked.           │
│   ─────────────────                  │  AGVs blocked.               │
│   add another AGV.                   │  Bottleneck station.         │
│                                      │  No. of AGVs too low.        │
│                                      │  Wrong product mix.          │
│                    ┌─ PLEASE CHOOSE─ │ ──── Possible Suggestions ────│
│                    │  Accept         │                              │
│                    │  Own suggestion │        ADD_AGV               │
│                    │  Next item      │                              │
│                    │  Next symptom   │                              │
│                    │  See last change│                              │
│   ─ Messages ─     │  INFO           │                              │
│                    │  HELP           │                              │
│                    │  QUIT           │                              │
└──────────────────────────────────────┴──────────────────────────────┘
```

```
┌─────────────────────────── F A S I M ──────────────────────────────┐
│  Simulation time : 600.              Statistics clear time : 120.0  │
├─────────────────────────────────────────────────────────────────────┤
```

Station #	Block %	Utilization %		AGV #	Idle %	Block %	Load utilization %
1	0.0	21.8		1	0	1.0	65.3
2	2.9	45.9		2	0	4.4	66.5
3	4.1	63.4		3	0	3.6	60.8
4	1.8	45.0		4	0	9.5	54.9

Part #	Finished number	At time	Number in system	Average time in system	Cycle time
1	33	587	4	52.5	14
2	36	595	1	56.5	13
3	36	590	5	97.2	13

Total number of parts in system now : 10   Maximum Allowed : 10
Total number of finished parts   : 105   Last time   : 596

FIGURE 4.24 Screens from FASIM and FASDT to illustrate the application of the goal-directed, data-driven simulator.

provided a coherent discussion on some aspects of material flow simulation and its use for aiding the evaluation of FAS designs. The goal-directed approach applied by FASDT is a comparatively new concept, and more research is necessary to validate it.

In conclusion, we believe that material flow performance evaluation is an essential constituent of robot-based cell and system design. The role of material flow simulation lies in the accurate evaluation of the cell/system operation design. Generalized simulators for single robot cells, multirobot cells, and the flexible assembly systems are necessary to reduce the long lead time and the great expertise required to build useful simulation models. We have presented an integrated framework of generalized data-driven simulators to allow a designer to evaluate a wide spectrum of potential robot-based assembly cell designs and also FAS designs. The designer does not need simulation or programming expertise to use these simulators.

ACKNOWLEDGMENTS

The work described in this chapter is part of research being carried out in the Republic of Ireland as part of the European Strategic Programme for Research and Development in Information Technology (ESPRIT) project 623, led by IPK in Berlin, FRG. We acknowledge J. Maguire and Tom O'Donnell from the University College Galway for their contribution in this work. We also acknowledge Frank Bieler and Christoph Felix from Universität Karlsruhe, Institut für Informatik III, who worked with us as visiting research students.

REFERENCES

1. K. E. Wichmann, "Trends in the Development of Simulation Software Tools for Analyzing Manufacturing Systems," in *Proceedings of the Third International Conference-Simulation in Manufacturing*, G. F. Micheletti, Ed., 1987, Turin, Italy.
2. Catalog of Simulation Software, *Simulation*, pp. 165–181, October 1987.
3. J. L. Peterson, "Petri Nets," *Comput. Surveys* (September 1977).
4. A. Alan B. Pritsker, *Introduction to Simulation and SLAM II*, 2nd ed., Wiley, New York, 1984.
5. Floyd H. Grant and James R. Wilson, "Material Flow Evaluation in Robotic Systems using Simulation," *Mater. Flow* 3, 83–97 (1986).
6. S. Wadhwa and J. Browne, "Collision Avoidance Analysis in Multirobot Cells Using Petri Nets," *Robotersysteme, Zeitschrift fur Informationstechnologie und Handhanbungstechnik*, Springer-Verlag, Berlin, Vol. 4, 107–115 (1988).
7. B. R. Fox and K. G. Kempf, "Opportunistic Scheduling for Robotic Assembly," *IEEE*, 2152–2157 (1985).
8. S. Wadhwa and J. Browne, "A Goal-Directed Data-Driven Simulator for FAS Design," paper delivered at European Simulation Multiconference July 8–10, 1987, Vienna.

9. S. Wadhwa and J. Browne, "Modeling FMS with Decision Petri Nets," *International Journal of Flexible Manufacturing Systems (IJFMS)*, Kluwer Academic Publisher, Vol. 1, 225–280 (1989).

10. Wilbert E. Wilhelm, "An Approach for Modelling Transient Material Flows in Robotised Manufacturing Cells," *Mater. Flow* 3, 55–68 (1986).

11. Shimon Y. Nof and Zvi Drezner, "Part Flow in the Robotic Assembly Plan Problem," *Mater. Flow* 3, 197–205 (1986).

12. P. Alanche, K. Benzakour, F. Dolle, P. Gillet, P. Rodrigues, and R. Vallete, "PSI: A Petri Net Based Simulator for Flexible Manufacturing Systems," Lecture Notes in Computer Science 188, *Advances in Petri Nets*. Springer-Verlag, Berlin, 1984.

13. H. Alla, P. Ladet, J. Martinez, and M. Silva, "Modelling and Validation of Complex Systems by coloured Petri Nets, Application to a Flexible Manufacturing System," Lecture Notes in Computer Science 188. *Advances in Petri Nets*. Springer-Verlag, Berlin, 1984.

14. F. Martin and H. Alla, "Discrete Event Simulation by Timed Coloured Petri Nets," in *Modern Production Management Systems*, A. Kusiak, Ed. Elsevier, Amsterdam, 1987.

15. J. Duggan and J. Browne, "ESPNET: Expert System Based Simulator of Petri Nets," *IEE Proc. D Control Theory Appl.*, Vol. 135, No. 4, 239–247, (July 1988).

16. Jerry W. Saveriano and William A. Gruver, "Optimization of Material and Information Flow Within Robotic Workcells and Systems," Technical Paper MS 82-131, Society of Manufacturing Engineers, Dearborn, MI.

17. S. Wadhwa, J. Maguire, and J. Browne, "A Design Evaluation Procedure for Robot Based Flexible Assembly Systems," *Proceedings of the 13th ISIR*, Brussels, 1986.

18. Daniel N. Hall and Kathryn E. Stecke, "Design Problems of Flexible Assembly Systems," *Proceedings of the Second ORSA/TIMS Conference on Manufacturing Systems: Operation Research Models and Applications*, K. E. Stecke and R. Suri, Eds., 1986.

19. John E. Lenz, "MAST: A Simulation Tool for Designing Computerized Metalworking Factories," *Simulation* 40:2 (February 1983).

5

Sensors and Grippers in Robot Work Cells

THEO DOLL   *University of Karlsruhe, Karlsruhe, Federal Republic of Germany*

Sensors and grippers constitute the interface between a robot and its environment. The modern intelligent robot systems need to be supported by various advanced gripper and sensor systems. This chapter describes some of the recently developed grippers and sensors, explaining their working principles in detail.

5.1 SENSORS IN ROBOT WORK CELLS

Sensors have a key function in the development of future robots. They are employed to reduce uncertainties concerning the robot itself as well as its environment. Use of intelligent robots will be possible only if one succeeds in developing low cost sensors and manages to integrate them in the robot's control system. Complex tasks for robots (e.g., assembly operations) need a combination of sensors in a manner similar to the use of multiple sensors by human beings for the same tasks. Sensors that simulate the human ability of vision or touch do not yet exist. For this reason, task-specific sensor configurations have to be developed for many applications.

Computer-integrated manufacturing cells are the basic unit of future manufacturing plants. They make it possible to do away with traditional mass production and to build configurable manufacturing installations to produce customer-specific products in small amounts.

In manufacturing environments, sensors can be used to measure information about materials, amounts, geometry, place/location, and time [1].

Material-dependent sensor information is related to the quality of
raw parts or to the behavior of parts during assembly operations by
industrial robots. Sensor information on amounts is needed in relation to quantity of pieces as well as to quantity of flow. Geometry-
based sensor information normally controls shape of parts and position accuracy, as well as surface properties. Information about the
location of parts helps for monitoring moving objects, translation,
and orientation. Time-based sensor information reports points of
time and precedence as well as such time-dependent measuring values as frequency, velocity, and acceleration.

In all these cases sensors are employed to exclude uncertainties—
that is, to detect them and react appropriately. In the context of assembly by robots, these uncertainties can be traced back to two factors: uncertainties from tools (robots, feeders, and fixtures) and
uncertainties in the parts themselves (manufacturing tolerances).

There are basically two ways to handle these uncertainties in a
manufacturing environment:

Avoid all uncertainties in the assembly planning phase.
Detect uncertainties with the aid of sensors.

The first possibility is the traditional approach, which gives rise
to inflexible, expensive, and time-consuming systems, especially the
construction of specialized feeding devices. Such a solution makes
quick product changes uneconomical because of long change over
times.

The development of advanced sensors specially related to the sensor-based detection of uncertainties will be a major step toward flexible manufacturing. In the future, sensors will more and more be employed in process control and process monitoring to guarantee the operation of a computer-integrated manufacturing installation and, if
required, to trigger the appropriate reaction of the connected control
computer.

5.1.1 Classification of Sensors

Sensors are, in principle, transducers that transform physical properties (the input signal) into electrical (output) signals. These properties can be electrical, magnetic, optic (surface reflection or transmission), or mechanical (distance, position, velocity, acceleration).

Sensors, used for measuring, have a twofold function: transforming physical properties into electric signals and processing these electric signals.

With the increasing integration of microelectronics and the miniaturization of electronic components, measuring units have been integrated with processing hardware. The term "sensor systems" is used

here for these complex devices. Sensor systems can be classified either according to the type of sensor itself or according to the type of information delivered by the sensor.

Classification According to Type of Information

Sensor information can in general be classified according to whether the information is internal or external.

Internal information is concerned with the state of the robot itself— for example, with such quantities as forces at various joints, joint velocities, and the states of various subsystems (e.g., gripper open or closed). External information is concerned with the environment of the robot, the location and orientation of objects relative to the robot, and other physical properties of objects (color, surface texture, etc.). Sensor information can also be classified according to its complexity:

Simple binary information for detecting the presence or absence of objects with a proximity sensor.
Complex binary information for detecting the position of an object with a tactile sensor array.
Scalar information (e.g., object temperature).
Vectorial information (e.g., when measuring forces with a wrist force sensor).

The sensor information may be further categorized as static or dynamic, depending on the rate of change of the information. Dynamic information is concerned with the changing structure of the environment, whereas static information describes nonchanging parameters of the environment.

Classification According to the Sensor Itself

Sensors can be categorized according to operating distance, type of interaction, or working principle. Three different operating distance may be classified:

1. Contact sensors (also called tactile or touch sensors) must be in direct contact with the object to pick up parameters describing interaction with the object. These parameters can be forces and torques, or such physical properties as temperature or mass of objects.

2. Short-range (or proximity) sensors operate very close to the object but not in contact with it.

3. Range sensors operate at distances greater than those of short-range sensors. The threshold value for robotic applications to distinguish between range and short-range sensors is approximately 20 mm.

The type of interaction distinguishes between active and passive sensors; thus the former emit energy (e.g., an optical range sensor

emits a light beam and picks up the reflected signal) and the latter
rely on energy from the object or from the environment. Typical
passive sensors are inductive or capacitive proximity sensors. An
inductive sensor produces a magnetic field and measures the change
of inductance. Camera systems for gray scale picture processing are
also passive sensors.

Sensors are usually based on one of the following physical principles:

Magnetic
Capacitive
Optic
Acoustic

or on a combination of these principles. The next section discusses
how the physical sensor principles influence the selection of a suitable sensor.

5.1.2 Criteria for the Selection of Sensors

In industry, sensors are needed mainly to detect uncertainties in the
manufacturing or assembly process. Before selecting an appropriate
sensor configuration for a given assembly task, one should determine
whether the uncertainties may be reduced by simple changes in the
assembly machinery or in the design of the parts. Such simple changes
may offer a more economic solution. On the other hand, monitored systems are more flexible. Strategies for monitored systems are presented
by Eversheim [2], who also lists the most significant stages in the planning of sensory applications and explains how to extract a measuring
task from a given problem.

Measuring Task

The parameters that describe a measuring task are type of measurement (distance, temperature, etc.), measuring range, accuracy,
and required measuring frequency. These parameters give a rough
specification of the measuring task.

Other than the measuring task itself, the following groups of parameters are important in selecting a sensor: integration of the sensor, object to be measured, and environmental conditions.

Integration of the Sensor

A sensor is a component of a more complex system into which it is
integrated. This integration process imposes restrictions that must
be considered when choosing a sensor. Mechanical restrictions are
size, shape, and weight of the sensor. The required power supply

and the kind of output signals (analog or digital, voltage or current) determine the effort necessary to integrate the sensor into the system electronically.

Object To Be Measured

The sensors that can be used for a certain task are also restricted by such specific object properties as material, surface, shape, and size. Inductive sensors can be used only for ferromagnetic objects, whereas, when using optical sensors, attributes of the surface must be considered. (Is the surface to be measured flat or cylindrical? Is the object transparent or opaque?)

Environmental Conditions

Environmental conditions influence the measuring accuracy of sensors and also decide whether the sensor would function. Sensors can be influenced by temperature, electrical and magnetic noise, vibration, light, humidity, and pollution (dust, smoke, oil). Interfering environmental conditions affecting the four physical principles of range sensors are shown in Figure 5.1 [3]. Further restrictions resulting from the use of a specific sensor are mentioned in the sections on range, short-range/proximity, force—torque, and touch sensors, where the working principles are explained and various developments are presented.

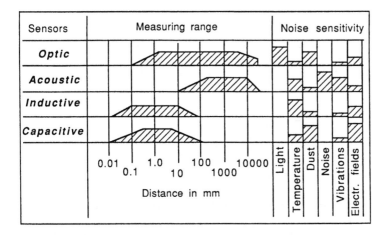

FIGURE 5.1 Interfering environmental conditions.

5.2 RANGE SENSORS

Basically, a range sensor is a device for determining the line-of-sight distance from a reference point, usually on the robot, to a target point on the surface of an object.

Applications for range sensors in the robotics area include:

1. Detecting the presence or absence of objects
2. Determining the distance between the end-effector or gripper of a robot and the object
3. Surface location
4. Surface orientation
5. Surface inspection
6. Collision avoidance
7. Navigation for mobile robots

The most important physical principles employed in range sensing are optics and acoustics. Other physical principles (electrostatics, etc.) are discussed in the section on proximity sensors, because of their short range.

Four methods are used for measuring distances:

Time-of-flight technique
Geometrical triangulation technique
Sensitive volume technique
Focusing technique

Based on these measuring methods, one can not only estimate the distance of an object surface but also determine the orientation of a surface, the three-dimensional position of an object, or the distance information can be used to create gray scale distance images. In this section the fundamental range sensing techniques are discussed and a few sophisticated sensor systems are described for measuring the range and orientation of surfaces and systems that can create range images.

5.2.1 Time-of-Flight Technique

Both optical and acoustic range sensors use the time-of-flight technique to determine the distance of an object surface. Two basic principles are used to determine the distance:

Measure the *round-trip time* of a transmitted pulse of light or ultrasonic burst.
Detect the *phase shift* associated with the transmission and reflection of a modulated light beam.

The principle of the time-of-flight technique is to transmit a wave burst over a short period of time and measure the time interval between the outgoing impulse and the returning impulse. The distance of the surface is calculated from the measured time interval and the speed of the pulse in a specific medium (for light, 300,000 km/s; for ultrasonic, 300 m/s). The other possibility is to measure the phase shift of the reflected light. A phase shift detecting system is described by Staugaard [4]. An amplitude-modulated laser beam is transmitted to an object, and the reflected light signal is out of phase with the original signal. With the detected phase shift, the distance of the object can be calculated. Such a phase-shifting system has demonstrated depth resolution within 1 cm.

Ultrasonic Sensors

An ultrasonic sensor generates a short burst of ultrasound in the transmit phase and detects the returning sound like a microphone during the receive phase. The sound is produced by applying an ac excitation voltage to piezoelectric or electrostatic elements. The ac excitation causes a vibration, which is transferred to a thin metal plate or diaphragm to produce the ultrasonic waves.

Today, several types of ultrasonic sensor are available as integrated modules that have the advantage of small size of transmitter and receiver, simplicity of measuring procedure, and low cost. Typical sensor elements have a maximum measuring distance ranging from 1.5 m at a frequency of 200 kHz to 10 m at a frequency of 40 kHz. Limitations in resolution (detection of small objects) arise from the transmitted beam pattern of the sensor element. The beam angle of different transmitters can vary from 1° to 30°. A typical acoustic beam with a beam angle of 10° and secondary zone lobes is shown in Figure 5.2. The measuring values depend on the temperature and pressure of the air and on air movements. With small air movements and temperature compensation, an accuracy of 1 mm within a range of 0.2–1 m is possible. The time required for measuring a distance of 1 m is 6 ms.

Detailed information on problems and solutions in airborne ultrasonic sensors in assembly and handling systems may be found in Ahrens [5]. A number of range measuring and edge detection experiments have been carried out to examine the influential factors under realistic conditions. Table 5.1 shows the results of the range measuring experiment. The sensor frequency used in the experiment was 120 kHz at an angle of 6°.

Measuring accuracy was determined depending on the size of a round steel reflector, the angular deflection of the reflector from the vertical, and several measuring distances. The measuring value written down is the average of 10 sampling values. To determine

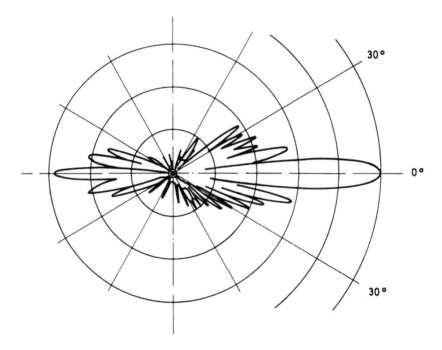

FIGURE 5.2 Ultrasonic lobe pattern.

the position of an object unambiguously, the positions of at least three edges must be determined. At a speed rate of 0.1 m/s, a positioning accuracy of ±4 mm with respect to the object was achieved.

Ultrasonic measuring systems are being employed in the field of robotics for collision avoidance during robot motions, for measuring stack heights when unloading pallets, for navigation of mobile robots, and for localizing three-dimensional objects in the robot's environment. Simple ultrasonic systems use the same sensors for transmission and reception. This configuration is not very useful for short-range sensing, since no reception is possible during transmission time, and the transducer needs some time to switch from transmission to reception. This difficulty can be overcome by the use of separate transmitter and receiver, which however has the disadvantage of size enlargement by a factor of 2. For short-distance range measuring, which is possible with just one transducer, an active method for damping the sound transducer is suggested by Miller et al. [6]. In experiments, distances down to 25 mm were measured.

TABLE 5.1 Results of Measuring Experiment

Reflector	Angular deflection	Measuring range (mm)	Measuring accuracy (mm)
1 cm^2	0°	150	±0.3
		500	±0.4
		1000	±0.5
	5°	150	±0.6
		500	±0.9
		1000	±0.8
10 cm^2	0°	150	±0.6
		500	±0.5
		1000	±0.8
	5°	150	±1.0
		500	±0.7
		1000	±0.8

Figure 5.3 shows a sensor configuration that can be applied to primitive environment recognition for robots. The sensor system consists of nine single sensors. One is a transmitter, the other eight are switched as receivers. The device works with a frequency of 200 kHz and has an accuracy of 2 mm in a range of 50 mm to 2 m. With this sensor even simple objects, edges, or the angle of surfaces can be detected.

Optical Sensors

Optical sensors coupled with the time-of-flight concept work in a manner analogous to the ultrasonic principle. Time-of-flight light detection systems are used in civil engineering and military applications for measuring line-of-sight distances up to 10 km with an accuracy of 10 mm. The problem associated with time-of-flight sensors is that very short time intervals have to be measured. Accuracies of 10 cm require measuring time intervals of nanoseconds. Such systems are not very well suited for robotic applications, where relatively short distances are involved [7], and this timing restriction exceeds the capability of all but the most expensive systems. Laser-based phase measurement systems present an alternative approach, but they require a

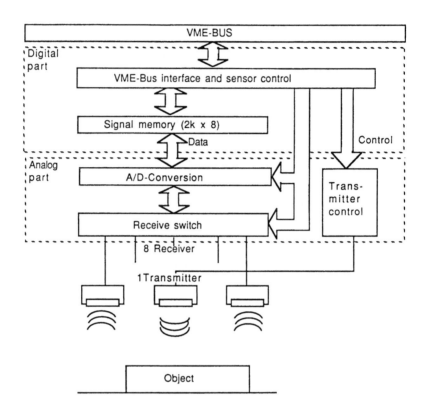

FIGURE 5.3 Ultrasonic sensor system for primitive environment recognition.

retroreflecting object surface and are quite expensive. A pulsed laser system is described by Fu [8].

5.2.2 Geometric Triangulation Technique

The second group of optical range sensors consists of the geometrical triangulation type. Two different approaches may be distinguished: single spot triangulation and complex pattern triangulation.

Single Spot Triangulation

An infrared light-emitting source projects a small spot of light onto the surface to be measured. This light is diffusely reflected onto a position-sensitive detection system such as PSD (position sensitive

detector) or CCD (charge coupled device) elements. A surface displacement causes a proportional displacement of the reflected light spot
on the position detection system. Figure 5.4 shows two possible geometric arrangements of the position detector and the light source. It

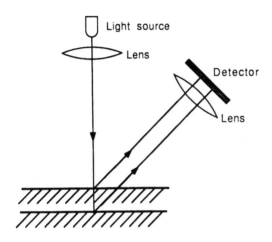

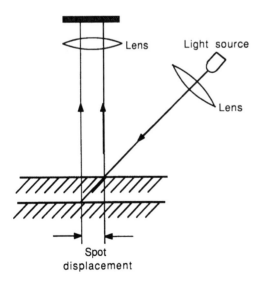

FIGURE 5.4 Two possible arrangements of the light source.

may be noted that in the bottom arrangement an object displacement will cause a sideways motion of the measuring point. Several types of triangulation range finder are available. Typical technical data are:

Measuring distance   40--100 mm
Measuring range   10-- 40
Accuracy $\geq$0.01 mm

Triangulation-based systems hold great promise for robotic applications. Sensor heads are small and lightweight. Variations in the reflectivity of the object and variations in the surrounding illumination can be compensated automatically by controlling the infrared light source.

Figure 5.5 shows a simple sensor built employing the triangulation principle. Dimensioning of the parameters can be done as follows. The laser produces a light spot on the object surface, which is mapped onto a CCD array under the angle α by the lens L; C is the distance between the lens and the intersection point of the optical axis of the lens with the laser beam, and x_1 and x_2 are distances between this intersection point and the points X and Y on the laser beam. Let f

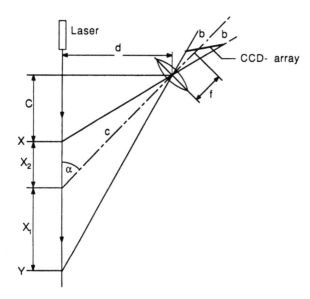

FIGURE 5.5 Triangulation measurement.

be the focal length of the lens and $C \gg f$. For a first approximation, the mapping can be treated like a mapping from infinity, and the distances x_1 and x_2 can be obtained by applying elementary geometric transformations to the image sizes b_1 and b_2:

$$x_1 = \frac{b_1 C}{f(\sin \alpha) - b_1(\cos \alpha)}$$

$$x_2 = \frac{b_2 C}{f(\sin \alpha) + b_2(\cos \alpha)}$$

If a measuring range from X to Y on the laser beam is desired, the CCD camera is tilted with respect to the lens until the images of X and Y are focused and placed at the edges of the CCD array. This procedure is best suited for distances $C \gg f$. For a CCD array with 256 elements with a distance of 13 μm, the maximum values for b_1 and b_2 are 1.65 mm.

If $C = 100$ mm, $f = 10$ mm, and $\alpha = 15°$, the distance C' of the intersection point is 26.7 mm, $X_1 = 165.5$ mm, and $X_2 = 40.2$ mm, resulting in a measuring range of 36.7–262.2 mm. The distance d from lens to laser beam is 25 mm.

Kanade [9] describes a sensor for measuring surface position and orientation based on the triangulation principle (see also Fig. 5.6).

The sensor head consists of a multiple light source, a system of lenses, and a sensor chip acting as a spot position detector. The light sources are placed in a circular pattern 60° apart, pointing toward the optical axis of the objective lens. The light sources are aligned so that all six light beams cross the optical axis at the same point. The light beams thus form an optical cone. The tip of the cone is placed at approximately the center of the range of operation of the proximity sensor. Each light source is independently controllable.

The lens system and the sensor chip form a camera arrangement. When an object intercepts the light beam, the resulting light spot is imaged onto the sensor chip through the lenses. The sensor chip then gives the spot position on its active two-dimensional surface. Knowledge of the spot position on the sensor chip (together with the camera optics) and the trajectory of the light makes it possible to find the coordinates of the point in three-dimensional space performing triangulation. From the measurement of multiple points in the vicinity, the surface orientation can be calculated.

If the triangulation range finder is moved in a fixed plane, it is possible to get an image with intensities proportional to the distance between the plane and the scanned objects. These distances can consequently be used to compute the geometry of an object. Another way to get the image of information is to use a triangulation laser scanner.

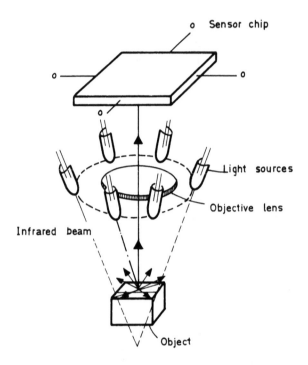

FIGURE 5.6 Working principle of the sensor head.

Such a scheme is shown in Figure 5.7. The light beam is emitted by
a 5 mW He-Ne laser and is deflected by a mirror to an intensity modulator, where the light is modulated by a control circuit that turns the
light on and off. The modulated light is directed at a rotating polygonal mirror with 20 facets, each of which deflects the light by an angle
of 12°. At the edge of two neighboring facets, the diffused light may
be detected by a photodiode and reported to the control circuit, producing the signal v_r, which starts and controls the generation of scanning lines by actuating the deflection mirror. The light beam is now
aimed at the object. The modulation voltage v_h allows scanning of a
selected picture area, by aiming the light beam at any desired point
of the object under investigation. For a pictorial representation of
the intensity image on the range array, see Figure 5.8.

Complex Pattern Triangulation

Another approach to the triangulation principle is to project a more
complex pattern—for example, a strip of light or ring beam—to an object

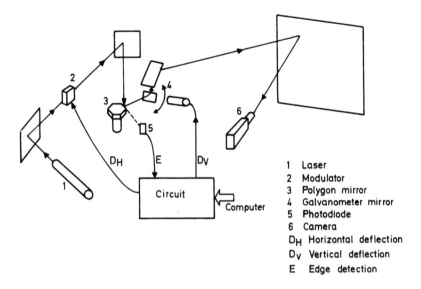

1 Laser
2 Modulator
3 Polygon mirror
4 Galvanometer mirror
5 Photodiode
6 Camera
D_H Horizontal deflection
D_V Vertical deflection
E Edge detection

FIGURE 5.7 Schematic diagram of a triangulation laser scanner.

surface. Complex pattern triangulation is normally used for surface tracing, where it is desirable for the range finding system to detect not only the position but also the inclination of the surface. A surface tracing range sensor called RORST (Riken Optical Range Sensing scheme for surface Tracing) using ring pattern projection is described briefly below. A more detailed description is available in Reference 10. Figure 5.9 shows the working principle of the RORST sensor; the symbols have the following meanings:

θ = projection angle of the axially symmetrical sheet beam
α = distance between the observation plane and the objective lens
R_0 = radius of the sheet beam at the lens positions
$r(\theta_1)$, $r(\theta_2)$ = radial position of the ring pattern image on the observation plane

An axially symmetrical light beam is projected and produces a ring pattern on the surface of the object to be investigated. The ring pattern image is observed through a lens system, and the radial position of the ring pattern image is measured for each azimuth θ. Based on the triangulation, range information in cylindrical coordinates $z(\theta)$, $R(\theta)$ is determined as follows:

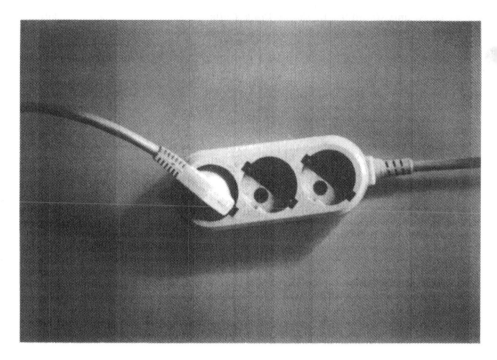

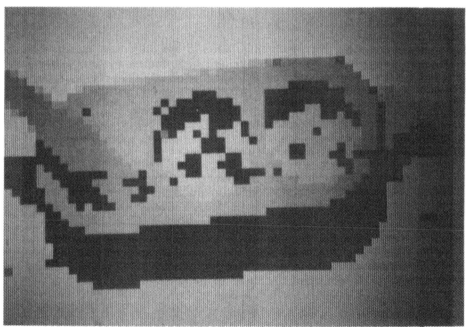

FIGURE 5.8 Photograph and distance image of an ac socket.

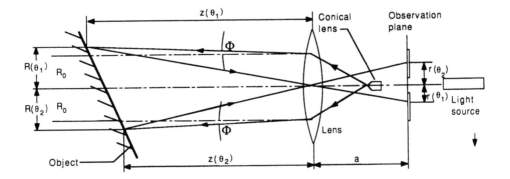

FIGURE 5.9 RORST

$$z(\theta) = \frac{aR_0}{r(\theta) - a \tan \phi} \qquad (5.1)$$

$$R(\theta) = R_0 \frac{1 + a \tan \theta}{r(\theta) - a \tan \theta} \qquad (5.2)$$

Thus, three-dimensional coordinates (θ, z, R) of points along the projected ring pattern can be obtained. The mean gradient of the surface for each azimuth $g(\theta)$ can now be estimated as follows:

$$g(\theta) = \frac{z(\theta) - z(\theta - \Pi)}{R(\theta) + R(\theta - \Pi)}$$

5.2.3 Sensitive Volume Technique

The sensitive volume technique uses the amount of reflected light to determine distances. A beam of light originating from an infrared light-emitting diode (LED) illuminates a surface. The closer the surface is to the sensor, the greater is the intensity of the reflected light reaching the sensor. A photodetector generates a voltage corresponding to the distance. Figure 5.10 shows the arrangement of the light source and the detector.

The main advantage of the sensitive volume technique is that very simple optical range sensors can be built. The problem associated with this technique is that the detector output is sensitive not only to the distance of the surface, but also to such other paramters as irregularity, reflectivity, and inclination of the surface. One way to overcome these sensitivities is to use one light source and an array of photosensors, or vice versa.

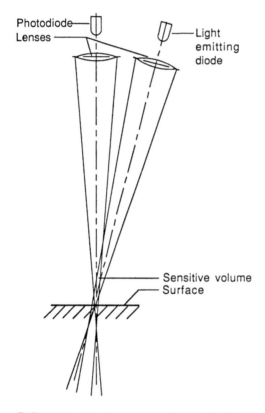

Photodiode
Lenses
Light
emitting
diode
Sensitive volume
Surface

FIGURE 5.10 Range sensing with the sensitive volume technique.

Okada [11] developed a sensitive volume sensor that can determine the distance of irregular surfaces reflecting optically diffuse light. Figure 5.11 shows the sensor head, consisting of a light source and an array of photosensors. The light beam B is guided through a pinhole HS and can reenter the sensor housing only through an aperture HR; thus it cannot reach the array of photosensors before having been reflected by the object. While the surface of the object is illuminated by the light from S, the photosensors of the array R are scanned to detect the light path used for making the output signal maximum. This effective path provides decisive information for determining the distance between the sensor head and the surface of the object. With both the light source S and the array sensor R situated on the same line, and where both the front and back lines of

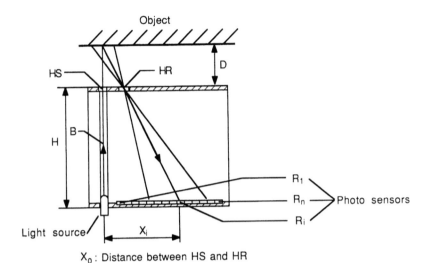

FIGURE 5.11 Configuration of a sensor head.

the sensor head are parallel, two homologous triangles are found to satisfy the following relation:

$$\frac{D}{D + H} = \frac{X_0}{X_i} \tag{5.3}$$

It follows from Equation 5.3 that

$$D = H \frac{X_0}{X_i - X_0} \tag{5.4}$$

Therefore, the distance D can be measured by detecting the position X_i where the photosensor array shows a maximum.

5.2.4 Focusing Technique

Another approach to range measuring is based on the focusing property of a lens. Autofocus is already well known as the distance sensing method in the camera industry. This optoelectronic technique can also be adapted for robotics applications. Two developments described elsewhere [10,12] are presented here. Kinoshita's principle [10] is shown in Figure 5.12. The light source S emits a light beam, which

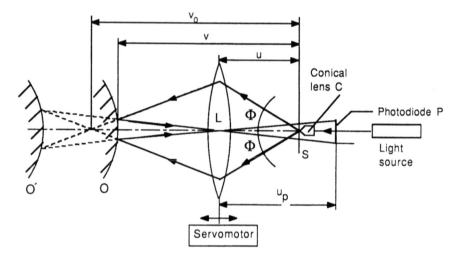

FIGURE 5.12 Range sensing by focusing.

is widened in a conical lens C and projected onto the object surface
O. The projection of the light beam on the object is a round spot,
which is reduced to a point when focused. The intensity of the spot
is picked up by a photodiode and, when a maximum brightness is
reached, the spot is assumed to be reduced to a point and the distance of the object surface can then be calculated from the focal
length of the lens and its current position.

Let f be the focal length and u the distance between the vertex
of the conical light sheet and the lens. The distance of the object
surface from the vertex of the conical light sheet can then be given
by the following equation:

$$v_0 = \frac{u^2}{u - f}$$

Let u_p be a distance between the observation lens and the observation plane and r_p the radius of the light spot on the observation
plane. The actual distance of the object surface from the vertex of
the conical light sheet can be given by the following equation:

$$v = \frac{u(u_p\, u \tan \psi - r_p f)}{u_p (u - f) \tan \psi - r_p f}$$

The measuring accuracy of this method may be influenced by the detection error of optimum imaging conditions.

Experiments were done with a conical lens with 170° vertex angle that produces a spot with 27.5 mm in diameter at a position 300 mm away from the vertex of the conical lens. The conical light beam is projected onto the surface of an object through the objective lens of a diameter of 100 mm and a focal length of 80 mm. The range detection error of this system is estimated less than 1 mm in the range of 420–490 mm.

Another interesting development is described by Rösler [12]. This method uses a laser beam, which is projected out-of-focus on the object surface. The diameter of the spot of light produced is inversely proportional to the object distance. Figure 5.13 shows a schematic diagram of this method.

A semiconductor laser produces a light spot on the object that is projected on a CCD array by a semitransparent mirror. The diameter of the spot (detected by the CCD array), the focal length of the lens system, and the aperture are used to calculate the distance of the object. For example, let the focal length be 100 mm, the aperture 25 mm, and the distance between lens and projection plane 180 mm. The diameter of the light spot will decrease from 10 mm to 2.5 mm if the object distance is increased from 150 mm to 200 mm. The difference of 7.5 mm is equivalent to a difference of 577 CCD cells if we use a

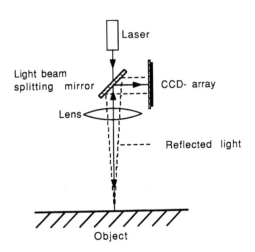

FIGURE 5.13 Range sensing by measuring light spot size.

cell distance of 13 μm. This results in a measuring accuracy of 0.1 mm.

The measuring range can be adapted to the current task by changing the three parameters focal length, aperture, and viewing distance. The rule of thumb for measuring range is 1.5–3 times the focal length, because accuracy drops with the square of the distance. The light spot must not be too big, because its intensity could become too low for the CCD array to detect its size unambiguously. The main advantage of this system is that no mechanical focusing is necessary.

5.3 PROXIMITY AND SHORT-RANGE SENSORS

Most of the range sensing techniques discussed so far are relatively sophisticated. Next we consider three fundamental approaches to proximity and short-range sensing:

Inductive sensors
Hall effect sensors
Capacitive sensors

These sensors work in close proximity to the objects to be sensed but not in contact with them. Proximity sensors are generally defined as sensors with binary output, which indicates the presence of an object within a specified interval. Employing each of these three principles, versions for short-range position sensing have been developed. A more detailed description of these sensors can be found in Trietly [13].

5.3.1 Inductive Proximity Sensors

An inductive sensor is based on the change of induction due to the motion of a metallic object in a magnetic field. If a metallic object is brought into a magnetic field, energy is drained from the oscillator, which results in damping. The functional relationship between the distance to be measured and the magnitude of damping created is unfortunately nonlinear. Figure 5.14 shows the schematic diagram of an inductive sensor. If the sensor works as a proximity switch, a defined signal is produced by a comparator circuit.

Inductive sensors are the most widely used industrial proximity sensors. The measuring range varies from 1 to 20 mm, depending on the coil size and on the material of the object to be measured. The main advantage of the inductive sensor is its repeatability of 0.001 mm under constant environmental conditions. Environmental factors (e.g., temperature, electrical fields, vibrations, and orientation of the sensor

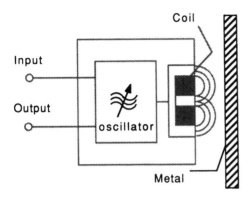

FIGURE 5.14 Principle of an inductive sensor.

relative to the object surface) have an influence on its signal output. The major application in automatic assembly for inductive sensors *so far* is the detection of metallic objects. They are also suited for edge detection or position determination of cylindrical parts (e.g., applications where an inductive sensor is guided over a part to compute a centered position by analyzing the analog signals). Several experiments with varying sensor distances and cylinder diameter proved that a clear signal minimum could be detected with parts of diameter up to 80 mm.

Highly developed inductive sensor systems are used for positioning welding robots. For seam tracing, multicoil systems usually are used. Figure 5.15 shows a multicoil seam tracing system with the output signal over the welding seam. Voltages 180° out of phase are induced in ring coils 1 and 2. If the sensor is exactly centered, the resulting induced voltages add to zero. If the sensor is not exactly centered, different voltages are induced because of the differences of the eddy current fields. From the sum of the induced voltages, a seam position signal can be derived, which delivers information about distance and direction of the seam.

The system described also allows height monitoring of a welding torch. The height of the sensor above the seam has a significant influence on the usability of the signal and, therefore, on the accuracy of the seam tracing.

The output signals are highly interdependent and must be considered together in the realization of a seam tracing system.

A more detailed description of these systems is given in References 14 and 15.

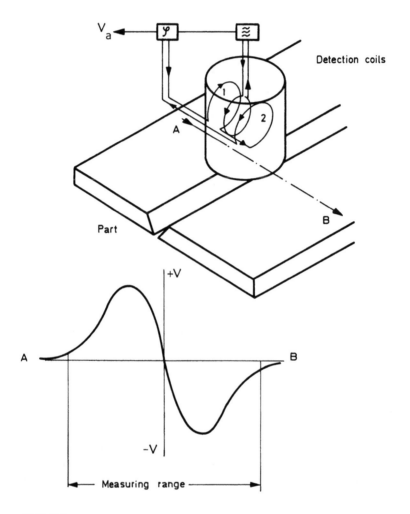

FIGURE 5.15 Multicoil seam tracing system.

5.3.2 Hall Effect Sensors

A Hall effect sensor detects magnetic fields or, when used in conjunction with a permanent magnet, is capable of detecting all ferromagnetic materials. Hall effect sensors are available as solid state components with the size of a small solid state transistor. They are in-
expensive and operate as switches, changing from off to on when the
field strength exceeds a certain threshold, or as linear devices,

delivering output proportional to the field strength of the magnetic field. Hall effect sensors can be used only for very close sensing from 3 to 12 mm using standard permanent magnets. The typical application for Hall effect sensors is to sense the linear or angular position of elements. If a permanent magnet is embedded in the jaw of a robot gripper, the magnetic sensor can be used to detect the opening of the gripper.

5.3.3 Capacitive Sensors

A capacitive sensor is based on detecting a change in capacity induced by a surface that is brought into the measuring range of the sensing element. A capacitive sensor consists of a sensitive electrode, which forms a parallel-plate capacitor together with the object surface.

The capacitance C is proportional to the effective plate area A and inversely proportional to the plate distance d. The constant ε represents the dielectric material:

$$C = \varepsilon \frac{A}{d}$$

There are a number of electronic approaches for detecting proximity based on a change in capacitance. Measurements are made by determining ac impedance, charging time, or the resonance of a tuned circuit. Capacitive sensors are available in ranges from 12 to 40 mm. They are generally cylindrical, with an integrated capacitance measurement circuit. The typical technical data (measuring range and accuracy) and applications are similar to those of inductive sensors. Capacitive sensors, however, have the advantage of detecting not only ferromagnetic materials but also other solid and liquid materials, with different degrees of sensitivity.

5.4 FORCE SENSORS

In industrial tasks that require mechanical interaction with the environment, such as assembly, grinding, or milling, it is necessary to control the interaction forces if the task is to be performed successfully.

The forces and torques encountered by a robot arm can be measured by joint force sensing, wrist force sensing, and finger force sensing.

The advantages and disadvantages of joint force sensing are discussed by Staugaard [4]:

The advantage of measuring arm joint forces indirectly is that a
separate system of force sensors is not required. The joint forces
are simply determined by measuring load variables that already exist in the system. The major disadvantage of sensing arm joint
forces in a manipulator is that the resulting force measurements
do not always provide an accurate indication of the exchange of
forces between the robot end-effector and its surrounding objects.
To get accurate force information, one must account for things like
joint friction, the load of the arm itself, and the inertial forces
created through arm movement.

Most of the commercially available force sensors are wrist force
sensors, described in the next section. Finger force sensors have
the advantage of being in direct contact with the object, so that very
small changes of forces can be detected. Finger-based force sensors
are discussed in Section 5.5.

Basically, a wrist force sensor is a device with some compliant sections and a few transducers, which measure the deflection of the
compliant sections. The most common transducer is the strain gage
type.

Careful attention must be paid to a good design of the mechanical
structure of the sensor element. It should have the following properties.

High stiffness: High stiffness of the sensor ensures that disturbing
   forces are quickly damped out to permit accurate readings after
   short time intervals. It also reduces deflections, which might add
   to the positioning error of the hand.
Compact design: The wrist force sensor is mounted between the robot and the end-effector to detect interactions between the manipulated object and the environment. Compact sensor design minimizes the probability of collisions between the sensor and the environment. Moreover, a large sensor adds to the lever arm and
   leads to a reduction of the maximum load.
Low weight: Additional weight decreases the maximum load of the robot and impairs its dynamic features. The weight of the sensor
   might create signals that must be compensated for, to avoid a decrease in accuracy.
Linearity: Good linearity of the sensor allows resolving the forces
   and torques with simple matrix operations.
Low hysteresis: If the measuring device consists of several parts,
   friction between the parts can lead to hysteresis effects.
Low coupling of signals: Low coupling of the signals simplifies signal
   processing but requires a more complicated design of the sensor
   element.

5.4.1 Basics of Force—Torque Measuring

A strain state in Cartesian space can be described by the force—torque vector $[F_x, F_y, F_z, M_x, M_y, M_z]$, which consists of three forces and three torques in the axes of an orthogonal coordinate system. The following assumptions are made to be able to measure this strain state.

The external forces and torques lead to an internal strain in the measuring device, which leads to an elastic deformation. The strain state of the measuring device is linearly dependent on the external forces and torques; in addition, the deformation is linearly dependent on the internal strain. From the type of deformation, conclusions can be drawn about the external strain.

The deformation consists of a change in length and angle (or consists of elongations and shearings, where a shearing always leads to an elongation). The relation between elongation and the strain at any point of the measuring device can be described as follows:

$$\varepsilon_i = c_{i1} F_x + c_{i2} F_y + c_{i3} F_z + c_{i4} M_x + c_{i5} M_y + c_{i6} M_z \tag{5.5}$$

For several measuring points, the following system of equations is obtained:

$$
\begin{aligned}
\varepsilon_1 &= c_{11} F_x + \cdots + c_{16} M_z \\
\varepsilon_2 &= c_{21} F_x + \cdots + c_{26} M_z \\
&\;\;\vdots \\
\varepsilon_n &= c_{n1} F_x + \cdots + c_{n6} M_z
\end{aligned}
\tag{5.6}
$$

This system of equations can be written in a matrix form as follows:

$$
\begin{bmatrix} \varepsilon_1 \\ \varepsilon_2 \\ \cdot \\ \cdot \\ \cdot \\ \cdot \\ \varepsilon_N \end{bmatrix}
=
\begin{bmatrix}
c_{11} & c_{12} & c_{13} & c_{14} & c_{15} & c_{16} \\
c_{21} & c_{22} & c_{23} & c_{24} & c_{25} & c_{26} \\
\cdot & \cdot & \cdot & \cdot & \cdot & \cdot \\
\cdot & \cdot & \cdot & \cdot & \cdot & \cdot \\
\cdot & \cdot & \cdot & \cdot & \cdot & \cdot \\
\cdot & \cdot & \cdot & \cdot & \cdot & \cdot \\
c_{N1} & c_{N2} & c_{N3} & c_{N4} & c_{N5} & c_{N6}
\end{bmatrix}
\begin{bmatrix} F_x \\ F_y \\ F_z \\ M_x \\ M_y \\ M_z \end{bmatrix}
\tag{5.7}
$$

or in a shorter form:

$$\varepsilon = C(f, m) \tag{5.8}$$

The desired force–torque vector (f,m) can be obtained by solving this matrix equation as follows, where C^{-1} is called the decoupling matrix.

$$(f,m) = C^{-1}\varepsilon \qquad (5.9)$$

The coupling matrix C is determined empirically for each new sensor device by loading the sensor with six linearly independent strain situations. The deformation values are picked up by strain gages, transformed into voltages, preprocessed, and transformed into digital values for further processing.

5.4.2 A Strain Gage Wrist Force Sensor

A strain gage wrist force sensor built at the University of Karlsruhe is shown in Figure 5.16 [16]. It is designed so that a gripper can be mounted in the sensor. To obtain a signal level as high as possible, four strain gages are switched together to form a bridge. Under the assumption that $\Delta R \ll R$, we get for each bridge (see Fig. 5.17):

$$\frac{V_A}{V_E} = \frac{1}{4}\left(\frac{\Delta R_1}{R_1} - \frac{\Delta R_2}{R_2} + \frac{\Delta R_3}{R_3} - \frac{\Delta R_4}{R_4}\right) \qquad (5.10)$$

For strain gages, the relation between elongation and change in resistance is:

$$\frac{\Delta R_i}{R_i} = \varepsilon_i k \qquad (5.11)$$

The k factor, which is determined by the manufacturer and supplied with each strain gage, is a characteristic quantity of the strain gage. Substituting Equation 5.11 into Equation 5.10, one obtains:

$$\frac{V_A}{V_E} = \frac{k}{4}(\varepsilon_1 - \varepsilon_2 + \varepsilon_3 - \varepsilon_4) \qquad (5.12)$$

Altogether, eight measuring bridges, each consisting of four strain gages, have been attached to the measuring device. The positions of the eight bridges are marked with numbers 1–8 in Figure 5.16. They have been arranged so that only the components of the force–torque vector that actually account for bending produce a signal. This leads to the system of equations:

$$
\begin{bmatrix} e_1 \\ e_2 \\ e_3 \\ e_4 \\ e_5 \\ e_6 \\ e_7 \\ e_8 \end{bmatrix} = C \begin{bmatrix} F_x \\ F_y \\ F_z \\ M_x \\ M_y \\ M_z \end{bmatrix} \tag{5.13}
$$

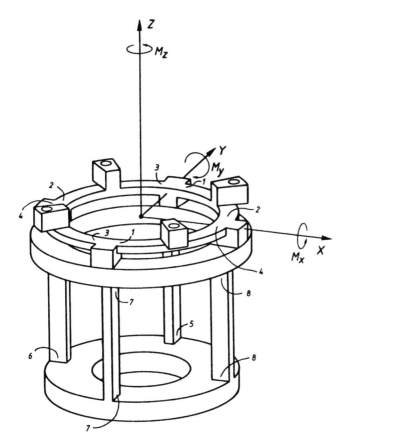

FIGURE 5.16 Mechanical design of the wrist force sensor; bridge positions are indicated by numbers 1–8.

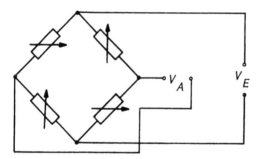

FIGURE 5.17 A bridge with four variable elements.

This may be written in a short form as follows:

$$e = C(f,m) \tag{5.14}$$

The elongations $\varepsilon_1, \ldots, \varepsilon_8$ from Equation 5.6 are replaced by the vector of sampling values, with the k factor, the basic bridge voltage, and the amplification factor of the analog preprocessing already included. The use of eight instead of the necessary six measuring values (see Eq. 5.5) promises a higher accuracy of the sensor, because random errors can be filtered out. A disadvantage is the high computational effort, because the coupling matrix C in this case is not a square matrix and cannot be inverted. The decoupling matrix must be obtained as follows: multiplying Equation 5.14 by C^T, the transposed matrix of C, gives:

$$C^T e = C^T C(f,m) \tag{5.15}$$

where $(C^T C)$ is a 6×6 nonsingular and invertible matrix, provided C is regular with regard to its columns. This is tested for each case, because C represents individual properties of the measuring device that are initially not known. Multiplying Equation 5.15 by $(C^T C)^{-1}$ gives the desired force–torque vector as follows:

$$(f,m) = (C^T C)^{-1} C^T e \tag{5.16}$$

where $(C^T C)^{-1} C^T$ is called the psuedoinverse C^I of matrix C. Using the C^I notation, Equation 5.16 may be written as follows:

$$(f,m) = C^I e \tag{5.17}$$

This equation in the expanded form may be expressed:

$$
\begin{bmatrix} F_x \\ F_y \\ F_z \\ M_x \\ M_y \\ M_z \end{bmatrix} = \begin{bmatrix} c_{11}^I & c_{12}^I & c_{13}^I & c_{14}^I & c_{15}^I & c_{16}^I & c_{17}^I & c_{18}^I \\ \cdot & \cdot & \cdot & \cdot & \cdot & \cdot & \cdot & \cdot \\ \cdot & \cdot & \cdot & \cdot & \cdot & \cdot & \cdot & \cdot \\ \cdot & \cdot & \cdot & \cdot & \cdot & \cdot & \cdot & \cdot \\ \cdot & \cdot & \cdot & \cdot & \cdot & \cdot & \cdot & \cdot \\ c_{61}^I & c_{62}^I & c_{63}^I & c_{64}^I & c_{65}^I & c_{66}^I & c_{67}^I & c_{68}^I \end{bmatrix} \begin{bmatrix} e_1 \\ e_2 \\ e_3 \\ e_4 \\ e_5 \\ e_6 \\ e_7 \\ e_8 \end{bmatrix}
\qquad (5.18)
$$

A rather simple relationship is now established between the signals $e_1, \ldots, e_8$ and the force and torque components $F_x, \ldots, M_z$, which can be resolved with eight multiplications and seven additions for each of the six components.

The sensor has been designed for the following maximum loads:

$F_x = F_y = 60$ N
$F_z = 180$ N
$M_x = M_y = M_z = 4$ Nm

The sensor distinguishes itself through very good linearity and high accuracy (error 1%) for loads along the x- and y-axes. For loads along the z-axis, errors are only slightly higher.

5.5 TOUCH SENSORS

The purpose of developing touch sensors is to give robots a tactile sense similar to that of a human hand. Touch sensors work in direct contact with the object and, therefore, the ideal place of operation for a touch sensor is on the fingers of a robot gripper.

Touch sensors can be used to:

Identify objects by measuring their geometries.
Detect position and orientation of a grasped part.
Detect slipping of a part in the case of collision.
Measure the clamping force and resulting forces.

According to an inquiry by Harmon [17], the following features are requested by potential users of touch sensors:

10 × 10 taxel on a surface of 2.5 × 2.5 cm
All taxels should have response times around 1–10 ms
Threshold value at least 1 g
Maximum load per taxel 1000 g
Low hysteresis
Robust design for industrial applications
Taxel need not be linear

This section discusses the working principles and materials of touch
sensors and presents a number of developments in this area.

5.5.1 Conductive Elastomer Touch Sensors

A conductive elastomer is a rubberlike material with embedded conductive strips, fibers, or particles, which contact if a force is applied to the elastomer.

The principle of a simple switching sensor with embedded carbon
fibers is shown in Figure 5.18. If a force is applied to the elastomer,
the fibers will have contact at the intersection points, and a simple
shape detection can be realized by testing the contact points.

The position of the contact points and the resulting force can be
measured if a pressure-sensitive elastomer is used. This material
consists of an elastomer with embedded metal or carbon particles,
which give it a certain conductivity. Figure 5.19 shows the distribution of conducting particles under load and no-load conditions. With
increasing load, more and more particles get into contact, which results in a decreasing resistance of the material.

Conductive elastomers offer a number of advantages. Sensors
based on the conductive elastomer principle are cheap, easily produ-
cible, and based on a simple measuring principle. Despite the high
number of materials tested, a good linear sensitivity, low hysteresis,
and high dynamic range may always be achieved at the expense of

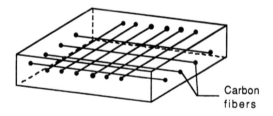

Carbon
fibers

FIGURE 5.18 A simple switching sensor.

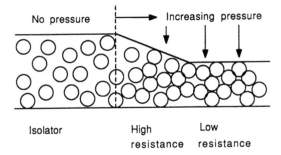

FIGURE 5.19 Distribution of conducting particles.

such mechanical properties as elasticity, abrasive resistance, and life span.

The change in resistance cannot directly be used as force value, because the relation between pressure and resistance is strongly nonlinear and hysteresis occurs because of lasting deformations of the material. A number of different designs of pressure sensitive sensor arrays are described by Overton [18].

Figure 5.20 shows the structure of a sensor pad containing bands of carbon-filled rubber separated from one another by bands of insulating rubber. The variable resistor appears as the variable contact resistance at the intersection of lines. An alternative configuration (Fig. 5.21) uses conductive columns of carbon-filled rubber in an insulating block of silicone rubber. This design eliminates electrical interaction between the transducers themselves.

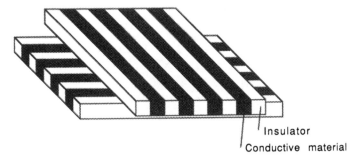

FIGURE 5.20 Sensor pad with strips of conductive rubber.

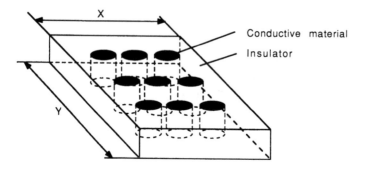

FIGURE 5.21 Sensor pad with conductive columns.

A digital VLSI tactile array sensor based on conductive rubber is described by Raibert [19]. The goal of this design is to use VLSI techniques to provide the sensor with specialized local computing facilities and also to provide the mechanical microstructures necessary for transduction. The same photolithographic process can be used to build the computing elements and the mechanical elements. A prototype chip has 48 tactile elements in a 6 × 8 array. Each element measures 0.32 × 0.64 mm.

Figure 5.22 shows the working principle of a single element. A layer of elastic conductive material is pressed against a plate with a triangular hole in it. Each hole has 15 contact electrodes. An electrode in a narrow region of the hole is touched only when the

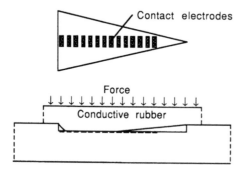

FIGURE 5.22 Working principle of a single element of the very large scale integrated tactile array sensor.

pressure is high, while an electrode in the wide region is touched also when the pressure is low. With different shapes of holes, an element may have different response characteristics. The primary advantage of this design is that the elastic material can be designed with optimal mechanical characteristics in mind, without considering its conductivity response to pressure, while the geometry of a sensing element could be manipulated to change the sensitivity resolution and dynamic range of the response.

5.5.2 Capacitive Touch Sensors

A similar approach is to use the capacitive effect to get a tactile array sensor. Boie [20] describes an 8 × 8 capacitive element array built into a gripper finger. Figure 5.23 shows a 4 × 4 section of this sensor array. The transduction of mechanical forces to representative electrical signals uses a three-layered sandwich structure. The top layer is columns of compliant metal strips over a central elastic dielectric sheet. The bottom layer is a flexible printed circuit board with rows of metal strips and multiplexing circuits. Electrically, the sensor is a capacitor array formed by the row and column crossings, with the middle layer functioning as a dielectric spring.

A capacitive-based tactile sensor measures an applied force by detecting changes in the distance between two parallel plates of a small capacitor. As the force is increased, the dielectric material between the plates gets compressed, leading to an increased effective capacitance. The dielectric material translates the force into positional changes and forms the dielectric gap between the capacitive plates.

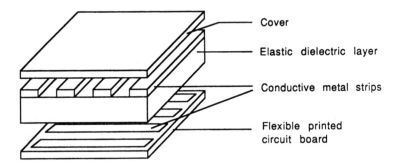

Cover

Elastic dielectric layer

Conductive metal strips

Flexible printed circuit board

FIGURE 5.23 A capacitive tactile array sensor.

Ideally, we can formulate a linear relation between force and compression after Hooke's law:

$$F = K \, \Delta x \qquad\qquad (5.19)$$

where F is the applied force, K is the effective spring constant, and Δx is the positional change produced by the force. The choice of a material that behaves closely according to this equation makes it easy to translate the output of the sensor back to the applied force. The positional change Δx can be computed with the following formula for the capacitance of two parallel plates:

$$C = \frac{\alpha A}{4\pi d} \qquad\qquad (5.20)$$

where α is the dielectric constant, d is the distance between the two plates, and A is the size of the effective plate area.

Capacitive sensor arrays have some advantages over conductive sensor arrays because the elastic materials need not be modified and desirable mechanical properties are generally consistent with low dielectric loss.

On the basis of this principle, Siegel et al. [21] developed an integrated tactile and thermal sensor for use with the Utah/MIT four-fingered dextrous hand. The tactile sensing portion is composed of an 8 × 8 array of force-sensing cells with 1.8 mm center-to-center spacing. In a lower layer a 4 × 4 matrix of surface-mounted thermistors is placed on a flexible printed circuit board. A performance analysis of the tactile sensor evaluating pressure sensitivity, measuring repeatability, and spatial selectivity is presented by Siegel et al. [22].

5.5.3 Optical-Based Touch Sensors

Optical tactile sensors transform a tactile image (strains or pressure) created by an object to an optical image. This principle provides an effective tool for converting optical images into conveniently processed signals. A commercially available optical array sensor is described by Rebman [23]. The sensor uses an 8 × 8 array pattern on an active area of 5 cm × 5 cm. The mechanical transduction is provided by an elastomeric touch surface. A mechanical edge, which is connected to the elastomeric surface, is used to partially or fully shade a light beam. The light beam is detected by a phototransistor; the deflection on the touch surface causes some shading and in turn a corresponding electrical response from the detector. This tactile sensor is used as a stationary element close to the robot, so the part to be examined must be brought to the sensor. The sensor provides information that can be used to locate the target relative to a reference point on the touch surface or to compute the size and the shape of an object. The sensor pad and the working principle of a single sensor element are shown in Figure 5.24

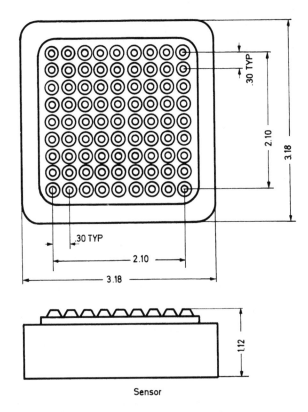

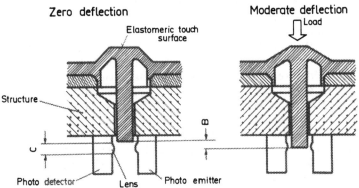

FIGURE 5.24 An optical touch sensor.

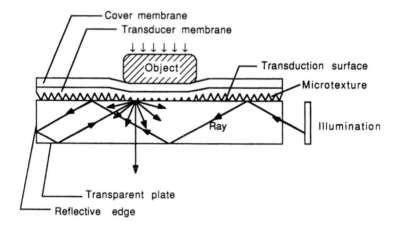

FIGURE 5.25 A tactile sensor based on total internal reflection.

Begej [24] describes the development of a tactile sensor based on
the principle of total internal reflection at an optical surface caused
by an opaque elastic membrane. An optical image is created in which
the intensity is monotonically related to the strains or pressures created by an impressed object.

The principle of this sensor is shown in Figure 5.25. A transparent plate is illuminated from one or more edges. Most of this light is
confined to the interior of the plate by total internal reflection of the
two surfaces. Now, an elastic transducer membrane is placed in contact with the transducing surface. The transducer membrane is opaque
and is textured on one side, for example, with cones. If an object is
pressed against the transducer membrane, it comes into contact with
the transducing surface of the transparent plate. At the contact
points of the opaque transducer membrane, diffused light is reflected
back into the transparent plate. Some light rays no longer satisfy the
conditions for total internal reflection within the transparent plate, and
pass through the opposite side. This mechanism produces an imprint
in which local light intensities represent the magnitude of the normal
strains or forces associated with an object in contact with the sensor.
This tactile imprint is converted to a digital form by a circular intensity difference (CID) camera.

Two prototype sensors were built. A longer type with an active
area of 7 cm × 12 cm and a 128 × 128 pixel CID camera is used to
image a 3.3 cm × 3.3 cm section of the area. A compact unit is designed for use on a robot gripper equipped with a cable of optical fibers that convey the strain image to a remotely located CID camera.

The active area is 2.2 cm × 2.5 cm with a density of 54 tactile elements per square centimeter. The design of the compact sensor also allows the development of nonplanar sensor geometries.

Schneiter and Sheridan describe an optical tactile sensor design based on fiber optics [25] that allows extremely high resolution (325 spots/cm^2). With the optic fibers the sensor signals can be routed from the sensor to a video camera; thus the device is immune to electromagnetic noise and may be used in noisy environments. The dynamic range is about 18:1 at a minimum reaction time of 1/30 second.

The working principle of the sensor is shown in Figure 5.26. From an emitter fiber, light is emitted in a cone and reflected from a surface. For a given light flux F, the intensity of the reflected light is dependent on the distance of the fiber to the surface. From receiver fibers, the light intensity I of the reflected light cone is measured. The relationship between the light intensity I and the distance to the surface is given by the formula:

$$I = \frac{\alpha F}{4\pi H^2 \tan^2 (\theta/2)} \tag{5.21}$$

where α is a constant for the reflectivity of the surface and θ is the cone angle of the emitted light beam. Different designs have been built employing this principle.

The arrangement of the sensor pad shown in Figure 5.27 consists of fiber bundles, a deformable clear elastomer, and a deformable reflector made of white silicone rubber. In one design light is directed from a light source into a separate emitter bundle and reflected back into a separate receiving bundle to a standard TV camera. A differ-

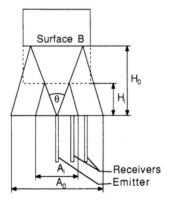

FIGURE 5.26 Reflected light intensity with receivers.

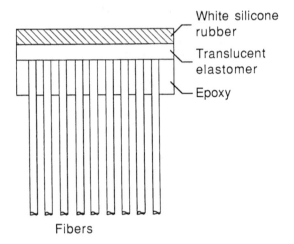

White silicone rubber

Translucent elastomer

Epoxy

Fibers

FIGURE 5.27 Arrangement of the sensor pad.

ent design uses each fiber bundle in conjunction with a beam splitter, for both emitting and receiving light beams.

5.5.4 Piezoelectric Touch Sensors

Certain materials show an electric polarization when subject to mechanical strain. This is called the piezoelectric effect. Polarization means that positive and negative charges in a material are shifted. On the outside of the material, a voltage proportional to the charge can be detected under no-load conditions, and if short-circuited, a charge proportional to the mechanical strain can be measured.

The polarization is proportional to the elongation of the material. A negative elongation or compression results in a negative polarization. This effect even works in the other direction (i.e., an electrical polarization brought into the material causes a mechanical elongation or tension).

The piezoelectric effect was first discovered in natural quartz and later in ceramics. In 1969 Kawai discovered that polyvinylidene fluoride (PVDF) can be poled and then exhibits piezoelectric properties. PVDF that has just been melted gets its piezoelectric properties only after being stretched and poled in an electric field at about 150°C. The piezoelectric constants generated are direction-dependent; that is, the direction of a mechanical strain must be considered when evaluating the charges and voltages.

Available PVDF foils are 9–40 µm thick and are coated with aluminum or nickel on both sides for contacting reasons. Their pyroelectric coefficient is about 4×10^{-5} C/m$^2 \cdot$K and the charge constant is approximately 10×10^{-9} C/N.

A piezoelectric sensor can be modeled electrically as a capacitor proportional to the mechanical stress if the electric charge does not move. The basic piezoelectric equations can be derived as follows.

For a piezoelectric material exposed to constant mechanical stress:

$$D = \varepsilon_\sigma E \tag{5.22}$$

where D is the areal density of charge, ε_σ is the dielectric constant, and E is the electrical field (V/m).

For a constant electrical field, we get:

$$S = s_E \sigma \tag{5.23}$$

where s_E stands for the reciprocal of the elasticity modulus, σ for the mechanical stress (N/m^2) and S for the relative elongation (m/m).

Thus the relationship between electrical and mechanical quantites may be described by means of the following linear equations:

$$S = s_E \sigma + dE \tag{5.24}$$

$$D = d \sigma + \varepsilon_\sigma E \tag{5.25}$$

The ratio of energy transformed to the energy input yields the square of another important constant called the coupling factor k. It can be calculated with the formula:

$$k^2 = \frac{gd}{s_E} \tag{5.26}$$

Piezoelectrically produced charges can be processed with a charge amplifier (Figure 5.28). The piezoelectric element is run in short-circuit mode to prevent voltages between the electrodes. This has the advantage that the charges cannot leak through the finite isolating resistors, which would lead to a measuring error. The short-circuit current delivered by the piezoelement is integrated to the charge Q in the capacitor C. Assuming an infinite no-load amplification of the operational amplifier, the output voltage may be expressed as follows:

$$V_a = -\frac{1}{C} \int_{t_1}^{t_2} i \, dt = -\frac{Q}{C} \tag{5.27}$$

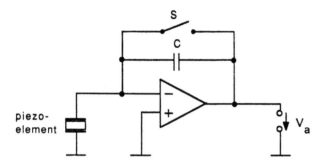

FIGURE 5.28 A charge amplifier.

The switch S is open between the points of time t_1 and t_2. For small charges, the measuring time $t_2 - t_1$ is limited by the parasitic input current of the operational amplifier. This input current is also integrated and causes a measuring error, which can be estimated as follows, under the assumption that the measured force is 1 N, which causes a charge of about 10 pC, and the input leakage current is about 50 fA.

The measuring time should be limited so that the measuring error caused by the input leakage current of the operational amplifier does not exceed 10%.

$$Q_1 = 0.1Q_p$$

where Q_1 is the charge caused by the leakage current and Q_p is the charge from the piezoelectric sensor. Using the formula for electric charges as

$$Q = it$$

one obtains

$$i_1t = 0.1Q_p$$

or

$$t = 0.1\left(\frac{10 \text{ pC}}{50 \text{ fA}}\right) = 20 \text{ s}$$

If the measuring time is 20 seconds and the measured force is 1 N, the resulting measuring error will be around 10% and can be reduced

by software measures, because the leakage current of the amplifier
can be assumed to be constant and the resulting parasitic charge
can be computed.

The potential of piezoelectric polymers for robotic applications was
investigated by Dario [26]. Polymers can be molded or thermoshaped
with conventional technologies, permitting transducers of complex
shapes to be built. Dario developed a composite transducer with a
skinlike structure. The bottom of the sensor consists of a printed
circuit board with a pattern of electrodes. A PVDF film is bounded
to the circuit board, and when the polymer film is pressed, a voltage
can be sensed at the electrodes. The second layer is a conductive
rubber to measure the static pressure. At the top, a third layer
made of a second film of PVDF with lower spatial resolution works primarily as a membrane and is sensitive to small forces. Some of the
PVDF sensor elements have a thin layer of resistive paint that can
heat the PVDF sensors when connected to a dc power supply. In
this case, 80 temperature variations can be detected by the polymer
sensor using the pyroelectric effect.

5.6 ROBOT GRIPPERS

The gripper of a robot is the only part that has mechanical contact
with the object; its main functions are to grip objects, hold objects,
and release objects.

Gripping establishes a defined position and orientation of the object relative to the robot. *Holding* secures the defined position and
orientation relative to the robot during the material transfer route
and assembly operation. When *releasing* the object, the relationship
between gripper and object is given up at a specified point.

The possibilities of executing different handling tasks are limited
mainly by the flexibility of the gripper system. The gripper system
can, therefore, be seen as an independent system within the robotic
workcell, which has a key function with respect to the flexibility of
the overall system.

Generally, a conventional gripping device used for industrial robots
is a specialized device that handles only one or few objects of similar
properties (shape, size, weight, etc.). When a single gripper alone
cannot cope with the variety of parts to be handled, a multiple gripper, a gripper change system, or a jaw change system can be used.

Grippers can be equipped with sensors for monitoring the gripping functions. An integrated sensor system can monitor the internal state of a gripper (e.g., jaw distance), and the structure of the
environment (e.g., object distances).

The integration of sensors leads to more intelligent gripper systems. Integrating sensors can provide the robot system with the

flexibility needed to cope with uncertainties. This could in many cases reduce the requirements of feeding devices, lead to simpler assembly systems, and help reduce system cost.

Various gripper systems have been developed that can adapt to the shape of objects. Such grippers are built with flexible fingers with several passive joints, which close around the object to grasp it (Fig. 5.29).

The design of the so-called dextrous hands, which imitate the versatility of the human hand, is still in the research phase. These devices feature multiple fingers with three or more programmable joints each. Figure 5.30 shows the mechanical configuration of such a system with three fingers and three individually programmable joints per finger. This kind of hand allows gripping objects of different geometries (cube, cylinder, ball, etc.); fine manipulations of the object can also be realized by the gripper system iself. This fine manipulation ability results in a task separation similar to that of the human hand—arm system.

A compromise between these highly sophisticated and therefore expensive systems and conventional grippers can be achieved by mounting some sort of compliant device between a conventional gripper and the robot. These compliant devices consist of two metal plates connected by elastomeric shear pads and are known as *remote center compliance devices* (RCCs). Remote center compliance devices can compensate positioning faults resulting from reaction forces in the assembly phase.

Active systems are equipped with force sensors to measure resulting contact forces, with computing hardware to calculate the required correcting movements and with a drive system to actually execute the correcting movements. Van Brussel describes such systems [27].

We concentrate next on standard grippers and on sensor-integrated grippers, for which two examples are presented. These gripper types

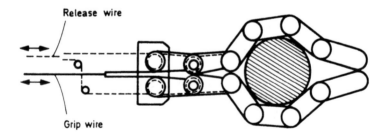

FIGURE 5.29 Gripper with passive joints.

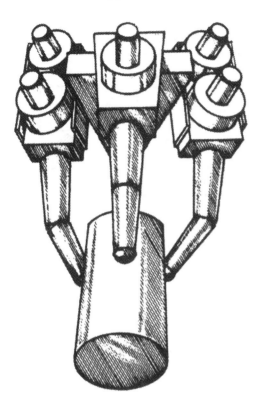

FIGURE 5.30 Three-fingered gripper.

are classified, and a list of influential parameters for selecting standard grippers is compiled.

A functional model (Fig. 5.31) of a robot gripper can be set up
according to the basic functions. Energy is transferred to the object by the jaws through the gripping force. Reacting forces and
moments are routed back to the robot via the gripper housing. A
control—sensory system can guard and control all the actions of a
gripper. The sensing capabilities may help the performance of a
gripper to a considerable extent. Not all parts have to be explicitly
present in all types of gripper.

5.6.1 Gripper Classification

The classification of grippers presented here is meant to serve as an
introduction for the selection of gripper systems.

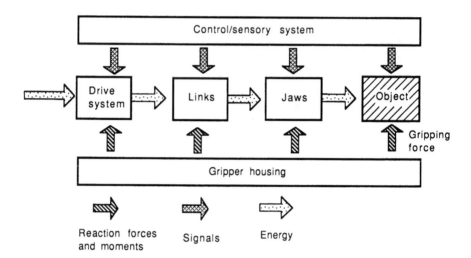

FIGURE 5.31 Functional model of a sensor–gripper.

One must first distinguish between standard grippers without sensors and sensor-integrated grippers. Standard grippers can be classified according to the following criteria:

Working principle
Drive concept
Number of clamping or suction elements

A classification from the first two criteria leads to a tree, as illustrated in Figure 5.32. For mechanical grippers, the characteristics are the number of links and their type of motion.

Sensor-integrated grippers can be distinguished by the kind of sensors they are equipped with or the controllability of the fingers or suction cups. Simple sensor-integrated systems can detect the presence or absence of objects or test the jaw position (open or closed). Complex systems (see Section 5.7.3) can, for example, determine the position or orientation of objects and measure the position of the jaws.

The jaws of a gripper can be controlled by position or force (see Section 5.7.2), and the jaws can be controlled either separately or together. If a two-fingered gripper has individually controllable jaws, an additional degree of freedom is gained in the direction of jaw motion, which can be used to correct positioning faults with the gripper itself. An example for a controllable vacuum gripper can be found in

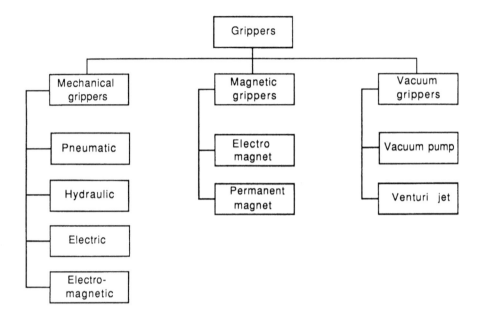

FIGURE 5.32 Gripper classification.

Tella et al. [28]. This gripper consists of several suction heads, which are connected to separate piston cylinders. This lifting mechanism allows the suction heads to be positioned independently.

5.6.2 Criteria for the Selection of Grippers

After various gripper concepts have been introduced, the necessary gripper selection parameters are described and some criteria for the selection of a suitable robot gripper are given. Detailed material about gripper selection can be found in Cardaun [29].

Figure 5.33 gives an example of a workcell and points out some of the factors that must be considered when selecting a gripper. In this workcell a gearbox consisting of housing, toothed rack, and pinion is assembled. The housing is taken from a box by robot 1 and mounted in a fixture device. The toothed rack is taken from a feeding device, transported to the fixture, and placed in the housing. Robot 2 takes the pinion from a feeder, transports it to the fixture, and places it in the housing. The assembled rack-and-pinion gear is taken from the fixture by robot 2 and placed on a conveyor belt.

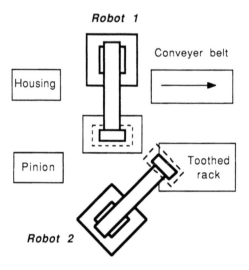

FIGURE 5.33 Example of a robot workcell.

The parameters to be investigated in detail for gripper selection can be grouped as follows:

The object
Feeding, storage, and fixture devices
Sequence of operation
The robot
Environmental conditions

The Object

The objects to be grasped are the first thing to be considered when choosing a gripper. The influence the object has on the selection of the gripper can be grouped into geometric, physical, and technological parameters.

Geometrical parameters are:

Object size
Number of potential gripping surfaces
Position of potential gripping surfaces
Distance of gripping surfaces
Geometry of gripping surfaces

Physical and technological parameters are:

Mass
Material
Inherent stability
Delicacy of surface
State of surface
Temperature

It may be noted that all these object properties are not static, but are subject to changes during the manufacturing process. Milling and assembly operations change object weight, size, and geometry. The variations of these factors influence the required gripper flexibility.

The object size, its mass, the orientation of potential gripping surfaces, and friction between gripper and object influence the force that must be applied by the gripper to guarantee a safe grip.

The geometry of potential gripping surfaces, the delicacy of surfaces, and the expected object temperatures influence the design and the material of the gripper jaws.

Feeding, Storage, and Fixture Devices

The accessibility of parts, their minimum distances, and any positioning uncertainties influence the design of the jaws and the smallest and widest possible jaw distances.

Depending on the device, additional external resulting forces must be considered while computing the necessary gripping forces in the storage or assembly phase.

Sequence of Operation

The definition of the sequence of operations is a major point in the design of the workcell. It influences the cycle time of the workcell, and consequently its productivity, and it determines the required flexibility of the gripper(s).

In our earlier examples robot 1 has to grasp two different parts. If a mechanical gripper can be found to meet the requirements of jaw distance and maximum gripping force, the technically and economically optimal solution has been achieved. If none of the available grippers is flexible enough, such additional devices as gripper changers or multiple gripper configurations must be considered. These additional devices add weight to the robot and require extra movements (to change tools, etc.). This leads to an increase in cycle times and loss of performance.

The Robot

The gripper must be adapted mechanically with flanges to the robot. The maximum gripper weight allowed for an application is equal to the load capacity of the robot minus the weight of the object to be gripped.

When choosing the robot gripper, the robot's drive system should also be considered, to ensure that no extra energy-supplying devices are needed.

Environmental Conditions

Environmental conditions of the operational area may require the gripper system to operate at temperatures beyond the usual ambient temperature.

Two kinds of requirement can be distinguished. The gripper may be exposed to extreme conditions (e.g., water or oil), or the gripper itself may produce inadmissible pollution, as when pneumatic grippers produce oil-loaded waste air. According to these requirements, the choice of the gripper can depend on the operational area.

5.7 STANDARD GRIPPER TYPES

5.7.1 Magnetic Grippers

For magnetic grippers, electromagnets and permanent magnets can be used, with use of the latter being more complicated because the gripping force, which is equal to the attractive force of the magnet, cannot be turned off, and the object must be released (pushed off) mechanically. The design of magnetic grippers is generally simple; different configurations can be built from standard components.

The objects' gripping planes should be clean, plane, and dry to achieve safe contact, and of course magnetic grippers can be used only for ferromagnetic objects.

Handling metal sheets is a typical application for magnetic grippers, but more than one sheet tends to be attached to the gripper. Lundström shows a gripper construction to get around this problem [30].

5.7.2 Vacuum Grippers

Vacuum grippers are mainly suction grippers that use a vacuum pump or a venturi jet to generate a (70—90%) vacuum between a suction cup and the gripping plane. The version that pressed out the air is hardly used any more and is not analyzed here. The suction cups are made of rubberlike materials (silicone, neoprene). Depending on the design and on the material, suction cups are suitable for workpiece temperatures up to 200°C. To adapt to local variations in surface angle, the vacuum cup can be equipped with a ball joint.

The exerted gripping force F_G depends theoretically on the contact area A between suction cup and gripping plane and the pressure difference Δp between suction cup and environment. The pressure difference that can practically be reached within the required cycle time is influenced by the roughness of the gripping surface, by air leakage, and by the performance and type of vacuum pump used. The maximum gripping force depends on the roughness of the surface and on the permeability to air of the grip surface.

Suction grippers are available in a variety of shapes and sizes. They offer low weight, and custom configurations can easily be built from standard elements. Suction grippers need only one gripping surface and are therefore first choice for stacking and unstacking operations. Suction grippers also offer the advantage of adapting to local variations in surface angle. Figure 5.34 shows an example of a surface adapting suction cup.

Other applications are handling such large, lightweight parts as cardboard boxes and styrofoam sheets, or parts that require careful handling. Vacuum grippers are unsuitable for precise positioning tasks, for assembly tasks with high forces, and for components with curved surfaces or with holes and cavities. A gripping surface should, in general, be even, airtight, dry, and clean.

5.7.3 Mechanical Grippers

Mechanical grippers, due to their manifold variations and universal applicability, are among the most common gripper types.

Drive Concepts

The most common drive concepts for mechanical grippers are pneumatic, hydraulic, electric, and electromagnetic.

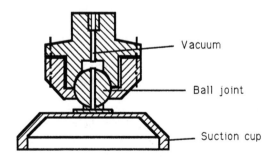

FIGURE 5.34 A surface-adapting suction cup.

In pneumatic systems compressed air is used to transport energy. Pneumatic systems are characterized by simple construction, low cost, easy maintenance, short reaction times, and a very good performance-to-weight ratio. Depending on the type of piston motion, two types of pneumatic drive system may be distinguished: linear piston and rotary piston.

The type of drive used has a major influence on the design of the transmission. The output of a linear piston system is a piston rod with linear movement, whereas for a rotary piston system the output is a rotating shaft.

With a pressure reducer, different gripping forces can easily be realized. The major drawback of this concept is the complicated control of the jaw position due to the compressibility of the medium. For certain applications, the oily waste air can lead to an undesired pollution.

Because of their many advantages, pneumatic drive systems are used in a wide variety of applications in the field of robotics. The medium air is usually available in the environment of a robot and can even be used in explosive environments.

Hydraulic drive systems are generally more costly. Hydraulic oil must be supplied and oil pressure must be generated. High forces are possible with small units, but in comparison to pneumatically activated grippers it is more difficult to adjust the level of the clamping force. The hydraulic gripper requires flow control and relief valves.

Oil is incompressible and facilitates better control of hydraulic drive systems, compared with pneumatic ones. Hydraulic systems are capable of handling high forces and torques and are therefore preferred for handling heavy objects.

For electric gripper systems, mainly dc and stepper motors are used. The advantage of electric drive systems lies in their compatibility with electronic control devices. This guarantees simple signal transmission and offers a good basis for position and force control. Electrical energy used in such systems is generally nonpolluting. The complete drive and control system of an electric gripper is a lot more complex than that of simple pneumatic grippers, but it offers extended functionality.

Electric motors require reducing gears. Gripping forces are generally lower than in pneumatic and hydraulic systems and control times are longer. The typical application for electric gripper systems is handling and assembly of sensitive, lightweight parts. Examples are grippers for complementing electronic boards with special chips or handling parts with no inherent stability.

Electromagnetic drive systems are characterized by their simple design and offer all advantages of electric energy supply. Position or force control of electromagnetic systems is generally not provided for. The gripping force is strongly dependent on the jaw travel distance, and only small jaw widths can be realized. Typical for electromagnetic grippers are applications that want to take advantage of electricity as energy source but set no requirements on the programmability of the system.

Transmission

The transmission must transmit forces produced by the drive system to the jaws. Depending on the drive concept used, rotary motion is transformed into linear motion, or reducing gears are used to increase the attainable output forces and torques.

In mechanical grippers, the transmission is the component that determines the characteristics of the gripper system. The type of transmission used determines the maximum jaw distance, the type of jaw motion, and the correlation between the maximum jaw distance and the gripping force.

The basis for a classification of gripper units is a classification of types of jaw motion. The jaws can move along a linear or a circular path. The jaws can be parallel all the way along the path or only at one point. The type of jaw motion is crucial for the usefulness of the gripper.

It may now be worthwhile to take a closer look at the characteristic of jaw motions. The following types of jaw motion can be distinguished.

Linear jaw motion with parallel jaws (Fig. 5.35a): The jaws move along a linear path with the jaws always being parallel. This type of jaw

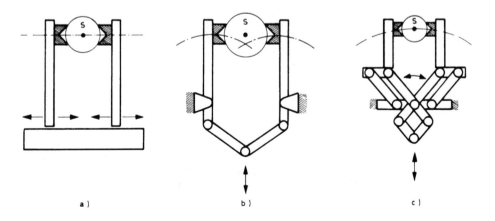

FIGURE 5.35 Types of jaw motion.

motion has the widest range of uses. Parts can be grasped alternatively from inside and outside, and an arbitrary jaw distance
can be chosen.

Circular jaw motion with parallel jaws (Fig. 5.35b): The jaws move
along a circular path and are always parallel. The gripping center changes with the jaw width, for the same gripper reference
position.

Circular jaw motion with parallel jaws only in one position (Fig. 5.35c):
The jaws rotate about one axis. The jaw surfaces are parallel only
in one position. In conjunction with plane jaws, this type of jaw
motion is useful only for applications with constant grip widths;
but if the jaws have notches, it can also be used for gripping cylindrical parts with different diameters.

Another characteristic feature is the functional correlation between
jaw motion type and the gripping force. Depending on the transmission design, the gripping force can be constant (or almost constant) along the jaw trajectory, or the maximum gripping force is
reached only in a small range (as in the toggle gripper design).

Toggle grippers can exert very high forces but can be employed
only if the required grip widths lie within a very small range. For
gripping parts of varying size a linear or nearly linear correlation
between jaw motion and gripping force is preferred.

From the correlation between jaw travel and gripping force, the
usable clamping capacity can be calculated to serve as a measure for
the flexibility of grippers. A wide usable clamping capacity is the
precondition for gripping objects of different sizes.

The Jaws

The jaws of a robot gripper are especially important. They have
direct contact with the object, and their shape and material have a
strong influence on the gripping forces required. The more contact
points there are, the smaller the required forces and the more safely
the object will be grasped.

Figure 5.36 shows how to grasp cylindrical parts with three different parts of jaws. The first type (Fig. 5.36a) has a plane gripping
surface. Contact between the gripper and the object occurs along two
lines of the jaw surface. This type of grip is called a force closure
grip because only the friction force keeps the object in its position.

A pure form closure grip requires contour-adapted jaws as shown
in Figure 5.36b. This type of contact is best suited to resist gravitational and inertial forces, because the object is supported in all possible motion directions. The disadvantage of this jaw type is its limited ability to grasp objects of different geometries. The third type
of jaws has a V-shaped gripping surface (Fig. 5.36c). Fourfold line

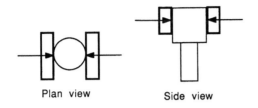

Plan view   Side view

a) Plane jaws

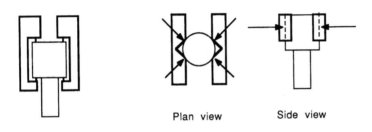

Plan view   Side view

b) Contour adapted jaws   c) V-shaped jaws

FIGURE 5.36 Different jaws for gripping cylindrical parts.

contact between object and gripper keeps the object in its place. This is a mixed form—force closure grip.

Gripper jaws are usually tailor-made for a specific application. Design guidelines for designing jaws can be found in *Robot Grippers* [31--33].

5.8 SENSOR-INTEGRATED GRIPPERS

It is the sensory ability that makes the human hand a versatile and precious tool. Similarly for a robot gripper, sensing is of primary importance for extending versatility. In fact, a sensitively controllable gripper can be only as good as its sensory system, which delivers sampling values for computing reference input for the drive system.

The art of designing an industrially usable gripper--sensor system consists of covering a maximum number of problem cases with a minimum of sensors. The entirety of all sensors used constitutes the sensory system of the robot and should consist of the fewer simple sensors possible, with respect to small size, weight, robustness, and easy maintenance.

Gripper and sensory system should be an integral part, and yet modular. The design of a sensory system should focus on the versatility of each sensor, sufficient immunity to interference, low probability of failure, and compact integrability into the gripper, which in turn requires low power dissipation.

In spite of intensive research activities in this field, the industrial use of sensor-integrated robot grippers lags expectations. Important subproblems in the development of a gripper—sensor system are:

Compilation of a suitable sensor configuration.
Development of a suitable gripper mechanics for integration of sensors
    and hardware for signal preprocessing.
Selection of a signal transmission system free of interference.
Development of a suitable interface for communication of the gripper—
    sensor system with the robot control unit.
Development of a suitable user interface for programming gripping and
    monitoring tasks.

Two examples of sensor-integrated grippers developed for two distinct applications are now presented.

5.8.1 A Pneumatic Gripper for Monitoring the Grip and Transfer Phase

A sensor configuration for monitoring the grip and transfer phase is shown in Figure 5.37. The CCD—sensor array between the jaws is of central importance when monitoring gripping actions. This sensor in conjunction with four tunable infrared LEDs is used for edge detection. With this arrangement, the presence of an object can be tested. An object can be identified and a centered position between the jaws can be calculated. The combination of a semiconductor laser and the sensor array results in a range measuring device according to the triangulation principle, as described in Section 5.2.2. The maximum measuring accuracy depends on the resolution of the sensor array and on the distance between object and sensor. The system can be fine-tuned to a suitable measuring range by rearranging the laser and sensor array or by exchanging the lens. The CCD array consists of 256 elements and has the dimensions of 16 mm × 16 mm × 20 mm, including lens. The accuracy is 0.1 mm over a range of 30–50 mm. If well designed, the signal processing hardware can be mounted directly in or on the gripper.

The jaw position is measured by a second CCD—sensor array. The jaw distance is calculated from the position of a black-to-white transition

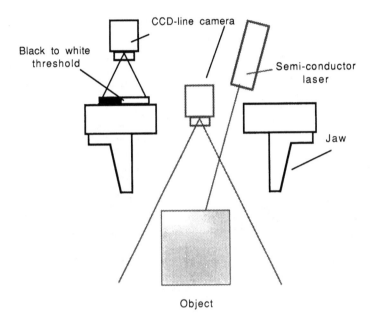

FIGURE 5.37 A sensor configuration for monitoring the grip and transfer phase.

marker attached to one of the jaws by illuminating it with infrared LEDs. All this is installed inside the gripper housing to be well protected against environmental influences.

A special pneumatic gripper mechanics was developed for integrating the sensory system and data preprocessing (Fig. 5.38). The drive kinematics was designed to provide a maximum free space in the center for the CCD-sensor array and the illuminating equipment. The hardware for the sensor data processing is separated into an analog and a digital part and placed to the left and to the right sides of the gripper housing, in two separate cases. The system described was developed at the University of Karlsruhe, in collaboration with the University of Kaiserslautern.

5.8.2 A Gripper—Sensor System for Sensitive Assembly Tasks

A robot hand for assembly is shown in Figure 5.39. It has force-controlled fingers and a force—torque sensor. The drive and sensor systems were designed to be as compact as possible, and the force-torque sensor, described in Section 5.4.2, is built around the motor and the sensor components. The hardware for processing of analog values, the microcontroller, and an optical fiber interface are placed

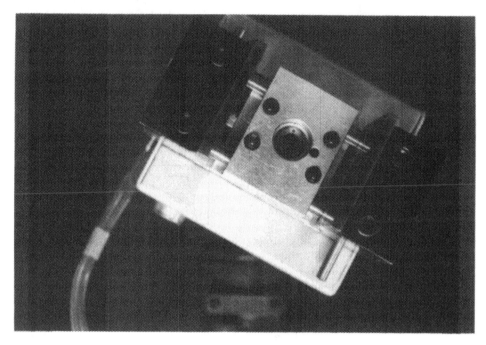

FIGURE 5.38 A special pneumatic gripper.

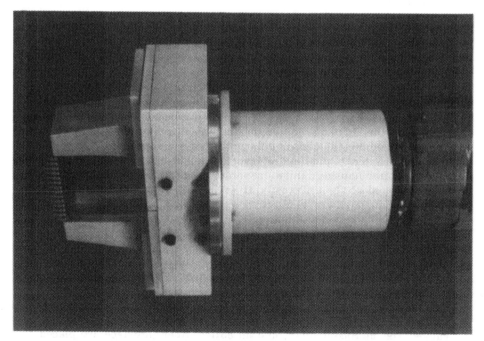

FIGURE 5.39 A robot hand for assembly.

between the supporting legs of the force—torque sensor. The gripper is a two-fingered parallel jaw design with a dc motor drive. The
jaw distance is measured by a potentiometer and the gripping forces
by a force sensor using strain gages, which has been integrated in
the jaws. The basic functions of the gripper are controlling the jaw
distance and controlling the gripping force.

The gripper is connected to a Puma robot, and both the gripper
and the robot are controlled by a VME—bus system via a serial interface. The high transmission rate of sensor data from the force—torque
sensor requires a fast link, which is immune to interference. It was
realized using an Intel MCS 51/96 and optical fibers. This combination allows serial transmission up to 187.5 kb/s. The transmitted
sensor data are processed on a 68000 board in the VME—bus system.
Other CPU boards realize trajectory interpolation, coordinate transformation, and I/O to and from the Puma joint controllers. The structure of this system is shown in Figure 5.40.

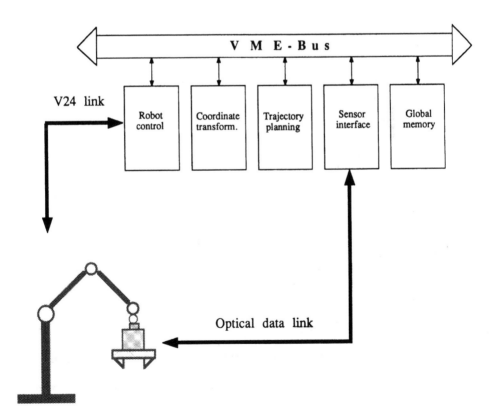

FIGURE 5.40 Overall structure of the Karlsruhe system.

Processes on all boards have access to sensor data in a dual-ported RAM via a sensor interface unit, and they can send commands and parameters to the sensor module. Process management and communication of all monitoring and control processes is done with a multiprocessor, multitasking operating system called HEROS (Hierarchical Executive Robot System). In the HEROS system, each processor has its own multitasking real-time system kernel, and the processes communicate interrupt-driven via global memory using semaphores. This concept guarantees task switching times of less than 100 µs, which permits the realization of control loops with extremely short delay times.

REFERENCES

1. U. Rembold and P. Levi, "Entwicklung von Sensoren für die Montageautomatisierung, atp-Sonderheft Fertigung und Automatisierung, 1988.

2. W. Eversheim and A. Hausmann, Strategies for Sensor Utilization in Assembly. Journal Industrial Robot No. 9, 1985.

3. S. Kämpfer, *Roboter, die elektronische Hand des Menschen*, VDI-Verlag, Düsseldorf, 1984.

4. A. Staugaard, *Robotics and AI: An Introduction to Applied Machine Intelligence*, Prentice Hall, Englewood Cliffs, NJ, 1987.

5. U. Ahrens, "Möglichkeiten und Probleme der Anwendung von Luft-Ultraschallsensoren in der Montage- und Handhabungstechnik," in *Robotersysteme* No. 1 (1985).

6. G. L. Miller, R. A. Boie, and M. J. Sibilia, "Active Damping of Ultrasonic Transducers for Robotic Applications. *International Conference on Robotics 1984*.

7. W. D. Koenigsberg, "Noncontact Distance Sensor Technology," in *International Conference on Robot Vision and Sensory Control*, 1983.

8. K. S. Fu, *Robotics, Control, Sensing, Vision, and Intelligence*. McGraw-Hill, New York, 1987.

9. T. Kanade, "Optical Proximity Sensors," Workshop on Intelligent Robots: Achievements and Issues. SRI International, Menlo Park, CA, 1984.

10. G. Kinoshita and M. Idesawa, "Optical Range Finding System by Projecting Ring Beam Pattern," *ICAR 1985*, pp. 177–184.

11. T. Okada, "Development of an Optical Distance Sensor for Robots," *Int. J. Robotics Res.* 1:4 (Winter 1982).

12. M. Rösler, "Preisgekröntes Laser-Distanzmeßgerät, *Elektronik* 19 (1986).

13. H. L. Trietly, *Transducers in Mechanical and Electronic Design*, Dekker, New York, 1986.

14. F. Blume and W. Schaffrath, "Induktive Sensoren zur Auto-
    matisierung von Schweißfertigungsprozessen," *Schweißtechnik*
    51, 1984.
15. P. Drews and G. Starke, "Sensoren fur das Lichtbogenschweißen
    mit Handhabungssystemen," Seminar: Integration von Sensoren
    in Handhabungssysteme, Hannover, 1985.
16. W. Meier, "Entwicklung und Implementierung eines Sensorver-
    arbeitungssystems für eine Roboter-Kraftmeßdose mit sechs
    Freiheitsgraden," Master's thesis, Karlsruhe University, 1984.
17. L. Harmon, "Automated Tactile Sensing," *J. Robotics* 1:2 (May
    1982).
18. K. J. Overton, "The Acquisition, Processing, and Use of Tac-
    tile Sensor Data in Robot Control," COINS Technical Report
    84-08.
19. M. H. Raibert, "An All-Digital VLSI Tactile Array Sensor, *Inter-
    national Conference on Robotics 1984*, pp. 314–319.
20. R. A. Boie, "Capacitive Impedance Readout Tactile Image Sensor,"
    in *International Conference on Robotics 1984*, pp. 370–378.
21. D. Siegel, I. Garabieta, and J. Hollerbach, "An Integrated Tac-
    tile and Thermal Sensor," *International Conference on Robotics
    and Automation 1986*, pp. 1286–1291.
22. D. Siegel, S. Drucker, and I. Garabieta, "Performance analysis
    of a Tactile Sensor," *International Conference on Robotics and
    Automation 1987*, pp. 1493–1499.
23. J. Rebman and K. A. Morris, "A Tactile Sensor with Electro-
    optical Transduction, in *International Conference on Robot
    Vision and Sensory Control*, 1983.
24. S. Begej, "An Optical Tactile-Array Sensor," COINS Technical
    Report 84-26.
25. J. L. Schneiter and T. B. Sheridan, "An Optical Tactile Sensor
    for Manipulations," *Robotics Comput. Integrated Manuf.* 1:1,
    65–71 (1984).
26. P. Dario et al., "Piezoelectric Polymers: New Sensor Materials
    for Robotic Applications," *Proceedings of the 13th ISIR*, 1983,
    pp. 14-34–14-49.
27. H. Van Brussel, "Robotics 1986—State of the Art," *Proceedings
    of the 16th ISIR*, 1986, pp. 1–15.
28. R. Tella, J. Birk, and R. Kelly, "A Contour-Adapting Vacuum
    Gripper," in *Robot Grippers*, D. T. Pham et al. (eds.), Springer-
    Verlag, 1986.
29. U. Cardaun, "Systematische Auswahl von Greiferkonzepten für
    die Werkstückhandhabung," Ph.D. Thesis at University of Han-
    over, 1981.
30. G. Lundström, *Industrial Robots — Gripper Review*, IFS Publi-
    cations, Bedford, UK, 1977, p. 43.

31. T. Arai and M. Asada, "Properties of Rotational Hands with Versatility, in *Robot Grippers*, D. T. Pham et al. (eds.), Springer-Verlag, Berlin, 1986.

32. A. Baz and J. Vossoughi, "Stress Considerastions in Robot Grippers," in *Robot Grippers*, D. T. Pham et al. (eds.), Springer-Verlag, Berlin, 1986.

33. D. T. Pham and S. H. Yeo, "Concentric Gripping of Cylindrical Workpieces Using Quasi-Parallel Grippers," in *Robot Grippers*, Springer-Verlag, Berlin, 1986.

6

Programming of the Robot Cell

KLAUS HÖRMANN   *University of Karlsruhe, Karlsruhe, Federal Republic of Germany*

6.1 INTRODUCTION

Robots and other numerically controlled machines are being used more and more to automate manufacturing processes. Unfortunately the programming of these machines is very time-consuming and expensive. Hence it is very important to provide the users of these machines with powerful and comfortable programming tools.

To program a robot workcell, one usually must coordinate various machines. Besides the robot, this typically includes numerically controlled (NC) machines, feeders, and fixtures. All these machines must be synchronized with each other and with the robot. For smaller workcells, this synchronization task normally is done by the robot control itself. For greater cells, a separate cell control (PLC, for programmable logic control, shortened to PC) is being used.

Hence, as a matter of fact, the programming of the robot workcell does not consist merely of robot programming but also of programming of the cell control and of NC machines. Nowadays, most of these machines may already be programmed using an advanced programming technique, such as textual programming or graphical interactive programming. Since the emphasis of this book is on robotics, and since there exists already quite a lot of good literature on NC programming, we will not treat these topics in any more detail.

6.2 OVERVIEW OF ROBOT PROGRAMMING
   METHODS

The purpose of a programming system is to generate a robot control program in a user-friendly way. Compared with the programming of other numerically controlled machines, robot programming has some peculiarities:

Robot programming is concerned with the generation of very complex motions. These motions are very hard for human beings to imagine without the help of the robot itself or a graphical simulation system.

The geometrical data of the real world (e.g., positions and orientations of the parts to be handled) have significant deviations from those data used in the model of the real world, which is used for off-line programming. Hence very often sensors are used to measure these geometrical quantities. These readings are the basis of a conditional program to compensate the deviations.

Typically, the robot must be synchronized with peripheral devices (sensors, end-effectors, feeders, fixtures, etc.).

Obviously, this list implies a considerable number of requirements for a robot programming system. The robot programmer must be able to specify the spatial operations of the robot in a simple way. The robot programming techniques must be adapted to the way of thinking of the user, to facilitate the description of spatial operations. The user wants to program the robot in a problem-oriented way without the need to measure coordinates in the workcell and to specify these coordinates explicitly within the program. Furthermore, the system must provide methods for real-time programming. The system has to support interrupt handling and the interaction between the robot controller and peripheral machines. To ensure that sensor data can be processed, means are necessary to program loops and conditional branches.

Summing up, a robot programming method must conform to the following criteria:

It is simple enough to be easily learned and handled.

It is problem oriented: the programming method must be adapted to the imagination of the user and to the particular application.

It supports off-line robot programming to avoid using the expensive robot workcell for programming.

Today's robot programming methods fulfill only parts of these requirements. The historical development of robots, starting with simple pick-and-place devices and leading to the present sophisticated devices, has been accompanied by the concurrent development of programming methods with increasing power and complexity. The first robots were programmed with plug boards. This method was later improved by on-line programming methods.

We use the term "on-line programming" if the programming process directly involves the robot itself. The main activity of on-line programming is the definition of trajectories. Usually trajectories are defined by specifying points (i.e., positions and orientations) with respect to the end-effector. The type of interpolation determines how the different limbs of the robot are moved and synchronized and how long this motion may last. The programmer may specify several types of interpolation together with such parameters as time, velocity, and acceleration.

To specify the points of a trajectory, the robot is guided to the desired points and the corresponding joint values are recorded. The most common ways to specify these points are by specifying the motions on a teach pendant ("teach-in programming") and by moving the robot via a master–slave linkage ("master--slave programming). Sometimes the robot may be moved manually while the trajectory is recorded as a series of closely spaced points. The latter method is especially useful for spray painting.

Further improvements of this method were achieved by the incorporation of velocity, time, program branching, and many other special functions. On-line programming is still the most widely used robot programming method; even so, it is awkward for more complex problems. These methods are already well known, hence are not treated in detail in this chapter.

Increasingly off-line programming methods are used, especially for complex applications. Off-line programming involves the construction of a program text, which is translated without the actual presence of the robot and is then translated or interpreted. A problem with this textual programming method is how to specify the movement points without the robot. Therefore, almost all textual programming systems are provided with a teach-in method, too.

In textual programming, the operating sequence of the work trajectory of the robot and effector is written with textual instructions. During the initial development phase such languages were obtained by extending existing programming languages with robot-specific instructions. The special robot programming languages were developed later. Section 6.3 is concerned with textual programming techniques, whereas Sections 6.3.1 and 6.3.2, respectively, give a general introduction into this area and a detailed treatment of the subject, using the SRL programming system as an example.

To simplify off-line programming further and to shorten the time needed for program development, computer graphic tools are applied. Computer graphics can be used, for example, to simulate the effects of an off-line written program on a graphical screen. Still more user-friendly are graphical interactive programming systems. These systems allow one to specify motions in an interactive way and to immediately observe the effect of a command. Section 6.4.1 gives a very short introduction to graphical programming and simulation.

An even more advanced programming technique is the task level programming method, also called implicit programming. The term "implicit" refers to the fact that these systems do not specify the robot operations in an "explicit" way, as is the case with the methods already mentioned. Instead, the specification is given implicitly by defining the goal of the operations rather than the operations themselves. Hence the system automatically plans the action sequence needed to achieve this goal.

Task-level programming systems are still in the research stage. Section 6.5.1 gives an introduction to this exciting research area, and Section 6.5.2 presents the GRIPS prototype system for assembly sequence planning.

6.3 TEXTUAL PROGRAMMING

6.3.1 Introduction

Even though programming languages for robots have been under development for almost two decades, they have not been generally accepted, as was the case with NC, scientific, and commercial languages. Present robot programming languages and methods are tailored to the application problem, the design of a specific robot, or the type of sensor system used by the robot control system.

In textual programming, the operating sequence of the work trajectory of the robot and effector is described with textual instructions. During the initial development phase such languages were obtained by extending existing programming languages with robot-specific instructions. The special robot programming languages were developed later.

A program written in such a language must be translated by a system program, which may be either an interpreter or a compiler. With an interpreter system, a program is decoded, instruction by instruction, and is executed stepwise by the robot control. In a compiler system the complete program is translated (compiled) into an intermediate code and is subsequently executed by an interpreter program. Direct interpretable languages do not achieve the power of compiler languages; however, they result in less expensive hardware and software for the robot. Today, textual programming languages are mostly based on the interpreter principle. An essential characteristic of textual programming is that the program can be developed off-line, without the use of the robot. The trajectories are described textually and translated by a program development system.

Two different methods are being used for movement specification. With the first method the specification of the coordinates for the trajectories must be defined explicitly in the program in numerical form as constants or variable values. This means that the coordinates must be either measured or read from a design drawing or obtained from a data base and inserted into the program. This method is very complicated and is therefore seldom used.

With the second method the trajectory is described only through symbolic identifiers. The values of the variables used for the start, intermediate, and end points, as well as the effector orientation, are not necessarily known at the time of programming. Concurrently with the programming, the move parameters are taught to the robot with the help of a teach-in system and are inserted in the textual program during the program execution (Fig. 6.1). This method is a pure textual programming procedure. Unlike the first method, it offers a great advantage in its degree of comfort and visibility of geometric relationships.

In addition to the motion commands, the robot languages are frequently supplemented with the following features:

Elements for the description of the geometry of the workpieces and
　　their geometric interrelations.
Data types and language elements for geometric calculations.
Language elements for the interaction with sensors and other peripherals.
Some real-time attributes and concepts for the realization of parallel
　　processes.

There is a great difference between the power and versatility of the languages that have been developed to date. The scale ranges from the simplest assembler level to modern high level languages.

A classification of some typical languages according to their capability is shown in Figure 6.2 [1]. Most of the languages are designed for explicit programming; that is, each movement must be explicitly described by the programmer. Tables 6.1 and 6.2 show the attributes of selected robot programming languages. The type of the arm configuration, the number of axes, and the type of sensors used are of interest. In some cases the language may be used to program several different arms or arm configurations. At present there exist more than 100 different kinds of robot programming language. Almost all are designed for special robot types and configurations. An advantage of textual programming is that the program, which is readable by human beings, can easily be changed, documented, and extended. A second major advantage is the high power of the textual languages. In case of such very complicated applications as sensor-oriented assembly, it must be possible to react instantaneously to changing

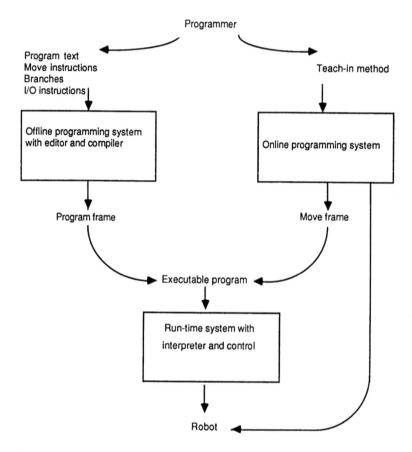

FIGURE 6.1 Combined on-line and off-line programming system for robots.

environmental conditions. This usually results in program internal calculations and program branches based on the results of these calculations.

A disadvantage frequently mentioned is that the textual programming process is hard to follow, since the writing and testing of the code are done independently of the robot. Hence a highly qualified programmer is needed. This objection can be rebutted, since textual programming is mainly used for the solution of complex tasks that cannot be solved by other means. With structured programming languages that are based mainly on Algol and Pascal, all essential

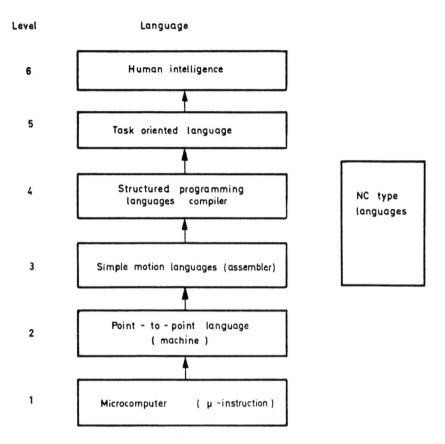

FIGURE 6.2 A classification of robot programming languages according to their capabilities.

functions of the basic language are retained. In addition, all extensions are provided which are necessary for robot programming. Some typical additional attributes are:

Geometric data type as vector, rotation, and frame.
Parallel, cyclic, or time-bounded processing of program parts.
Move instructions with different interpolation and move parameters
(speed, acceleration, etc.).
Parallel instructions for two cooperating robots.
Specification of different trajectories.
Instructions for the manipulation of the end-effectors and the tool
system.

TABLE 6.1 Various Implemented Robot Programming Languages

Robot	Manufacturer	Arm configuration[a]	Number of axes	Languages
T3	Cincinnati Milacron	R R R R R R	6	T3
Puma	Unimation	R R R R R R	6	RPL VAL
Stanford	Sheinmann	R R P R R R	6	AL PAL
IBM arm	IBM	R P P P R R	7	Funky Emily Maple Autopass
PACS arm	Bendix	R P P R R R	6	RCL
Allegro	General Electric	P P P R R R	6	Help
Anomatic	Anorad	P P P R	4	Anorad
Sigma	Olivetti	P P P P	4	Sigla

[a]R = rotational joint ; P = prismatic joint.

Processing of sensor signals.
Instructions to control the signal flow of input/output interfaces.

Since most manufacturers of robots use a proprietary programming language, the presently available languages differ in various aspects concerning syntax, program structure, and features. This fact results in the following disadvantages to the user of different robots.

1. Every robot system used in a facility needs a specially trained programmer.
2. The process of transferring an existing robot program to another system is equivalent to implementing the robot task a second time, even if the target robot has similar kinematics.
3. There is no possibility of transferring taught locations from one robot control to another one.

A solution to this problem could be obtained by a standardized interface between the robot programming system and the robot control system, as shown in Figure 6.3. This would provide the following advantages:

TABLE 6.2 Capabilities of Different Languages

Programming language	Language type	Number of arms	Other arm configurations	Sensors	Vision
Funky	Point-to-point programming			Touch	
T3	Assembly	1		Limit switches	
Ancrad	NC program	1			
Emily	Assembly	2	X	Touch Proximity	
RCL	FORTRAN	1			
RPL	FORTRAN	1		Touch Vision	Location Orientation
Sigla	Assembly	1—4	X	Force Torque	
VAL	Assembly	1		Vision	
	ALGOL	2	X	Force Torque	Recognition
Help	Pascal	1—4	X		
Maple	PL/1	1	X	Force Proximity	
MCL	APT	1	X	Touch Vision	Modeling Recognition
PAL	Transformations	1	X		

Programs written with one particular programming system could be used to control robots of different types.

A robot could use programs or data from any programming system.

The resulting flexibility in the use of a robot system and the possibility of easily reconfiguring a robot cell would improve both the efficiency and the performance of a robot installation. In Germany such

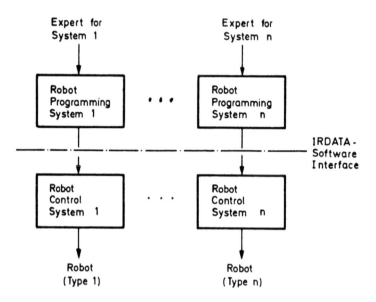

FIGURE 6.3 The concept of the IRDATA interface.

a standardization effort has been undertaken (VDI 2863, [2]), which is currently being considered to form an ISO standard.

6.3.2 An Example of a Modern Textual Programming System: The SRL System

Off-Line Programming

Industrial robots are being applied more and more in the areas of assembly operations and for complex handling tasks. In this particular area, programming with high level programming languages is quite helpful. This type of programming facilitates the evaluation of sensor signals and enables the use of CAD data. Also the off-line programming of complex operations is supported.

In spite of the development of high level programming languages and modern programming techniques in the last several years, only a few are being used in the industrial practice. This slow acceptance is primarily due to a considerable resistance in industry against the rather complex high level languages. The SRL programming system presented here is the result of research and development carried out at the University of Karlsruhe and may form the basis of a robot programming language for industrial use.

The objective of an off-line oriented programming system is to program the robot without its direct physical presence. Of course some portions will always have to be programmed on-line with the help of the robots in the real working environment. The most important element of off-line programming is a language that enables a description of the problem to be solved. To use a high level language, it is necessary to develop a compiler to check the correctness of the program and to translate the program into a code suitable for the execution by the robot. The type of definitions of the geometrical data of the robot's movements plays an important role. This is because otherwise these data are quicker and easier for the programmer to handle with the help of an on-line teach-in method, or off-line through graphic assistance.

The High-Level Programming Language SRL

In designing a programming system for robots, the general practice has been to develop the hardware first and then the programming language. In this case the programming language will support only the capabilities of the particular robot under consideration (e.g., through special movement or gripper commands). In such systems, however, the commands not needed for the particular hardware used are missing. This results in the development of a highly specific programming system that cannot be transferred to other robots.

The language SRL (structured robot language) is a versatile programming language with many únique features [3,4], utilizing many important concepts of the existing robot programming languages and scientific languages. Pascal and AL (assembly language) form the essential basis of this development. Until recently, the AL explicit programming language for robots, implemented worldwide in many institutions, had been the most powerful language developed.

The general program concept of SRL was taken from Pascal. Its features include a declaration part, loops, program branchings, assignment statements, procedures, and functions. SRL offers the data types shown in Table 6.3. A second group of data types was taken from AL. Spatial positions and orientations can be represented more easily in a Cartesian space than in a robot coordinate system. Hence, the physical space is represented in Cartesian coordinates, using a reference point of the robot as the origin. All specifications are referred to in terms of this coordinate system. Points in space are described with a vector. The gripper and/or tool located on the robot's arm is described by the position of its so-called tool center point and its orientation.

The orientation of an object in space may be described through two equivalent data types. The programmer uses the one that solves the problem in the most efficient way. One possibility is the use of the

TABLE 6.3 SRL Data Types

Integer	Standard data types as they are
Real	declared in Pascal
Char	
Boolean	
Vector	Position specification
Rotation	Orientation specification
Orientation	Orientation specification
Frame	Position and orientation
Array	Array of same elements
Record	Composition of different elements

"rotation"; this describes the orientation through the use of an axis about which rotations are performed and an orientation angle. This form of description may conveniently be used, for example, when using CAD data. In this case the axis may represent a normal to the surface. The data type "orientation" describes the orientation through the specification of three angles of rotation about the three axes of the coordinate system. Such a method of describing a position and orientation is generally known as the "frame concept." The arithmetic of SRL was extended to include this data type to enable calculations of trajectories, orientations, and positions.

One of the most important requirements in considering the application of a robot programming language is the power and flexibility of its move commands. The move commands in SRL support the frame concept as well as the programming at the robot joint angle level. In addition, various modes of motion control such as point-to-point control, linear Cartesian, and circular interpolation are expressed through different commands so that the programmer becomes aware of the different resulting movement trajectories. Table 6.4 gives an overview of the available move commands.

Each movement may be specified by different parameters. Such parameters include the time, speed, acceleration, and force exerted at a movement, as well as intermediate points during the movement and sensor and interrupt conditions. Combining the single move commands with the corresponding parameters enables every desired complex trajectory to be described.

A further feature of SRL is its system specification part. Here the programmer defines the industrial robots used for the manipulation

TABLE 6.4 SRL Move Commands

Command	Definition
PTMOVE	Point to point move; all the axes move with maximum acceleration and speed
SYNMOVE	All the axes start and stop simultaneously (axis synchronization)
SMOVE	Movement of the gripper in a straight line
LANEMOVE	Movement of the trajectories, which are calculated in higher degree polynomials
VIAMOVE	Movement is smoothly transferred to the next move
CIRCLEMOVE	Movement in an arc
MOVE	Movement with user-specific definition
DIRMOVE	Movement in one direction until step condition is achieved
DRIVE	Movement of one or more robot axes

task. It also describes the end-effector (i.e., the tool system or gripper the robot arm holds), sensors, I/O ports, interrupts, and other hardware facilities and assigns to them freely selectable symbolic names. In the program, the hardware is addressed via these names. On one hand, this increases the self-documentation of a program, and on the other hand, it allows an existing program to adapt to another environment by merely changing the specification part. Moreover, the data structures of the sensor information are described in the specification part and are made accessible in the program.

It is often advantageous to store environment models or geometrical object models in a database or to construct and change data with the help of a planning module (e.g., CAD system). The robot can thereby be informed about the configuration of its environment with a database. After a change in the environment (e.g., through the movement of an object) the data are updated. This capability was already considered in SRL. The user has several functions to use in working with a database.

SRL has a special multitasking concept to satisfy the real-time requirements of robot programming. Thus, a validity range and a life duration are assigned to a program section range that is parallely executable in SRL, as is also the case with procedures and variables in

Pascal. This further aids the reliability of the program. Relatively simple commands allow the user to have the sections executed in a parallel, cyclic, or time-shifted manner. To prevent unknown and undesired temporary side effects, SRL offers capabilities for the synchronization of parallel processes.

The SRL Compiler

The principal tasks of a compiler are to check the syntactic and semantic correctness of programs and to generate an executable machine code. Since the machine codes are normally different, it would be necessary to implement an individual code generation for every type of industrial robot. To avoid this expense, the software interface IRDATA (Industrial Robot Data: see Ref. 2) is used. The SRL compiler first generates the IRDATA code. This is then executed by the control computer, and thus the specific control instructions are generated for one or several industrial robots. Thus, IRDATA can also serve as a programming interface to other robot controllers or to other programming systems.

To develop the SRL compiler, several requirements first had to be met. One requirement was that changes in the language should be easy to modify in the compiler without high cost. Therefore the GAG system (Compiler Generator based on Attributed Grammars) developed at the University of Karlsruhe was used in the development of SRL [5].

This software tool generates compiler modules based on the syntactic and semantic description of a programming language provided by the user. These modules are complete and consistent according to the input description. The module for the generation of IRDATA code and the lexical analysis, however, were not developed with the GAG system (Fig. 6.4). They must be programmed by hand. The prerequisites for the use of the GAG compiler generator is the syntactic and semantic description of SRL with the help of an attributed grammar. The advantages of this description are its high grade readability and the necessity to define the language in a precise and formal manner. In addition, the completeness and consistency of the language definition are enforced, as they are both automatically checked by GAG. Constructing a formal and correctly working compiler with GAG takes only approximately 20% of the usual development time for such a software system.

The SRL compiler has the following modular construction:

Scanner (lexical analysis)
Parser (syntax analysis)
Semantic analysis
Listing generation

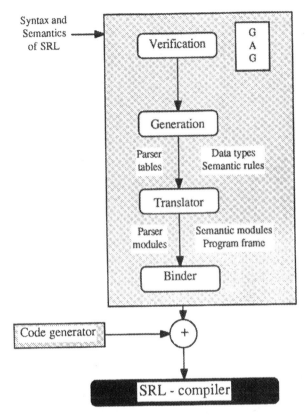

FIGURE 6.4 SRL compiler design with the help of the GAG generation system.

MOPS (Module for the Optimization and Parallelization on the Source
    code level)
IRDATA code generation

Except for the parser and the semantic analysis, all modules are implemented as independent programs (Fig. 6.5). The scanner is constructed on the basis of a program frame provided by GAG. It prepares the SRL input program for the syntactic analysis. Thus, the
pre- and self-defined symbols, compiler options, numbers, and program comments are recognized and converted into optimal form for
analysis and also are stored as a data file of program symbols.
    The syntax analysis attempts to analyze the program symbols using
the syntactic rules specified in the grammar. The errors detected are

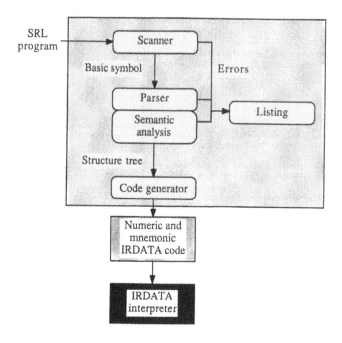

FIGURE 6.5 Structure of the SRL compiler.

collected in a data file and documented. Parallel to the analysis, the individual steps are recorded in a *structure tree*, which represents the syntactic structure of the SRL input program. It also serves as an interface to the next module, the semantic analysis. It supplements the structure tree with missing information, resulting in an attributed structure tree. This tree is analyzed according to the semantic rules of the attributed grammar, and the deviations from these rules are stored as an error file. The attributed structure tree is the basis of further modules. The optimization module MOPS can next be called. It allows an optimal utilization of the available resources (multiprocessor system) through parallelization of sequential program statements.

The IRDATA code generation is now called. It transforms the information contained in the structure tree into IRDATA code. Through this, the IRDATA code is generated in two different forms: mnemonic and numeric. Although in practice only the numeric code is processed, the transformation from SRL into IRDATA by means of the mnemonic code is essentially easier for the programmer to understand.

Parallel Processing Through MOPS

There are also some simply structured programming languages for industrial robots that contain language constructs for describing parallel program executions. The parallelism is mostly restricted to program parts, which are processed parallel to each other. Internally, however, these parts must be processed sequentially. A simpler possibility to describe parallelism on language levels is the specification of a special Begin and End mark within which all the statements may be executed parallel to one another.

Apart from this programmed parallelism, a program may contain sections or statements that have been specified as sequentially executable but show an inherent parallelism. This means that they would be executable parallel to each other without the program's effects being changed.

The module developed at the University of Karlsruhe for optimization and parallelization on the language level (MOPS) automatically determines sections of the user program for parallel processing in the control computer [6]. However, the effects of the user program must be maintained and the efficiency possibly be even enhanced. The only program sections examined are those in which the sequence of statements is not exactly defined through the language description. The statements are normally executed in the sequence in which they have been written. This, however, is not mandatory; it *is* necessary that the effect of the execution be the same as if the statements had been executed sequentially.

Since the compiler generates an executable code, it is necessary to map the parallel structure onto this code. Normally one is not able to find the appropriate number of processors to enable the assignment of a one-to-one relationship between processors and parallel statements. This assignment and the necessary distribution of the work that goes with it cost computation time. At the very worst, the parallelized program runs slower than the original one.

The execution of an automatically parallelized program requires the individual parallel sections to be synchronized. The parallelization module must generate a kind of synchronization code.

Because of its design, the SRL system offers an optimal interface for an automatic parallelization module. The internal representation of the input program (the structure tree) results from the SRL compiler after the syntactic and semantic analysis. This structure tree fulfills all conditions already listed. A module for automatic parallelization therefore has a definite software interface in the structure tree and the language description. It underlies only the constraints as the implementation language of the SRL compiler and the internal representation of the structure tree.

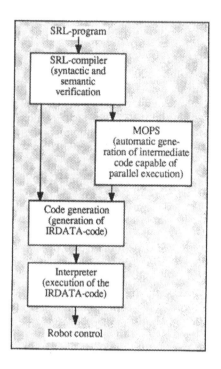

FIGURE 6.6 Integration of the parallelization module into the SRL system.

The processing and parallelizing of a program by the SRL system
follow the scheme shown in Figure 6.6. The module for the parallelization MOPS is an extension of the SRL compiler which can be optionally called after the syntactic and semantic analysis. The result
is that the modified structure tree will be passed back to the SRL
compiler for code generation.

MOPS has made it possible to use future multiprocessor structures
for a more efficient processing of the user program. In addition, the
user-defined parallelism is tested, which adds to the program's safety.

Programmable IRDATA (PR-IRDATA)

The IRDATA code was conceived as an interface between compilers
of higher level languages and the robot controllers. The definition of
the mnemonic code in addition to the numerical code, however, still
allows direct programming in IRDATA code. The mnemonic code is
legible for the programmer and shows a certain comprehensibility.

The application area of this type of programming can be on-site pro-gramming or program correction.

The structure of IRDATA does not support direct programming in IRDATA. Therefore, it appears logical to change the definition of IRDATA to include and support direct programming by mnemonic code. To avoid confusion, this programmable version of IRDATA is desig-nated as PR-IRDATA (programmable IRDATA). At the same time, a module was developed to transform the PR-IRDATA program into mnemonic and numeric IRDATA code [7].

In PR-IRDATA, the following changes or extensions to IRDATA were made: the line number of each command was abandoned; and each statement may optionally be preceded by a symbolic label, which can be addressed by program branchings.

Procedures are defined with a name. A procedure call is executed by <<CALL,proc-name>> and no longer with the specification of the line number. The specification of the block nesting depth is optional. In IRDATA variables are addressed relative to the blocks, while in PR-IRDATA, a name and a type may be assigned to the variables. They are addressed in the program with these names and without the type indication. This, however, necessitates a different struc-ture of the PR-IRDATA program. In PR-IRDATA, nested procedure definitions corresponding to the program structure are allowed. This type of nesting is comparable to that of Pascal. Moreover, comments are possible which are not processed in the translation. The possibil-ity of arithmetic expressions greatly facilitates the programming of calculations and formulas.

One can generally assume that a compiler-generated code is syn-tactically and semantically correct. This, however, is not necessarily guaranteed by direct programming with PR-IRDATA. Hence, the translation also must perform compilerlike functions, which are re-flected in the structure of the translator. It must first be verified that the individual PR-IRDATA statements of an input program cor-respond to the required syntax. In addition, it must be determined analytically whether these commands are in a suitable context to each other. For example, the beginning of every loop must have a corres-ponding end, and loops may be nested in each other only, not crossed. After this syntactic analysis, the required semantics must be verified. This comprises the correct usage of variables, data types, labels and procedure calls. When all these criteria are met, the translator still must convert the nested program structure into a sequential structure and resolve the arithmetic expressions. Moreover, the line numbers must be generated and the labels must be made to refer back to these. Addressing by symbolic variables must be changed to block-relative addressing, and missing type specifications must be added. When all this is done, the IRDATA code may be generated in numeric and mne-monic form.

Integration of the Teach-in

It would be quite troublesome and time-consuming if in each and every application one had to define the frames off-line in the source program or with the help of the frame editor. The user would not only have to determine the exact corresponding position but also be able to imagine the more abstract orientation specification for any arbitrary gripper or tool orientation and then be able to transform them into angle and rotation axis specifications. Since this can be managed much better using the industrial robots themselves, the robot is used as a kind of measuring device, enabling one to define the destinations of moves or the intermediate points of a move trajectory through the interactive definition of the desired position and orientation.

Frames defined in this manner can now be integrated into the user program in different ways. One possibility is to enter the values in the first execution of the program, whereby the program stops at each undefined frame. With the help of the teach-in, the user commands the robot to the desired positions and orientations. The robot control records the corresponding joint values and enters them into the program code. Then the robot control proceeds with the program execution until the next undefined frame occurs.

The advantage of this method is that because of the loading of frame lists, no additional management cost is required for subsequent program executions. But on the other hand, the following, more flexible method is preferred for use in the SRL system. With this method the frames are defined by teach-in independently from the program execution and stored in a frame list. Therefore, they may also be corrected or supplemented as needed with the frame editor without the program having to be newly translated.

The frames defined by teach-in are stored in a list or in a data file and may be used as parameters of move commands, for example, as the destination or intermediate points of a motion trajectory. Normally, the frames are internally represented in a Cartesian system. In special cases, when a very exact execution is required and the numerical deviation of the coordinate transformation is too large, the frame equivalent must be stored in robot coordinates. In this case the robot control maintains two representations of this frame because the normal representation in Cartesian coordinates is required for further calculations, too. In a movement command, however, the robot coordinates are used as destination.

In addition to the movement destinations, it may also be required to describe some additional parameters as well (e.g., "welding current on/off" or "gripper open/closed"). The user will find it more convenient to record these specifications on site during the teach-in phase than to note them and enter them later off-line into the program. Moreover, the data required for description of objects (e.g., workpieces or tools) may also be recorded.

Frame Editor

The frames used for the description of the position and orientation of objects in space can be described in the program text through a virtually incomprehensible numerical description. In this case a change of these values would require a new compilation.

Another possibility is to refer to a frame merely by a symbolic name and to separately define the frame, using the teach-in method, for instance. As an alternative to the teach-in module, a frame editor, developed at the University of Karlsruhe, allows the definition and correction of frame values without requiring the operation of the robot. The frame editor operates through a frame list in which frames that were already defined by the on-line teach-in method possibly can be found. The programmer assigns a freely selectable frame name; the frame editor then searches in the frame list to determine whether an entry was already stored under this name and displays the individual frame components. If the frame name is not yet present in the list, it will be newly entered and the value "0" displayed. After correction, the frame values are stored on command. They can be referenced by a frame variable using an SRL command of the user program.

In the current implementation, the representation of the frame value consists of three real numbers for the position vector, a real number for the rotation angle, and three real numbers for the rotation axis. From these entries a Denavit–Hartenberg matrix representation is internally calculated. Work in progress allows a graphic representation of the frames. Along with the processing of individual frames, complete motion trajectories can be compiled and then simulated or actually executed.

Simulation and Debugging

Almost every programming system provides the user with testing tools for detecting errors. Among the most important of these procedures are methods that allow the program statements to be executed in a stepwise manner. After each step of the program, the user may display parts of the memory, inspect the register and variable values, and alter them if required. He may also continue the program at any arbitrary statement. These three techniques allow an investigation of the program data at critical program positions and thereby specific search for errors.

Simulation systems are more and more being used for program development. These systems allow one to display the robot motions on
a computer graphic screen. Hence, the possibility is increased by
the robots still being able to remain in use during program development. Modern robot simulation systems allow the definition of arbitrary industrial robots and objects. This allows one to configure
and test various models to guarantee an optimal adaptation between
the robot and its environment. Most of the systems have an interface that permits transfer of the already calculated and simulated
motion trajectories directly to the industrial robot. The programming usually is done using a robot programming language.

Through slow motion representation techniques, collisions and errors may be detected. Simulation systems are not suitable for final
error removal, however, since the running charges of the environment (e.g., the reaction of sensors) may not be simulated properly.

How To Use the System

Figure 6.7 summarizes the structure of the SRL system and the interfaces between the modules. According to the order of complexity,
there are three different levels of programming:

1. High level programming language (SRL)
2. Programmable IRDATA
3. Teach-in method and frame editor

The SRL compiler and the PR-IRDATA converter produce executable
IRDATA code, whereas the frame editor and the teach-in components
construct a frame list with movement data.

According to the requirements of the problem to be solved, the programmer may use all four software tools for construction of the user
program or only one or two modules. The robot control, in general,
consists of three components: IRDATA-interpreter (robot-independent part), move control, and sensor processing.

A short example of two SRL instructions will illustrate the coordination of the components. Using a normal text editor, the programmer
enters two instructions into the programming system: a command for
reading a sensor value and a move command. The next step is to start
the SRL compiler. The compiler reads the program text, carries out
the lexical, syntactic, and semantic analysis, and generates among
other things five IRDATA statements (see Fig. 6.8). The first SRL
statement is mapped onto the first IRDATA statement, whereas the
following three statements generate the calculation and storing of the

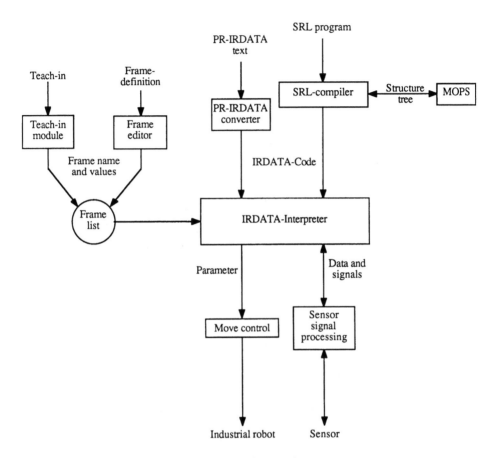

FIGURE 6.7 Components and interfaces of the overall SRL programming system.

component of a frame. The variables are represented through the block relative addresses (BR) and the intermediate results are stored in a stack (ST). The robot movement is initiated by the last IRDATA statement.

Figure 6.8 shows the IRDATA statements in mnemonic form. Its numeric equivalent is being read from the loader of the IRDATA interpreter and stored in the program memory in compressed form. These statements are decoded before being processed; the parameters are prepared, and then the respective function module is called.

SRL - Programm

A:= INPUT SENSOR (3)
SMOVE puma TO FRAMEC(ROTC(YAXIS, 180), VECTORC(A,0,50);

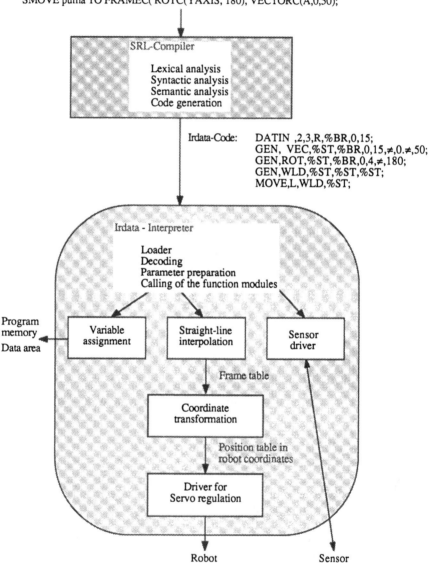

SRL-Compiler

Lexical analysis
Syntactic analysis
Semantic analysis
Code generation

Irdata-Code: DATIN ,2,3,R,%BR,0,15;
                GEN, VEC,%ST,%BR,0,15,≠,0.≠,50;
                GEN,ROT,%ST,%BR,0,4,≠,180;
                GEN,WLD,%ST,%ST,%ST;
                MOVE,L,WLD,%ST;

Irdata - Interpreter

Loader
Decoding
Parameter preparation
Calling of the function modules

Program
memory
Data area

Variable
assignment

Straight-line
interpolation

Sensor
driver

Frame table

Coordinate
transformation

Position table in
robot coordinates

Driver for
Servo regulation

Robot   Sensor

FIGURE 6.8 Translation and execution of SRL instructions.

Executing Off-Line Generated Programs

Industrial robot control has so far concentrated mainly on the hardware-oriented functions and the control of robot movements. This is not surprising, since the developers were influenced more by the knowledge of mechanics and control theory than by that of computer science. Even today for some manufacturers (and users), robot control is nothing more than an NC control.

The present controls for industrial robots are, despite their differences, based on a certain basic structure. Of central importance here is the control computer with a 16- or 32-bit processor and the memory consisting of at least several hundred kilobytes of random access and programmable read-only memories (RAM and PROM). There is normally also a bus and several serial and parallel interfaces. The latter are needed for the processors of the digital control of the robot axes. In addition to the usual interpretation of the user program, the processor of the control computer normally must execute the trajectory planning with interpolation and coordinate transformation, as well.

In the future, even more components and functions will be added to the robot control. This will require much more powerful computer hardware which might also be realized as a multiprocessor system.

Control Concepts

Until now, the main emphasis of movement control has been on industrial robots with the following functions:

Trajectory planning and interpolation (straight lines, circular interpolation, complex curves by means of polynomials)
Variations in the trajectory parameters (speed, duration, acceleration, precision)
Changes of the orientation during the movement
Special functions (e.g., palletizing)
Integration of additional parameters (controlling welding current or controlling additional axes)

Besides these motion control tasks, new robot controls will contain functions to perform the following:

In/output with digital or analog ports
Modifying movements through sensors
Simple program flow control
Program management (e.g., selecting programs or cyclic starting of programs)

This, however, is not sufficient for future tasks. The control functions must therefore be extended to include:

Arithmetic calculation of geometries and trajectories
Explicit (i.e., formulated by the user's program) processing of sensor data in the robot control
Structured complex control of the program flow
Input/output by terminal and of data files

The IRDATA standard [28] for a software interface between the programming system and control exerts an essential influence in this direction. The IRDATA interface concerns only the formulation of functions the control is to execute, not the hardware interfaces or protocols. Usually the IRDATA code is executed by an interpreter implemented on the robot control. The statements contained in IRDATA (all of which, of course, must not be implemented on the respective control) cover nearly all functions that were required previously and are recognized to be sensible for the future. These include:

Movement commands
Specification of technological parameters
Robot and environment description
Program flow control
Arithmetic and Boolean functions
Input/output commands
Test data and test functions

This standard goes beyond the narrow range of the manufacturer and will support the extended use of advanced programming languages and programming systems. The control functions are divided into a large robot-independent section, such as arithmetic or program flow control, as well as into a robot-dependent part to which all the movement controls belong. The robot-independent part can be implemented on a supervisory computer with an interface to the respective control(s) of the industrial robot(s). It is therefore possible to define a hierarchy of interfaces between the elements of the robot control as shown in Figure 6.9.

Robot-Independent IRDATA Interpreter Section

The total set of IRDATA statements is divided into statements of different types. These include, for example, statement type 22000 for program flow control or type 5000 for movement commands. In every statement type, code numbers specify which particular action of this statement type should be executed. The statement types and

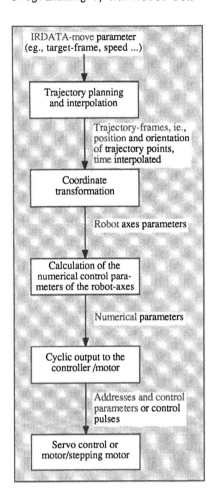

FIGURE 6.9 Functions and interfaces for move control.

the code numbers are combined by adding them to the so-called main-word of an IRDATA statement. Every IRDATA code piece consists of a series of statements, separated by semicolons. Each statement in turn consists of a series of words separated by commas. In the IRDATA interpreter, each IRDATA statement is processed through a series of procedures. These procedures are arranged on different levels as shown in Figure 6.10 [8].

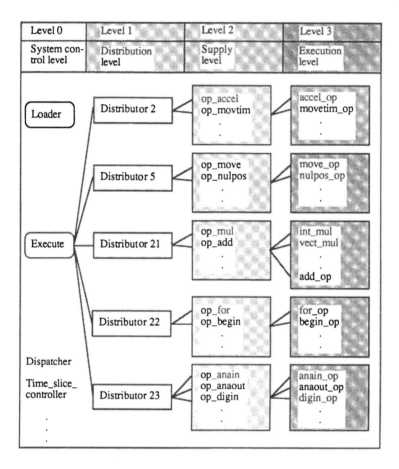

FIGURE 6.10 Hierarchical model of the IRDATA interpreter.

Level 0 contains the routines necessary for the control of the interpreter. The loader, for instance, converts the IRDATA code (provided as ASCII file from the program) into an internal, less memory-
intensive representation, which is thus faster to process. Along with
this, the IRDATA code to be interpreted is also checked for syntacti-
cal correctness, to relieve the interpreted from this task during the
execution of the program. This is quite sensible, as the user can already at this phase detect errors in the program. Otherwise, errors
might not be detected until later. For example, the industrial robot

might have been working correctly for weeks when suddenly it encounters an erroneous statement that must be interpreted, thus causing the interpreter to have to abort the program. Errors that cannot be recognized at this early stage (e.g., the program branches to a destination described by a variable, which later will contain a nonexistent statement number) lead to the calling of a central error routine during the interpretation, which initiates a sensibile error handling.

Level 1 consists of the so-called distribution procedures. It assures that for every statement type, the corresponding preparation procedure of level 2 is called, which then places the parameters for the just-processed statement onto the stack. The last levels construct the already mentioned processing procedures, whereby in general one processing procedure corresponds to each type of IRDATA statement, which is made clear by the naming of the procedures (Fig. 6.10).

Movement Control

In the Karlsruhe implementation of the interpreter, it was possible to achieve robot independence in the higher levels of control. This was possible through a trajectory planning module, also developed at Karlsruhe. The interpreter communicates with this module through a defined interface. When calling a robot movement, as with the other statements, the above-described levels of the interpreter are also executed down to the processing routines. On this level, the motion parameters are passed on to the trajectory planning (Fig. 6.9). For test purposes, an additional option can be set in the interpreter which allows the movement parameters to be displayed on the screen only.

According to the IRDATA definition, several parameters are specified directly with the move command, whereas others are introduced through special IRDATA statements. These last items include, for example, the speed of the motion, the duration of motion, or the acceleration. An actual unit of measurement may thereby also be selected. The specification may either be given as absolute or in percentages. These predefined move parameters must be "noticed" by the interpreter up to execution of a move command.

6.4 GRAPHICAL PROGRAMMING
   AND SIMULATION

Graphical programming uses computer graphics to illustrate the programming process. In contrast to the textual programming, it is

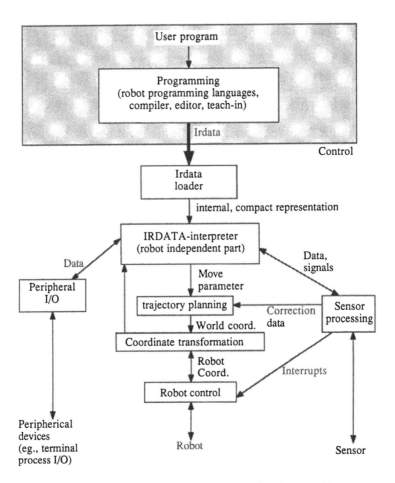

FIGURE 6.11 Structure of the IRDATA implementation.

possible to define also the robot motions and to simulate the program execution using a graphical screen.

Graphical programming uses models of the real objects in the work-cell and the robot itself. The methods of modeling these geometrical components range from simple vector editors to comfortable CAD systems. There are also different methods of modeling the functions of the robots and peripheral devices. More comfortable systems allow one to model the characteristics of particular controllers. This, for instance, makes it possible to predict program execution times.

There are different ways of presenting an object on a computer graphic screen. Most of the systems use wireframe models to represent the objects, especially for objects in motion. Some systems allow the elimination of hidden lines at the cost of lower visualization speeds. This means that the presentation more or less flickers.

For the definition of motion points (i.e., positions and orientations), normally two methods are available. The first method is similar to the on-line programming methods and uses dials or other devices to move the robot on the screen. The second method allows one to specify motion points via the definition of geometrical relations between the robot and its environment. For instance, the contact between a gripped workpiece on a table implicitly defines the joint quantities of the robot needed to achieve this situation.

The definition of the motion points is completed by textual commands to specify the motion parameters and the control flow of the program. Some systems offer one general-purpose robot programming language, which can be translated into a particular proprietary robot programming language by a specialized postprocessor. This has the advantage of easier exchangeability of robots. Other systems have specialized programming modules for each type of robot.

Common to all graphical programming systems is that the resulting programs need some fine corrections before the program can be executed in reality. This is due to inevitable differences between the environment of the robot and the programming system's model of this environment.

As an example, the simulation system ROSI was described in Chapter 2. It is well suited for graphical programming of robots.

6.5 TASK-LEVEL PROGRAMMING

6.5.1 Introduction

Task-level programming (also called implicit programming) is an abstract method of programming a robot by specifying the task to be performed, instead of directing the robot itself. The transformation from the abstract representation of the task specification to the concrete representation of a robot program is done by the system. Task-level programming systems are especially useful for difficult applications such as assembly tasks.

To accomplish an assembly task with the help of robots, many different problems must be solved. First, the necessary equipment and the layout for the assembly cell are determined by an assembly expert. Then the robot programmer must commit himself to a sequence of single actions of the resources of the robot cell. He must also think about many geometrical problems (e.g., how to grip a workpiece in performing a certain parts-mating operation; how to move a workpiece so that no collision occurs during motion; how to use sensors to bring uncertainties under control).

The task of an assembly planning system for robots is to find automatically a sequence of actions in order to solve the specified assembly task.

Since planning is one of the basic techniques of artificial intelligence (AI), considerable research on this topic has been underway. Early research had been devoted to non-domain-specific planning or to rather abstract problems (e.g., block world problems). Only recently real-world problems have received increasing attention. This development has probably been influenced by the research on expert systems.

Aside from this AI-oriented research, there has not been much work on real-world robot planning. Early works include References 9–11. Taylor [9] synthesized sensor-based programs of the robot programming language AL by parametrizing prototypical strategies (so-called procedure skeletons). Such skeletons contain a framework of motions, error checks, and computations for a particular type of task. The planner performs geometrical computations and error computations and decides which strategy to apply and how to parametrize this strategy. A similar approach based on procedure skeletons is taken in the LAMA system. For AUTOPASS, the syntax and semantic of a task-level robot programming language was defined. An emphasis of this research was an algorithm for collision-free path-planning for a robot.

More recent approaches are the TWAIN system [12] and the SHARP system [13,14]. The TWAIN system proposal includes modules for layout planning, fine motion planning, grasp planning, gross motion planning, and selection of feeders and fixtures. The task specification is hierarchically decomposed into plan islands. Constraint propagation of symbolic expressions of plan variables is used to communicate between plan islands to find instantiations for the plan variables. If the plan variables cannot be instantiated, backtracking is used to modify the decisions made up to this point.

The SHARP system proposal consists of modules for grasp planning, fine motion planning, and gross motion planning. It also uses constraint propagation to coordinate the modules.

6.5.2 Overview of the GRIPS System

This section gives an overview of a Graphical Implicit Planning System represented by the acronym GRIPS, a system that allows automatic generation of robot programs. The system was designed especially for automatic parts assembly. Input to the system are geometrical descriptions of the workpieces and their geometrical goal relationships (Fig. 6.12). The robot and gripper actions necessary to

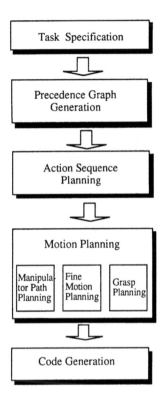

FIGURE 6.12 Structure of the GRIPS system.

establish these goal relationships are planned by the system. Output of the system is a sequence of program statements in an internal language that may be easily translated into languages understandable by commercially available robot controllers.

The geometrical descriptions of the workpieces are obtained using a CAD system. The system provides a geometrical editor to model the geometrical goal relationships among the workpieces. The planning process is then performed hierarchically in several planning levels. In the first step a set of precedence graphs is generated which represents the restrictions with respect to the order of the assembly operations. This is performed by a simulated disassembly of the final assembly product. The system disassembles the product step by step. At each step the parts that may be disconnected from the rest of the assembly with one single translational motion are identified. The direction for such a motion is computed from the type of contact using some straightforward heuristics.

At the next level, a linear sequence of assembly operations is planned which takes into account the order restrictions of the precedence graph. At this planning level the plan consists merely of symbolical actions. The executability of the assembly operations cannot yet be completely considered. Using some simple heuristics, however, such actions are selected that have a high probability of being executable.

The last planning level is concerned with the planning of the particular motions necessary to execute the symbolical actions. The three types of motion necessary for assembly are as follows:

Collision-free motions to grasp or release a workpiece
Collision-free fine motions to establish a goal contact between two workpieces
Collision-free manipulator gross motions to transport workpieces from one location to another

Finally, a code generation is performed to produce an executable robot program. The code generator module translates the internal form of the action sequence into a robot programming language. One of the target languages currently supported by the system is the programming language MCL of our simulation system ROSI (see, e.g., Ref. 15). This allows one to simulate the robot programs graphically. Code generation for other languages such as VAL and IRDATA are possible.

Task Specification

Input to our task planner is a specification of the assembly task on a very high level. This specification includes both the geometry of the parts themselves and the geometrical relations between the parts of the final assembly. One way to specify these geometrical relations between the parts is to define the positions and orientations of the parts with respect to a global coordinate system. This method, however, is rather tedious, because the coordinates of the parts necessary to establish a particular type of contact must be calculated.

A better way to define these contacts is to specify a set of symbolical spatial relations between the geometric features of the two objects. In the case of a polyhedra, such features are faces, edges, and vertices, and auxiliary elements such as axes of symmetry. Examples for symbolic relations between these features are two faces are in contact, and two edges are parallel with a particular distance. This is a sort of *implicit* specification, since the coordinates of the parts are inferred by the system instead of being given *explicitly* by the user.

To facilitate this type of specification and to avoid textual descriptions of contacts, a graphical editor was implemented [16,17]. This

system models a scene composed of different objects, which are read
from a library. The user can select the object features using a graphic
screen and dials or a light pen. The user may model the relations between these features step by step in an interactive way. The result
of each step is shown immediately on the screen.

Internally, the system uses a term rewrite system [18] to infer the
position of the parts. A list of spatial relations is considered as a
word of a language. The symbols of this language are the type of relation, the specification of the features of the objects involved, and
the distance between these features. The term rewrite system tries to
reduce a set of spatial relations to one particular relation without any
loss of information. The system repeatedly tries to apply rewriting
rules to a list of initial relations until no further simplification can be
achieved. The system is able to detect two types of error. One type
is an inconsistency between two spatial relations. The other type occurs if the situation is not specified in enough detail (i.e., too many
degrees of freedom are left). In both cases the error is reported to
the user and the user is asked to correct the specification.

Precedence Graph Generation

Precedence graphs are one method of representing constraints for
an assembly. Each node of the graph represents one particular assembly operation and the edges represent the precedence relations
between two operations. Unfortunately, a precedence graph is not a
suitable means of expressing all possible assembly sequences. An alternative to precedence graphs consists of the AND/OR graphs used
by Sanderson and Homem-de-Mello [19]. In the system described here
[20,21,39], a set of alternative precedence graphs is used instead.

The system automatically computes the ordering constraints from a
geometrical model of the parts and their spatial goal relations. The
system performs a disassembly of the final product step by step. At
each step the possible departure directions for each of the parts are
determined using some simple heuristics based on the contact surfaces
involved. Each departure direction is then investigated to determine
whether the part can be removed in this direction without colliding
with any other part. If there is a collision between two parts, a
precedence constraint between these two parts has been detected.
If there is no collision, the part is removed together with all other
parts that can be removed simultaneously and independently of each
other. In the next step, each part is again tested as to whether it
can be removed. This computation proceeds until all parts of the assembly have been removed.

In the next planning step, two additional sets of rules are applied
to each of the precedence graphs. One type of rules concerns the
stability of the assembly. An unstable situation during the assembly

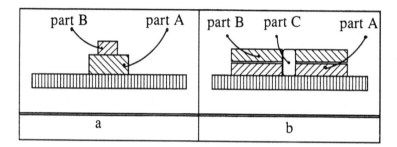

FIGURE 6.13 Criteria for precedence constraints: (a) stability and (b) side effects.

creates a need for an additional precedence constraint. In Figure 6.13, for example, part A must be assembled before part B; otherwise the assembly would not be stable. Another type of rule incorporates heuristic knowledge into the system, to avoid faulty interactions ("side effects") between parts of the assembly. Figure 6.13 shows an example of this: if the discs A and B were assembled before the shaft C was mounted on the base plate, the discs could slip on the surface or on each other while C was being assembled. A heuristic rule then imposes a precedence constraint between the three parts so that part C would be assembled first.

Figure 6.14 shows a simple assembly, and Figure 6.15 shows the resulting set of precedence graphs. Level 0 in Figure 6.15 shows the initial situation with no precedence constraints at all. Level 1 shows the four alternative precedence graphs after the disassembly method has been applied to the assembly. The graphs on level 2 result from applying the stability tests and those on level 3 from the side-effect rules. Some of these graphs contain cycles (those in Fig. 6.15 with a heavy border line). Since a cycle represents a contradiction, these graphs need not be further considered.

Figure 6.16 shows another example, the Cranfield Assembly Benchmark Kit [22]. The resulting precedence graph is shown in Figure 6.17. This example requires considerable computational effort: the total CPU time on a VAX 8700 computer was approximately 2932 seconds.

Action Sequence Planning

After the ordering constraints have been computed with respect to the geometry of the assembly itself, a sequence of executable operations must be planned. This process takes into account the actual task environment (i.e., the geometry of the workcell and the resources available, such as robots, grippers and fixtures).

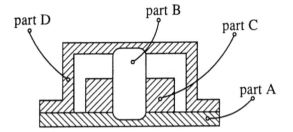

FIGURE 6.14 Example of a simple assembly.

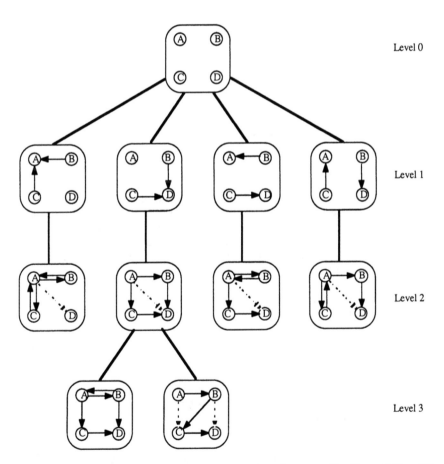

FIGURE 6.15 Precedence graphs resulting from application of the different sets of rules to the example of Figure 6.14.

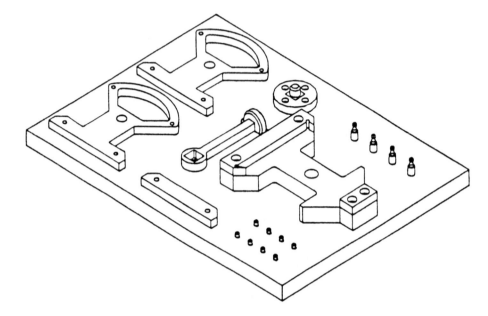

FIGURE 6.16 The Cranfield Assembly Benchmark kit.

On this planning level the term "goal" denotes an assembly operation that must be achieved. In the first step, a goal is selected that
has no preceding goals (i.e., no other assembly operations to be executed prior to it). This goal is investigated to determine whether it
can be achieved within one operation cycle. An operation cycle is a
sequence of robot actions as follows:

Moving an arm without holding a workpiece
Approaching the workpiece with the gripper
Grasping the workpiece
Separating the workpiece from a base plate, a fixture, or other object
Departing
Moving an arm that holds a workpiece
Approaching the workpiece to another object
Establishing a defined contact to this object
Releasing the workpiece
Departing

Goals that can be achieved by this sequence or by a subset of it are
called *simple goals*. If the selected goal is not a simple one, the strategy

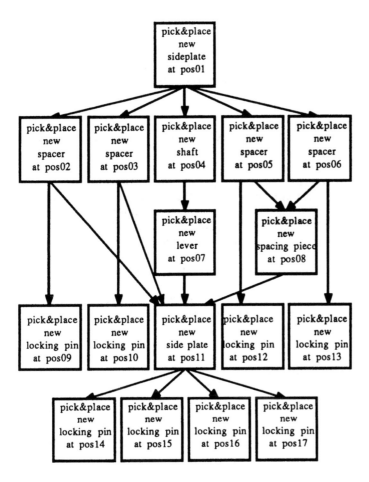

FIGURE 6.17 Precedence graph resulting from the Assembly Benchmark.

planner tries to construct a subgoal that makes the selected goal achievable by one operation cycle. This means that the selected goal becomes simple from the point of view of the new subgoal. With this backward chaining process a sequence of goals is searched and developed so that every goal is simple with respect to the preceding goal.

The following examples demonstrate different conditions under which a goal could not be achieved by one operation cycle. Let us assume the goal: "Connect workpiece A to workpiece B such that the

surfaces a of A and b of B are in contact," which will be expressed
by "Connect A.a to B.b."

In case 1 object A cannot be grasped because it is obstructed by
object X. One way to deal with that problem is to remove X and connect it to another object Y. Thereby, the following conditions must
be fulfilled:

Y must be distinct from object A.
If Y is equal to object B, then X may not be connected to Y (i.e., B)
  such that surface b is obstructed.

Case 2 differs from case 1 in that object X cannot be removed. The
grasp surface A.g, which is needed for the specified connect operation, is obstructed. We assume that A can be gripped in another way
using some other unobstructed grasp surfaces. An object Y must then
be found, which allows A to be put down such that the grasp surface
A.g is free. Cases 3 and 4 may be dealt with accordingly. Table 6.5
summarizes these different cases and the heuristic rules for subgoal
construction.

The next planning step is to assign the real resources of the physical robot cell to these goals. This depends on the capabilities of the

TABLE 6.5 Four Action Sequence Planning Cases

Case	Obstacle	Subgoal
1	The grasp surfaces of A, which are needed for the grasp, are obstructed by an object X, which is removable.	Connect X.x to Y.y ($Y \neq A$, $Y.y \neq B.b$)
2	The grasp surfaces of A, which are needed for the grasp, are obstructed by an object X, which is not removable.	Connect A.x to Y.y ($A.x \neq A.g$, A.g free)
3.	The surface b of the workpiece B is obstructed by an object X, which can be removed.	Connect X.x to Y.y ($Y \neq A$, $Y.y \neq B.b$)
4.	The surface b of the workpiece B is obstructed by an object X, which cannot be removed.	Connect B.x to Y.y ($B.x \neq B.b$, B.b free, $Y \neq A$)

available resources. For a grasp operation, a gripper must be selected that can grip the object in such a way that the necessary parts-
mating operation can be executed by the robot. The decisive consideration in the selection of the robot arm is whether the arm can reach,
with the selected gripper, all positions necessary to execute the intended operation (e.g., the start and goal position of a pick-and-place
operation). If no resources that provide all necessary capabilities can
be found, a subgoal must be determined such that the change of the
actual state to the subgoal state and the subgoal state to the goal
state may be done with the available resources. This is shown in
the following example.

Let us assume a robot cell with two robot arms. A workpiece must
be moved from position x to position y. If x and y cannot be
reached by the same arm, the goal cannot be achieved with one
operation cycle. Assuming that the workspaces of the two arms
are overlapping, a rendezvous position z in the common workspace
of the two robots can be found. With this heuristic rule, a subgoal "Put workpiece to z" may be established.

After a sequence of robot actions has been found, the geometrical parameters of the different operations must be determined exactly. This
is described next.

Motion Planning

Input to the geometric planning level is the framework of operations
of one operation cycle, which can be considered as a sort of skeleton
plan. The task of geometric planning is to fill this skeleton with motion parameters or to establish that no parameters can be found. Geometric planning is concerned with three different problems:

Parts-mating planning (or fine motion planning): finding a fine motion to connect or disconnect two parts.
Grasp planning: finding a fine motion to establish a stable grasp of
an object or to release an object.
Gross motion planning: finding a path for the robot and the gripper
and the payload.

Together, these motion elements result in a combined complex motion.
An example for a pick-and-place operation of this type is shown in
Figure 6.18. The main problem associated with these combined motions is to coordinate the solutions found by subplanners for each of
the individual motions. Typical coordination constraints are the following.

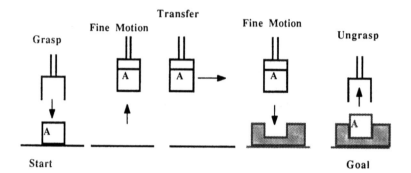

FIGURE 6.18 Different types of motion necessary for a pick-and-place operation.

1. The motions for grasping, releasing, and parts-mating must be collision free not only for the gripper with respect to the object, but also for the robot arm connected to the gripper.

2. Both the grasp operation and the release operation must (trivially) use the same geometrical relation between the gripper and the object to be grasped. Hence this relation must take into account the obstacles for the gripper at both the start and goal locations. In this way a relation will be avoided that, for example, would make sense at the start location but would cause a collision at the goal location.

3. There exists a mutual dependence between grasp planning and fine motion planning in such a way that the solutions found for one of the two problems constrain the possible solutions for the other problem. For instance, does one particular grip constrain the fine motions possible with this grip? Or, does one particular fine motion strategy constrain the parameters of a grip that is necessary to execute this fine motion?

4. Once solutions have been found for the grasp and the fine motion problem, one must search for a suitable gross motion to connect the end points of the grasp and the fine motion trajectory.

These problems are treated in the following way.

1. All motions are planned with respect to one common reference point located in the wrist of the robot arm. This enables a propagation of configuration space obstacles from the gross motion planner to both the fine motion planner and the grasp planner.

2. The grasp and release operations are planned using an artificial situation that contains both the obstacles at the start location as well as the obstacles at the goal location. That is, a logical OR is performed for both types of obstacle, and the planning takes into

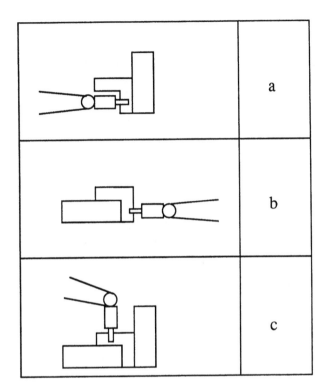

FIGURE 6.19 To be able to plan the grasp and the release operation, the start (a) and the goal environment (b) must be combined (c).

account the sum of obstacles (Fig. 6.19). Figure 6.19a and 6.19b contain the start and the goal scene, respectively; Figure 6.19c shows the combination of both scenes.

3. To resolve the problem of the mutual dependence between fine motion planning and grasp planning, the faces of the workpiece that are in contact with other workpieces at the goal location must first be identified. Then these faces must be communicated to the grasp planner, to ensure that they are not considered as grip candidates. With this approach we try to avoid grips that are likely to lead to failures during fine motion planning.

4. The sequence in which the planning is performed is: grasp planning, fine motion planning, gross motion planning. First, we generate a list of possible grip candidates, ordered by quality. Each grip candidate is described by the geometrical relation between the

object and the gripper's fingers. Then, using the best grip candidate, we plan the motions necessary to grasp the object and to release it. This planning takes into account the configuration space obstacles of the robot arm. Hence the resulting motion is collision-free both for the gripper and for the robot arm. Then, again using this grip candidate, we plan the two fine motions necessary to disconnect the workpiece from its environment in the start situation and connect it to other workpieces in the goal situation. If no suitable fine motions have been found, backtracking is initiated to try another grip. In the last step, if the fine motions have been found, a gross motion is planned which connects the end points of the two fine motions. If no gross motion can be found, another pair of fine motions must be planned. The search space constructed by this approach is shown in Figure 6.20. In general, if the system fails to find a solution, it backtracks to the next higher level and tries an alternative. So, in the worst case, the complete tree is searched.

There already exist a large number of algorithms for each of the particular motion problems. What is unique in this approach is that a Cartesian configuration space representation is used as a sort of communication medium between the different subplanners. These subplanners are presented in the sections that follow.

Gross Motion Planning: Many of the motions of a robot manipulator are gross motions without sensor interaction—for example, to transfer

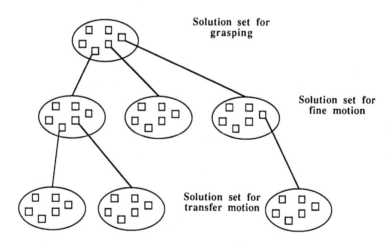

FIGURE 6.20 Search space in which the overall motion planning is performed.

parts from one position to another (*pick and place*). Hence an important function of a robot action planner is the ability to plan collision-free motions among obstacles in the work space.

The *configuration* of an object is a set of parameters that specify completely the positions of all points of the object. These parameters make up the *configuration space*, the *C space*, of an object. A point in this C space defines a configuration of the object. A typical C space, for instance, is the joint space of a manipulator.

An obstacle in the workspace now corresponds with a C space obstacle in the configuration space. A point within a C-space obstacle represents a configuration of the robot that would cause a collision with an original obstacle. The space outside the C-space obstacles is called the free space.

Once the original obstacles have been transformed into C-space obstacles, the problem can be reduced to finding a path for a point through free space which connects the start and the goal location.

This problem has received considerable attention in robotics research. Unfortunately, this problem is inherently difficult for robots with revolute joints. Most of the approaches known use a free space search in the configuration space of some of the links of the robot. All these C-space approaches (see, e.g., Refs. 13, 23--26) use the joint space of the robot as configuration space. The main disadvantage of these methods from the point of view of task planning is their poor interface to other geometrical planning methods (i.e., grasp planning and fine motion planning). Another interesting approach [27] uses freeways for the hand and payload, a C-space representation for joints 1 and 2, and constraint propagation to communicate between these two representations. It solves the problem for four joints of a Puma-class robot. Unfortunately it has rather severe limitations concerning the representation of the objects.

The approach taken here [28--30] is primarily based on a Cartesian free space representation for the first three joints of the manipulator. This allows one to decouple the limbs of the robot so that the shape of a limb does not depend on the configurations of preceding links in the kinematic chain. It also allows separate computation of the free space obstacles for each link.

The Cartesian free space is constructed with respect to the coordinates of a reference point in the wrist of the manipulator. Unfortunately, the Cartesian position of such a reference point does not unambiguously specify the configuration of the robot. To solve that problem, *kinematic states* of the robot are introduced and different free space obstacles are computed for each state of the robot. When searching for a path for the reference point, the only configuration space obstacles considered are those that are relevant for the current kinematic state of the robot. The path may also require switching between different states. This is possible at certain points in the configuration space where the joint angle intervals of the different kinematic states overlap.

The boundary of a free space obstacle is computed approximately by contact analysis between a real obstacle and the limbs of the robot. A space grid consisting of cubes of uniform dimensions is used to represent the Cartesian configuration space. The free space obstacles are mapped into this cube space by projecting their boundary into the cube space and filling the interior of this hull. All cubes that are partly or completely occupied by a free space obstacle are marked. In addition, the real obstacles are mapped into the cube space. The free space outside all the free space obstacles is called *free configuration space* (FCS), and the free space outside of all the real obstacles is called *free space* (FS). Figure 6.21 shows a block-shaped obstacle and the resulting configuration space obstacle after it has been mapped into the cube space.

A configuration for the arm is safe if the reference point of the arm is placed within an FCS cube. The *cube space skeleton* (CSS) is defined as the orthogonal space grid defined by the centers of the FCS cubes (i.e., CSS vertices) and their connecting edges (CSS edges). The reference point of the arm and the hand is the point

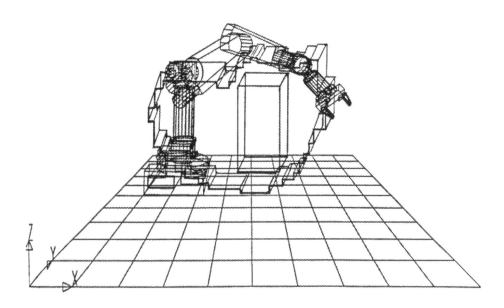

FIGURE 6.21 An original obstacle and the resulting Cartesian configuration space obstacle.

where all three rotation axes of the hand intersect. The orientation range of the hand is subdivided into discrete orientations. If the reference point is placed at a CSS vertex, one can map the hand and payload into the cube space. This results in a pattern for each discrete orientation, which corresponds to the cube space occupancy of the hand for this particular orientation. Similarly one can compute a pattern for each rotational transition between two adjacent orientations. This computation is performed for the entire set of discrete orientations and transitions.

A free configuration for the arm *and* the hand is now represented by a CSS vertex and a free pattern for the hand configuration at that vertex. A translation between two CSS vertices is safe if the two orientation patterns are safe. A rotation between two discrete orientations is safe for one particular CSS vertex if both orientation patterns and the transition pattern are safe. A state-space graph with nodes representing free configurations and arcs representing free translations and free rotations is constructed. The graph is searched using the A* algorithm. The heuristic function is simply the sum of the shortest orthogonal distance and the number of rotations between the start and the goal orientation.

Grasp Planning: One problem with task-level programming of industrial robots is to find feasible grasping configurations. The known approaches in the literature for grippers with two parallel jaws may be classified as follows.

1. A number of approaches (see, e.g., Refs. 31--34) reduce the task to a two-dimensional problem. They investigate grasping configurations for a gripper with two parallel jaws. First, pairs of parallel faces are investigated to see whether they are suitable for a grip. Then, for each pair of faces, the obstacles of the object itself and the obstacles of the local environment are projected onto an intermediate plane (the *gripping plane*) between the two faces. Within this plane a path is searched for the projection of the jaws and the palm from the initial configuration to the intersection of the two faces.

2. A second approach [35] computes an explicit free space description in the configuration space of the gripper. For this free space description, a free space graph is built and searched for a free path for the reference point of the gripper.

The approach taken here [34] is similar to the first method mentioned above and was partly influenced by Laugier [31,32]. In a first phase (which may be off-line), feasible grasping configurations are computed based on the geometrical descriptions of the object and the gripper. In the second phase (which may be on-line), additionally the knowledge of the environment of the object is used to select one of the grasping configurations and to plan a collision-free motion for the gripper.

A grasp configuration is defined as a position and orientation of the gripper with respect to the object. To achieve a *good grip*, some constraints must be taken into account.

1. The gripper must not collide either with objects in the environment or with the object itself.
2. The grip must be stable; that is, the object must not change its geometrical relation to the gripper during any gripper motion.
3. The grip must also take into account the geometrical relations between the object and the goal environment, to avoid impossible grasping configurations.
4. The motion necessary to grasp and release the object must be collision-free not only for the gripper and the payload, but also for the robot manipulator.

The grasp planner allows the processing of polyhedra. The geometrical model employed for object representation is the boundary representation form, which may be generated by the usual CAD systems. The gripper model was specifically dedicated to grippers with two parallel jaws. It employs a functional model of the *palm* and the *fingers*, which may be parametrized to represent different gripper geometries.

A potential grasping configuration is a pair of object faces for which the following conditions must hold:

1. The faces are reachable; that is, each face must have at least one convex edge.
2. The two faces are parallel, and their normal vectors point into opposite directions.
3. The distance between the faces is between the minimum and maximum opening width of the fingers.
4. The overlap between the two faces is large enough. This overlap is referred to as *gripping face*. The plane parallel to these faces and in the middle between them is called the *gripping plane.*

Figure 6.22 shows an object X1 to be grasped and the gripping plane parallel between the two faces A and B. When the gripper approaches the gripping face, the obstacles for the fingers and the palm must be analyzed. These obstacles may be imposed by the object itself or by its environment. They are the bodies that intersect the translational volume of the fingers and the palm when the reference point of the gripper is restricted to the gripping plane. These obstacles are computed and projected onto the gripping plane.

A path for the gripper is now collision-free if the projection of the fingers and the hand onto the gripping plane avoids the appropriate gripping plane obstacles. A configuration space is constructed with

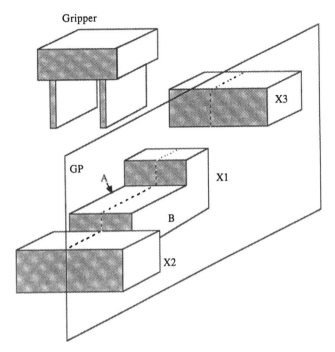

FIGURE 6.22 Example scene showing an object X1 to be grasped using faces A and B and two obstacles X2 and X3; the gripping plane lies in the middle of A and B and is parallel to them.

the two translational and one rotational degree of freedom of the reference point. In addition, the gripping plane is intersected with the configuration space obstacles of the cube space of the robot manipulator. These additional gripping plane obstacles represent configuration space obstacles for the robot manipulator with respect to translations of the reference point within the gripping plane. Hence they are invariant to the gripper orientation within the gripping plane and can be easily transformed into three-dimensional configuration space obstacles. A path outside the configuration space obstacles corresponds to a safe motion for the arm and the gripper. The free space outside the configuration space obstacles is partitioned into cubes of equal dimensions. A heuristic search technique is used to find a path consisting of a chain of safe cubes connecting the start and the goal location.

Fine Motion Planning: The current implementation [36] allows one to plan, with respect to the reference point in the wrist, pure translations that consist of a sequence of free motions and compliant motions. The resulting path is collision-free for the gripper, the payload, and the robot manipulator. The basic technique employed is a heuristic search in a partitioned configuration space description.

The fine motion planner module is able to process polyhedra that model the gripper, the payload, and the obstacles. The polyhedra must be described in boundary representation form. Nonconvex

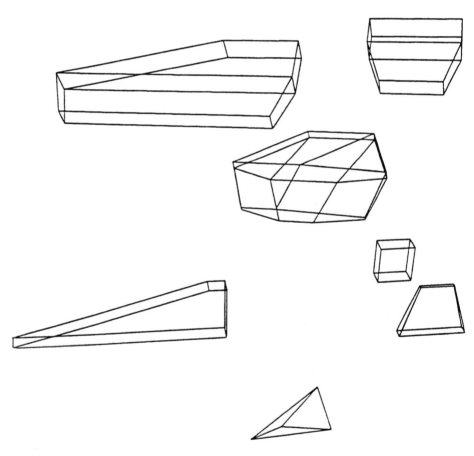

FIGURE 6.23 Original scene with an object X to be moved and the resulting configuration space.

polyhedra are currently replaced by their convex hulls. Work is under way to divide nonconvex polyhedra into disjunct subpolyhedra. For each moving object—obstacle pair, a configuration space obstacle is computed using an algorithm similar to the one described by Lozano-Pérez [37]. Figure 6.23 shows an example with an original scene and the resulting configuration space scene. The object X to be moved is shrunk to a point, while the obstacles are grown accordingly.

The resulting set of points is used to compute the convex hulls of the configuration space obstacles using an algorithm based on work by Preparata and Hong [38]. Figure 6.24 shows a nonconvex object and its convex hull computed by the system. The configuration space obstacles are described in a boundary representation form and are mapped into the cube space described above. The mapping algorithm (see Ref. 28) performs a hierarchical subdivision of the object until either the cube size is reached or the particular part of the space is empty or completely filled with an object (Fig. 6.25).

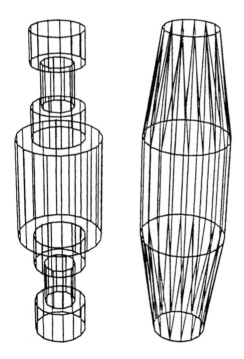

FIGURE 6.24 A nonconvex object and its convex hull after application of the algorithm.

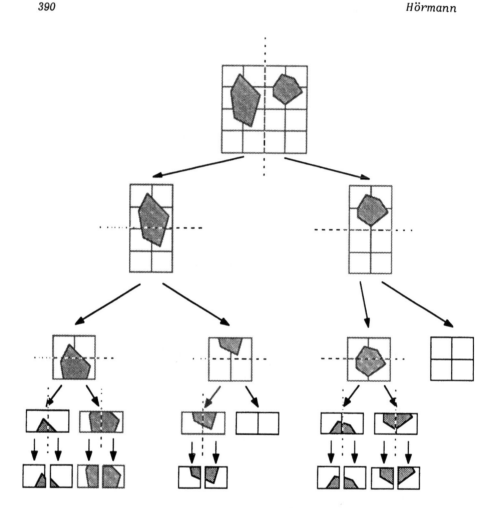

FIGURE 6.25 The mapping algorithm performs a hierarchical subdivision either until the cube size has been reached or until the particular part of the space is either empty or completely filled with an object.

In this cube space a heuristic search using the A* algorithm is performed from the start location to the center of a cube near the goal configuration space surface. Then a guarded motion is planned to establish an initial contact between both objects. Afterward a sequence of compliant motions through several configuration space surfaces is searched, which connects the landing point with the goal point. Figure 6.26a shows a situation of a goal point that can be reached without

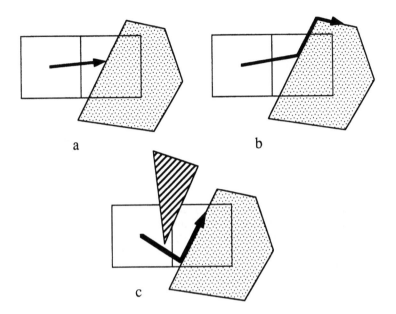

a   b

c

FIGURE 6.26 Different possibilities of reaching the goal contact: (a) with a guarded motion; (b and c) with a combination of guarded motion and compliant motion.

any need for intermediate compliant motions, which are in turn needed in Figure 6.26b. Figure 6.26c shows how the landing point is constructed to avoid local obstacles between the center of the cube and the configuration space surface.

In the last step, a smoothing algorithm is used to improve the path found by the algorithm, since the path leading through the centers of subsequent cubes may be rather jagged.

REFERENCES

1. S. Bonner and G. Kang, "A Comparative Study of Robot Languages," *Computer*, December 1982.
2. VDI 2863: IRDATA, Part 1, final version. VDI (Society of German Engineers), Düsseldorf, December 1987.
3. C. Blume and W. Jakob, "Design of Structured Robot Language (SRL)," *Proceedings of Advanced Software for Robotics*, Lüttich, 1983.

4. C. Blume, *Konzeption eines Programmiersystems für Industrier-*
    oboter. VDI-Verlag, Düsseldorf, 1987.

5. B. Westermann, "Syntaktische und semantische SRL-Beschrei-
    bung und Generierung des SRL-Compilers mit Hilfe von GAG,"
    Diploma thesis, University of Karlsruhe, 1984.

6. B. Heck "Entwurf und Implementierung eines Moduls zur Op-
    timierung und Parallelisierung auf Sprachebene (MOPS)," Dip-
    loma thesis, University of Karlsruhe, 1985.

7. H. Franke, "Definition einer Irdata-Programmstruktur und Im-
    plementierung des zugehörigen Irdata-Umsetzers," Studienarbeit,
    University of Karlsruhe, 1985.

8. M. Beck, "Spezifikation des SRL-Codes im Irdata-Rahmen und
    Entwurf und Implementierung des roboterunabhängigen Inter-
    preterteils." Diploma thesis, University of Karlsruhe, 1984.

9, R. H. Taylor, "The Synthesis of Manipulator Control Programs
    from Task-Level Specifications," Ph.D. dissertation, AI-Labora-
    tory, Stanford University. Rep. AIM-282, July 1976.

10. T. Lozano-Pérez and P. H. Winston, "LAMA: A Language for
    Automatic Mechanical Assembly," *Proceedings of the Fifth UCAI,*
    Cambridge, MA, August 1977.

11. L. I. Lieberman and M. A. Wesley, "AUTOPASS: An Automatic
    Programming System for Computer Controlled Mechanical Assem-
    bly," *IBM J. Res. Dev.* 21:4 (1977).

12. T. Lozano-Pérez and R. A. Brooks, "An Approach to Automatic
    Robot Programming," in *Solid Modelling by Applications.* J. W.
    Boyse and M. S. Picket, Eds.). Plenum Press, New York, 1984.

13. C. Laugier and F. Germain, "An Adaptive Collision-Free Tra-
    jectory Planner," *Proceedings of the International Conference*
    on Advanced Robotics, Tokyo, September 1985.

14. C. Laugier and J. Pertin-Troccaz, "SHARP: A System for
    Automatic Programming of Manipulation Robots," *Third Inter-*
    national Symposium on Robotics Research, Paris, October 1985.

15. R. Dillmann and M. Huck, "A Software System for the Simula-
    tion of Robot Based Manufacturing Processes," *Robotics* No. 2,
    3-18 (1986).

16. G. Werling, "An Editor for the Graphical-Implicit Modeling of
    Scenes" (in German), Diploma thesis, Institute for Real-Time
    Computer Control and Robotics, University of Karlsruhe, 1988.

17. B. Frommherz and G. Werling, "Specifying Configurations of
    3D Objects by a Graphical Definition of Spatial Relationships,
    ESPRIT No. 623," Working Paper IP-UKA-01.88/1 (also sub-
    mitted to the AIENG 88, August 1988, Stanford, CA).

18. W. J. Clesle, "Interpretation of Spatial Relationships among 3D
    Objects" (in German), Studienarbeit, Institute for Real-Time
    Computer Control and Robotics, University of Karlsruhe, 1987.

19. A. C. Sanderson and L. S. Homem-de-Mello, "Task Planning and Control Synthesis for Flexible Assembly Systems," in *Proceedings of the NATO Advanced Research Workshop on Machine Intelligence and Knowledge Engineering for Robotic Applications.* A. K. C. Wong and A. Pugh, Eds., Springer NATO ASI Series, Vol. F33, 1987.

20. B. Frommherz and J. Hornberger, "Automatic Generation of Precedence Graphs, ESPRIT No. 623," Working Paper IP-UKA-!2.87/1 (accepted by the 18th ISIR, 1988, Lausanne, Switzerland).

21. J. Hornberger, "A Method to Compute Precedence Constraints from a CAD Model Automatically" (in German), Diploma thesis, Institute for Real-Time Computer Control and Robotics, University of Karlsruhe, 1987.

22. K. Collins, A. J. Palmer, and K. Rathmill, "The Development of a European Benchmark for the Comparison of Assembly Robot Programming Systems, Robotics Technology and Applications," *Proceedings of the First Robotics Europe Conference* (Brussels, June 27–28, 1984), Springer-Verlag, Berlin, 1985.

23. B. Faverjon, "Obstacle Avoidance Using an Octree in the Configuration Space of a Manipulator," *Proceedings of the IEEE International Conference on Robotics*, Atlanta, March 1984.

24. L. Gouzenes, "Strategies for Solving Collision-Free Trajectory Problems for Mobile and Manipulator Robots," *Int. J. Robotics Res.* 3:4 (1984).

25. T. Hasegawa, "Collision Avoidance Using Characterized Description of Free Space," *Proceedings of the International Conference on Advanced Robotics*, Tokyo, September 1985.

26. T. Lozano-Pérez, "Automatic Planning of Manipulator Transfer Movements," MIT AI Memo, December 1980; also in *IEEE Trans. Sys., Man, Cybern.* SMC-11 (October 1981).

27. R. A. Brooks, "Planning Collision-Free Motions for Pick-and-Place Operations," *Int. J. Robotics Res.* 2:4 (Winter 1983).

28. K. Hörmann, "An Algorithm for Computing Collision-Free Paths for Industrial Robots" (in German), Ph.D. thesis, Department of Computer Science, University of Karlsruhe, 1987.

29. R. Sperl, "Implementation of a Path-Planning Algorithm for Industrial Robots with Six Degrees of Freedom" (in German), Diploma Thesis, Institute for Real-Time Computer Control and Robotics, University of Karlsruhe, 1988.

30. H. Wildner, "Generation and Visualisation of Configuration Space Obstacles for Industrial Robots with Six Degrees of Freedom" (in German), Diploma thesis, Institute for Real-Time Computer Control and Robotics, University of Karlsruhe, 1988.

31. C. Laugier, "A Program for Automatic Grasping of Objects with a Robot Arm," *International Symposium on Industrial Robots*, Tokyo, October 1981.

32. C. Laugier and J. Pertin, "Automatic Grasping: A Case Study in Accessibility Analysis," *Proceedings: Advanced Software in Robotics*, Liege, Belgium, May 1983.

33. J. D. Wolter, R. A. Volz, and A. C. Woo, "Automatic Generation of Gripping Positions," *IEEE Trans. Sys. Man Cybern.* SMC-15:2 (March/April 1985).

34. A. Hörmann, "A Method for Automatic Grasping of Parts with an Industrial Robot" (in German), Diploma thesis, Institute for Real-Time Computer Control and Robotics, University of Karlsruhe, 1986.

35. T. Lozano-Pérez, "Automatic Planning of Manipulator Transfer Movements," *IEEE Trans. Sys. Man Cybern.* SMC-11 (October 1981).

36. V. Werling, "Planning of Fine Motions for Assembly Operations" (in German), Diploma thesis, Institute for Real-Time Computer Control and Robotics, University of Karlsruhe, 1987.

37. T. Lozano-Pérez, "Spatial Planning: A Configuration Space Approach," *IEEE Trans. Compt.* C-32 (February 1983).

38. F. Preparata and S. Hong, "Convex Hulls of Finite Sets of Points in Two and Three Dimensions," *Commun. ACM* 20, 87–93 (February 1977).

39. B. Frommherz, "A Concept for a Robot Action Planning System" (in German), Ph.D. thesis, Department of Computer Science, Karlsruhe, 1989.

7

Path-Planning Methods for Robot Motion

C. BUCKLEY   *Integrated Systems Laboratory, Federal Institute of Technology at Zürich Switzerland*

7.1 INTRODUCTION

Robot manipulators are used principally to move payloads from one position to another. Each motion these manipulators make must be carefully planned in advance, and often improved iteratively, incurring much cost and loss of time. This chapter deals with the topic of *path planning* that is, the automatization of motion generation.

7.1.1 Applications Versus the Theoretical Forefront

There is a great deal of theoretical research in this area, and many results have been generated. Not so many of these, however, have found their way into actual robot systems sold for application, nor are they applied to these systems after sale. To some extent, this is because manufacturers feel that these advanced techniques are costly to implement and should therefore command a price higher than most customers are willing to pay. By and large, however, this research simply has not been able to produce results robust enough for implementation in an industrial environment.

This does not mean that the quality of the research concerned has been inferior. On the contrary, it has provided many fundamental insights and has served as a valuable stimulus in the investigation of many promising theoretical topics that otherwise might have been neglected. This is especially true in the field of symbolic computation.

No, the path-planning problem is simply a difficult one. Per-haps, because of the challenge it presents, it attracts much research interest. However, as more and more people have gotten involved, the problem has grown in complexity instead of skrinking, and it has taken on a legitimacy of its own. Today there is far less pres-sure to develop a working solution—everyone knows it's a hard prob-lem, so it has become acceptable to be engaged in solving a little piece.

Needless to say, this development is bad news for those wanting to apply generated results to real-world problems.

7.1.2 The Various Meanings of Path Planning

As if the complexity of the subject weren't trouble enough, there are quite a number of different, only loosely related problems being investigated under the general heading of "path planning." Among these are:

1. *The position path-planning problem*: how to move a body or manipulator from an arbitrary initial position to an arbitrary fi-nal one, without colliding with anything in between. This is the primary concern of this chapter.
2. *The trajectory generation problem*: how to take the results of an algorithm of the first type and optimize the trajectory to make it smoother, or more easily executable. This is treated only briefly here.
3. *The fine motion planning problem*: how to move around near objects and apply forces to them. This is also treated only briefly here.

To cope with this polyvalence of interest, a certain amount of jar-gon has been developed to differentiate among the various versions, which in turn makes keeping up harder still.

7.1.3 The Thrust of This Chapter

This chapter has three purposes:

1. To provide a short overview of the issues being discussed in current path-planning literature. This is intended to make it easier to skim the reams of literature that appear yearly.
2. To provide a review and synthesis of all major past research done on the position path-planning problem.
3. To outline a scheme for developing a robust position path-plan-ning system based on the results of this research in the not-too-distant future.

An effort has been made to keep the presentation as simple as possible. Mathematical formulas and developments have been purposely avoided. This is *not* based on a presumption of simplicity on the part of the readership. However, working through such material takes time, and this book is supposed to be a digest, which should be easy to read, and even entertaining. Complete references have been given; if you are interested, you will find more than enough formulas and abstractly expressed ideas to keep you occupied for months.

The point of view represented in this chapter is somewhat unique. You will most likely find it to be quite different from those of others working in the field. Perhaps if others saw things similarly, progress would be made more rapidly. At least it is hoped that you may profit from what is written here and that you may understand what has been done, perform research that complements it, and develop successful applications soon.

7.2 READING THE PATH-PLANNING LITERATURE

For a variety of reasons, most path-planning research is done in universities. The results of such work more often appear in the form of articles or reports than as useful industrial prototypes. Path planning in general, and position path planning in particular, is a very good topic for academic research—while it is intellectually challenging, it is also simple to understand intuitively, and consequently an easy thing to motivate students to work on. This means that there are a lot of results, and therefore a lot of reports to read. When, along with the sheer quantity, one considers the unwritten rules of gamesmanship that apply to academic writing, just keeping up with report contents can become a difficult task indeed.

Keeping up with this literature is one goal that can never be met, although you have to try anyway. It is certain that as this chapter goes to press it is already out of date. Furthermore, different readers come from different parts of the world, and the literature that one knows the best is usually dependent on what is being worked on locally. It is virtually guaranteed that some reader will say "How can he pretend to write an overview on path planning? He hasn't even talked about <*reference x*>, and 'everybody' knows about that one."

Practically the only solution to both these problems is to get the reader actively involved in the literature reading process, by providing "meta-level knowledge," as opposed to "leaf-level knowledge." That is what this section is about. It consists of a very brief, top-level discussion of the author's conception of the central issues in path planning, giving for each issue an estimation of the extent to which progress has been made and what remains to be done. This is the kind of summary that anyone who had read the literature, spoken with its authors, gone to the conferences, and spent some time considering the matter would develop, although perhaps only in passive, mental form.

By writing such a thing down, it is sought nonetheless to make it possible for those who have neither the time nor interest to go through these steps themselves to obtain a useful ersatz. It is hoped that, having worked through these next few pages, when subsequently reading articles (for example, try it out by reading the papers of your favorite local "stars"), you should find it easier to quickly determine:

Whether it is worth reading an article thoroughly.
How much progress is represented by what is reported.
How much remains to be accomplished before the results can be used.

7.2.1 Categorizing the Issues

Issues in path-planning research arrange themselves into three major categories:

1. The types of constraints the research algorithm was capable of dealing with.
2. The type of information the method used in the planning process.
3. The conceptual model around which solutions were organized.

The first two categories are classical enough. The third appears to be less orthodox, and some might question whether it belongs in the discussion at all. Section 7.3 will show, however, that this is perhaps *the* most important factor in determining the results achieved.

7.2.2 Constraints

Constraints pertaining to path-planning methods are of two types. The variety we will call type I consists of constraints a method may handle; constraints on the method itself are referred to as type II. The more of the former, the better, since the method is then more general. However, the latter have played a decidedly more important role in path planning, since often many of these have been needed to implement a working algorithm.

Geometric Constraints

Position path-planning is concerned with the development of methods that correctly handle geometric constraints of type I. These are simple to describe: they state that of all the bodies that make up a robot manipulator and its surroundings, none may collide during motion.*

This description is deceptively simple. If there are n three-dimensional bodies, each of which has two-dimensional boundaries, there can be at worst n^2 pairs of collision points[†], each of which may roam anywhere in n^2 four-dimensional (2×2) spaces. That's a lot of points to keep track of. When the bodies concerned are nonconvex, there can be multiple contact points, and things get messy.

Therefore, it is useful and important to differentiate among path-planning efforts according to the *degree to which geometric constraints were imposed on the method.* Many early quite serious efforts concerned themselves with planning motions for:

A single point
A manipulator approximation consisting of a (small) set of points on
    the manipulator
A single line segment
A "stick-figure" manipulator (multiple line segments)
Lower dimensional projections of manipulators (typically in the plane)
Parts of manipulators

to be able to even hypothesize a solution. While such approximations are valid for demonstrating computational principles, they can be tolerated for applications only to the extent that the approximations do not cause information to be lost. This has almost never been found to be the case.

For example, many manipulators of the "Scara" type are used in assembling components into circuit boards. These manipulators have 3 degrees of freedom, which serve to translate and orient a separately controlled plunger in a plane parallel to the circuit board surface. Since the manipulator motion per se is essentially planar, it is convenient to model it in a 3 degree-of-freedom configuration space. However, this works only if it may be safely assumed that the plunger motion and changing payload size will never cause a collision, so it still would be possible to "hook" a neighboring fixture with an average payload (such as a flyback transformer) on its way to being inserted. To accurately model this, the dimensionality of the problem must simply be increased.

---

*An advanced variant requires that collisions be foreseen and orderly.
[†]If the collision can be modeled as such.

Methods for realistic problems (no stick figures or points, with the full number of degrees of freedom of the mechanism represented) have been published. As of this writing, they are almost always more complicated to implement and slower, albeit in principle more reliable, than restricted methods.

On the other hand, methods making these convenient geometric approximations have been found to be unacceptable unless carefully watched over; beware of anyone claiming otherwise.

Kinematic Constraints

Limitations of how bodies may move with respect to one another are referred to as kinematic constraints. These include limit stops and nonholonomic constraints (e.g., rolling constraints, pivot constraints). Such constraints are type I constraints when they reflect the characteristics of the manipulator for which a path is to be planned, without special consideration of the path-planning method to be used.

Such constraints have often been applied as type II constraints on path-planning methods. That is, the following restrictions are often found:

No nonholonomic constraints: This is often called the "free body" path-planning problem. Motion planning for mobile robots has often been done this way, even though it takes special, nonrobust drive trains to allow the results to be executed. In fact, most mobile robots found in university laboratories have specially designed drive trains to allow them to carry out motions generated under this assumption.

"Cartesian" constraints: It is far easier to *build* rotary mechanical joints—conventional (and therefore cheaper) motors turn endlessly around. It turns out to be computationally easier, however, to require that only translational motions (also called linear) be used.

Complete circular arc assumptions: If a joint must turn around, if it can be allowed to spin completely and freely around, computation also becomes simpler, since a degree of freedom can be removed from the problem by replacing the link by this swept volume.

Again, although many of these assumptions are not realistic, they are often made just to get results out.

Dynamic Constraints

These dynamic constraints bring in the notions of force, and indirectly, time. Precise position control systems on servo-controlled robot manipulators work by using high forces to get actuators on track. This can be hazardous, especially when working with objects that cannot be moved by a manipulator, only damaged by it. Path-planning problems with constraints of these kinds have been called "compliant motion problems."

The way in which time may plan a role is best demonstrated by an example. Given the planar two-link arm in Figure 7.1, with joint angles θ_1 and θ_2 as labeled, the relationship between the end point x and the joint angles is given by Equation 7.1:

$$\begin{pmatrix} x \\ y \end{pmatrix} = \begin{pmatrix} l_1 \cos \theta_1 + l_2 \cos (\theta_1 + \theta_2) \\ l_1 \sin \theta_1 + l_2 \sin (\theta_1 + \theta_2) \end{pmatrix} \tag{7.1}$$

Partial differentiating with respect to θ_2,

$$\begin{pmatrix} \dfrac{\partial x}{\partial \theta_2} \\ \dfrac{\partial y}{\partial \theta_2} \end{pmatrix} = \begin{pmatrix} -l_2 \sin (\theta_1 + \theta_2) \\ l_2 \cos (\theta_1 + \theta_2) \end{pmatrix} \tag{7.2}$$

When the arm is fully extended (the elbow is straight), $\theta_1 + \theta_2 = 0 \rightarrow \sin (\theta_1 + \theta_2) = 0$. Therefore,

$$\frac{\partial x}{\partial \theta_2} = 0 \tag{7.3}$$

and

$$\frac{\partial \theta_2}{\partial x} = \infty \tag{7.4}$$

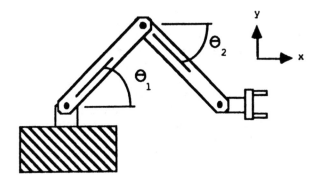

FIGURE 7.1 A manipulator with a dynamics problem.

and joint 2 must move infinitely fast to effect a reasonable speed in x. In manipulator kinematics, this is called a "singularity problem." No actuator capable of exerting a finite force can accelerate a mass at infinite rates. Consequently, to avoid position errors, the acceleration time must be lengthened.

Much good work in compliant motion planning has been done. However, owing to the difficulty of this topic in and of itself, algorithms for compliant motion do not handle geometric constraints well.

Path-planning methods have been reported that deal with the time problem described above. These, however, also ignore geometric constraints, or at the very best accept as input paths that already satisfy geometric constraints, and perturb them to satisfy dynamic constraints while (hopefully) not reintroducing geometric constraint violations.

On the other hand, the type II approach to the problem of dynamic constraints has often been employed. This says, "If you move the manipulator slowly enough, dynamics are not a problem, right?" This constraint today still makes even some of the more impressive position path-planning systems look shabby, as they execute painfully slow motions to avoid damaging the manipulators for which they have been implemented.

7.2.3 The Use of Available Information

The availability of dimension has to do with the types of information path planners take advantage of to operate.

Environmental Modeling

A very clear separation can be drawn between planners that model the environment and those that don't. Planners that don't rely mainly on sensor information, and they prove useful only in very restricted situations. For example, much work has appeared in which mobile robots solve path-planning problems using a wall-following technique. This works only in two dimensions, where the dimension of the path is the same as the dimension of the obstacle boundaries.

Planners that do model their environment either get the environmental model delivered them (e.g., from a CAD database) or build it up from data they collect from their sensors while exploring. The best results with the latter kind have also been obtained for two dimensions only—higher dimensional path planners almost exclusively use preprepared environmental models.

The principal issue of importance is the means by which the environmental model is stored. The following issues should be raised:

Is the model represented in a form the path-planning method can use?

Does the model represent the environmental situation accurately enough? Can the model be incrementally updated?

Note that environmental modeling is distinct from algorithm memory as discussed below. The former has to do with knowledge of the shapes of *static* objects with the manipulation space. The latter has to do with allowable *configurations* in the configuration space, and their connectivity.

Sensor Information

Many different sensors may be used to provide input to a path planning method. They measure force, velocity, and actuator position, as well as geometric information about the environment, the latter usually optically. It is not appropriate to go into this area here.

Of significance, though, are the two ways in which sensors can be used: (a) they can be used simply to provide *supplementary* information to be used on top of trajectories generated based on the environmental model, or (b) they can be used to update the geometric model itself. The first application is quite common, while the second, although often proposed, has been implemented seldom. Sensors will not be a big issue in the results reported here.

7.2.4 The Planning Model

By far the most significant advances in path-planner capabilities have been obtained through changes of the path-planner model.

Configuration Space Versus Operation Space

Apart from earlier considerations concerning the number of degrees of freedom used for a model, it is also important whether the modeling is done in *configuration space*, a Euclidean space whose axis dimensions correspond to joint degrees of freedom, or in *operation space*, a Euclidean space corresponding to translation and rotation coordinates of the moving payload part of manipulator (its "hand").

It can be advantageous to use configuration space because the results can be directly sent to actuator hardware; however, it is hard to hypothesize desired trajectories in this space. Reasonable end-effector trajectories look very funny in configuration space. When configuration space is treated as Euclidean, it is often discontinuous where it should not be. If you think describing *trajectories* in configuration space is hard, consider how hard it is to describe entire *obstacles*.

Operation space offers advantages insofar as goal trajectories are very easy to hypothesize. Most modern step-by-step manipulator programming is done in this space. Operation space, however, does not accurately represent the manipulator configuration: for manipulators with more than one rotary joint, one hand position corresponds to several possible joint positions. This shortcoming in the case of redundant manipulators is even more extreme.

Lately, trends have been toward working exclusively with configuration space and trying novel, very complex techniques to overcome the associated problems of constraint modeling.

Trajectories Versus State Transitions

As they are physically executed, trajectories of planned paths consist of a continuous evolution of the several degrees of freedom of the manipulator involved. Computers are finite-state machines, but continuous trajectories may be approximated by a series of state transitions. This category has to do with how that information is represented.

Two alternatives are used, called the C^0 and C^1 alternatives. With C^0 all the information for a trajectory is stored in its end points. No continuity higher than zeroth order is maintained. To achieve an arbitrarily fine approximation of a trajectory, additional points must be stored. In working in a high dimensional space, this can get expensive.

Mechanics is inherently second order: acceleration is the second derivative of position. This means that even if only zero-order information is specified, to implement it, at least first-order continuity is implicitly required. The C^1 approach takes advantage of this to represent more complicated trajectories using fewer data.

There has been good work done in specifying trajectories that do enforce first-order continuity (e.g., Refs. 1 and 2). However, much remains to be done to combine this with the more extensive position path-planning results.

Algorithm Memory

To give the results of a path-planning algorithm, it is at least necessary to output the path. The question to be posed here is: How much more information needs to be stored? This question has been by far the most controversial question in path planning, although no one has identified it in exactly this way. Because it is so central, a good deal of attention is paid to it in Section 7.3, where claims are backed up with references to published results.

Here, it suffices to say that it is not hard to write a path planner that will be guaranteed to find a path if one exists (to a particular resolution). What *is* hard is to find one that does so without having to create an enormous computer model of the entire continuous configuration space, which exceeds the capacity of modern computers.

Subdivision Versus Trajectory-Based Methods

The distinction between subdivision- and trajectory-based models has to do with the fundamental conceptual model used to develop the path-planning algorithm.

With subdivision algorithms, the entire continous path-planning space is assumed to be subdivisible into a finite number of chunks. Points in one chunk differ from points in another in some important way—for position path planning, the key criterion is whether the points satisfy geometric constraints. There are three questions of importance here:

1. Is it necessary to represent the whole configuration space?
2. Is it indeed possible to do so accurately?
3. If so, how much does it cost?

Again, these matters are covered in much more detail in Section 7.3. Here we content ourselves with delivering the punchline—the answers are: usually not, yes, and way too much, respectively.

In contrast, trajectory-based methods* forego representing an entire space, and simply represent one or more trajectories in that space. This is not as universal a strategy as the subdivision method, since a configuration space can never be covered this way.

For more on this distinction, see the subsection entitled "Local Versus Global Methods" and the beginning of Section 7.3.2.

Evolutionary Versus Ensemble Development of Trajectories

Trajectories are elaborated using an evolutionary or an ensemble strategy, regardless of the algorithm's position vis-à-vis the basis of the conceptual model, as just discussed.

Evolutionary development involves beginning at one end of a trajectory and developing it sequentially by moving a configuration state toward the opposite end point (Fig. 7.2a). This is virtually the only method used by subdivision algorithms, because the graphical search algorithms used to solve the intermediate problem they develop operate this way. Many trajectory-based methods also work this way.

---

*In the author's Ph.D. thesis [3], these methods were called "continuum methods." The idea is the same, but the name has been changed, since experience has shown that "trajectory-based methods" conveys the idea better.

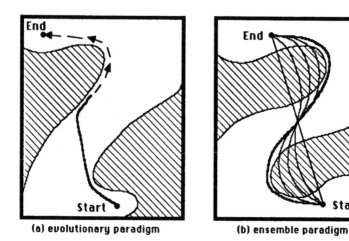

(a) evolutionary paradigm   **(b) ensemble paradigm**

FIGURE 7.2   Two trajectory development paradigms.

Ensemble development is shown in Figure 7.2b. Using this method, a trajectory is hypothesized in advance, and *transformed as a whole*. Very few algorithms use this development method.

This topic also is treated extensively in Section 7.3.

7.2.5   Other Classifications

The classification discussed in the scheme of Figure 7.2 is the author's own; the reader will certainly encounter others in the path-planning literature.   Some of the better known examples are discussed in this section.   With respect to the classification presented above, these characterizations fall into two categories:

1.   They have nothing to do with path planning.
2.   They are groupings of attributes of the above-described categories.

This section is included as a kind of cross-reference between the two sets.

Findspace Versus Findpath

The findspace/findpath distinction really doesn't have a lot to do with path planning.   However, since many people seem to think it does, it is important to dispense with it right away.

The term *findspace* is used to describe the problem of fitting a number of oddly shaped geometric objects into a given volume.   This is not really a path-planning problem, because its solutions need not be realizable—it is acceptable to have collisions while moving the various pieces of the environment into place.

The counterpart problem to findspace is *findpath*. This actually corresponds to the solving of a path-planning problem. Constraints used are typically geometric only—no (or few) kinematic and dynamic constraints. Put another way, the findpath problem is what is here called the position path-planning problem.

Gross Motion Planning Versus Fine Motion Planning

Gross motion planning refers to solving path-planning problems with mainly geometric constraints. The majority of early work, which simply called itself findpath algorthims, got shifted to this category as the need arose to draw a distinction between the two. Again, kinematic and dynamic constraints play a very small role, except that considerable effort has been made to account for "standard" industrial manipulator kinematics. Gross motion planning would be exactly like the position path-planning problem discussed above, except that it is also thereby implicitly understood that since final trajectories never come close to constraints, one doesn't have to be too careful about how these constraints are modeled. This assumption, of course, breaks down in regions of a configuration space in which the acceptable zones are very flat or small.

Fine motion planning is *not* really a complementary concept to gross motion planning, although the two are often presented contrastively. While the magnitude of the motions in fine motion problems might be expected to be small, this is not necessarily the case. Rather, fine motion planning problems are those in which geometric constraints play a relatively minor role. Usually they are manifest only as converted to dynamic constraints ("You can't drive the manipulator through this surface, because that would require larger forces than are allowed."). The goal in fine motion problems is not so much to evolve a twisted, tortuous path that avoids randomly placed obstacles as it is to make short motions that separate parts or bring them into contact without harming them. Typical fine motion planning problems are inserting pegs into holes and grasping.

Complete Versus Incomplete Algorithms

Complete algorithms are those for which it may be proven that, when given a soluble problem, they always find a solution. When given an insoluble problem, they figure out that it is insoluble. Complete algorithms are almost always subdivision algorithms.

Incomplete algorithms are those that are not complete.

In the author's classification presented above, the idea of completeness is included under the concept of memory. See Section 7.3.

Local Versus Global Methods

Local methods are those in which an evolutive trajectory development paradigm is used and for which path-planning memory contains
only the trajectory being developed.

Global methods are those that have more memory than local methods. While such methods could conceivably use for trajectory development paradigms other than the evolutive one, many do not.

7.2.6 Retrospective: Why Methodological Diversity?

As predicted at the beginning of this section, our enumeration of
issues shows that the points of difference among various path-planning algorithms have far more to do with how the *solutions* to the
problems are formulated than with the problem statements themselves.

Such a diversity at the beginning of study of a subject is to be
expected. However, work on path planning has been going on for
some 18 years now, and it is unusual for such a divergence of opinions to persist for so long. Why has this come about?

Some insight into this question may be gained in the next section.

7.3 A SURVEY OF PATH-PLANNING
RESULTS

From a historical viewpoint, progress in path-planning research can
be viewed as the confluence of two "schools" of thought of separate
origin. One of these schools is composed primarily of computer sci-
entists, whose work is based on the idea of subdivision. The other
school is composed primarily of numerical analysts and mechanical
modelers, whose work is based on the generation of trajectories,
without subdividing space.

7.3.1 Subdivision Path Planners

The idea behind the subdivision path planners is that the entire
configuration space of a path planning problem can be *subdivided*
into a finite number of *equivalence classes*, each of which contains
only points that may be treated equally. Therefore, by determining
what these classes are, and how they fit together, the path-planning problem may be transformed into one of simple graph search,
for which there exist many well-understood algorithms.

The first algorithms described here are the very earliest ones.
None of these really developed into works of lasting usefulness—all

were too simplistic. They are mentioned here though, since many of the ideas that are so important in currently popular subdivision methods were already present in these early works.

Whitney's Work

The first path-planning algorithm based on the subdivision method was actually implemented as a true finite-state machine. This was reported by Whitney [4], who addressed a highly simplified problem, the planning of motions of a manipulator gripper in a plane parallel to a table top. Collisions involving other moving parts of the robot manipulator and the environment were not considered. The state space of the problem was discretized to a constant mesh. Permitted positions were spaced at equal distances a few inches apart, and only four rotation angles were allowed. Some care was required to ensure the validity of these assumptions. Permissible state transitions (i.e., the states reachable from any given state) were simply programmed explicitly into a graph. Figure 7.3 is a sketch of a sample path-planning world used by Whitney.

Whitney evaluated several combinational algorithms for searching this graph, including dynamic programming and the A^* algorithm. The A^* algorithm is similar to other neighbor search graph algorithms (e.g., breadth-first, depth-first), but it uses a heuristic cost function to order the entries in the visit stack [5]. Whitney preferred A^* because "it frequently was able to avoid searching the

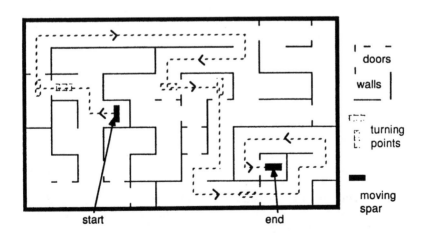

FIGURE 7.3 Whitney's simplified path-planning world.

entire graph." Practically every other subdivision path planner since has used A^* to search the problem graph.

Whitney pointed out that as the dimensions of the problem domain increased, the storage and time requirements to search the corresponding graph became "staggering." Consequently, the demonstration examples that Whitney gave were quite small*. This problem has also plagued practically every subdivision algorithm since, and it remains the main problem with algorithms of this type.

The true finite-state machine nature of Whitney's work was simultaneously its best and its worst attribute. To its credit, it captured the imagination of the people at that time who were not at all predisposed toward the use of computers in this area. On the other hand, the simplicity of the approach made it really useful for demonstration purposes only.

As might be expected, the path-planning algorithms that immediately followed Whitney's work were concerned primarily with rectifying this particular shortcoming of the Whitney algorithm.

Widdoes's Work

Widdoes's work in obstacle avoidance was done using Paul's computer system [6] for the Stanford manipulator [7] at the Stanford Artificial Intelligence Laboratory [8].† Due to the time constraints within which he had to work, much was unimplemented, and his methodology was really applicable only to the special geometric configuration of the Stanford arm (see Fig. 7.4).

However, the work is important because it described the strategy of subdividing continuous space into subsections that *cover* the original space. Instead of states of the finite-state machine, one has (hopefully homogeneous) regions. The problem of what to do when a region is *not* homogeneous was not considered.

---

*They were computed on a Digital Equipment Corporation PDP-8, a real "pioneer." Approximately 3000 12-bit words were used.

†A brief note on the availability of this document is in order. It has been cited by several authors (e.g., in Ref. 9), as a Stanford University Computer Science Report, number 227. This citation is in error; there is no such report. The document in question was a term paper done for the introductory robotics *course* at Stanford with course number CS227 (!). Although the report is very hard to find, the now elementary nature of what is important there makes this no longer essential. However, any one interested may write to the author for a copy.

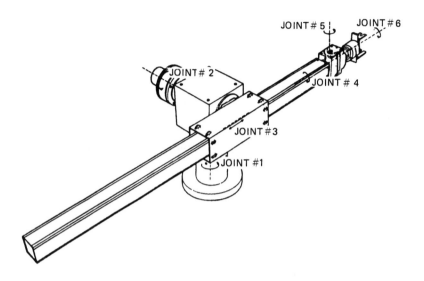

FIGURE 7.4   The Stanford arm.   (From Ref. 7, with permission.)

Lewis's Roving Robot with Arm

Lewis's path-planning work, done about the same time as that of Widdoes [10], was concerned with the installation of a Stanford manipulator (the same as used by Widdoes) on a mobile cart.   The primary obstacles to be avoided were parts of the cart (wheel wells, etc.), which did not move with respect to the manipulator base.

To plan paths in this fixed environment, Lewis introduced a method different from that proposed by Widdoes, called the *freeway method*.   The name of the method is very suggestive of how it works: a finite number of "good" (presumably) collision-free paths are chosen through the manipulator's configuration space between areas in which a lot of activity is foreseen.   Then, instead of searching a graph composed of a complete subdivision of the manipulator space, a planned motion consists of the following.*

---

*It should not escape notice that Lewis did this work at the Jet Propulsion Laboratory, near Los Angeles.   This area has one of the most developed freeway networks in the world, and at the time of Lewis's work, these still more or less fulfilled their purpose of efficiently moving people around.

1. Straight-line motion to the nearest "freeway entrance."
2. Motion through the "freeway system" to the "freeway exit" nearest the goal point.
3. Straight-line motion from the "freeway exit" to the goal point.

This scheme was particularly suited to Lewis's problem: for the most part, the obstacles with which he had to contend were various body parts of the cart on which the manipulator was mounted. These were few in number and unchanging, so it was a small matter to maintain freeway descriptions by hand.

Of course, the disadvantages of the method are clear: when new obstacles are introduced, there is a lot of reprogramming to do, and there is no way of determining how sure the method is—you still have to stand up over it and be sure it doesn't make a mistake.

Udupa's Planner

Two more ideas important for subdivision path planning were introduced in yet another algorithm developed specifically for the Stanford manipulator. This is the idea of the "grown obstacle," and it was presented by S. M. Udupa [11].

As mentioned above, a major difficulty in keeping track of geometric constraints in path planning is that *pairs* of points on different geometric object surfaces must be tracked. When the configuration space of the problem is organized naively, this means that there are a *variable number of degrees of freedom in the configuration space*, depending on the number of obstacles.

This is not correct; there are only as many degrees of freedom as there are independent actuators in the manipulator (unless something falls!). To have a configuration space with only this many degrees of freedom, Udupa implemented a scheme whereby "the manipulator was shrunk to a point, and the obstacles were grown to compensate."

Recalling the shape of the Stanford manipulator from Figure 7.4, Udupa's strategy was based on the assumption that the major problem in doing path planning for the Stanford arm was keeping the boom (> 1.0 m long) from colliding with something, especially since collisions could occur at the end to which the hand is not attached.*

Udupa's transformation procedure worked as shown in Figure 7.5. Figure 7.5a represents a view of the Stanford arm among a field of

---

*This is somewhat of a simplification to what Udupa actually did, but the details don't contribute much to the discussion.

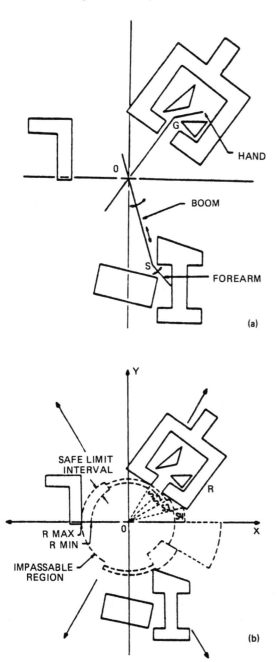

FIGURE 7.5   Udupa's growing transform.

obstacles, after the first transformation, in which the width of the boom has been shrunk to a line segment and the obstacles have grown porportionately. There is some ambiguity here, since how much bigger to make the obstacles depends entirely on which point of the manipulator contacts them.

The second stage of the transform changes the robot representation to a point. This point is chosen to be the tip of the boom. Then, the space is broken up into regions in which there are no boom collisions, inspecting both the front and back ends of the boom. This is shown in Figure 7.5b. Clearly this strategy is limited to the Stanford arm or similar type of manipulator.

The other idea Udupa introduced that was to have lasting influence is that of refinability. The subdivisions obtained after the second phase of the transformation are also subdivisions when they are broken into smaller pieces, as long as all the parts are unions of spherical patches of varying lengths. Udupa's software provided, under certain conditions (not clearly stated in the report), for the selective refinement described in the preceding section.

The major obstacle to the general application of Udupa's work was its specificity to the Stanford manipulator.

Lozano-Pérez's Work

By far the most intensive effort at subdivision path planning has been carried out by Tomás Lozano-Pérez, along with those who have worked with him at the Artificial Intelligence Laboratory of the Massachusetts Institute of Technology. Most people publishing in this field compare their work in some way with work reported out of this group.

The first paper published by Lozano-Pérez was with Wesley at IBM [12]. It drew much attention because it was the first serious attempt to separate path-planning theory from its application to a specific manipulator. The authors considered the conceptually simpler problem of planning the collision-free motion of a polyhedral body through a maze of polyhedral obstacles in the plane. This problem had been posed in terms of a moving point in earlier work [13]. This result fits well with the subdivision ideals, since it is easy to show that all shortest paths are made up of a finite number of segments between vertices which are "visible" from each other, hence the term VGRAPH (visibility graph).

Lozano-Pérez and Wesley observed that by taking the set difference* of all the polyhedral obstacles with the moving body fixed in

---

*Also called the Minkowski difference: between two convex sets A and B, the set difference is defined as $A \ominus B \equiv \{x - y \mid x \in A, y \in b\}$.

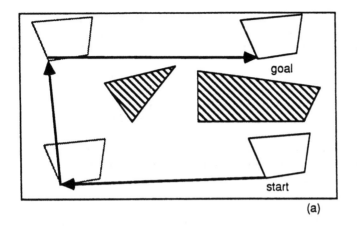

(a)

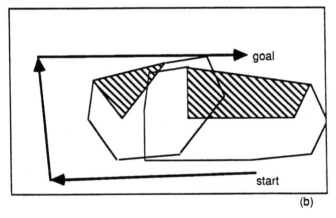

(b)

FIGURE 7.6 Lozano-Pérez and Wesley's growing transformation.

a given orientation, the problem reduces to the moving point problem again, this time with somewhat bigger "grown" obstacles. An
untransformed example and its transformed counterpart are shown
in Figure 7.6a and 7.6b, respectively.

As the authors observed, while this is an intriguing fact, it falls
somewhat short of being sufficient for solving the path-planning problem in general, since: in many instances, a collision-free path exists when the object is rotated during translation, but not otherwise. A variety of heuristics were posed to overcome this. In addition, the VGRAPH algorithm is useful only in the plane, because
in higher dimensions, the boundaries of obstacles can be more than

one-dimensional. To extend this algorithm to three dimensions would involve adding the ability to determine the shortest path consisting of line segments between points on an ordered set of edges, such as that reported in Ref. 14.

To deal with parameters of rotation, Lozano-Pérez introduced what has since become his trademark, the *slice-projection method*. In its simplest form, this method is not difficult to understand.

First, the idea is based exclusively in configuration space, or a space that is a Cartesian product of the servo-controlled joints of a manipulator. As with any space used to describe the problem, this one will be split up into allowed regions (free space) and nonallowed regions (configuration space obstacles). If the manipulator has rotary joints:

The relationship between configuration space obstacles and obstacles
    in the space in which the robot manipulator moves will not be 1:1.
The surfaces of the configuration space obstacles will be curvy and
    hard to describe.

The key to the projection-slice method is that rather than trying to describe what these curves look like in n-dimensional space, one chooses some subset of the configuration dimensions of the problem and breaks them up into little pieces.

In the simplest case of one-dimensional reduction, it works this way.

Consider an n-dimensional configuration space, for which some of the coordinates are rotary joint coordinates, so that configuration space obstacles are curvy, unimaginable things, which are not even convex. To "dissect" the space, so to speak, you simply cut it up using a number of evenly spaced hyperplanes, all of which are parallel to each other. In and of itself, this operation does nothing special—the separate pieces are of the same dimensionality as what you started with.

However, if the cutting hyperplanes lie sufficiently close together, the separate pieces can be collapsed along the dimension of the hyperplane normal, and the resulting *slice projections* are one dimension simpler, yet represent a reasonable approximation to their higher dimensional origins.

In Figure 7.7, a two-dimensional example of this is presented. Figure 7.7a is the original space. In two dimensions, hyperplanes are simply lines. Figure 7.7b shows this space cut up into individual pieces. Figure 7.7c shows the resulting pieces reduced along the plane-orthogonal dimension, in this case into lines.

There's nothing really to prevent doing such cutting and projecting along more than one (mutually orthogonal) dimension at a time. The result is exactly the same as if one reduced the space

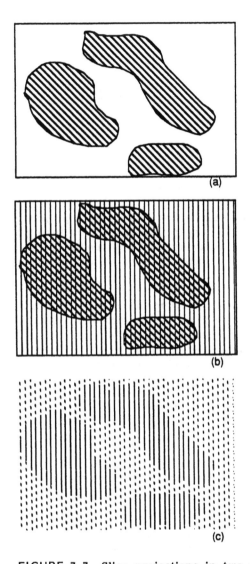

FIGURE 7.7 Slice projections in two dimensions.

one dimension, then reduced each of the resulting spaces along another dimension, and so on. Therefore, in three dimensions, reduction along two would produce a two-dimensional "forest" of line
segments, and reduction along three would produce a three-dimensional array of points.

In practice, reduction to the case where leaf elements are lines suffices, since these can be accurately represented by storing the boundary coordinates between blocked and unblocked segments, as shown in Figure 7.7c.

As yet, nothing has been said about how these projection slices are computed. Lozano-Pérez showed that if the following conditions are met:

All pairs of bodies that may possibly collide are convex.
When projection slices are simultaneously taken along multiple dimensions of a treelike manipulator, these dimensions are those most "outward" on the tree topology.

The configuration space obstacles making up a slice before projecting it onto the remaining dimensions may be constructed by taking set differences (as in VGRAPH), but *using the swept volume of the moving bodies as they move between cutting planes instead of the moving bodies themselves.*

Pulling all this together, the path-planning algorithm for a planar polyhedron among polyhedral obstacles *allowing for rotation* reported in Reference 12, consisted of cutting up the configuration space dimension corresponding to the single angle describing the orientation of the moving body into a number of slices, and constructing VGRAPH-style grown obstacle sets for each orientation. In this paper, this approach was mentioned more as a proposal than in terms of anything general—only a few orientations were used, sometimes even isolated orientations (trivial projection slices). Transitions between projection slices were taken care of "by hand," using such strategies as:

Find an intermediate position in which the moving object can freely turn between one of two orientations, and such that there exists each a path to each of the starting and ending points in the two orientations.

It was observed that each projection slice essentially replicated the storage and time complexity of an isolated nonrotating VGRAPH problem.

In a subsequent paper, Lozano-Pérez presents a slightly different application of his ideas [9]. In this one, planning is done for a so-called Cartesian manipulator, one for which the first three joints that move a comparatively smaller wristlike object around are translational. This allows the entire path-planning problem to be taken care of using one three-dimensional projection slice (along the wrist rotation dimensions), which is projected onto a single three-dimensional space, represented by a reference point mounted on the wrist.

Because of this simplification, a three-dimensional VGRAPH algorithm could have been used, except for the shortcomings mentioned previously.

Instead of doing this, all bodies (including the swept ones) were represented by (three-dimensional) polyhedral approximations. These were of course convex. The space was not, but it was broken up into convex polyhedral pieces. In this case, paths could easily be planned by describing straight-line paths between neighboring cells (through their common planar face), and straight-line paths within a convex polyhedral cell.

Importantly, the principle of selective refinement was first used here: free space need not be fully refined. Cells in space could belong to one of three types:

1. *EMPTY*: contain no configuration obstacles.
2. *FULL*: contain nothing but configuration obstacles.
3. *MIXED*: could contain anything.

Initially, all configuration space (the dimensions that were not projected away) is one *MIXED* cell. Since the planner cannot find any empty cells through which to specify a path, it decomposes the *MIXED* cell into a number of *convex EMPTY, FULL,* and *MIXED* cells and tries again. In principle, this cycle could be repeated until cells of a certain minimum size are reached. In this particular implementation, iterative refinement was carried out only until the start and goal configuration states were in *EMPTY* cells.

Brooks, working with Lozano-Pérez, incorporated the ideas developed in Lozano-Pérez's previous works into a complete algorithm for planning paths for a planar convex polyhedron among similar polyhedra [15]. There were two translation degrees of freedom and one rotational degree of freedom.

Lozano-Pérez's previous work placed a premium on a reasonably accurate representation of the configuration space at all times, using a combination of swept volumes, polyhedral transformations, and the grown obstacle transformation he had described in his early work. Brooks made a significant departure from this philosophy, in that the shapes represented in configuration space were always parallel-epipeds. Arbitrarily fine approximation was obtained on an as-needed basis by subdividing initially coarse cells, but only bisection cuts were used.

Instead of using polyhedral approximation of swept volumes, Brooks worked directly with the surface constraint curves of the polyhedral bodies—for two dimensions, the case analysis is not so bad. Initially, configuration space is cut into *EMPTY, MIXED,* and *FULL* cells by taking extrema progressively along each configuration space axis in turn. A path is searched from start configuration to goal configuration, which consists only of *EMPTY* or *MIXED* cells. If one is found, an attempt is made to refine the *MIXED* cells into *MIXED* and *EMPTY* ones, so that the path runs entirely through *EMPTY* cells. This is done by intersecting constraint surfaces with edges of the parallelepiped, and using heuristics so that one of the cells is completely empty if possible, and the biggest volume is given over to an empty or full cell. Paths between the cells are determined in a manner functionally equivalent to that described in Reference 9.

This algorithm attracted a lot of interest in that for the first time, a solution was presented to a general, theoretically grounded path-planning problem, which seemed to be reliable to arbitrary accuracy. It also attracted interest because it was written up in a reasonably understandable way. Schwartz and Sharir, whom we mention again presently, had published a planner with similar capabilities about a year earlier [16], and some might even consider it superior to that of Brooks, since it did not use any approximations (Brooks had to use slice projections for his rotation dimension, which could theoretically give the wrong answer). However, the algorithm of Schwartz and Sharir was so conceptually complicated that practically no one took the time to grind through the paper, and those who did had no interest in implementing it by the time they finished. It is to be added, though, that Schwartz and Sharir are mathematicians, so this probably did not bother them too much, and they themselves *did* report an implementation for their work.

To finish with the work of Lozano-Pérez & Company, one more algorithm should be cited. In Reference 17, Lozano-Pérez describes a version of his projection-slice algorithm that is valid for all manipulators composed of a single kinematic tree. It avoids the restrictions on slice projections mentioned above by using one-dimensional slice projections exclusively. These are organized into a hierarchy as deep as need be. As the reader might imagine, this can be *quite* expensive computationally. Examples reported are only two dimensional.

Schwartz and Sharir's General Case

After describing their algorithm for moving a planar convex polygon among others [16], Schwartz and Sharir wrote a follow-on report that has become something of a landmark for path planning in that it describes an algorithm which, if worthwhile or possible to implement, would be sufficiently general to meet virtually anyone's requirements [18].

Whereas earlier authors contented themselves to work with polyhedral objects, Schwartz and Sharir observed that if one only restricted the surfaces of the bodies modeled to be level sets of algebraic polynomials, and described rotational degrees of freedom, not by the use of trigonometric angles but by algebraic variables, using transformations such as

$$z = \tan \frac{\theta}{2}$$

or even

$$s = \sin \theta, \; c = \cos \theta$$

with the algebraic constraint

$$s^2 + c^2 = 1$$

to keep the dimensionality of the problem the same, then *all the constraint sets describing forbidden and permissible configurations of any manipulator so described could be written in terms of a Boolean combination of a set of algebraic polynomial constraints.*

Once this observation had been made, use could theoretically be made of a method due to Collins for topologically decomposing the configuration space into cells of admissible and inadmissible regions [19]. This fits right into the now familiar scheme of transforming the path-planning problem into one of graphical search, then searching it using an algorithm like A^*, only *this graph exactly characterized the problem—there would never be any need to refine it!*

There were certain problems associated with this approach, some of which the authors were able to solve, and some of which proved to be insurmountable. To be able to transform a feasible planned path expressed in terms of a suite of adjacent feasible cells back into a single parameter trajectory of points, previous authors took advantage of the polyhedral convexity of the cell. For this general case, the cells are in hyperspace, and neither polyhedral nor convex, and intuition is useless in attempting to derive a path.

Perhaps the major contributions of Schwartz and Sharir were to show that even in hyperspace, the connections between such cells could be found, that they could be made continuous. They gave algorithms on how to compute them and described the conditions under which this could be done. Unfortunately, these results seem to have been overshadowed by the results given in the next paragraph.

Problems that could not be overcome were principally ones of complexity. While they showed that a path-planning algorithm could be written in this framework in polynomial time, the growth rate is very large. Simply in terms of characterizing the number of cells in the Collins decompsition, the *power* of the polynomial term depends exponentially on the number of dimensions of the configuration space. It turns out that even under the most generous assumptions, it is cheaper to tesselate a manipulator configuration space along each of its dimensions into as many subchunks as there are bits of accuracy in the binary number system of your computer than it is to compute all the Collins cells required [3].

Nonetheless, these results are pleasing because of the insight they provide into the structure of the problem.

Other Subdivision Work

Many others have undertaken similar efforts, much more than can be instructively summarized here. Of all the published subdivision known to the author, the following observation applies:

To the extent that a published position path-planning work advertised to be "complete" to a given resolution does not suppose away certain degrees of freedom of the problem using some kind of reasonability heuristic, and depends on the slicing up of configuration space into cells of like class (admissible or inadmissible) and converting the problem into graphical search, there exists no algorithm that does not fall into one of the following two categories:

1. Its *worst-case* behavior is the equivalent of enumerating over all possible configuration states to some small fixed resolution.
2. Its representation scheme is incapable of representing the continuous problem accurately.

Put another way, at its worst and most general, subdivisioning does nothing short of developing an arbitrarily fine model of continuous space.

This includes all the algorithms discussed in this section so far, the algorithms described in References 20—27, as well as many others you may find.

Now fortunately, most algorithms, including all the ones reported here, rarely exhibit such behavior. To avoid doing so, they employ a variety of *heuristics* to accelerate their search. The ground rules employed by all the algorithms described up to now have been:

Searches begin at one of the end points of the motion and work sequentially toward the other.

Interfaces between the geometric aspects of the problem and the topological aspects of the search space are in the form of predicate query: "Is this [set of] state[s] admissible or not?"

The entire space is modeled to some degree of accuracy before searching of it begins.

Donald's Work

There remains one work to be discussed. Technically, Donald's work [28] is a subdivision algorithm. In a way, though, it is the exception that proves the rule, and it will serve as an excellent transition point into the results discussed in the next section.

In terms of capabilities, the algorithm is so much more powerful than any of those reported above that it can solve the *three-dimensional free-body path-planning problem*, which has 6 degrees of freedom (3 translational, 3 rotational). As stated in the preceding section, this algorithm also, in the worst case, could search at every point in a configuration space that is the Cartesian product of all the individual degrees of freedom (or, approximate continuous space).

However, its performance was reported to be significantly better than this, primarily due to the use of some powerful *heuristics*. Simple, two-dimensional path planners can easily be written which can solve arbitrarily complicated problems using an "English garden maze rose petal" strategy: one models the garden maze as a finite number of channels, connected together by junctions. To avoid traversing a channel twice, one marks its entrance and exit with rose petals. One can tell whether a junction has veen visited by looking to see whether any of the channels leading away from it have rose petals. This is essentially a seventeenth-century version of graphical search. The only problem lies in navigating within the channels, and this is simple—one essentially "follows the walls."

One can consider Donald's heuristics as a sort of wall following. However, while the English garden maze is for our purposes two-dimensional, the configuration space for a three-dimensional free-body problem has six dimensions. This makes things implementationally but not conceptually harder:

One must first compute the level sets of constraint surface. Constraint surfaces are *not* necessarily one-dimensional:
> It may be necessary to choose remaining degrees of freedom according to heuristic considerations (e.g., moving toward the goal point along the shortest path).

There may be more than one constraint active at one time.

Donald's contribution was to analytically enumerate the different types of constraint that are possible in the specific case of a three-dimensional convex polyhedron in collision with other three-dimensional polyhedra. He then expanded these symbolically and used numerical methods to move along constraint surfaces, to move around obstacles and to implement other heuristic strategies. This was quite an expensive undertaking: Donald reported that it took 12 hours for a dedicated VAX computer simply to put the various equations representing possible constraints into machine-calculable form. However, this may be considered to be a fixed overhead cost—these equations were subsequently loaded into another program, which planned the actual paths.

Using such heuristics, Donald found that it was rarely necessary to expand more than a small fraction of the multitudinous nodes corresponding to configuration space "points." Donald called his structure a lattice and actually implemented it as a single balanced binary tree, whose comparison key was the coordinate vector of the node, ordered lexicographically.

In fact, one is given to understand that the enumerative capacity of this lattice is rarely needed. Donald states [28, p. 79]:

> In practice we have had no problem in selecting a very fine resolution strategy for the lattice . . . [and making use of] "smarter" heuristics which attempt to exploit the coherence of [configuration space]. [In doing this] we obtain a planner which is not only complete, but which can solve complicated problems in a reasonable amount of time. We continue to find the lattice useful for recording the planner's explorations by [these heuristics].

Donald reported his system to be qualitatively superior in terms of running times, presumably when compared with other planners (principally those that had already been developed at MIT, where Donald did his work).

Summary of Subdivision-Based Algorithms

So, in retrospect, we see that as more and more accurate algorithms based on the subdivision approach were developed, the advantages associated with its use steadily diminished, as implementors were inexorably driven toward a representation that could potentially turn into an arbitrarily precise representation of a continuous space.

Heuristics have played an important part in the searching of these large spaces. By and large, the search paradigm used has been that of A^*, or something similar. The more successful algorithms (of the ones discussed, namely those of Brooks and Lozano-Pérez [15] and of Donald [28]) distinguished themselves from the less successful ones in that they:

Did not expand the entire space before starting to search it, but rather worked only with the parts that were in some sense "needed."
Paid considerably more attention to the heuristics used to search a configuration space than to the representation of the configuration space itself (in the case of Ref. 28).

In the next section, it will be seen that this is exactly the philosophy adopted by those developing an entirely different class of path planners, the *trajectory-based path planners*.

7.3.2 Trajectory-Based Path Planners

Unlike the case of subdivision path planners, in trajectory-based path planners no attempt is made anywhere to fully model configuration space in the computer doing the planning. Instead, these planners hypothesize a trajectory and then attempt to verify it. If it is not valid, another trajectory is generated, either based on the preceding one or at random. In practice, the number of attempts to find a new trajectory is limited, so the planner doesn't run forever, but gives up after a certain time and asks for help.

In much of the path-planning literature, algorithms described in the preceding section are characterized as "global," while the algorithms in this section are characterized as "local." In this author's opinion, this is an Orwellian exercise in Newspeak, since it does not fairly characterize the difference between the two classes.

While it is certainly true that the better subdivision (global) algorithms can study the entire configuration space in arbitrarily fine detail, these run very long* when they do so. Consequently, if you draw a "patience threshold" beyond which you are no longer willing to wait for results, these algorithms will amost certainly overstep it.

On the other hand, authors of subdivision algorithms regularly maintain that the only atlernative to using their approach is to use trial and error. Another thing they maintain is that the search directions chosen for a trial depend exclusively on conditions in the immediate neighborhood of a trial point, which must be moved from the starting configuration state to the goal configuration state. *The results of this section will show that in fact neither of these claims is true.*

Peiper's Work

The first path-planning algorithm of any kind for a general-purpose manipulator was reported by Peiper [30] at about the same time that Whitney published his work [4]. Peiper sought trajectories valid for *all* links of a 6 degree-of-freedom articulated robot manipulator. Peiper used the manipulator-generalized coordinates (i.e., the joint angles) as a configuration space. Manipulator links and obstacles were represented using set unions of different-sized instantiations of one of three different object primitives:

---

*For example, as reported by Donald [29], the algorithm described in Reference 15 can run for 2 days on certain problems. Similar results were personally experienced by the author.

Half-planes
Spheres
Cylinders

This limited set of primitives made interference detection easy. Three-dimensional objects with circular cross sections may be effectively modeled by their center points or axes if the object radii are subsequently subtracted from any interobject distance computed.

Peiper planned paths by generating hypothetical ones and testing them. Testing of a path was done by moving along it in small increments and checking for collisions. If a collision occurred, the path was rejected and another was attempted. Four main heuristics were used to convert unacceptable trajectories into acceptable ones. They are listed in the order in which they were tried.

1. Form a straight line in generalized coordinate space between the initial and final positions.
2. Compute two intermediate positions near the starting and initial points, which place the end-effector of the manipulator "up and then over" the objects being avoided.
3. Generate intermediate positions that "fold" the manipulator up by moving all joints toward their positive limit stops.
4. Generate intermediate positions that "fold" the manipulator up by moving all joints toward their negative limit stops.

If any of the intermediate positions also were found to be unacceptable, they were displaced slightly and checked again. There were also heuristics to check for visitation of the same configuration state more than once; had this process been made recursive, therefore, it would have been very similar to the configuration state graph described later by Donald [28]. There were also a number of minor heuristics for special cases.

Peiper reported good success with his approach, although he specifically did not claim it to be perfect. In particular, he reported a tendency toward failure when the manipulator was started in a configuration in which the joints were near their physical stops and when objects were so placed that the arm could get through only by going between two objects. A more precise evaluation is unfortunately not possible—none of trial examples were saved. It will be found that this is true of *most* published path-planning algorithms. More will be said about this in the last subsection of Section 7.4.1 "Testing."

From a qualitative point of view, it is interesting that some 15 years earlier, Peiper was undertaking problems of the same scope as Donald (6 degrees-of-freedom, three-dimensional space) with an algorithm that was much simpler, and was getting results out. It could well be that Donald's algorithm is capable of more, but his examples don't show this conclusively.

On the other hand, even if Donald's algorithm were demonstrably better, the question naturally presents itself: Without worrying so much about being able to build a complete configuration space model, what could be accomplished if the quality of the heuristics used in trajectory-based path planning could be improved? The answers prove to be quite interesting.

Myers's Work

Myers's contributions were to combine Peiper's basic idea of hypothesizing trajectories, testing them for validity, and generating alternate hypotheses with what might be called the first nonlocal, nonsubdivision search technique. This worked in the following way.

1. Initially, the end points of a trajectory to be found were checked and determined to be valid. If not, the problem returns failure.

2. When, during checking of the validity of a trajectory, an invalid region was found (by searching along the trajectory), an attempt was made to find the "deepest" intersection point. This deepest point was then moved in some heuristically chosen direction (e.g., using an "upward" heuristic) until it became feasible again.

3. Having found a new intermediate point, the entire path-planning algorithm was called again on the two halves of the path so generated, beginning from step 2.

4. Termination of the recursion happens when the subpath being considered is verified to be collision-free.

The net effect is something in between the evolutionary and ensemble trajectory development methods. An example of such an algorithm as applied to a very nonconvex set of obstacles is shown in Figure 7.8.* A purely evolutive algorithm, even one that used wall following, would have a hard time making any headway at all with this problem.

As an aside, Myers also added the use of a box volume primitive to the spheres and planes used by Peiper.

Myers also reported qualitative good results. However, since his test problems did not at all resemble those of Peiper, it is again difficult to make a more exact comparison.

Descent-Based Methods

Up until this point, all of the trajectory-based methods (and even most of the subdivision methods) described have in common that the

---

*Never mind, for the moment, where the valid intermediate points come from; this is covered in the subsection entitled "Buckley's Work."

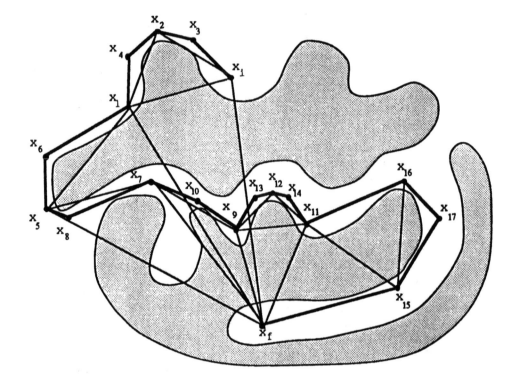

FIGURE 7.8 A very difficult two-dimensional example solved by
Myers's approach.

basis for generating alternating trajectory hypotheses was quite informal. For example, the idea of moving the manipulator "up and
over" obstacles played a very important part of the strategies of
both Myers and Peiper. However, this strategy is inappropriate
when obstacles are suspended.

In both cases, we need a more robust source for a *direction* in
which to move in configuration space in order to clear a collision.
Several had observed that in optimization theory, the shape of the
configuration space constraint could be used to provide this information.

In fact, why not treat the path-planning problem as a nonlinear
proramming problem?

While the goal in nonlinear programming is really to find a single
acceptable end state, problems usually are solved through the hypothesis of an initial guess, followed by the successive relaxation

of this guess until it stabilizes at (hopefully) the desired goal. *The desired trajectory is generated as a side effect of this relaxation.* In nonlinear programming problems, there are constraints, and these might as well correspond to the noninterference constraints in path planning. This means that only the nonlinear programming solution methods that maintain their constrints feasible throughout the course of the relaxation may be used, but there are many of these.

Bar far the best known work of this type is that of Khatib [31]. Khatib initially implemented his system as an added feature to a dynamic model-referenced robot manipulator servo control system.* He modeled the influence of obstacles as forces acting on the manipulator, which is what would actually happen if a collision did take place. To do this, he defined a set of continuously differentiable, analytically computable potential functions based on the *minimum distance* from smooth surfaces called n-ellipsoids, which approximated workspace obstacles. For example, to approximate a cube with dimensions $(2a, 2b, 2c)$,† the surface is represented by the equation:

$$\left(\frac{x}{a}\right)^{2n} + \left(\frac{y}{b}\right)^{2n} + \left(\frac{z}{c}\right)^{2n} = 1$$

Such a shape is plotted using a standard three-dimensional function plotted in Figure 7.9, where x, y, z are the Euclidean space coordinates, and n is a positive integer power. As n is increased, the approximation to the cube increases, but the gradients of the equation also become more poorly conditioned.

The gradient of these n-ellipsoids could be used to *simulate* repulsive forces of obstacles in the environment on certain distinguished points fixed on the surface of the moving robot manipulator. These forces, along with a similar attractive force acting on other points of the manipulator, and related to its goal position, have the net effect of having the manipulator control system solve a penalty function nonlinear programming problem, as well as maintaining servo.

This method of obstacle avoidance through the exertion of simulated forces represents an important contribution. In contrast to the finite-state machine algorithms, the continuous nature of the problem space is an advantage rather than a nuisance, since it allows the generation of well-behaved perturbations. The idea of descent-

---

*It might be said that Khatib solved a fine motion planning problem in which the movements were not fine.
†Centered at the origin—the equation may be transformed, but it is not nearly so understandable.

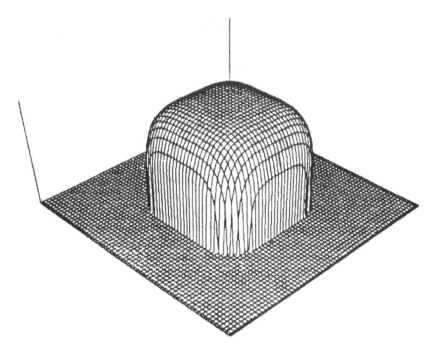

FIGURE 7.9   An *n*-ellipsoid with *n* = 4.   (From Ref. 31, with permission.)

based trajectory generation has been raised before [32].   This algorithm, though, depended on the use of special shapes (sticks and circles) to run.   Local constraint representation had been used by Grechanovsky and Pinsker [33] to position extra degrees of freedom of a redundant manipulator following a *fully prescribed* trajectory.*   Khatib's work, however, represents the first attempt to make a global representation of interobject constraints.

   There was considerable room for improvement on Khatib's algorithm.   Given the inherent nature of his technique, it is only to be expected that problems of nonconvexity would arise, as they did.   Furthermore, Khatib's potential functions were not defined for interaction between two general objects, only for interaction between a certain subclass of objects and a point.   Therefore, accurate simulation

---

*The work described in Reference 33 was done in the USSR.   Although not reported in the west until the summer of 1983, there is every indication that this work was carried out well before 1980.

of interference forces between two general objects would require enumeration of a large number of points on one of them. This shortcoming was partially corrected in a later implementation.

The Problem of Minimum Distance

Generally, though, the problem of calculating minimum distances remained. Too much accuracy was simply lost when using Khatib's approximate environmental models. It is quite common for solid modeling in general, and manipulator modeling in particular, to be done using polyhedral boundary representation models [34], and so interest developed in the computation of the minimum distance between these.

Actually, this very problem had been addressed at about the same time that path-planning research got started. Comba looked at the problem of detecting whether two polyhedral objects intersected through the solution of a nonlinear programming problem [35]. Again, for this to run smoothly, the function he minimized was required to have continuous derivatives, and so on. The function he developed to solve the problem turned out to be asymptotically equal to the minimum distance between the sets, as two smoothing constants were taken to zero [3].

Comba's representation of polyhedra was simply a set of hyperplanes. This is not a very efficient way of working, especially when quite a bit more information is available in a typical boundary representation model: the edges, vertices, and indirectly other faces neighboring on a face.

Toward the end of the 1970s, considerable interest in the field of computational geometry arose in its own right, and this problem was taken on in earnest. An early article by the same Schwartz who authored the path-planning algorithm discussed earlier produced a time-optimal algorithm for planar polygons. The minimum distance function was also used in checking collisions in an off-line robot motion simulator [36].

Buckley's Work: The Minimum *Directed* Distance Δ

However, even given an optimum algorithm for computing the minimum distance between convex polyhedra in three dimensions, we still lack an adequate base for generating trajectory perturbations using either Myers's method or the full ensemble trajectory method. Here the minimum distance is virtually useless, *because its gradient drops to zero identically in all places where the trajectory is inadmissible* (i.e., where the gradient is needed most).

To solve this problem, Buckley developed a constraint function called the *minimum directed distance*, which is equivalent to the minimum distance when the two convex bodies between which it is

computed do not intersect, and equal to the shortest distance in any
direction the bodies must be moved to clear each other if they do intersect [37]. The formal definition of the function is given in terms
of a two-stage extremization problem, one of the stages of which is
in a dual, conjugate indicator space. This formulation has several
advantages.

It makes it easy to prove the function continuous.

It allows for the development of an analytic expression for the function gradient. This gradient is also dependent on rotational degrees of freedom, and so on, to the extent the standard Jacobian
matrix is nonsingular, allows for expressions of gradients in any
configuration space.

It facilitated the development of time-optimal algorithms for its computation between two- and three-dimensional polyhedral objects
stored in boundary representation form.

It can be shown that the Δ minimization problem is exactly equivalent to the constraint case analysis described previously [15 or 28].
Put another way, when Δ functions between two convex objects are
used as constraint functions, all are applicable all the time. (This
is the way it must normally be for constraint functions used with
descent algorithms.)

The Δ function provides a convenient framework for: dividing
the computational workload up between different computing agents
(one object pair per agent), and moving the question of trajectory
perturbation onto a more familiar theoretical basis.

Note that using this function to compute gradients of zero-level
sets (which was not possible with the plain minimum distance), a
conventional hemstitching algorithm can provide essentially the same
functionality as Donald's numerically sliding along symbolically derived constraint surfaces, if you want to use it that way.

In somewhat of a departure from usual practice in path-planning
research, an attempt was made to do more than mere qualitative
evaluation of a path-planning method. A two-dimensional path-planning algorithm was written that was essentially equivalent to that of
Myers. The only differences were as follows.

1. Instead of being detected on a trial-and-error basis, collisions were discovered through the minimization along the hypothesized trajectory of any Δ function that came close to zero. This
automatically yielded something that could be used as the "deepest"
point in the collision.

2. This deepest point was relaxed out using a hill-climbing
method based on the Δ function. (This is how the intermediate
points in Fig. 7.8 were generated—Δ is computed between the configuration point and the configuration space obstacle.)

This algorithm was compared against the two-dimensional algorithm published at MIT [15] *on a common set of test problems.* The Δ algorithm was a supersimplistic one, being hacked together in a couple of weeks. Nonetheless, the results turned out to be interesting.

The test problems were for the most part generated at random and were put into the test battery when they turned out not to be trivial for both algorithms to solve. One was adapted from one of the "hard" problems in Reference 15. A graphical representation of this "hard" problem is shown in Figure 7.10. The idea is to move the white rectangle from its starting position at the bottom left to its goal in the upper center. Shown is Brooks's solution to the problem—the Δ solution looks similar, only somewhat smoother.

In four of the nine tests studied, the Brooks algorithm ran faster than the Δ algorithm. This includes one instance in which the Δ algorithm failed altogether to find a solution. In the other five cases, the Δ algorithm ran faster. In two instances, including the problem

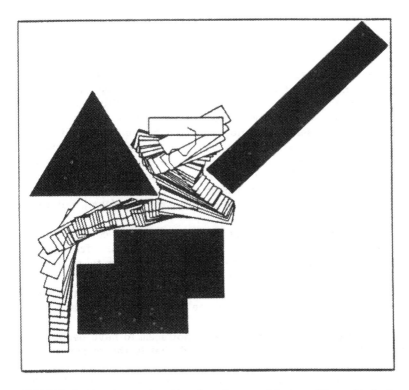

FIGURE 7.10 A hard path-planning problem. (From Ref. 3, with permission.)

of Figure 7.10, the Δ algorithm ran more than an order of magnitude faster, with the Brooks algorithm requiring tens of minutes to find a solution. It is unclear to what extent it is useful to distinguish this execution "to exhaustion" from the case of not findng a solution within some specified time limit.

Qualitatively, the Brooks's algorithm seemed to perform somewhat better for simple problems, while the Δ algorithm did demonstrably better for certain difficult problems, especially those involving passage through a restricted opening or channel. By contrast, the case in which the Δ algorithm failed completely is typical of its behavior in instances in which large initial path deviations are required to find one that is collision-free.

Based on these few tests, we can say ony that the algorithms gave comparable accountings of themselves. That is, there was no clear winner. Further testing clearly should be done. For a hint of the direction in which this has since been taken, see the research proposals in Section 7.4.

7.3.3 Bringing the Two Paradigms Together

Thus at the end of both the subdivision and trajectory streams of development, we see evidence on both sides that in spite of the importance commonly attached by subdivisionists to developing computer representations of configuration space, the most advanced algorithms of both groups effectively dispense with this. Although Donald's algorithm uses a structure that could turn into a configuration space representation under worst-case conditions, it would become very unwieldy should this occur, and practical observation shows that it does not. Donald's memory model more closely resembles something like a "scratchpad" in which previously visited locations were stored. Buckley's simple algorithm turned in performances roughly equivalent to those of Brooks's algorithm, but using no configuration space model at all.

To be sure, the worst-case security of having a configuration space model is worth something. The use of memory-free algorithms is *not* being promoted here. The conclusions to be drawn, however, are that both approaches have something to offer, and that perhaps a better memory model has rather less resemblance to the true geometric configuration space than it does to, say, the human memory.

By and large, these observations do not seem to have been picked up on by others working in the field, and that is the reason for this chapter. Fortunately, there has not been a total eclipse—there remains one, quite recent work to be discussed which does seem to take these considerations into account. One can only hope this is an omen of better results to come.

Faverjon and Tournassoud's Mixed-Strategy Work

Although the mixed-strategy idea is young, it is quite appealing in that it combines aspects of the subdivision and trajectory paradigms of Faverjon and Tournassoud [38] (the authors call them "global" and "local").

Insofar as the subdivision side of the algorithm is concerned, these authors divide the configuration space of the path-planning problem into uniform cells, in accordance with predictions summarized above (see subsection "Other Subdivision Work" of Section 7.3.1). However, these cells are relatively large.

These cells are linked together in a graph, again as usual. Initially, the weights on the graph are set to be uniform. Consequently, to plan a path, you need only:

1. Determine which cells contain the starting and ending goal.
2. Plan a path through the graph leading between these configuration cells.
3. Execute the path as a sequential series of steps between neighboring cells.

However, no account has yet been taken of obstacles. In a traditional subdivision algorithm, the stepwise transitions between neighboring cells are taken for granted—this is one of the underlying principles of the method.

On this point, the authors diverge in their approach. *In executing the stepwise transitions between neighboring cells of their subdivision, the authors make use of what they call a "local" planner, which is essentially the trajectory-based planner described by Tournassoud [39].* As with other trajectory methods, it is not at all certain that this will succeed. When it does not, it is blocked by an obstacle at a false local stationary point.

This condition, however, can be detected—one need only compare the configuration state achieved with the one desired. If they are close enough, the local step can be declared a success; otherwise it is declared a failure. Once a failure has been determined, a new global path must be computed.

In trying to solve one or more problems, several attempts can be made to pass between two adjacent cells (these do not always yield the same result, since the authors also propose a sort of smoothing to straighten out the path). The results of these repeated trials are accumulated, and they influence the path weights of the transition for the $A*$ algorithm. Transitions that succeed more often receive lower costs, while transitions that succeed less often receive higher ones. Consequently, zones in which there are many problems tend to be avoided by the subdivision planner.

Recalling that this method was based on a relatively coarse subdivision of the configuration space, the question of how this space may be selectively refined remains to be treated. This is also simply answered: if a cell is too large, the results of repeated trials through the cell will lead to success and failures, hopefully in equal distribution. Consequently, the variance on the sample size is large. When this variance exceeds a threshold, the cell is subdivided; otherwise it is left alone.

This approach is very dependent on a number of previous attempts to plan paths through the space. Consequently, as with the subdivision methods (which treat every problem as a new one), the initial calculation times may be rather long. However, this model also has a kind of "memory"—it learns as it goes along. What's more, it can be *taught*: if you ask it to execute paths you know are successful, it will afterward tend to prefer those paths in relation to others. Here then, at last, is a reasonable framework for integration of the freeway concept proposed early on [10].

In the research cited, only results pertaining to planning paths for a simple disc in a two-dimensional maze (2 degrees of freedom) are reported. It is sincerely hoped that this project is still being actively pursued.

7.4 FUTURE WORK IN PATH PLANNING

Since the appearance of the works described in this chapter, perhaps the greatest effect on the research community has entailed convincing people that the position path-planning problem, as it has been posed in the works described here, is intrinsically a hard one and not likely to be solvable in the near future.

A few results continue to appear, still proposing solutions to the path-planning problem based on the subdivision of configuration space. Almost invariably, though, when it comes to implementation, these approaches draw the line at a few degrees of freedom, lest the computations become too complex.

As far as is known to the author, no path planner has since appeared that does not fall into one of the categories previously described—in particular, many are precisely characterized by the observation "Other Subdivision Work" (see Section 7.3.1).

It is not surprising, then, that recent results from the more esteemed people in the field have all had to do with simplified versions of the path-planning problem, about which it is still possible to say something theoretically coherent. Examples of this kind of work may be found in References 40—43.

In and of themselves, these results are not unwanted: the topics undertaken are very exciting, and it is wonderful that such well-defined, challenging problems are made available to be solved by graduate students who will later assume leading roles in the research sector. After all, college and university professors are there primarily for the purpose of educating students.

In terms of *applied* industrial research in path planning, though, this bodes ill, because it is exactly the kind of development of a research topic into an end in itself that can significantly slow down the delivery of good "engineering" solutions. In the case of position path planning, this is not at all necessary, as results discussed later in this chapter have shown. Consequently, the next section sketches out the author's prognosis of how a research program leading to a working position path-planning system robust enough for use in an industrial environment might be set up, to be completable in a reasonable length of time.

7.4.1 Proposal for Future Research

There need not necessarily be a correlation between complexity of an implemented path-planning system and its success—some of the most successful results reported here have been quite simple. Consequently, it is to be expected that reasonable results are attainable within the framework of a moderate expenditure (we are not talking about another 10-year "race to the moon"). This proviso as to magnitude aside, the following issues must be addressed.

Memory

The results reviewed in this chapter have been mainly divided into two parts: subdivision-based algorithms and trajectory-based algorithms. As previously observed, the same two groups are characterized by many "subdivisionists" as proponents of "global" versus "local" methods—the former look everywhere, while the latter look only at the current configuration state ("where one is"). This characterization has been shown to be invalid.

Behind this entire issue is really the question of memory. If one were to distinguish between algorithms that can remember more than just a single path (or even worse, just part of one), and algorithms that can remember having been several places in a configuration space and do not necessarily have to do with the path currently being planned, the subdivision algorithms can be neatly and aptly described as memory algorithms that receive an extremely large (albeit poorly organized) initial memory charge. True, reported trajectory algorithms have had memory capacity for one or two trajectories at the most, but this does not mean that things could not be done differently, or that the method in general is wanting.

There is no question that memory of some form is necessary in path planning. True, memory-deficient algorithms have been published that nonetheless could compete with memory-rich subdivision algorithms. However, this only goes to show the importance of correctly describing *motions*—if you describe motions in more detail than blind heuristics based on end points (e.g., straight lines), you can do a lot even without memory.

The form this memory is to take is another matter entirely. Spatial subdivision is only one idea, which has been around far too long and appears to have been only moderately useful. This is not to say that spatial subdivision is not useful for *theoretical analysis* of path-planning problems. However, for practical implementations, other representations appear to be more apropos.

"Sensing" the Modeled Environment

Although an effective means of representing spatial memory remains to be developed, substantial progress has been made in interacting with geometric databases to determine whether a configuration state is valid. For simply detecting intersections on a predicate basis, very efficient means are available from computational geometry.

For following along the configuration space surfaces of polyhedra in contact, the methods of either Donald or Buckley may be used. While Donald's methods offer the advantage of symbolic computation, should this be important, Buckley's methods are simpler to implement and are useful for a wider range of configuration spaces. These methods may also be used simply to *detect* intersections—for an application of this, see Reference 44, but this kind of application does not begin to take full advantage of intersections.

Still to be developed are the means of describing the proximal behavior of two convex bodies that are modeled using something other than polyhedral surfaces. Furthermore, while nonconvex bodies may be treated through convex decomposition and separate treatment of the convex subparts, this introduces certain inefficiencies, since surfaces separating the different subparts are treated twice. Finally, rapid implementation using parallel architectures could bring considerable speedup if properly addressed. This author's current research lies in this direction.

Types of Heuristics

As yet, only very few types of motion heuristics have been explored. Mainly, these are based on the evolutionary paradigm, in which the trajectory is determined by its end points and is checked by being stepped through sequentially.

The flexible trajectory has received very little attention. No true flexible trajectory algorithm for path planning has been reported. An early such algorithm, reported as a means of solving an optimal control problem, required that zones of constraint activity be enumeratively identified—only the configuration space points where constraight activity shifted were computed [45].

There may be other types of useful heuristics.

Testing

Finally, we come to the issue of testing. Accepting the conclusion that no analytically provable solution to the general position path-planning problem that is also implementable is likely to be developed soon, but that heuristic algorithms are promising, *the need to perform evaluation through testing becomes unavoidable.*

However, there is no standard method of evaluating path-planning problems, and no benchmarks published in sufficient detail to be transferred from one research effort to another. Although meager, the comparison run between the works reported in References 37 and 15 stands virtually alone in this regard.

Many different kinds of problem might be included in a standard test case database. This definition of such a database would be best performed by a broadly based committee within the framework of one of the international engineering organizations—the IEEE seems a good choice. Here is a set of issues to be addressed by test case definition:

Free-body versus articulated chain problems
Planar versus spatial free-body problems
Free-body problems requiring the use of rotational degrees of freedom versus those that do not
A range of articulated chain problems graded along various degrees of freedom
Problems involving shapes of a simple type (e.g., spheres) versus problems with arbitrary shapes
Problems with polyhedral shapes versus problems with more general ones
Problems that are configuration space convex versus those that are not

This list comprises traditional position path-planning problems only; it would have to be expanded considerably to encompass the path-planning problems of the types mentioned in the first part of this chapter.

7.5 CONCLUSION

It is hoped that the reading has been easy and agreeable, but that nonetheless the reader has come away not only with a better understanding of the origins, development, and state of the art in position path planning, but also with a good idea of what remains to be done to put the research progress that has been made to productive use, and this in the near term.

To summarize the position taken in this chapter, if we can manage to overcome certain prejudices stemming from confusion of the desire to *use* computers to solve path-planning problems and the desire to map the path-planning problem onto a conceptual framework resembling a computer (the subdivision approach), we can begin to develop useful path-planning algorithms without waiting for the resolution of fundamental theoretical issues in the representational geometry of manifolds in hyperspace.

ACKNOWLEDGMENT

The author acknowledges the time and material resources granted him to write this chapter by the Integrated Systems Laboratory, Federal Institute of Technology, at Zürich, Prof. Dr. Wolfgang Fichtner, Director.

The research for this work was carried out while the author was with the following institutions:

Stanford University, Stanford, California
Veteran's Administration Hospital, Palo Alto, California
University of Montpellier II, Montpellier, France
Federal Institute of Technology, Zurich, Switzerland

Thanks are also due to Schlumberger's Palo Alto Research Laboratory, where the author's experimental investigations were carried out.

Responsibility for the contents and conclusions of this work rests solely with the author.

REFERENCES

1. J. Y. S. Luh and C. S. Lin, "Optimum Path Planning for Mechanical Manipulators," *ASME J. Dynam. Sys. Meas. Control* 102, 142—151 (June 1981).

2. R. H. Taylor, "Planning and Execution of Straight Line Manipulator Trajectories," *IBM J. Res. Dev.* 23:4, 424—436 (July 1979).
3. C. E. Buckley, "The Application of Continuum Methods to Path Planning," Ph.D. thesis, Stanford University, Department of Mechanical Engineering, 1985.
4. D. E. Whitney, "State Space Models of Remote Manipulation Tasks," *IEEE Trans. Autom. Control* AC-14:6, 617—623 (December 1969).
5. P. Hart, N. J. Nilsson, and B. Raphael, "A Formal Basis for the Heuristic Determination of Minimum Cost Paths," *IEEE Trans. Sys. Sci. Cybern.* SSC-4:2, 100—107 (July 1968).
6. R. P. Paul, "Modelling, Trajectory Calculation and Servoing of a Computer Controlled Arm," Computer Science Report STAN-CS0-72-311, Stanford University, November 1972. Also available from NTIS.
7. V. D. Scheinman, "Design of a Computer Controlled Manipulator," Artificial Intelligence Memo AIM-92, Stanford University, June 1969.
8. C. Widdoes, "A Heuristic Collision Avoided for the Stanford Robot Arm, unpublished term paper for Stanford University course CS227, 1974.
9. Tomás Lozano-Pérez, "Automatic Planning of Manipulator Transfer Movements," *IEEE Trans. Sys., Man, Cybern.* SMC-11:10, 681—698 (October 1981).
10. R. A. Lewis and A. K. Bejczy, "Planning Considerations for a Roving Robot with Arm," *Third International Joint Conference on Artificial Intelligence*, Stanford, CA, 1973.
11. S. M. Udupa, "Collision Detection and Avoidance in Computer-Controlled Manipulators," *International Joint Conference on Artificial Intelligence*, Boston, 1973, pp. 737—748.
12. T. Lozano-Pérez and M. A. Wesley, "An Algorithm for Planning Collision-Free Paths Among Polyhedral Obstacles," *Commun. ACM* 22:105, 560—570 (October 1979).
13. M. B. Ignat'yev, F. M. Kulakov, and ?. ?. Pokrovskiy, *Robot Manipulator Control Algorithms*, Document 59717, U.S.—U.S.S.R. Joint Publications Research Society, August 1973.
14. J. Y. S. Luh and C. E. Campbell, "Collision-Free Path Planning for Industrial Robots," Technical Report CH1788-9/82/0000-0084, IEEE, New York, 1982.
15. R. A. Brooks and T. Lozano-Pérez, "A Subdivision Algorithm in Configuration Space for Findpath with Rotation," AI Memo 684, MIT Artificial Intelligence Laboratory, December 1982.

16. J. T. Schwartz and M. Sharir, "On the Piano Movers' Problem. I. The Case of a Two-Dimensional Rigid Polynomial Body Moving Amidst Polygonal Barriers," Courant Institute of Mathematical Sciences Report 39, New York University, October 1981.

17. Tomás Lozano-Pérez, "A Simple Motion Planning Algorithm for General Robot Manipulators," *National Conference on Artificial Intelligence.* AAAI, Philadephia, 1986, pp. 626—631.

18. J. T. Schwartz and M. Sharir, "On the Piano Movers' Problem. II. General Techniques for Computing Topological Properties of Real Algebraic Manifolds," Courant Institute of Mathematical Sciences Report 41, New York University, February 1982.

19. G. Collins, "Quantifier Elmination for Real Closed Fields by Cylindrical Algebraic Decomposition, in *Lecture Notes in Computer Science,* Vol. 33, Springer-Verlag, Berlin, 1975, pp. 134—183.

20. R. A. Brooks, "Solving the Find-Path Problem by Good Representation of Free Space," *IEEE Trans. Sys., Man, Cybern.* SMC-13:2, 190—197 (March/April 1983).

21. B. Faverjon, "Obstacle Avoidance Using an Octree in the Configuration Space of a Manipulator," *IEEE Conference on Robotics and Automation,* Atlanta, 1984, pp. 504—512.

22. K. Hörmann, "A Cartesian Approach to Findpath for Industrial Robots," *NATO ASI Series F,* Vol. 29, Springer-Verlag, Berlin, 1987, pp. 425—450.

23. R. T. Chien, L. Zhang, and B. Zhang, "Planning Collision-Free Paths for Robotic Arm (sic) Among Obstacles," *IEEE Trans. Pattern Anal. Machine Intell.* PAMI-6:1, 91—96 (1984).

24. E. K. Wong and K. S. Fu, "A Hierarchical-Orthogonal-Space Approach to Collision-Free Path-Planning," *International Conference on Robotics and Automation,* IEEE, St. Louis, 1985, pp. 506—511.

25. L. Gouzènes, "Strategies for Solving Collision-Free Trajectories (sic) Problems for Mobile and Manipulator Robots," *Int. J. Robotics Res.* 3:4, 51—65 (1984).

26. M. Herman, "Fast, Three-Dimensional, Collision-Free Motion Planning," *International Conference on Robotics and Automation,* IEEE, San Francisco, 1986, pp. 1056—1063.

27. S. Kambhampati and L. S. Davis, "Multiresolution Path Planning for Mobile Robots," *IEEE J. Robotics Autom.* RA-2:3, 135—145 (September 1986).

28. B. R. Donald, "Motion Planning with Six Degrees of Freedom," Technical Report 791, MIT Artificial Intelligence Laboratory, Cambridge, MA, 1984.

29. B. R. Donald, "Hypothesizing Channels Through Free-Space in Solving the Findpath Problem," Memo 736, MIT Artificial Intelligence Laboratory, Cambridge, MA, 1983.

30. D. L. Peiper, "The Kinematics of Manipulators Under Computer Control," Ph.D. thesis, Stanford University, Department of Mechanical Engineering, 1968.

31. O. Khatib, "Real-Time Obstacle Avoidance for Manipuators and Mobile Robots," *International Conference on Robotics and Automation*, IEEE, St. Louis, 1985, pp. 530—535.

32. L. A. Loeff and A. H. Soni, "An Algorithm for Computer Guidance of a Manipuator in Between Obstacles," *ASME J. Eng. Ind.* August 1975, pp. 836—842.

33. Y. Grechanovsky and I. Sh. Pinsker, "An Algorithm for Moving a Computer-Controlled Manipulator While Avoiding Obstacles," *Eighth International Joint Conference on Artificial Intelligence*, Karlsruhe, West Germany, August 1983, pp. 807—813.

34. A. A. G. Requicha and H. B. Volcker, "Solid Modeling: A Historical Summary and Contemporary Assessment," *IEEE Comput. Graphics Anim.* March 1982, pp. 9—24.

35. P. G. Comba, "A Procedure for Detecting Intersections of Three-Dimensional Objects," *J. Assoc. Comput. Mach.* 15:3, 354—366 (July 1968).

36. W. E. Red, "Minimum Distances for Robot Task Simulation," *Robotica 1*, 231—238 (1983).

37. C. E. Buckley, "A Foundation for the Flexible-Trajectory Approach to Numeric Path Planning," in *Languages for Sensor-Based Control in Robotics*, NATO ASI Series F, Springer-Verlag, Berlin, 1986.

38. B. Faverjon and P. Tournassoud, "The Mixed Approach for Motion Planning: Learning Global Strategies from a Local Planner," *Tenth International Joint Conference on Artificial Intelligence*, Milan, 1987.

39. P. Tournassoud, "On Motion Coordination," Rapport de Recherche 549, Institut Nationale de Recherche en Informatique et en Automatique, 1986.

40. J. T. Schwartz and M. Sharir, "On the Piano Movers' Problem. III. Coordinating the Motion of Several Independent Bodies: The Special Case of Circular Bodies Moving Among Polygonal Barriers," Courant Institute of Mathematical Sciences Report 52, New York University, September 1982.

41. M. Sharir and E. Ariel-Sheffi, "On the Piano Movers' Problem. IV. Various Decomposable Two-Dimensional Motion Planning Problems," Courant Institute of Mathematical Sciences Report 58, New York University, February 1983.

42. G. Ramanathan and V. Alagar, "Algorithmic Motion Planning in Robotics: Coordinated Motion of Several Disks Amidst Polygonal Obstacles," *International Conference on Robotics and Automation*, IEEE, St. Louis, 1985, pp. 488—493.

43. C. Ó'Dúnlaing, M. Sharir, and C. K. Yap, "Retraction: A New Approach to Motion Planning," *Annual Symposium on the Theory of Computing*, ACM, Boston, 1983.

44. S. A. Cameron and R. K. Culley, "Determining the Minimum Translational Distance between Two Convex Polyhedra," *International Conference on Robotics and Automation*, IEEE, San Francisco, 1986, pp. 591—596.

45. A. E. Bryson, W. F. Denham, and S. E. Dreyfus, "Optimal Programming Problems with Inequality Constraints. I. Necessary Conditions for Extremal Solutions," *AIAA J.* 1:11, 2544—2550 (November 1963).

8

Automatic Error Detection and Recovery

MARIA GINI   *University of Minnesota, Minneapolis, Minnesota*

8.1 INTRODUCTION

Robots are operating today on a wide variety of tasks, such as object handling, painting, and welding. Even though assembly is still considered to be a difficult area, more and more robots are used in manufacturing to do assembly tasks. New areas outside manufacturing, such as exploration of unknown environments and medical applications, are being considered. Future growth areas will center on highly complex tasks [1].

The reliable execution of a robot task depends on the accuracy of the robot, the correctness of its task description, and the certainty of its environment. Real-world uncertainties often interfere with the robot task. Unexpected events such as misplaced or missing parts cause the robot to fail completely at its task, no matter how minor the disturbance may be, unless the robot has the ability to detect changes in the environment through its sensors and to replan its task.

Many error events can be anticipated by the production engineer. For instance, missing components may be detected by simple presence tests or by vision. However, there is always the possibility of events that have not been considered. To protect the robot and the workcell from damage, all errors must be detected. A major difficulty in robot programming derives from the need to foresee all cases requiring different processing.

Failures in achieving a task are the result of errors, but not every error produces an immediately detectable failure. Errors can

occur at many levels: at the mechanical level (a joint becomes
locked), at the hardware level (a sensor does not function properly so that the arm crashes into the table), at the controller level,
in the computer controlling the robot (either at the hardware or the
software level), and in the environment.

To give some flavor to the problem, imagine a simple scenario.
Consider a robot in a factory performing pallet loading operations
with various objects arriving on a conveyor belt. As each new object arrives, the robot is able to identify what object it is and to
decide how to grasp it. The robot continues executing its task as
long as there are objects. Suppose that at some point, an object
falls from the hand of the robot or off the conveyor belt, possibly
because it was not in the right place to be grasped properly. Unless the robot has been explicitly programmed in advance to detect
or handle such an error, it may blindly continue its task as if the
error had not occurred. Depending on the circumstances, the results of the error will range from negligible to catastrophic. If the
object lands over an already completed pallet, for example, the error is much more serious than if the object falls on the floor.

This chapter deals only with errors in the environment. They
tend to be more unpredictable and difficult to characterize with
mathematical models. The unpredictability makes them good candidates for the development of knowledge-based approaches. As Ger-
ardo Beni said [2], robots differ from computer peripherals, such
as disk drivers and printers, because robots communicate with each
other in physical space. A printer never needs to know where, in
real space, another printer is. It never needs to see or touch the
disk driver that sits next to it. From this necessity of physical
communication arises the complexity of robot programming.

Recovery from environmental errors is far more complicated than
recovery from simple controller errors, since even expected errors
can manifest themselves in unexpected ways. Instead of anticipating
errors, we will show how to use knowledge-based programming techniques so that the robot can autonomously exploit its knowledge to
detect and recover from failures.

The process of dealing with failures begins with *error detection*.
Once an error has been detected, the goal is *error recovery*. Error recovery has been defined [3] as "an ability in intelligent robotic systems to detect a variety of errors and, through programming, take corrective action to resolve the problem and complete
the desired process."

It is difficult to detect when something significant has occurred.
Many things that happen are usually reported to the computer controlling the robot, but not all of the events are significant. The
same event may be important in some circumstances and almost irrelevant in others. Deciding when something is important is the

first step in the error detection process. It also is difficult to detect the cause of the error and its effect on the robot environment. Not all errors produce immediate failures. When errors appear some time after what caused them happened, it is more difficult to detect the cause. Detection of error requires interpretation of sensor data. Sensors provide a huge amount of potentially useful data that must be interpreted to be of any use. Knowledge of what the program does helps greatly in the interpretation process. For example, the same error might call for a different recovery procedure depending on the circumstances and on the task being executed. Losing a part over an open grid might require human intervention if there is no replacement for the part or if the lost part will cause jams, but the error could simply be ignored if a large supply of parts is available and no damage is expected.

Another interesting facet of error recovery is that it must be sensitive to the recent history of the robot. Consider the following example. A robot is taking parts from a conveyor belt and putting them in boxes, six to a box. Suddenly, parts stop coming down the conveyor. The robot detects an error and traces it back to a missing part. The obvious recovery strategy is to wait a specified time for the next part to come and then continue the work. However, if the same error continues to occur, the robot should eventually stop waiting for parts, since the problem might be with the conveyor belt or with the supply of parts.

Several critical problems must be solved to do recovery in the most general cases. A sophisticated planning system is needed. The planner must be able to reason about available resources, their usage characteristics, their purposes, and their assignment. The same error might call for a different repair strategy depending on circumstances such as availability of spare parts. Different strategies might have to be explored and compared. For many problems the system must be able to acquire additional information from sensors [4]. Planning to acquire information is difficult because it requires the system to know what situations are critical to a task and how to detect them. Since the environment might have changed by the time the recovery procedure has been generated, methods of dealing with time and moving objects are needed. This includes, for instance, finding a free path for the robot in a changing environment and reasoning with time.

The interest in off-line programming makes automatic error recovery even more essential. The need for automatic and general-purpose error recovery will grow with the complexity of the applications and the software developed. A study done in 1982 by Arthur D. Little Company for the ICAM project [5] examined expected developments in robotics. The study grouped the developments according to whether they can be expected within the short term (2—4 years) or in the long term (3—6 years). The study suggests that real-time I/O access to external databases, integration with factory networks, and graphics capabilities should be expected in the short term. Automatic collision avoidance, recovery after unexpected events, and force sensing are expected in the long term.

Good error detection and recovery will improve robot performance, reduce damage and overheads, and thus increase global production per unit cost [6]. Operating automatic error recovery techniques would avoid the need of identifying all the relevant cases, thus leading to a simplification of the programming techniques. In addition, it could significantly reduce the amount of engineering required to run robots.

8.2 ROBOT RELIABILITY

Reliability is a serious problem in robot programming and a difficult one as well. Parts slip, fall, jam, and get misplaced; surfaces become wet or slippery; and operations fail. The range of potential problems is so vast that there are no standard engineering procedures that address them.

Many ideas used to improve the reliability of robots have been taken from the fields of software reliability and software safety [7—9]. A common approach to failure analysis and diagnosis is to apply techniques based on fault trees, event trees, or cause-consequent diagrams. It has been proposed to combine fault tree based failure analysis with rule-oriented reasoning [10—12] or to use fuzzy logic [13].

Techniques in computer system reliability deal with states of information, not states of the world, and are primarily directed toward restoration of internal data states, not physical conditions. Unfortunately, robots operate in the real world and the errors of interest are those manifested in the real world, not within the robot's software.

8.2.1 Problems with Robot Programming

Further expansion of robot use is impeded by several problems inherent in robot programming [14—16].

First, the robot must be provided with a model of the environment. The robot interacts with a real world that is too complex to be represented by a complete, exact model. Thus, it must be decided what information to include in the model. This decision depends on the task to be performed. For example, the operation of inserting one object into another could be described more appropriately by specifying a motion strategy than by providing a geometric description of parts. Knowledge about opposing faces of an object and nongeometric properties such as the coefficient of friction of the faces, or the center of gravity of an object, could be more important for grasping the object than a complete description of its geometry. To move parts with a robot, we need to know little more than their grasping position and destination. To do error recovery, however, additional qualitative knowledge might be necessary. For instance, before attempting to generate a strategy for grasping a part, it is important to know whether the object is small enough to be grabbed by the robot hand. The exact weight of a part is seldom important, but we might need to know if a part is light enough to be carried by the robot without causing the robot arm to break. A precise description of a surface property (e.g., roughness) is not as important as the fact that the part is slippery. The problem of modeling centers not only on the decision of what to model, but also the manner of representing this knowledge. Although the introduction of robot programming languages has represented a breakthrough in industrial robots, most languages are poor at describing models of the robot environment.

A second problem in robot programming involves the ability of the robot to detect and recover from errors. This process is difficult because there are discrepancies between the world model and the real world. To make matters worse, the robot itself has inherent inaccuracies (actions are not exactly reproducible), and there are real-time constraints. It is tricky, then, to determine when a slight discrepancy constitutes an error as opposed to normal variation, and whether a minor miscalculation will result in disruption or go unnoticed. Inappropriate decisions may be costly. Failures to detect an incorrect arm position may result in the arm crashing into a wall. On the other hand, a program that has worked well hundreds of times may fail because of a small difference in the size of one part, a difference that should have been insignificant. Intelligent programs should make the appropriate decisions in cases such as these.

A third area of concern in robot programming is the need to do as much work as possible off-line—that is, "off-line with respect to the robot," not "off-line with respect to the computer," as commonly intended in computer science. To control development costs, many robot users use robot programming languages to develop robot task descriptions off-line. When possible they test their programs off-line using simulators to further avoid tying up production line robots with nonproduction work. Off-line programming also allows robot programs to be developed before the parts that the robot will manipulate are available. This makes it possible to modify the design of the parts to ease the assembly process. Another advantage is that off-line programming can be performed in an environment more suited to programming than the factory floor. Larger computers can be used because they are not tied to a single robot. This opens the area of robot programming to applications of artificial intelligence techniques. Unfortunately, programs developed off-line tend to be unreliable and error prone. Most of the problems come from the lack of real sensor data, since sensor interaction can only be simulated [17]. Despite the potential problems, off-line programming is here to stay. A Delphi forecast of markets and technology published in 1982 [5] projects an increase in off-line programming from an estimated 6% in 1981 to 28% in 1990. Integration of multiple sensors and multiple robots into a total manufacturing system will certainly require powerful off-line programming capabilities.

A final concern in robot programming is the reduction of the burden on the programmer. Currently, programmers of robots must depend on their experience, intuition, and common sense in deciding how to describe the task and how to ensure its proper execution [18]. The programmer is responsible for anticipating all the possible errors and for determining the actions to take to recover from them. Since not all the possible errors can be considered, a long testing cycle is needed. The most common approach is to program the robot to recognize and handle any specific errors observed during the testing phase. This does not guarantee that the robot will be able to deal with every error it encounters while in production.

The problems we address in this chapter involve programs that fail in spite of being formally correct. We are not interested in the issue of errors in programming. We are not interested in the errors of the kind that any reasonable robot simulation system would detect.

8.2.2 Techniques for Error Recovery

There are two general approaches to error recovery: forward recovery, in which the system is transformed into a correct state following the point of the error, and backward recovery, in which the system is restored to an earlier state that is known to be correct.

Backward recovery is commonly used in data processing systems. A typical example is a system in which the program's state is saved periodically. If the system fails while the program is running, it can be restarted at the point of the last successful checkpoint. Another technique is the use of recovery blocks. Critical instruction sequences in a program are each placed in special blocks in which they are executed. The results are checked for validity, and if the validity check fails, an alternative instruction sequence retries the operation. Backward error recovery is very simple to implement because it requires no knowledge about the failure to work. It is attractive also because it can be used with unanticipated faults. However, in general it is difficult to restore a robot to an earlier state after an error because it is impossible to undo physical actions.

Forward recovery is a more complex method. The system must first diagnose the failure and determine the difference between the current state of the system and a desired state that the system should have reached. Then it must determine how to get the system from its current state to the desired state. Exception handling is the most widely used technique for forward recovery. Some programming languages, such as PL/1 and Ada, have specific instructions for exception handling. Forward recovery is the technique most often used in robotics [19].

Even though forward recovery is more complex to implement, it is easier to control a robot to perform actions to reach a desired state than to undo actions that have already been performed. In the real world most actions are irreversible, or there is no obvious way of reversing them.

8.3 DEALING WITH ERRORS: BACKGROUND

8.3.1 Dealing with Errors in Robot Programming

Errors are dealt with in different ways in the current state of the art in industrial robots [20].

If no sensors are used, the robot performs its task regardless of success. Even though this approach might appear to have little value, it is appropriate when the robot is very accurate or the task is very simple.

When sensors and programming are available, the programmer can incorporate into the program checks for every likely error and can program how to recover from it [21—23]. This is expensive both in engineering and in robot computational resources. It is also easy to forget something. Assume, for instance, that the program checks to ensure that a part has been grasped. During a transfer motion, the part could be lost. Even though this might happen only rarely, if no additional checking were done, the successive process could be severely disrupted. No matter how complex is the error checking, the robot might produce severe damage to itself or the workcell every time it encountered something the programmer did not expect.

Since it is difficult to consider all possible errors, many of which might never happen, another method is to generate all possible error conditions during the generation of the program from a task-level

description. This works well for most likely errors or for specific
tasks [24], but in general it is impossible or impractical to anticipate all possible errors.

Latombe and coworkers [24,25] use inductive learning from experiments. They save execution traces of several attempts to carry
out a task and generate from them a program that performs successfully the same task. For a given task, an initial sequence of actions
to achieve it is given by the user. This is called the ground plan.
After each motion in the ground plan has been executed by the robot, the system checks position and force sensory data to determine
whether the planned situation has been achieved. If that is the
case, the next motion in the ground plan is performed. Otherwise
rules are called to propose a corrective plan to achieve the desired
situation. The correction is patched into the ground plan and execution is resumed. Uncertainties and variations in the parts produce different execution traces. The ground plan may work well a
few times and then stop working because of a variation in the parts.
To be able to catch most of the errors, several executions of the
same task are usually required. Traces are described by linear
graphs, in which the nodes represent motions and the arcs represent situations. The induction process proceeds through iterative
transformations of the traces by merging nodes and arcs that are
equivalent. The result is a graph with branching points and cycles,
which represent the flowchart of the program. This method is certainly interesting but it requires the robot to be used for a considerable amount of time, since there are many conditions to be checked,
hence many trials. There are no guidelines for finding the conditions
and the cases to be checked, to identify all the relevant cases that
could produce errors. An additional drawback is that this method
requires large amounts of rules specific to the task. So far the
method has been applied only to the insertion of parts.

A different approach to the same problem uses skeletons of procedures and numerical methods to deal with uncertainties. Taylor
[26] developed a robot programming system that used the notion of
procedure skeletons to represent typical motion strategies in a parametrized fashion. The system produces sequences of motions by
selecting procedure skeletons appropriate to the task and filling in
parameters such as object locations and destinations. Then model
errors and tolerances are propagated numerically forward through
the robot task, checking for unacceptable levels of uncertainty.
The method computes the error bounds that guarantee correct execution of the task. Brooks [27] has developed a similar method that
we describe in the next section.

The group at the National Bureau of Standards [28,29] has developed a hierarchical control architecture called AMRF (automated
manufacturing research facility), for a small-batch metal machining

shop. The system is divided into three hierarchies: task decomposition, world model, and sensory processing. At each level, goals are decomposed into simpler goals for the next lower level. The sensory system updates the world model as rapidly as possible to keep the model consistent with the physical world. It remains to see how well the architecture can be applied to different, more complex domains.

8.3.2 Planning in Artificial Intelligence

Planning is a fundamental requirement of an error recovery system. Once an error has been detected and interpreted, the system must plan how the robot is to recover from the error.

Artificial intelligence research in planning makes a careful distinction between planning activities and the execution of the plan, often ignoring the latter. Current systems that combine planning with execution monitoring and replanning almost always do so only in simulation [30,31].

Traditionally planning systems include a plan generator and a plan executor [32]. The plan generator constructs a sequence of actions, called a plan, before starting to execute it. The plan is a sequence of primitive operations that are directly executable by the system. Usually execution is monitored to ensure that each operation produces the desired effects. If not, the system can return control to the plan constructor so that the plan is modified appropriately.

Cheeseman [33] describes a spectrum of planning systems depending on the amount of a priori knowledge and off-line sensor planning.

1. Most *classical planning* systems assume a perfect model of the world. They concentrate on the solution to the planning problem, which, as Chapman [34] has shown, is undecidable. Early attempts that made use of a planning system to control a robot [35] showed unexpected complexities even in carefully constrained environments. Uncertainties introduced by sensors and actuators, as well as errors in the world models used, cause plans to fail. In addition, the robot environment can change before the plan is executed. Even a simple change in the world can make the plan totally unusable, requiring extensive replanning. Since it is difficult (perhaps even impossible or at least impractical) to know all the consequences of an action, and since the world model can never be accurate, a different approach is to develop a plan assuming that everything goes well and to monitor its execution.

2. When planning systems were first used to control real robots, the need for *execution monitoring and replanning* became apparent immediately. Techniques have been developed to reuse as

much as possible of the original plan during replanning. This approach requires developing a model of how the world is supposed to be and a way of recognizing when it differs from expectations. The triangle tables used by STRIPS [35] are a method for encoding this type of knowledge.

3. *Conditional planning* has been developed for cases in which the number of alternatives in the plan is limited and the world remains the same no matter which part of the conditional branch is followed.

4. More recently, interest has developed in *sensor planning*. Monitoring the execution introduces the need of having to deal with sensors and to perform sensor interpretation. Since sensor interpretation is complex, it can be facilitated by sensor models and probabilistic sensor interpretation.

5. In many domains much of the information needed to generate the plan can be obtained during the execution. In these cases it is appropriate to use *deferred planning*. Parts of the plan are developed not at planning time but at execution time. This leads to interleaf planning and execution [36]. Calls to the planner are allowed in the plan. This approach seems to be appropriate to control real robots in real time. Integrating planning and execution requires representing the relationship between high level goals and information gathering needs, describing the sensors and their ability to satisfy the information gathering needs, maintaining a world model for use by the planning system, and deciding when to stop planning and start executing, and vice versa.

6. *Reactive planning* is based on precompiled procedures that are scheduled for execution depending on the circumstances at the time of execution. This allows the system to react to unexpected changes without becoming overcommitted to a specific plan. Recent work [31,37] in this area appears promising, despite many open problems in interpreting the sensor data.

Planning research has produced a number of relevant ideas and solutions (see survey, Ref. 32). Hierarchical planning techniques demonstrated in the NOAH system [38] and in SIPE [30] allow the planner to operate on several levels of abstraction. Nonlinear planning systems [39] generate plans to achieve a conjunction of goals without forcing more ordering on the steps of the plan than needed. The executor can then take advantage of the parallelism in the plan to optimize execution.

Some of the earliest experiments with automatic robot error recovery took place with Shakey, the mobile robot project at the Stanford Research Institute (SRI) between 1966 and 1972 [35,40]. Shakey could execute simple tasks involving a few objects in its environment of several connected rooms. Shakey was programmed using the STRIPS system, a planning system that took descriptions of what existed and what was desired and produced a sequence of instructions describing how to achieve the goal(s). STRIPS provided a language for describing the robot's environment and a desired configuration. The description produced a map of Shakey's environment, specifying usable paths among locations and the positions of relevant objects. When given a goal in terms of a desired configuration of the world, STRIPS would automatically produce a sequence of commands to Shakey to achieve the goal. STRIPS would produce task descriptions for Shakey in the form of triangle tables that specified plan steps to perform and preconditions to check beforehand. If Shakey encountered an error or some other unexpected event in its environment, the precondition checks would fail. At that point Shakey would stop and wait for STRIPS to replan its actions according to new knowledge derived from the unexpected event. The experiments showed unexpected complexities in dealing with the real world, even when the environment was carefully constrained. However, STRIPS is still one of the very few systems that have been tested on a real robot.

Research with the Jet Propulsion Laboratories' robot also addressed robot error correction [41]. As part of this work, Srinivas [41] developed a formalization of techniques to explain robot failures and to generate expectations of the effects of failures. The first of these techniques, failure reason analysis, used knowledge about why robot actions fail to generate a fault tree identifying possible reasons for a particular failure. The second technique, multiple outcome analysis, would generate a set of possible outcomes from a given failure along with an indication of sensor information that would remove ambiguities from among the possibilities. Unfortunately, Srinivas's system has been tested only in simulation.

More recent planning research is beginning to address plan execution issues. Research involving SIPE has examined the problems of monitoring execution and recovering from detected errors [30, 43]. Still the execution of plans is simulated, and the errors are simulated. Recent work in contingency planning also addresses problems related to error detection and recovery [44]. Progress has been confined to very general solutions and simulation-based systems.

These studies all make a number of assumptions: knowledge about events is correct, each action produces precisely defined postconditions, there are no uncertain data, correct predicates are generated from sensor data every time they are needed, and sufficient knowledge is provided to take into account all the possible states of the environment [30,45]. To be useful for robotics, a planning system must be able to deal with the inexactness of the real world, its geometry, and its noise. Reasoning and planning in AI have been developed in conjunction with cognitive tasks, mostly abstract and idealized. These assumptions are unrealistic for even simple robot tasks [46].

To be used in robotics, the traditional approach to planning requires sophisticated models of the environment (not only the geometry of the objects is important but also the frictions involved when objects come into contact, the uncertainties in the robot's and objects' positions, etc.) and models of the sensors (which sensors are available, what is their uncertainty, etc.).

The use of models allows the planner to complete planning before execution. Also by modeling uncertainties, the planner can generate plans that will succeed in spite of real-world uncertainties [27]. This approach has yet to prove its usefulness for real robot tasks. So far it has been applied only to simple tasks, such as insertions of pegs into holes [47,48] and in simulation [27]. When the environment is complex, it might be unrealistic to plan in advance for every possible event.

Doyle [4] has developed a system that expands any given plan by adding appropriate perception requests to it. The idea is to include in the plan only the perception requests that are important to verify the execution of the plan. He uses knowledge of what the sensors can do and decides which sensors are important to guarantee the correct execution of the task. The system has been implemented for a simulated robot.

An approach to programming robots and dealing with errors is to generate from the task-level description a program that is guaranteed to be correctly executed even in the presence of uncertainties in the environment. This requires models of robot kinematics and dynamics, and models of physical properties of objects such as friction. This approach has been applied to simulated fine motions for specific tasks such as insertion operations [47]. Handey [48] is a task-level programming system that produces robot programs and performs a graphic simulation of their execution. The system assumes an ideal robot and environment and does not deal with potential task uncertainties. Simulated sensors are used to locate the required parts in the simulated workcell. An experimental version of Handey has been implemented and tested on a real robot.

Brooks [27] deals with uncertainties and has a symbolic method to propagate errors. His method can be applied not only in the forward direction to propagate errors, but also in the backward direction to compute acceptable ranges of values on initial positions. Brooks is able to add sensing operations to the initial plan to provide more accurate information to the robot controller. His method is not intended to be a method to detect execution errors. His goal is to automatically determine whether a robot plan is feasible. If the program is feasible, there is no need for additional engineering; otherwise methods of reducing the uncertainty must be found. One method is to add appropriate sensing operations to reduce the uncertainty enough to guarantee success. Brooks performs the computations needed to propagate uncertainties in a symbolic way by using a system of constraints. The advantage of symbolic computations is that computations may be performed in two directions: the initial constraints can be fixed and the result computed, or the desired result can be given and the required initial constraints computed. So far his technique is applicable only to linear programs. Another problem is that errors could be easily overestimated because compensation phenomena are not taken into account.

Donald [49] studied the formal problem of error detection and recovery within the motion planning paradigms used by Lozano-Pérez. Modeling uncertainties [27] and taking into account errors in the model [49] help, but still the results will be as good as the models used. The real world is so complex that it may prove impossible to develop adequate models of it.

The fundamental problem with most AI approaches is in assuming that perception and action can be decomposed into two independent activities that interact only through a symbolic world model. To be more precise, the assumption most researchers have made is that after an action is executed, the perception system can update the world model to maintain its consistency; the planner than looks at the updated world model and decides what to do next or how to fix the current plan. Unfortunately, perception is too complex and the world too unpredictable. It is unrealistic to expect to maintain a consistent world model except in trivial cases.

8.3.3 Dealing with Errors During the Execution of Robot Tasks

A few research groups have attempted to design realistic systems that work with real robots.

The group at the University of Toulouse [50–52] designed and implemented a system, NNS, to control a manufacturing cell that includes planning and some error recovery. They have almost no sensing in their system except an industrial vision system, and the operations they perform are limited to pick and place. They update the workcell model after each failure, which is computationally expensive except for trivial cells. The system concentrates on using assembly knowledge during the assembly task itself. Tasks are specified as a sequence of changes in the state of the workcell. The structure of the system makes the intentions of the robot's actions clear during task execution, making it possible to automatically detect and recover from errors. NNS does not address the problem of task specification; tasks are apparently coded directly in terms of workcell state descriptions. While such a representation has clear advantages, it is at odds with customary industrial practices in robot programming.

The group of Lee and Hardy [53] at the University College of Wales have studied the problem of error recovery in conjunction with industrial robot tasks. They are working on an experimental system, called AFFIRM, for representing knowledge about the robot's task, workcell, and sensors. Tasks are encoded using the frame-based knowledge structure shared by the rest of the system. It appears that the work on AFFIRM concentrates on error recovery for a specific robot task rather than for a class of tasks [53].

Automatic error recovery work has been done by Nelson [54] on the sheep shearing robot. The recovery system for the sheep shearing robot is designed to apply specific recovery strategies to specific sensor readings. The system does not attempt to represent knowledge about errors and recoveries in a flexible fashion or to support inferencing. While the technique of directly coupling sensor readings to error handling greatly simplifies the system, it also restricts the system to handling errors that can be explicitly recognized by sensors.

Weisbin and coworkers [55] at Oak Ridge National Laboratory have recently developed a system that allows their mobile robot to respond to unexpected situations as it navigates an unstructured environment. The system is implemented using the real-time expert system shell Picon. It has been tested on simple navigation problems, and it remains to be seen how it will scale up to more complex tasks and more sophisticated sensors.

A method that combines task planning with reaction to sensor data, developed by Georgeff [37], uses knowledge about the task to guide the interpretation of sensor data. Georgeff [37,56] and his group have designed the PRS (procedural reasoning system) system to allow reactive reasoning. PRS keeps procedures instead of generating plans every time, so reducing the work the planner has to perform. This allows the system to operate in real time. Other planners take too long to replan, which means that reactions to unexpected events may come too late. The PRS system in a sense is more like an operating system than a planner.

Firby [31] has developed an interesting model based on the use of RAPs (reactive action packages). Actions are executed in a way that is adaptive, interruptible, and self-correcting. Plan selection is done entirely at execution time and is based only on the situation existing then. The problem with this approach is that the decision on what to do next is based on the assumption that sensors can update the world model after each action has been completed. Assuming that it is always possible to generate logical descriptions from sensor data is unrealistic.

Even these recent important contributions do not solve the problem of sensor interpretation of unexpected events. In fact, most unexpected events are simulated by human interaction via the keyboard, not interpreted by sensors [37].

Brooks [57] argues for robustness based on a highly distributed control and on a task-achieving behavior. His approach promises to be effective at least for some tasks. It is not yet clear how his architecture can incorporate high level planning, but using a task-oriented decomposition of tasks helps greatly in solving problems of sensor interpretation.

8.3.4 A System for Error Detection and Recovery

We proposed the application of artificial intelligence techniques to the problem of robot error analysis and recovery [58]. The basic notion is to use symbolic reasoning techniques to analyze the robot's activities and relate them to a symbolic model of the robot's workcell. Errors would then be detected by watching for sensor readings that indicate the unsuccessful completion of some required action. Automatic recovery could be effected by analyzing the differences between the expected and actual results of the robot's actions and applying planning techniques.

The approach to error detection and recovery that we propose here is more similar to the way people handle errors and unexpected events. By relating events to general knowledge, human beings can identify unexpected situations; similarly, by applying common sense and domain-specific knowledge, they can find solutions to situations never seen before. The key to human performance is in the knowledge about the environment and about the specific task at hand. We want to do something similar for assembly robots. Since the domain is limited and reasonably constrained, the amount of knowledge needed can be managed by using present technology [59]. This approach is different, even though not incompatible, with approaches based on generating robot programs that take into account every possible outcome of robot actions.

In our approach we start from a description of the robot task and we generate from it an executable description, that we call an augmented program (AP). The AP describes the sequence of actions the robot must execute, the sensor readings necessary to verify the performance of the robot, and information about how the robot's actions affect the environment. The AP is a finite-state machine in which state transitions are driven by events generated by the robot controller or by the sensor processing system.

The AP is executed by the AP executive, which interacts with the robot and the processes that handle sensors. The AP executive initiates physical actions, tells the sensor handlers what to look for next, and waits for significant sensor readings. Depending on the sensor readings, it decides what the next state will be. Sensors are used to verify the robot's proper operation and to detect errors. The AP executive maintains a history of the robot's activities called the event trace. The event trace is used for reasoning about errors and to keep logs of robot activities to improve future performance by predicting likely errors.

When an error is discovered, the recoverer takes control of the situation until a recovery procedure is generated and submitted for execution. The recoverer examines the state of the environment and updates the internal world model. To keep some consistency between the internal world model and the real world, tests can be run through the robot and the sensors. The recoverer examines the world model and the original plan to see what has been affected by the error and replans the recovery strategy.

The architecture of our system is shown in Figure 8.1. The architecture is simple, and we consider it to be appropriate as a general-purpose architecture for error detection and recovery. More details are given later in this chapter.

8.4 EXECUTION MONITORING, ERROR DETECTION, AND ERROR INTERPRETATION

Knowledge-based, multisensor-dependent task planning and task execution are, according to many researchers, the most important research problems in robotics [60,61]. Sensors should be included in the execution of the control structure for procedure parameter feeding, for feedback control loops, and for fault/emergency interruptions [62].

8.4.1 Dealing with Sensors

To perform an intelligent task, a robot needs to interact with its environment through its sensors. Often information about the environment must be obtained before decision making, planning, and verification of high level tasks can proceed. Since the environment is dynamic, it is impossible to store complete information a priori into the robot. Thus the robot must gather the necessary information through its sensors. The gathering of information could be accomplished by explicit information gathering goals, or it could be performed as part of the process of verifying the correctness of the execution of an action. Proper identification of the information gathering goals requires the ability to represent the relationships between high level goals and their implicit information gathering needs. Also sensors and their ability to satisfy the information gathering needs must be described.

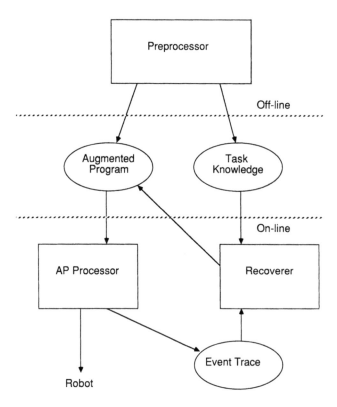

FIGURE 8.1 Error recovery system components.

Unfortunately, dealing with sensors is difficult. Most of the recognized work in AI has been done in computer vision. The research done in computer vision is so vast and complex that we will not even attempt to cover it in this chapter. Instead, we explore current trends dealing with sensors and with uncertainties, and in sensor interpretation [63].

A promising approach has been pursued at the University of Utah: Henderson and coworkers [64,65] proposed object-oriented models of the functions performed by physical sensors as well as the logical functions performed by the combination of sensors and programs. Sensors in their system are defined as abstract computational processes that operate on output from other sensors, which may be physical sensors or "logical" sensors. The definition of "logical sensors" provides a virtual interface between the planning and the sensor system, thus allowing changes to occur in the sensors without great effects on the planning system.

The basic research questions are as follows.

Decide which sensors to use: This requires knowing what sensors
   are available and what they can do [4]. Using the idea of defining logical sensors [53,64], each sensor (either physical or
   logical) can be described by its features (function performed,
   range, accuracy, repeatability, time needed to obtain the processed data, expected lifetime for data, etc.). The description
   of the sensors can then be used to select appropriate sensors for
   each task [4]. Particularly interesting tasks involve selecting
   sensors for normal execution (What are the most appropriate sensors? How can we combine values from multiple sensors? How
   can we detect inconsistencies in sensor data? What should be
   done when the appropriate sensors are not available?); selecting
   sensors for general monitoring and surveillance [What are the appropriate sensors for general surveillance (e.g., intrusion of a
   person in the working area of the robot) and for general safety
   (e.g., a drop in power?) How should these sensors be used to
   maintain consistency in the world model and to prevent failures?
   How often should we check them?]; and selecting sensors to obtain additional data during recovery (What are the appropriate
   sensors? How do we select them?).
Identify the relevant sensor values before executing the program:
   In particular, it is important to find out the relevant values for
   each sensor at each step during the execution and to determine
   the tolerance of the values. This is often called "expectation
   generation." How to obtain the values from the description of
   the task and of the objects is still an open research problem
   [66]. Absolute values and tolerance change during the task.
   For instance, the precision of reaching a point depends on having
   to perform an action there or on having just to go through an
   intermediate point, which depends on the fixtures and tools
   available. The problem of uncertainties is very important in robotics. Uncertainties are caused by the robot and the parts in
   the workcell. Precision of robots often is much smaller than
   their repeatability. The parts have tolerances, but most important is the uncertainty on position and orientations of parts delivered by feeding systems. Another important problem is that
   sensory data are also affected by errors. When a camera is used
   to identify a part in its field of view, we know that the position
   of the parts is affected by a measurement error.

8.4.2 Reasoning About Sensor Data

Sensors are fundamental to high performance robots, but they are
hard to work with. Because we want the robot to feel if it holds

an object in its gripper, or to see objects in the workcell, the robot is equipped with strain gages and cameras. But each camera
image is simply a matrix of dots that varies with lighting changes,
and the strain gages simply return analog force values along three
axes. It is difficult not only to select sensors and anticipate critical sensor readings but also to combine information from multiple
sensors in a meaningful way.

The fusion of data from multiple sensors is important. Specific
techniques have been developed for specific problems; certainty
grids, for example, have been used successfully to combine range
data from a variety of sensors to produce maps for a mobile robot
[67]. Chandrasekaran and Miller [63] approach the problem by em-
bedding it in a more general diagnostic process that reasons about
the sensed environment as a whole, leading to an application-specific solution. Paul et al. [60] propose a distributed architecture
in which sensors are independent agents that contain local expertise
and communicate perceptions through a blackboard.

Interpretation of sensor data is difficult unless thresholding techniques can be used. An example is the interpretation of grasping
forces. Depending on the object being grasped, it is possible to
select thresholds for grasping forces and to interpret the force values as meaning that there has been a touch, that the object has been
grasped, or that the object is being crushed. In many other cases
this one-to-one mapping between force values and their interpretation is impossible because the same forces have different interpretations.

Even more difficult is the interpretation of sensor data when it is
not known what they mean. An interesting approach is to use techniques from qualitative physics, advocated among other by Forbus
[68], but much remains to be done.

8.4.3 Error Analysis

During assembly, errors can be detected through a variety of sensor
readings or status indications. The range of readings indicating errors depends on the task at hand and the kinds of sensor and status
information the robot has available.

When errors are detected, an error interpretation process must
start to discover the source of the problem. A precise diagnosis is
important for an accurate, appropriate recovery. The raw data indicating that an error has occurred have little meaning without some
context. After detecting the error, the system must decide what
really occurred. For example, the real cause of the error may be
the misorientation of a part, the absence of a tool, or inaccuracy
in the arm position. The interpretation can be produced by
combining raw sensor data, the context and the semantics of the

instruction just executed, and knowledge about the effects of each instruction.

To do precise reasoning we must have a good causal model. In the context of robots this means that we must have a model about the actions. An example of a technique developed for this purpose is the failure reason model of Srinivas [42]. Knowledge about the effects of instructions helps in constraining the number of possible sources of errors.

The error interpretation process must combine raw sensory data, context, and semantics to derive the qualitative meaning of the error situation. In general it is not enough to know what instruction was being executed when an error occurred. The role or context of the instruction in the task is important in constraining the set of possible errors. The context can also eliminate certain other errors from the set of possibilities. For example, suppose that the robot was moving its arm and the gripper's touch sensor unexpectedly came in contact with something. This is an error because it deviates from our expectations. If the robot's intention was to try to grasp a part, the sensor reading indicates several possible qualitative errors: temporary wobble of robot gripper, misorientation of expected part, or collision with an unexpected object. On the other hand, if the robot was physically carrying a cube when the touch sensor is activated, the error could be total slippage of cube. This shows how semantics and context play important roles in classifying errors based on raw sensor data.

Error interpretation becomes more difficult as the complexity of the task increases, when errors are not detected immediately, and when there are multiple errors. For example, consider a task where a robot moves cubes from a feeder to a shipping pallet, a dozen at a time. What might happen if a cube falls from the gripper and lands on the pallet, knocking another object off? The failure reason model applies only to the objects and situations directly related to the sensor reading indicating the error. The robot thus associates an error with a part only if it uses its sensors on the part and finds an error. The lost part won't be missed until someone down the line tries to unload the pallet and finds it one part short.

A promising approach is to apply techniques of qualitative reasoning about physical space and motion [69]. Qualitative techniques give a powerful way of deducing possible problems in the workcell implied by sensor readings. These techniques also give us the tools to decide where and how we might use sensors to find out exactly what has happened in the work cell. DeKleer [70] demonstrated a system that used qualitative reasoning to solve physics problems involving an idealized roller coaster. Hayes [71] suggested that this qualitative, symbolic approach to physics problems might provide computable answers in more general situations in which customary quantitative approaches would fail. Forbus [72] demonstrated a system that analyzed interactions between bouncing balls in a two-dimensional world; the system could answer questions about potential collisions that were intractable by straight quantitative methods. These techniques have since been applied to a range of problems in physics and electrical engineering [73].

The important issues are summarized in the next two subsections.

How to Interpret Sensor Data

Sensor data are needed for situation assessment and sensor fusion, but they have no value unless they are interpreted. Given an unexpected sensor reading, how can the system determine whether it indicates an error or just a normal variability? This is more difficult than it seems because sometimes unexpected sensor readings do not really signal an error. The ability to recognize really unexpected events is important to enable the robot to detect errors that have not been considered by the programmer; but since variations are normal in robotics, it is difficult to detect when a change is significant as opposed to being a normal variation. The problem of uncertainties is very important in robotics. Uncertainties are caused by the robot and the parts. Each robot has a specified repeatability. Its precision often is not defined, or it is much less than the repeatability. The parts have tolerances, but most important is the uncertainty on position and orientations of parts delivered by feeding systems. Another important problem is that sensory data are also affected by errors. When a camera is used to identify a part in its field of view, we know that the position of the parts is affected by a measurement error.

Interpretation is done at different abstraction levels. Knowledge about the task and the environment is critical to decide what are significant data. Sometimes a solution is to get additional sensor data, but it is difficult to decide when to get them and from what sensors. Having models of expected sensor data helps in the interpretation, but matching models with real data is still largely an unexplored area. In addition, it does not help in interpreting really unexpected events.

How to Determine the Likely Causes of the Error

Knowing the cause is essential to recovery. The failure reason analysis used by Srinivas could be appropriate for simple cases, but it leads to combinatorial explosion. Qualitative reasoning techniques seem to be appropriate to generate a smaller number of candidate causes for errors. Additional knowledge about errors can also be used. Some errors are more likely to happen than others, and they should be checked first. Dealing differently with common and uncommon errors may reduce the complexity of the process.

8.5 ERROR RECOVERY

As we said earlier, it is important for the planner to have the ability to cope with failures at execution time; this allows the plan to be tidied up by undoing the failure-dependent parts and removing the steps in the plan already executed. The planner itself can generate a repaired plan to achieve the original goals.

There are several important issues in the application of planning. The planner must be able to reason about available resources, their usage characteristics, their purposes, and their assignment. The planner must be able to reason about the available operators (applicable operations) at more than one level of abstraction [30].

Another important problem is that of determining the degree of repair needed to recover from an error. More research is needed on recognizing the precise point in the task from which to continue, patch, or stop the repair plan.

It will be nice to have sophisticated strategies for replanning that depend on both user-defined and general criteria. Often different replanning sequences can be generated. Some are expensive to generate but easy to execute; others might be easy to generate but expensive in execution (e.g., a simple replanning strategy for assembly is to restart from the beginning and discard the partial assembly). Availability of parts, time, and cost of recovery are some of the criteria the planner should be able to reason about.

The basic questions (in addition to the questions already described) are:

Decide what has been affected: Not all errors produce immediate failures. An error can cause other errors. A failure that does not appear until some time after the error that caused it is more difficult to detect. Most of the times only a partial knowledge of the environment is available. Acquiring additional knowledge (e.g., through a camera) is not always the best solution because of time constraints. Qualitative methods to reason about the workspace and the parts affected can provide suggestions about the nature of the problem, permitting the acquisition of new knowledge to be focused on a limited number of areas.

Determine how to fix it: This requires determining the initial feasibility of repairing or recovering from the error and, if feasible,
developing a strategy for repair. To determine the initial feasibility, we must evaluate the following: Is the error catastrophic?
Have any important tools been lost or put out of operation? What
is the qualitative nature of the error (motor burnt, wobble)? Are
we now short of new materials for another reassembly? What additional resources are needed? How do we use them? What is
the nature of the repair (patch or complete overhaul)? What repair strategies do we have? Are there any user-defined heuristics or criteria to select (or override) our repair strategies?
Some repairs could be faster in time, while others may be less
expensive in other ways.

    A program that implements the recovery strategy must be con-
structed and submitted for execution. The appropriate reentry
point in the original program must be identified. If another failure occurs during the recovery, we must ensure that it is handled without ill effect. Common errors should be recognized and
canned recovery plans used. It is not obvious what should be
considered to be a common error and what appropriate recovery
plans should be. If we have too many, the system becomes slow
and gigantic; if we have too few, the system has to do a lot of
replanning.

Update the world model: It is important to maintain some consistency
of the world model after errors. This can be fairly complex for
moderately complex models, since there is the need to maintain a
distinction between what the system "believes" to be true because
of its own reasoning and what the system believes to be true because of sensing.

Explanation facilities: These are important to interface the system
with human operators, to keep logs of the system functioning,
and to justify conclusions reached and actions taken.

8.6 A SYSTEM FOR ERROR DETECTION
AND RECOVERY

Our approach [74] relies on a model of the world in which the robot
is operating, methods to interpret sensor data, and procedures to
generate recovery actions. We have designed and implemented a
system that does automatic error detection and recovery during the
execution of assembly tasks [75—77].

    We are interested in physical errors and faults in the robot work-
cell. These include the robots themselves, tools, feeders, and components. The errors we aim to detect include collisions, jammed

parts, gripper slip, orientation and alignment errors, and missing parts.

Our method relies on an extensive use of knowledge. It performs monitoring, interpreting, diagnosing, and planning. We start from a complete description of the robot task, from which a symbolic model of the world in which the robot operates is created. In addition, we do symbolic tracking of objects during the execution of the task, and we generate and maintain expectations about the meaning of robot actions in the environment. Sensor interpretation allows the robot to identify error situations, and causal reasoning is used to find plausible interpretations of errors. Replanning techniques are then applied to generate patches to programs to recover from errors.

Enough knowledge must be available to the robot to enable it to detect errors and recover from them, but we must be concerned about real-time requirements. Our solution calls for a system organized in two parts, an off-line phase and an on-line phase. The activities of extraction of intentions and model construction are done off-line with respect to the robot. The monitoring and interpretation of sensor data are done on-line with a robot monitoring system that is simple, fast, and independent of the robot. The diagnosing and planning are done on-line every time an error is found.

During the off-line phase the *preprocessor* generates from the user program an executable program (the augmented program, AP) that contains the necessary error checking. At the same time, knowledge is collected about the robot environment, the parts being manipulated, and the task, to be used during the recovery process. This information is used to predict where errors might occur and their nature to generate the recovery strategy. The user can interact with the system and provide additional information if desired.

The *AP processor* operates on-line every time the robot program is executed. It monitors the execution of the task to take into account unexpected changes in the environment, as well as inaccuracies in sensors and effectors. Differences between planned and real states of the environment will always happen and should be expected. The AP processor is able of real-time response to sensor data, and it has decision-making capabilities. When an error is discovered, the *recoverer* takes control of the situation until a recovery procedure has been generated and submitted to the AP processor for execution. Sensors are used to detect errors. Qualitative knowledge about sensor data and their interpretation is used to sift through the sensor information that is available, extracting only the most relevant data. The trace of events, with the intent of the program, aids in interpretation of any error that occurs and in the progress of generating the appropriate recovery procedure.

In addition, we use both general and specific strategies for error identification and recovery. General strategies are part of a knowledge base that contains general knowledge about assembly tasks and recovery. An example of a general strategy is "If an operation failed on a part, try it again with a new part of the same type." Specific strategies are task dependent and based on knowledge extracted from the task description. The same error might call for a different recovery procedure depending on the circumstances and on the task being executed. Losing a part would require human intervention if there is no replacement for the part or if the lost part would cause jams, but the error would simply be ignored if a large supply of parts were available and no damage was expected.

Our system uses a manipulator-level robot programming language to specify the task the robot is to execute. This description is given in the AL robot programming language [18,78,79], though any other manipulator-level language should work as well [23,80]. Even though AL is not used in any commercial robot, it has many features that other languages have adopted. We have expanded AL to handle descriptions of objects; thus it can be used to drive our graphic simulation system. Since we intend to experiment with existing robots, we will develop techniques for handling different programming languages. In particular we are interested in AML and VAL II, which are available in our laboratory.

The system assumes that this task description is accurate and correct in the sense that a robot simulator would execute it reliably. Thus the only errors the system should expect will be introduced by real-world uncertainties.

Additional details on the design of the system, its current state, and examples of problems we have solved with it can be found in Reference 81.

8.6.1 Program Preparation

During the program preparation phase, the system analyzes the robot's program off-line to generate a specialized description of the robot's task. This task description has two parts: the program describing the robot's task, and additional facts about the task's intentions and the physical objects (robots, parts, tools, etc.) the task uses. The first part, the augmented program, gives the sequence of actions the robot must execute, the sensor readings necessary to verify the performance, and information about how the robot's actions affect objects in the robot's workcell. The second part of the task description contains detailed information about the workcell layout and attributes of the robot and the parts. The model of the task and of the environment is generated from the application program with the help of a knowledge base that contains general knowledge about assembly.

```
begin
  open minihand to 5;
  move miniarm to frame (rot(xhat,-90), vector(4,20,2));
  center miniarm;
  move miniarm to frame (rot(xhat,-90), vector(14,20,12));
end;
```

FIGURE 8.2   AL program to pick up a cube.

A simple AL program and the corresponding AP are illustrated in Figures 8.2 and 8.3. The task described in the AL program is a simple pick-and-place operation. As we said before, the AP program is a finite-state machine. The states are numbered. Each state contains a collection of entries. The first describes the action the robot is to do ("robot-do") and the meaning of the instruction in the physical world. In particular, "expect" shows what will happen at the end of the state if all goes well; "imply" shows logical deductions about objects or the robot that can be made from the intentions extracted during the preprocessing phase. Each other entry specifies a condition to be checked using sensor data and the next state if the condition becomes true. Since the preprocessor generates the conditions using knowledge about the intention of each robot action, the same AL statement usually generates different conditions. As an example we can look at the states 2 and 4 in Figure 8.3.

The example is trivial but it provides enough details to show some of our techniques. The same example appears in Reference 76, where additional details can be found.

8.6.2   Program Execution and Execution Monitoring

The program generated by the program preparation phase is executed by a processor (AP processor), which in turn interacts with the robot. The AP is a finite-state machine in which state transitions are driven by events generated by the robot controller or by the sensor processing system. The system uses sensor information to detect errors and other unexpected conditions. Because monitoring of sensory information is a crucial part of the system and because this must occur in real time, separate processes will be developed to handle sensors. The AP processor initiates physical actions, tells the sensor handlers what to look for next, and waits for significant sensor readings. Depending on the sensor readings, the

```
(setq grab.ap
   '([1 ((robot-do open mini 5.0))
        ((open mini 5.0) 2)
        ((hand-error mini) 6)]

     [2 ((robot-do move mini (-90 90 90 4 20 2)))
        ((reach mini (-90 90 90 4 20 2)) 3)
        ((joint-error mini) 6)

     [3 ((expect grasp mini obj_block) (robot-do center mini))
        ((center mini) 4)
        ((hand-error mini) 6)]

     [4 ((imply grasp mini obj_block) (expect carry mini obj_block)
          (robot-do move mini (-90 90 90 14 20 12)))
        ((reach mini (-90 90 90 14 20 12)) 5)
        ((joint-error mini) 6)
        ((a-untouch mini) 6)
        ((b-untouch mini) 6)]

     [5 ((imply idle mini) (imply done) (robot-do end))]
     [6 ((imply error) (robot-do end))]]))
```

FIGURE 8.3 Augmented program obtained from the program in Figure 8.2.

AP processor decides what the next state will be. Sensors are used to verify the robot's proper operation and to detect errors. Information in the AP about sensor usage is used to sift through sensor input and extract relevant data. Relevant sensor data are used to invoke subsequent robot actions and to trace robot activities. Each effector and sensor is managed at the lowest level possible by its own module. Each module is independent. No horizontal communication between robots and sensors is allowed. Every communication will take place through the AP processor.

The system monitors the execution of the robot program and maintains a history of the robot's activities called the event trace. The trace of events is a sort of log of the execution of the task. It is used for reasoning about errors and to keep logs of robot activities to improve future performance by predicting likely errors. Information about the objects being manipulated by the program and a trace of the robot's actions with recent sensor readings are essential to the recovery process.

An event trace from the successful execution of the example shown before is illustrated in Figure 8.4. The first element in each line shows the time at which the entry was generated. Entries of the type "sense" show sensor values that have been sensed, entries of the type "expect" shows expectations, and entries of the type "imply" show logical deductions that the system makes using its current knowledge.

8.6.3 Error Interpretation and Recovery

Once an error has been detected, the recovery process can start using a trace of relevant events and whatever information is available about the task to determine the causes and effects of the error. Only after the cause for the error has been identified or, at least, after alternative plausible causes have been found, can the

```
(setq event-trace
    '((0 initialize)
      (0 new-state 1)
      (0 expect (open mini 5.0))
      (10 sense open mini 5.0)
      (10 new-state 2)
      (10 expect (reach mini (-90 90 90 4 20 2)))
      (27 sense reach mini (-90 90 90 4 20 2))
      (27 new-state 3)
      (27 expect (grasp mini obj_block) (center mini))
      (36 sense center mini)
      (36 sense open mini 4.0)
      (36 new-state 4)
      (36 imply grasp mini obj_block)
      (36 expect (carry mini obj_block) (reach mini (-90 90 90 14 20 12)))
      (48 sense reach mini (-90 90 90 14 20 12))
      (48 new-state 5)
      (48 imply idle mini)
      (48 imply done)))
```

FIGURE 8.4 Event trace from the execution of the program in Figure 8.3.

recovery process start. This information is then used to build an appropriate recovery procedure. The steps in the recovery procedure are appended to the existing AP, and the AP processor is restarted with the first step of the recovery procedure.

Figure 8.5 shows the AP recovery steps used to recover after the part carried has been lost. The failure recovery involves a simple heuristic applied to dropped parts. The heuristic is that parts fall straight down; thus a dropped part can be found directly under the place where the gripper detected that it dropped.

We are interested in a recovery process that is "recoverable"; that is, if something unexpected happens during the execution of the recovery procedure, the system should be able to recover from the new error without losing track of what it was doing. We are also interested in a recovery process that is able to obtain additional sensor data. Our choice for the representation of the program allows us to generate patches to an existing program as many times as needed and to generate additional instructions to gather new information.

```
(setq patch.ap
    '([7 ((robot-do open mini 5.0))
        ((open mini 5.0) 8)
        ((hand-error mini) 6)]

      [8 ((robot-do move mini (-90 90 90 8 19 2)))
        ((reach mini (-90 90 90 8 19 2)) 9)
        ((joint-error mini) 6)

      [9 ((expect grasp mini obj_block) (robot-do center mini))
        ((center mini) 10)
        ((hand-error mini) 6)]

      [10 ((imply grasp mini obj_block) (expect carry mini obj_block)
          (robot-do move mini (-90 90 90 14 20 12)))
        ((reach mini (-90 90 90 14 20 12)) 5)
        ((joint-error mini) 6)
        ((a-untouch mini) 6)
        ((b-untouch mini) 6)]]))
```

FIGURE 8.5 Recovery steps added to the AP of Figure 8.3 after the error.

If an error is detected, the monitor passes the control to the
recovery process that interprets the error by using the event trace
and general knowledge about errors.   After the error has been identified, the recoverer devises a recovery strategy.   In the process of
error interpretation, the recoverer also can run some tests through
the robot.   If a plausible recovery strategy is found, it is executed;
otherwise the robot's operator is informed of the error and given information about its cause.   Figure 8.6 shows an example of an error
that requires user's intervention.

We have implemented a method similar to failure reason analysis
[41,42].   We will integrate it with qualitative reasoning to make it
more effective and to reduce the number of possible causes.   We
execute each step in the plan and check for its correctness by
using only the relevant sensors.   This minimizes the use of sensors and also reduces the complexity of sensor interpretation.   We
rely on the description of the task and the environment to recognize expected sensor values.   So, the interpretation of sensor data
can be driven by expectations.

To minimize extra processing, we do not guarantee a consistent
world model.   We will keep a model of the world as accurate as

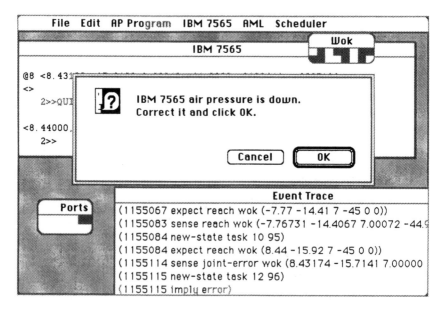

FIGURE 8.6   A request for manual intervention.

possible by using deductions and resorting to sensing when needed. Other groups [31,51] update the world model continuously. Computationally this can be a tremendous task. Since it is undesirable to synchronize the robot actions with the update process, it is also likely for the robot to use incorrect information. Updating the world model might require more time than it takes for the robot to do its task—for instance, when a vision system is used. Our system decides which sensor readings are important.

We do not volunteer information, in contrast with the approach taken by Drummond [44], who suggests, for instance, that whenever the robot carries an object for the first time, the robot should decide on its own to obtain the weight of the object itself. In general, it is unrealistic to assume that sensors can be read at any time without being calibrated and without knowing how to interpret the data. (The acceleration can affect the weight of the object, initial oscillations in the values of the force might be difficult to detect, etc.)

General solutions to problems of space and motion are complex and often intractable. We looked for effective results by restricting the problem in two ways. First, we deal with problems in the constrained environment of a robot workcell. Information is available at many levels of detail about the objects in the workcell. The number and locations of objects are constrained by the task at hand. The second restriction is that we are not trying to get exact solutions. The physical reasoning techniques simply tell the system what to look for; sensors are then used to find more exact information about location and orientation.

Our approach revolves around a qualitative structural model of the robot work cell. The model starts with the empty workcell described by the task knowledge base and uses the event trace to determine the workcell's current contents and configuration. Sensor readings from the event trace are used to constrain the probable locations of objects. Models in the task knowledge base of idealized parts provide assumptions that fill in additional details about the workspace. Intentions given in the event trace let the model represent the differences, if any, between expected and actual sensor readings. Physical laws (e.g., gravity, friction, conservation of energy) help identify uncertainties in the workcell, particularly when errors are detected.

8.6.4 Current State of the System

We have an experimental working system that controls an IBM 7565 Cartesian robot. The system includes an executive written in C and executed on a MacIntosh, and a replanner running on a VAX 11/780. The system has been tested on a variety of examples of pick-and-place tasks requiring the use of the force sensors in the fingers of the robot. We have successfully detected errors caused by a part dropped during a transfer, a part missing during a pickup, collision with unknown obstacles during a movement, and collision with objects during insertion into a fixture. We have limited our attention to recovery strategies that do not require undoing physical actions (e.g., unscrewing or unfastening). This is because not every action can be easily undone by reversing it.

We have also developed prototype components of the error recovery system using Franz Lisp on the Unix time-sharing system at the Artificial Intelligence Laboratory at the University of Minnesota. This system includes a robot simulator for testing purposes. We have a working implementation of a large subset of AL, called MiniAl [82], which allows us to do experimentation. A simulation system (MnCell) available in the Productivity Center of the University of Minnesota is driven by programs in MiniAl and will incorporate our execution monitoring system with simulated sensors.

8.7 CONCLUSION

Many robots have limited sensory and computational resources, and none are able to continuously develop a complete world model. Our system decides which sensor readings are important and which readings enable the system to detect errors. The knowledge is relatively contained, and, if properly represented, its use will not affect in a significant way the real-time execution of robot tasks. After an unexpected situation has been detected, however, more time will be taken to analyze it completely and to decide how to recover from it.

Our architecture for the monitoring system is general enough to accommodate different needs in intelligent control. The knowledge needed during real-time execution can be centralized into a monitoring system that becomes the intelligent part of the robot. The monitor supervises the task, but it delegates specific sensing and motor tasks to robot controllers and sensor processors. Such a modular construction allows this design to be used with a wide variety of existing controllers or sensors.

We can envisage many future expansions to our work. A learning system could be added to learn about likely errors and about successful error recovery strategies. Manufacturing faults and trends could be discovered. By observing the behavior of a robot for some time, human experts learn how to program it and how to avoid errors. It would be useful to use learning to make it easier to develop new programs.

The programming language AL could be be replaced by a task-level language when available [83,84]. We trust that our research will help in identifying needs in task-level descriptions by identifying the important knowledge that the robot needs to take decisions.

Monitoring malfunctions of other manufacturing systems will be a relatively simple extension. There are plenty of problems in manufacturing in which use of knowledge on-line can help solve monitoring problems and take decisions about corrective actions without the need for human intervention. Considerable problems arise in trying to monitor a complex and varied pattern of jobs. The factory of the future must be not merely adaptable through human intervention, but self-adaptive.

REFERENCES

1. D. J. Atkinson, "Telerobot Task Planning and Reasoning: Introduction to JPL Artificial Intelligence Research," in G. Rodriguez, Ed., *Proceedings of the JPL Workshop on Space Telerobotics*, Vol. 1, JPL Publication 87-13, Pasadena, CA, January 1987, pp. 339—350.
2. G. Beni and S. Hackwood, "Editorial," *J. Robotic Sys.* 1:2, 119—121 (Summer 1984).
3. S. Y. Nof, *Handbook of Industrial Robotics*, Wiley, New York, 1985.
4. R. J. Doyle, D. J. Atkinson, and R. S. Doshi, "Generating Perception Requests and Expectations to Verify the Execution of Plans," *Proceedings of the Fifth National Conference on Artificial Intelligence*, Philadelphia, August 1986, pp. 81—87.
5. D. H. Smith and R. C. Wilson, "Industrial Robots: A Delphi Forecast of Markets and Technology," *SME* (Dearborn, MI), 1982.
6. B. Carlisle, "Key Issues of Robotics Research," in H. Hanafusa and H. Inoue, Eds., *Robotics Research, The Second International Symposium*, MIT Press, Cambridge, MA, 1985, pp. 501—503.
7. A. E. Green and A. J. Bourne, *Reliablity Technology*. Wiley, London, 1972.
8. Nancy G. Levenson, "Software Safety: What, Why, and How," *Comput. Surv.* 18:2, 125—163 (June 1986).
9. N. H. Narayanan and N. Viswanadham, "A Methodology for Knowledge Acquisition and Reasoning Failure Analysis of Systems," *IEEE Trans. Sys., Man, Cybern.* SMC-17:2, 274—288 (March/April 1987).

10. M. L. Shooman, *Probabilistic Reliability: An Engineering Approach*, McGraw-Hill, New York, 1968.

11. D. J. Williams, P. Rogers, and D. M. Upton, "Programming and Recovery in Cells for Factory Automation," *Int. J. Manuf. Technol.* 1:2, 37—47 (1986).

12. S. Y. Nof, O. Z. Maimon, and R. G. Wilhelm, "Experiments for Planning Error-Recovery Programs in Robotic Work," *Proceedings of the ASME International Conference on Computers in Engineering*, New York, August 1987.

13. Y. Tsukanoto and T. Terano, "Failure Diagnosis by Using Fuzzy Logic," *IEEE Proc. Decis. Control*, 2, 1390—1395 (1977).

14. S. Bonner and K. Shin, "A Comparative Study of Robot Languages," *Comput. Mag.* December 1982, pp. 82—96.

15. T. Lozano-Pérez, "Robot Programming," *Proc. IEEE*, 71:7, 821—841 (July 1983).

16. U. Rembold and W. Epple, "Present State and Future Trends in the Development of Programming Languages for Manufacturing," in U. Rembold and R. Dillmann, Eds., *Computer-Aided Design and Manufacturing*, Springer-Verlag, Berlin, West Germany, 1986, pp. 279—322.

17. A. P. Ambler et al., "An Experiment in the Offline Programming of Robots," *Proceedings of the 12th International Symposium on Industrial Robots*, Paris, June 1982, pp. 491—504.

18. G. Gini and M. Gini, "Dealing with World Model Based Languages," *ACM Trans. Program. Lang.* 7:2, 334—347 (April 1985).

19. I. J. Cox and N. H. Gehani, "Exception Handling in Robotics," *IEEE Computer* 22:3, 43—49 (March 1989).

20. J. Y. S. Luh, "An Anatomy of Industrial Robots and Their Controls," *IEEE Trans. Autom. Control*, AC-28:2 (1983).

21. G. Gini and M. Gini, "Explicit Programming Languages in Industrial Robots," *J. Manuf. Sys.* 2:1 (1983).

22. B. E. Shimano, C. C. Geschke, C. H. Spalding III, and P. G. Smith, "A Robot Programming System Incorporating Real-Time and Supervisory Control: VAL-II," in K. Ratmill, Ed., *Robotic Assembly*, IFS Publications and Springer-Verlag, New York, 1985, pp. 201—217.

23. R. H. Taylor, P. D. Summers, and J. M. Meyer, "AML: A Manufacturing Language," *Int. J. Robotics Res.* 1:3 (1982).

24. J. C. Latombe, "Automatic Robot Programming," *Proceedings of the IFAC Symposium on Artificial Intelligence*, Leningrad, U.S.S.R., October 1983, pp. 273—282.

25. B. Dufay and J. C. Latombe, "An Approach to Automatic Robot Programming Based on Inductive Learning," in M. Brady and R. Paul, Eds., *Robotics Research*, MIT Press, Cambridge, MA, 1983.

26. R. H. Taylor, "The Synthesis of Manipulator Control Programs from Task-Level Specifications," Ph.D. thesis, Report AIM-282, Stanford Artificial Intelligence Laboratory, Stanford, CA, July 1976.

27. R. A. Brooks, "Symbolic Error Analysis and Robot Planning," *Int. J. Robotics Res.* 1:4, 29—68 (Winter 1982).

28. J. A. Simpson, R. J. Hocken, and J. S. Albus, "The Automated Manufacturing Research Facility of the National Bureau of Standards," *J. Manuf. Sys.* 1:1 (1983).

29. J. S. Albus, R. Lumia, and H. McCain, "Hierarchical Control of Intelligent Machines Applied to Space Station Telerobots," in G. Rodriguez, Ed., *Proceedings of the JPL Workshop on Space Telerobotics*, Vol. 1, JPL Publication 87-13, Pasadena, CA, January 1987, pp. 155—165.

30. D. Wilkins, "Monitoring the Execution of Plans in SIPE," *Comput. Intell.* 1, 33—45 (1985).

31. R. J. Firby, "An Investigation into Reactive Planning in Complex Domain," *Proceedings of the Sixth National Conference on Artificial Intelligence*, Seattle, July 1987, pp. 202—206.

32. Austin Tate, "A Review of Knowledge-Based Planning Techniques," AIAI-TR-9, Artificial Intelligence Applications Institute, University of Edinburgh, June 1985.

33. P. Cheesman, "Uncertainty and Planning: A Summary," in W. Swartout, Ed., DARPA Santa Cruz Workshop on Planning, *AI Magazine* 9:2, 124—127 (1988).

34. D. Chapman, "Planning for Conjunctive Goals," *Artif. Intell.* 32, 333—377 (1987).

35. R. E. Fikes and N. J. Nilsson, "STRIPS: A New Approach to the Application of Theorem Proving to Problem Solving," *Artif. Intell.* 2, 189—208 (1971).

36. Edmund H. Durfee and Victor R. Lesser, "Incremental Planning to Control a Blackboard-Based Problem Solver," *Proceedings of the Fifth National Conference on Artificial Intelligence*, Philadelphia, August 1986, pp. 58—64.

37. M. P. Georgeff and A. L. Lansky, "Reactive Reasoning and Planning," *Proceedings of the Sixth National Conference on Artificial Intelligence*, Seattle, July 1987, pp. 677—682.

38. E. Sacerdoti, *A Structure for Plans and Behavior*. American Elsevier, New York, 1977.

39. E. D. Sacerdoti, "The Nonlinear Nature of Plans," *Proceedings of the Fourth International Joint Conference on Artificial Intelligence*, Tbilisi, USSR, 1975, pp. 206–214.

40. Nils J. Nilsson, Ed., "Shakey and Robot," Technical Note 323, SRI International, Menlo Park, CA, April 1984.

41. L. Friedman, "Robot Learning and Error Correction," *Proceedings of the Fifth International Joint Conference on Artificial Intelligence*, Cambridge, Massachusetts, 1977, p. 736.

42. S. Srinivas, "Error Recovery in a Robot System," Ph.D. thesis, California Institute of Technology, 1977.

43. D. E. Wilkins, "Domain-Independent Planning: Representation and Plan Generation," *Artif. Intell.* 22, 269–301 (1984).

44. M. E. Drummond, "A Representation of Action and Belief for Automatic Planning Systems," in M. P. Georgeff and A. Lansky, Eds., *Reasoning About Actions and Plans: Proceedings 1986 Workshop*, Kaufmann, Palo Alto, CA, 1987.

45. M. J. Schoppers, "Universal Plans for Reactive Robots in Unpredictable Environments," *Proceedings of the Tenth International Joint Conference on Artificial Intelligence*, Milan, Italy, 1987, pp. 1039–1046.

46. M. Brady, "Artificial Intelligence and Robotics," *Artif. Intell.* 26:1, 79–121 (1985).

47. T. Lozano-Pérez, M. T. Mason, and R. H. Taylor, "Automatic Synthesis of Fine-Motion Strategies for Robots," *Int. J. Robotics Res.* 3:1, 3–24 (Spring 1984).

48. E. Mazer, J. Jones, A. Lanusse, T. Lozano-Pérez, P. O'Donnell, and P. Tournassoud, "Using Automatic Robot Programming for Space Telerobotics," in G. Rodriguez, Ed., *Proceedings of the JPL Workshop on Space Telerobotics*, Vol. 35, JPL Publication 87-13, Pasadena, CA, January 1987, pp. 130–150.

49. B. Donald, "Robot Motion Planning with Uncertainty in the Geometric Models of the Robot and Environment: A Formal Framework for Error Detection and Recovery," *Proceedings of the 1986 IEEE Conference on Robotics and Automation.* San Francisco, April 1986, pp. 1588–1593.

50. Rachid Alami and Helene Chochon, "Programming of Flexible Assembly Cell: Task Modelling and System Integration," *Proceedings of the 1985 IEEE International Conference on Robotics and Automation*, St. Louis, March 1985, pp. 901–907.

51. H. Chochon and R. Alami, "NNS, a Knowledge-Based On-Line System for an Assembly Workcell," *Proceedings of the 1986 IEEE Conference on Robotics and Automation*, San Francisco, April 1986, pp. 603–609.

52. Ernesto Lopez-Mellado and Rachid Alami, "An Execution Monitoring System for a Flexible Assembly Workcell," *Proceedings*

of the 1986 International Symposium on Industrial Robots, 1986, pp. 955—962.

53. N. W. Hardy, D. P. Barnes, and M. H. Lee, "Declarative Sensor Knowledge in a Robot Monitoring System," in U. Rembold and K. Hormann, Eds., *Languages for Sensor-Based Control in Robotics*, NATO ASI Series, Vol. F29, Springer-Verlag, Berlin, West Germany, 1987, pp. 169—187.

54. M. J. Nelson, "Automatic Error Recovery for Adaptive Robots," Technical Report 86/3, Department of Computer Science, University of Western Australia, Nedlands, Western Australia, September 1986.

55. C. Weisbin, "Self-Controlled: A Real-Time Expert System for an Autonomous Mobile Robot," *Comput. Mech. Eng.* September 1986, pp. 12—19.

56. M. Georgeff and A. Lansky, "Procedural Knowledge," *Proc. IEEE* 74:10, 1383—1398 (1986).

57. R. Brooks, "A Robust Layered Control System for a Mobile Robot," *IEEE. J. Robotics Autom.* RA-2:1, 14—23 (1986).

58. M. Gini and G. Gini, "Towards Automatic Error Recovery in Robot Programs," *Proceedings of the Eighth International Joint Conference on Artificial Intelligence*, August 1983, pp. 821—823.

59. Committee on Army Robotics and Artificial Intelligence (CARAI) et al., "Applications of Robotics and Artificial Intelligence to Reduce Risk and Improve Effectiveness," *Robotics Comput.-Integrated Manuf.* 1:2, 191—222 (1984).

60. Richard R. Paul, Hugh F. Durrant-White, and Max Mintz, "A Robust, Distributed Sensor and Actuation Robot Control System," *Proceedings of the Third International Symposium on Robotics Research*, MIT Press, Cambridge, MA, 1986, pp. 93—100.

61. Russell H. Taylor, James U. Korein, Georg Maier, and Lawrence F. Durfee, "A General-Purpose Control Architecture for Programmable Automation Research," *Proceedings of the Third International Symposium on Robotics Research, 1985*, MIT Press, Cambridge, MA, 1986, pp. 165—173.

62. C. C. Geschke, "A System for Programming and Controlling Sensor-Based Robot Manipulators," *IEEE Trans. Pattern Anal. Mach. Intell.* PAMI-5:1, 1—7 (January 1983).

63. B. Chandrasekaran and Don W. Miller, "An Artificial Intelligence Approach to Sensor Conflict Detection," *Proceedings of the International Topical Meeting*, American Nuclear Society, City, Pasco, WA, September 1985.

64. T. Henderson and E. Shilcrat, "Logical Sensor Systems," *J. Robotic Sys.* 1:2, 169—193 (1984).

65. T. Henderson, C. Hansen, and B. Bhanu, "The Specification of Distributed Sensing and Control," *J. Robotic Sys.* 2:4, 387—396 (1985).

66. A. A. Requicha, "Toward a Theory of Geometric Tolerancing," *Int. J. Robot. Res.* 2:4, 45—60 (1983).

67. A. Elfes, "Sonar-Based Real-World Mapping and Navigation," *IEEE J. Robotics Autom.* RA-3:3, 249—265 (1987).

68. K. Forbus, "Measurement Interpretation in Qualitative Process Theory," *Proceedings of the Eighth International Joint Conference on Artificial Intelligence*, August 1983, pp. 315—320.

69. K. D. Forbus and D. Gentner, "Learning Physical Domains: Toward a Theoretical Framework," in E. R. Michalski, Ed., *Machine Learning*, Vol. 2, Kaufmann, CA, 1986, pp. 311—348.

70. J. deKleer, "Qualitative and Quantitative Knowledge in Classical Mechanics," AI-TR-352, MIT Artificial Intelligence Laboratory, Cambridge, MA, 1975.

71. P. Hayes, "The Second Naive Physics Manifesto," in R. Brachman and H. Levesque, Eds., *Readings in Knowledge Representation*, Morgan Kaufmann, Palo Alto, CA, 1985.

72. K. D. Forbus, "Qualitative Reasoning About Space and Motion," in Dedre Gentner and Albert L. Stevens, Eds., *Mental Models*, Erlbaum, Hillsdale, NJ, 1983, pp. 53—74.

73. D. G. Bobrow, Ed., *Qualitative Reasoning About Physical Systems*. MIT Press, Cambridge, MA, 1985.

74. M. Gini, "Symbolic and Qualitative Reasoning for Error Recovery in Robot Programs," in U. Rembold and K. Hörmann, Eds., *Languages for Sensor-Based Control in Robotics*, NATO ASI Series, Vol. F29, Springer-Verlag, Berlin, West Germany, 1987, pp. 147—167.

75. M. Gini, R. Doshi, R. Smith, and I. Zaulkernan, "The Role of Knowledge in the Architecture of a Robust Robot Control," *Proceedings of the IEEE International Conference on Robotics and Automation*, St. Louis, Missouri, 1985, pp. 561—567.

76. R. Smith and M. Gini, "Reliable Real-Time Robot Operation Employing Intelligent Forward Recovery," *J. Robotic Sys.* Summer 1986, pp. 286—301.

77. M. Gini and R. Smith, "Monitoring Robot Actions for Error Detection and Recovery," in G. Rodriguez, Ed., *Proceedings of the JPL Workshop on Space Telerobotics*, Vol. 3, JPL Publication 87-13, Pasadena, CA, January 1987, pp. 67—78.

78. T. Binford, "The AL Language for Intelligent Robots," in *Languages et Methodes de Programmation des Robots Industriels*, IRIA Press, Paris, France, 1979, pp. 73—88.

79. M. S. Mujtaba and A. Goldman, "AL User's Manual," Stanford Artificial Intelligence Laboratory Memo AIM-323, Stanford, CA, January 1979.
80. B. E. Shimano, C. C. Geschke, and C. H. Spalding III, "Vol. II: A New Robot Control System for Automatic Manufacturing," *Proceedings of the IEEE International Conference on Robotics*, Atlanta, Georgia, 1984, pp. 278—292.
81. R. Smith, "An Autonomous System for Recovery from Object Manipulation Errors in Industrial Robot Tasks," Ph.D. thesis, University of Minnesota, 1987.
82. M. Endo, "Implementation of MINIAL," Master project, University of Minnesota, 1984.
83. L. I. Lieberman and M. A. Wesley, "AUTOPASS: An Automatic Programming System for Computer Controlled Mechanical Assembly," *IBM J. Res. Dev.* 21:4, 321—333 (1977).
84. R. J. Popplestone et al., "An Interpreter for a Language for Describing Assemblies," *Artif. Intell.* 14, 79—107 (1980).

9
Databases for Planning and Manufacturing

KLAUS R. DITTRICH, ALFONS KEMPER, and PETER C. LOCKEMANN
University of Karlsruhe, Karlsruhe, Federal Republic of Germany

9.1 INTRODUCTION

The "integration" in computer-integrated manufacturing (CIM) is a challenge to both technical and organizational talents. On the technical side, integration is accomplished by two basic information technologies: communication and databases. Communication facilities serve the immediate interaction between the various, spatially distributed agents (persons and machines) within the manufacturing process, whereas databases serve as a kind of "buffer" to preserve information until it is needed for communication, often again and again, by unknown agents after unknown periods of time.

Because so many agents generate and utilize common information— design processes, planning processes, manufacturing processes, quality control, inventory control, marketing—and because of the length of time over which new information is produced, CIM databases tend to become gigantic repositories. Database technology must, therefore, provide the means for effective and efficient, timely and reliable maintenance and update of and access to databases. In addition, database technology must ensure that data bases remain effective communication media even if years pass between depositing a piece of information by one agent and picking it up by another. Furthermore, even though a database—like the agents—may be spatially distributed, it serves its communication function only if it acts as a single coherent database.

Robots are one kind of agent within the manufacturing process. One may argue that robots by themselves deal with relatively small amounts of data that are of more than passing relevance and, as such, have little motive for using large repositories and the techniques that these entail. But even if this were so, as participants in wider communication they will encounter the database facilities, and hence should influence the communication standards and technical solutions established by the corresponding database systems. Moreover, as we shall demonstrate in this chapter, there are some database tasks that are clearly identifiable with robotics and as such require specific solutions. Nor does database technology restrict itself to the management of large-volume sets of data. It also ensures long-term viability of the data, controls the consistency of data upon changes to the database, and keeps processes that simultaneously utilize the same data from interacting with one another. All these are capabilities that are important even if the database itself is relatively small.

What tasks in connection with robotics are suggestive of database support? Two areas come readily to mind.

1. *Planning*: Before a robot can be installed, it must be programmed to perform a predetermined set of duties. No matter how adaptive and flexible the resulting functions, a certain amount of preplanning is necessary. This is especially true for robot-based manufacturing cells that are configured by multiple robots, peripherals, and numerically controlled (NC) machines, all of them interacting to perform a manufacturing task. Planning is mostly done off-line and may be supported by sophisticated programming environments encompassing the following components.

A geometric design processor for interactive graphic design of objects such as mechanical parts, transport facilities, and layouts of manufacturing cells. This is a typical computer-assisted design (CAD) task; thus some or many of the data may originate from other design processes and be drawn from other CAD databases within the communication network.

A robot design processor for the description of the robots to be used including, beyond the geometries, the topologies and dynamic and kinematic properties. Again, these data may be fetched from a library within the network or even from the robot manufacturer. Data from both these design processes comprise what is called a "world model database."

A robot emulator system that allows one to plan a robot program and then to check the program off-line (i.e., without physically manipulating the robot) for correct behavior. This is done by simulating the robot operations, using the world model in the database.

A compiler that translates the checked-out robot program into a corresponding program directing the robot motions. Depending on

the built-in flexibility, fewer or more data from the world model
may be incorporated into the program. The compiled program
may again be deposited in a program library.

2. *Manufacturing*: Simple robots performing singular tasks may
be provided with a single program from the library. With increasing
intelligence and functional breadth of the robots, more flexible approaches will be needed:

Downloading of precompiled programs to establish one of several
function sets within the robot.
Provision on demand of data from the world model database, to permit the robot to recognize and differentiate situations that require a specific response; in this case the program incorporates
a fair amount of interpretation of world model data.
Acquisition by the robot of long-range data that are to be collected
in the world model database for further planning, quality control,
or error control, or to be sent on to other databases within the
network, where they may be utilized for statistical or accounting
purposes and the like.

The short summary clearly demonstrates that robotics cannot be
divorced from database technology. Robots communicate with CIM
databases directly or indirectly. We also notice, however, that a
particularly important role is assumed by the world model databases.
We shall concentrate our discussions in this chapter, therefore, on
this type of database. Moreover, because of the essential role the
planning of the robot actions plays, we confine our examples to robot programming; more precisely: we focus our attention on robot
programmers who impose their views of the robotics world on the
programs they write.
With high level language programming, then, both the language
constructs and the database view must be in terms of the terminology and structures that are natural to the world of manufacturing
processes. Extraneous details such as translating statements into
control signals or structural descriptions into appropriate storage
structures, placing data into and retrieving data from the proper
locations in main memory or on peripheral storage, and achieving
the necessary response times should largely be kept from the programmer. Likewise the programmer should be able to rely on such
typical database characteristics as maintenance of consistency, multiuser access synchronization, and database recovery, without explicitly taking the necessary precautions.
Hence, this chapter devotes much of its attention to structuring
and accessing the world model database in an application-specific
fashion. The other issues are covered only to an extent that permits

the reader to appreciate their importance and to be able to judge whether commercial database products that may be encountered meet these requirements. Unfortunately, no manufacturer of a database system can afford to gear a product to one particular application, but will instead provide general-purpose interfaces. Consequently, the chapter discusses how to adapt existing interfaces to the purpose at hand.

After a brief survey of database system characteristics and the introduction of a running example, we start with the adaptation of a classical state-of-the-art interface, the *relational* data model. We continue with a representative of modern approaches to *structurally object-oriented* data models and conclude with its extension to a level at which many of the operations within the manufacturing cell are directly incorporated in the data model in the form of operations on the corresponding data structures. This level of support is called *behavioral object orientation*.

9.2 DATABASE BASICS

9.2.1 Characteristics of Database Systems

Since we assume that most readers are not database experts, this section presents the basic database concepts that are referred to throughout. A *database system* (DBS) is a collection of stored data together with their description (the *database*) and a hardware/software system for their management, modification, and retrieval (the *database management system*, DBMS). Eight key features are commonly considered to make up the function and quality of a database system:

Data integration
Application-oriented data structuring
Consistency
Multiuser operation
Recovery
Transactions
Protection
Data independence

Database systems were conceived precisely because of the need for *data integration*. The data of an entire enterprise are administered under central control in a nonredundant way. Changes to the data by one application program become instantaneously visible to all other applications as soon as they have been released. Data may be accessed with equal ease by all applications, provided they are entitled to see them. The technical solutions for integration may vary—some data may be replicated for performance reasons, and the data may be distributed across a network—but these remain transparent to the applications. Take as an example a workpiece that has been completed by the design department. No matter where the data are physically located, a robot programmer could pick them up to incorporate them into a new robot action.

The way in which data are structured when stored into and retrieved from a database should come as close as possible to the concepts and notations of an application, without concern for such physical storage considerations as blocks, tracks, or bytes, or operating system parameters such as file access methods or buffer sizes. Unfortunately, there is a potential conflict between the requirements of a particular application and the need for an enterprise-wide communication standard. *Application-oriented data structuring* means finding a middle ground between these needs. Typical examples are determining structures that accommodate both boundary representations and solid models, or 2D and 3D representations.

As far as possible, a database system guarantees that the data it manages are as authentic as possible an image of (are *consistent* with) the application world. In other words, a database models the so-called *miniworld*, the facts that are considered to be of interest for a given application. For example, if a geometric piece has been declared to be a cube, the changes that can be performed on it are severely limited.

A consequence of integration is that many users must be expected to access the database at the same time. *Multiuser operation* gives the user the impression of currently being the only one on the system. Thus access to data must be free from interference from *concurrent* users; that is, it must be properly *synchronized*. Take two programmers correcting two successive assembly steps in a robot program. If they do it in turn—no matter what order—the second will build on the completed results of the first. However, if they act simultaneously, without knowing of the other, the outcome may be a disaster!

Large quantities of operational data represent valuable assets for the enterprise. Therefore, such data must be guaranteed to be *persistent*; that is, a database system should provide *recovery* mechanisms that prevent the loss of information in cases of hardware and software malfunction or when user operations are aborted. Consider a hard disk failure with a concomitant loss of parts of the world model data used to control the operation of a manufacturing cell!

Because such simple database operations as updating a single data item usually do not reflect meaningful application semantics, a concept is offered to define comprehensive units of work (called *transactions*). The database system then maintains certain properties with respect to consistency, concurrency, and persistency for

the entire sequence of operations that make up the transaction. For
example, a robot programmer constructing a particular path of a
gripper may not wish to preserve or make visible his design until
he has convinced himself of its correctness.

Protection mechanisms make sure that only properly authorized
accesses to the database can be executed; they also include a component to define the authorization rules desired for a specific application environment. Take again our world model database: only
a few persons may enter new geometric designs into it, even fewer
may modify them, and the robot programs are not permitted to make
changes to them, just read them.

Finally, database systems are constructed to provide data *independence*. With *logical data independence*, an application program
is restricted to exactly the data it needs, and in the desired format (we say that the program is provided with a specific *view* of
the database). In this way the program becomes immune to any
changes to the database outside its own view. Consider the world
model database, which certainly reflects just a small portion of the
corporate database but is not affected by changes, say, to the production schedules. *Physical data independence* refers to the way
data are physically stored and to the strategies by which they are
accessed. For example, robot programs need not be changed or re-
compiled when new storage media are added to the database system configuration, or a new access technique is included to speed
up retrieval.

9.2.2 Operation of Database Systems

As we noted before, it is by way of data integration that database
systems provide a common communication base. In turn, communication enforces certain standards—referred to above as application-
oriented data structures—on the agents. As a consequence, the
miniworld semantics need no longer be hidden in the application
programs. Instead, the semantics must be explicitly represented
in the database itself. All together, there are three distinct sources
that determine the database semantics:

1. A set of rules that determine the organization of the database
   items into data structures.
2. Rules for expressing miniworld semantics that go beyond mere
   data organization.
3. Data items as supplied by the users or their application programs, that are fitted into the organizational scheme.

Let us start with step 1, the fact that a DBMS interface should
be general enough to accommodate many diverse applications elicits

a two-phase process for setting up the data organization. During
the first phase an appropriate *data model* is selected. The data
model determines the DBMS interface, and it offers a set of basic
organizational principles in much the same way as a programming
language offers basic constructs for controlling the steps in which
a program is to be executed, or the structures into which data
items are to be assembled. Correspondingly, a data model defines:

Which *types* the elementary data items may have (i.e., which *value
sets* for data items the system recognizes) and how these data
items may be used; typical examples are strings and real numbers.

Into which *structure* types data items may be assembled; an example may be the principle of organizing items into rows and
these into tables.

Which *restrictions* may automatically be observed for structures;
consider again tables, where each row may be uniquely identified, or a table may not exceed a given length.

The *operators* that may be applied to the database to insert, modify, delete, and retrieve data items. Because the data model
deals with organizational principles only, the operators must be
defined universally for all data items and structures that obey
these principles. For example, the operators should be able to
deal with any kind of table. Operators with this generality are
called *generic*.

The set of operators forms the basis of the *data manipulation
language (DML)* of a database system. There are several options
for presenting the DML to the user: a stand-alone DML can be
used for interactive access; more important to robot programming,
however, the DML may be embedded in one or more *host languages*
(conventional programming languages like PL/I, FORTRAN, and
Pascal).

During the second phase of step 1, the organizational principles
are applied to construct the actual data structure types into which
the data items that go into the database are to be assembled. Take
again our data model from above, which permits the construction of
arbitrary table structures with columns whose values are elements
of some value set. We must now decide which specific tables we
need and what they will look like. A table is determined by giving
it a name, declaring the number of columns and their headings, and
specifying a value set allowed for every column. Each table sup-
posedly reflects a particular part of the miniworld semantics. For
example, in geometric design we may have tables for expressing
points in space, edges, surfaces, cuboids, and so on. The set
of data structure types thus defined forms a *database schema*.

Operators for the definition of schemas are combined into a *data definition language (DDL)* that resembles the declaration facilities of higher level programming languages. Although some DBMSs do not require that the entire schema be defined before any data manipulation takes place, DML operators are possible only for structures that have been properly declared.

Not all the miniworld semantics may be expressible using the principles of the data model. For example, in our tabular model we may formulate in the schema how each row is to be uniquely identified. But it usually is not possible to give an upper limit for the size of a table. Even worse, there is no way to enforce that once a cube has been entered as a set of surfaces, and these as a set of edges and vertices, its vertices may not be changed individually but only in totality, to preserve the properties of cube. Hence, in step 2 one would have to add explicit rules that govern restrictions of this kind, provided the DBMS offers the appropriate facilities. The rules are mostly formulated as predicates over data items or structures; we refer to them as *consistency constraints* or *semantic integrity constraints*.

Only after all this has been set up are we ready to proceed to step 3, insert data items into the database and later retrieve, modify, and delete them, using the DML. In our example, we now have reached the point where we may build up a database of geometric designs, robot descriptions, and the like, all put into the form of tables.

If we assume that all the CIM databases in our network look like a single centralized communication base, or if there is indeed only a single central database, then the database schema to be designed is the one that governs the central base. Seen from a different angle, the central database schema is *the* basic communication standard for all cooperating applications. However, as mentioned in Section 9.2.1, an individual application usually needs only part of the structures and data items defined in the schema. For example, the geometric processor may elect to ignore all the information on robots. The application may even want to see its part organized differently from what the central schema dictates (e.g., with some columns of a table omitted and the remaining ones reshuffled). To this end, database systems provide the concept of *subschema* or *view*. There are specific means in the DDL to define various subschemas for one schema, and in fact, applications are often forced by the DBMS to refer to a subschema rather than the schema itself.

9.2.3 Database Implementation

What the user sees in terms of the DDL, DML, schema, and operations must ultimately be translated into storage structures on peripheral storage media and programs manipulating them. This is a complicated process that is responsible for the large size of database systems. All together, a stored database contains various classes of data:

User data as supplied via DML operations.

Schema data as supplied via DDL operations.

System data that are created and maintained by the DBMS to handle all necessary transformations.

Administrative data (e.g., consistency constraints or authorization rules).

Access paths that are constructed and maintained by the DBMS to perform DML operations more efficiently (e.g., indexes to retrieve data according to particular, frequently used criteria without having to scan the whole database sequentially).

Whereas the user data comprise, from a user's point of view, the only data useful to him (*primary data*), the other classes (*secondary* or *auxiliary data*) often make up a significant portion of the entire database.

All these data are the subject of *storage structure management*. To a certain extent users may influence the translation of the logical data structure as seen by them to the storage structures, and they may also determine which access paths are to be maintained by the DBMS. The corresponding activities, referred to as the *physical database design*, as well as the task of specifying the database schema and the consistency constraints (*logical database design*) are usually the responsibility of a group of specialists called the *database administration*.

A second component of a DBMS is *transaction management*. It is responsible for all aspects associated with the user transactions including observance of consistency rules, synchronization of transactions, recovery of transactions and database contents, and database protection. Often the schema data are not maintained in the database system itself but are transferred to a separate system, a *data dictionary*, where they are kept together with other data concerning design and operation of the database system.

9.2.4 Further Reading

Readers whose appetite has been whetted are referred to several modern books [4,18,25,26] that give an in-depth introduction to database systems.

9.3 RUNNING EXAMPLE

From the vast amount of information that must be processed in engineering applications, let us just pick one representative example:
the *manufacturing cell*. Among other objects, a manufacturing cell
consists of robots and assembly items (i.e., rigid solid objects).
Clearly, a world model database supporting CAM applications must
store computer models of all objects present in a manufacturing
cell. Therefore, in this section we first analyze the data modeling requirements and then investigate the standard operations to
manipulate these objects.

9.3.1 Modeling Data in CAM Applications

Representation Schemes for Rigid Solids

The basis for any model of a manufacturing cell is a way to store
information about geometric objects in a computer. There are several—quite different—representation methods for solid objects [15,
20].

The two representation schemes for which database support is
feasible [15] are constructive solid geometry (CSG) and boundary
representation (BR).

The CSG scheme is, together with boundary representation,
the most widely used in existing systems. Whereas boundary representation supports in particular the automatic manipulation of objects,
the CSG representation is mostly used to support the input of geometric objects by the user of the geometric modeling system.

Constructive Solid Geometry

The CSG scheme is a volumetric representation of geometric objects. In this approach an object is described as a composition of
a few primitive objects. The composition is achieved via motional
or combinatorial operators. Example operators are the (regularized)
union, intersection, and difference of two solid objects. Motional
operators are, for example, rotate and scale. The description of
a geometric object in CSG format forms a tree where the root of
the tree represents the solid object being modeled.

In Figure 9.1 we show the CSG tree for an example object
"bracket with 4 holes." In the CSG tree each nonterminal node
represents an operation, either a rigid motion or a combinational
(set) operator. Terminal nodes represent either a motion parameter or a primitive object. Each primitive object is described by
its parameters, such as length, width, and height, as well as its
relative position. Since a geometric object must be decomposed into primitive objects (corresponding to the leaves of the tree), the

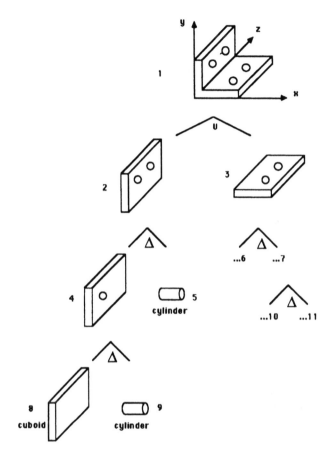

FIGURE 9.1   CSG tree of the bracket.

CSG tree of a complex mechanical part can become very high.   This might lead to inefficient data retrieval if there is no suitable data access support.

Boundary representation

In the boundary representation scheme a solid object is segmented into its nonoverlapping faces.   Each face, in turn, is modeled by its bounding edges and vertices.   We present the representation of a cuboid in Figure 9.2.

From a database point of view, we note that this representation scheme consists of different abstraction levels (i.e., faces, edges,

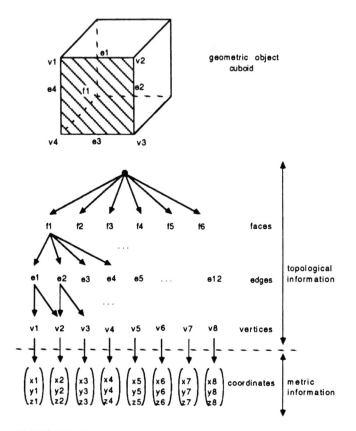

FIGURE 9.2   Boundary representation of a cuboid.

and vertices).   In contrast to the CSG scheme, the height of the tree is constant, that is, 3. A more complex solid object just leads to more nodes in the tree without increasing the height.

The lowest level of the tree stores, the metric information, that is, three-tuples $(x_i,\ y_i,\ z_i)$ for vertex v_i, for i in $\{1,\ldots,m\}$.   The second level of the tree stores the edges as combinations of the vertices.   Edge e_i is represented by the tuple (v_{i1},v_{i2}), where i is in $\{1,\ldots,n\}$.   On the next level of the tree each node describes a *variable* number of edges, which represent the boundaries of one face of the rigid object.

Modeling Robots

We examine the parameters needed to characterize a robot.   We shall try to give a classification of these parameters.   Since a robot is a highly complex device, we restrict the discussion of details to the robot axes.

Structure and Characteristic Parameters of an Industrial Robot

An industrial robot (often called a manipulator) can, from a mechanical standpoint, be divided into three components:

The base
The arm
The gripper

Whereas base and arm are permanent components of a manipulator, a gripper may be exchanged for another one. Moreover, each of these components may again be decomposed into smaller components. This is particularly true for the gripper containing sensors that are of further importance to the modeling task. These few remarks already indicate that robot models will turn out to be extremely complex, certainly going beyond what is needed for this presentation. Consequently, we restrict ourselves to the arm and gripper (without sensors). We discuss which information to keep in the database with the model for these components.

In modeling a robot, hence its components, four aspects must be covered:

Mechanical structure
Geometry
Kinematics
Dynamics

The mechanical structure, outlined above, needs some more refinement. An arm is composed of a number of arm segments (also called axes) that are interconnected via joints. Likewise, a gripper may consist of several axes and, additionally, of one or more tools. Examples of tools are facilities for drilling, welding, or screwing. Gripper fingers can also be considered to be tools.

The geometric model of a manipulator or its components is primarily intended for graphic representations. Both constructive solid geometry and boundary representations are suitable [16]. Geometric data (e.g., the axis dimensions) are also needed to plan the robot motions.

The kinematics of a robot are defined by number, types, and mutual position of the axes. The description is usually in the form of kinematic chains, where each axis has its own coordinate system whose position is determined relative to the coordinate system of the preceding axis.

The dynamic model describes the gravitational forces and torques influencing the robot and caused by its intrinsic dynamics, as well as the forces and torques caused by the robot motions. More formally, the dynamic model is a mathematical model consisting of higher order differential equations. Its parameters can partly be obtained from the kinematic and geometric models.

Modeling a Robot Axis

For the remainder we restrict ourselves even further. We examine in more detail how to describe one robot component, an axis, taking the aforementioned four aspects into account.

Each axis has its own geometry. In classical robot programming systems the corresponding information is maintained in separate geometry files. Usually, they keep the information redundantly both as a boundary representation for the purpose of internal computations and in constructive solid representation for communicating with the external world. For efficiency reasons one will often store with an axis additional values such as lengths (e.g., length of an axis) and volume, although these may be computed from the other parameters.

Following Dillmann and Huck [5], where kinematic chains are described for robots, the kinematic parameters of an axis are determined as follows. Place within each arm segment or with each joint a coordinate system (called a motion coordinate system) such that (see Fig. 9.3) the z_i-axis coincides with the axis about which joint $i + 1$ rotates, and the x_{i+1}-axis is perpendicular to the z_i-axis, pointing away from it, where i and $i + 1$ are the sequence numbers of two successive points.

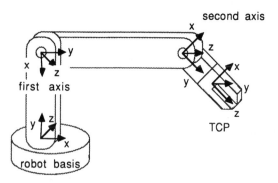

FIGURE 9.3 Motion coordinate systems in robot axes.

When the entire kinematics is being modeled, neighboring coordinate systems must be related to one another. Take an object described in coordinate system i. To express it in coordinate system $i + 1$, the following operations must be performed:

1. Rotation about the z_i-axis with angle ROT_Z such that axis x_i parallels axis x_{i+1}.
2. Translation along axis z_i by an amount TRANS_Z.
3. Translation along axis x_{i+1}, or alternatively along axis y_{i+1} by amounts TRANS_X or TRANS_Y, respectively, such that coordinate system i coincides with coordinate system $i + 1$.
4. Rotation about axis x_i or axis y_i by angles ROT_X or ROT_Y, respectively, such that axes z_i and z_{i+1} agree in position and direction.

In other words, the position of any coordinate system relative to a neighboring one can be described by rotational angles and translation lengths.

Characteristic for the description technique used here is that the kinematic parameters are expressed in so-called robot coordinates, that is, in values that can directly be interpreted by the robot controls and translated into motions. On the other hand, for the programming of robots the objects to be manipulated (picked up) by the robot are expressed in so-called world coordinates. Hence, transformations between robot and world coordinates are necessary.

All transformations may be described in the form of Denavit–Hartenberg (DH) matrices [5]. These are 4 × 4 matrices, where the first three rows and columns contain the information on the orientation of the coordinate system, and the last column describes the position of the origin of the coordinate system (i.e., the last column represents a translation vector with respect to the origin of some other coordinate system). The matrix elements are given by the rotational angles and translation lengths mentioned above.

DH matrices are a particularly elegant way to solve the coordinate transformations in robot programming. Since they are based on the concept of homogeneous coordinate systems, all transformations within the kinematic modeling of an axis may simply be expressed in the form of matrix multiplications.

Parameters needed to compute the dynamic behavior of an axis are, among others, its mass, its maximum acceleration, and the maximum velocity of the joint.

Figure 9.4 summarizes the discussion in the form of an entity–relationship diagram [2,18]. Entity-relationship (ER) diagrams are a widely used system analytic technique for structuring and

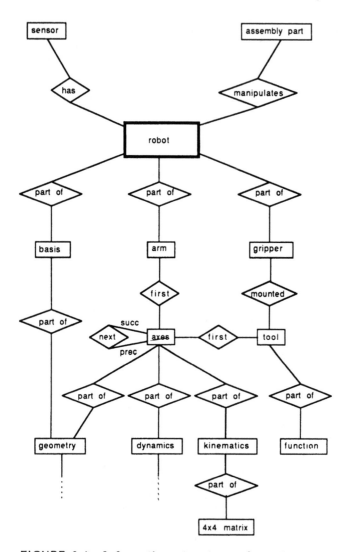

FIGURE 9.4   Information structure of a robot.

visualizing a miniworld before deciding how to organize a corresponding database schema.   In this conceptual data model, rectangles
represent entity sets and diamonds represent relationships among
entity sets.

9.3.2 Geometrical Transformation

To be able to move an object to different locations and to manipulate a particular object in size, one can apply three geometric transformations to the object stored in the geometrical database [12]. The three possible transformations are *translate, scale* and *rotate*. We will briefly explain how these operations are executed on an example object, say a cuboid represented by its eight vertices $\{v_1, v_2, \ldots, v_8\}$.

Translation

Translation corresponds to moving the geometrical object within the three-dimensional coordinate system relative to the origin. The orientation of the object in the 3D space is not altered by a translation. A translation is defined by its translation vector $T = (D_x, D_y, D_z)$. A single vertex is translated by adding the translation vector to the vector representing the vertex in the 3D system:

$$v_i = (x_i, y_i, z_i)$$

$$T = (D_x, D_y, D_z)$$

$$T(v_i) := v_i + T = (x_i + D_x, y_i + D_y, z_i + D_z)$$

To translate a geometrical object represented in boundary representation requires translating all vertices of the object that are stored in the BR schema.

Homogeneous Coordinates

The other two transformation operations (i.e., scaling and rotation) can be naturally defined as multiplications of the vertex (vector) with a corresponding transformation matrix, as we show below. To be able to combine different transformations of the same object (e.g., rotation and translation), one would like to also represent translation as a matrix multiplication. Then we would be able to combine different transformation matrices by multiplying them.

To represent the translation also as a matrix multiplication, the concept of homogeneous coordinates [12] must be employed, as is done in many graphic packages. This concept requires a vertex to be stored as a four-element vector instead of three elements. Then vertex v_i is represented as follows:

$$v_i = (x_i, y_i, z_i, 1)$$

Now the translation matrix is formed as a 4×4 matrix. The translation of the vertex v_i is then defined as follows:

$$T(v_i) = (x_i, y_i, z_i, 1) * \begin{pmatrix} 1 & 0 & 0 & 0 \\ 0 & 1 & 0 & 0 \\ 0 & 0 & 1 & 0 \\ D_x & D_y & D_z & 1 \end{pmatrix} = (x_i + D_x, y_i + D_y, z_i + D_z, 1)$$

Scaling

An important concept in viewing geometrical objects on a computer display is varying the size in form of scaling (or stretching). Vertices (as end points of vectors) can be scaled by S_x along the x-axis, S_y along the y-axis, and S_z along the z-axis, according to the scaling matrix S as follows (from now on we assume homogeneous coordinates):

$$S(v_i) = v_i * S = (x_i, y_i, z_i, 1) * \begin{pmatrix} S_x & 0 & 0 & 0 \\ 0 & S_y & 0 & 0 \\ 0 & 0 & S_z & 0 \\ 0 & 0 & 0 & 1 \end{pmatrix} = (S_x * x_i, S_y * y_i, S_z * z_i, 1)$$

Rotation

Rotations are used to change the orientation of geometric objects in the 3D space.

In three dimensions, we must distinguish rotations about the three different axes. Here we show only briefly the definition of rotation about the z-axis. The interested reader is, again, referred for more detail to Foley and van Dam [12].

A rotation about the z-axis is defined by the rotation angle ϕ. Corresponding to this angle, the rotation matrix $R_z(\phi)$ is constructed as follows:

$$\begin{pmatrix} \cos\phi & \sin\phi & 0 & 0 \\ -\sin\phi & \cos\phi & 0 & 0 \\ 0 & 0 & 1 & 0 \\ 0 & 0 & 0 & 1 \end{pmatrix}$$

The rotation of vertex v_i is then given as $v_i * R_z(\phi)$. Rotation of a geometric object that is stored in boundary representation is carried out analogously to scaling and translation; that is, each vertex must be rotated.

Simulation of Assembly Operations as Geometrical Transformations

As described earlier, the world model database forms the central part of a robot programming system. One major task in such a system is to simulate off-line robot operations (e.g., assembly operations) of the form:

mount cog wheel x on shaft y

The standard geometrical transformations described above are the operations used to model such an operation. We assume that object x (the cog wheel) exists at some location in the world of this robot application; that is, it exists in the world model database. The same holds for object y, the shaft onto which x is to be mounted. Simulating this assembly operation means, in terms of the world model database, changing the location of object x. In our particular case, this is achieved by the following (standard) geometric operations:

1. Translate x by T_1 (*pick up object x*).
2. Rotate x about the z-axis by $R_z(\phi)$.
3. Rotate x about the x-axis by $R_x(\Theta)$. } (*rotate x*).
4. Rotate x about the y-axis by $R_y(\Gamma)$.
5. Translate x again by T_2 (*mount object x on y*).

A program fragment that would achieve this transformation is given below:

```
for all vᵢ in {v₁,v₂,...} do
    begin
    vᵢ := vᵢ*T₁ ;
    vᵢ := vᵢ*R_z(Φ);
    vᵢ := vᵢ*R_x(Θ);
    vᵢ := vᵢ*R_y(Γ);
    vᵢ := vᵢ*T₂ ;
    end
```

This results all together in $5 * j$ multiplications of a vector with a matrix, where j is the number of vertices used to describe ob-object x. For example, if eight vertices are used to model the object, the method above would result in 40 multiplications. A much more efficient way is to combine the transformation matrices T_2, $R_z(\phi), R_x(\Theta), R_y(\Gamma)$ into one matrix M.

$$M := T_1*(R_z(\phi)*(R_x(\Theta)*(R_y(\Gamma)*T_2)))$$
```
for all vᵢ in {v₁,v₂,...} do
    vᵢ := vᵢ*M;
```

This method results in $4 + j$ multiplications; that is, for an object with eight vertices, 12 multiplications are necessary.

Modeling Robot Movements as Geometrical Transformations

Two different kinds of movements can be performed by an individual robot axis: translation or rotation of the axis.

As pointed out previously, all robot movements can be modeled by manipulating the kinematics description of each axis of the particular robot.   This is achieved by transforming the DH matrices that represent the current kinematic state of the robot's axes appropriately.   Since the kinematics description of each axis is based on its preceding axis, we do not have the global position of a particular part of the robot stored in the database.   Rather, the position must be computed from the kinematic chain.

For example, to compute the current gripper position, one multiplies all DH matrices that describe the current state of the robot's arm elements (for the notation see Fig. 9.4):

```
initialize GRIPPER_POS:
for i := 1 to #axes do
    GRIPPER_POS:=GRIPPER_POS *
                    robot.arm.axes[i].kinematics.DH_matrix
```

9.4   USING CURRENT DATABASE TECHNOLOGY

9.4.1   The Relational Data Model

After introducing our running example, we return to our point of departure at the end of Section 9.2.2.   Our first objective is to settle on a data model that constitutes a sensible compromise between the needs of individual applications such as robotics and enterprise-wide communication standards, which may additionally have to serve engineering design, manufacturing planning, and inventory control, to name a few.   As a matter of fact, if we turn to commonly available database technology, we are given few choices. The two predominant data models are the *network data model* and the *relational data model*.   Even though both were originally conceived with business and administrative applications in mind, they are a good point of departure for discussing the use of database technology in robotics.   The relational model is easier to start with for the database novice.

The basic assumption underlying practically all data models is that one describes an object of interest, say a vertex, by a set of properties, say an identifier and its x-, y-, and z-coordinates. Suppose now that given a particular vertex, we write down its values for these properties in a row.   The basic organization principle of the relational model is one in which we collect the rows of similar objects (i.e., of objects with the same set of properties) into a table.   Each column in the table, then, corresponds to a specific property. By adding column headers to the table, we may even decide in which order we want the columns to be presented to us. The table for vertices could thus look like this:

vertex			
vid	x	y	z
v3	2	3	0
v2	2	0	0
v6	2	0	3
v5	2	3	3
v1	0	0	0
v4	0	3	0

The notion of *relation* instead of table derives from two further assumptions:

No two rows are identical.
The rows may appear in any order.

These assumptions give a table all the characteristics of a mathematical relation as a set of rows, henceforth referred to as *tuples*. A column heading is called an *attribute*. Because of this, an extensive mathematical theory has been developed over the years for the relational model, and it has greatly contributed to the foundations of database technology in general. While we avoid a discussion of the theory in this book, we shall refer to some of its results from time to time.

The relational model in its pure form (and the one offered by commercial products) imposes one severe restriction: the values under any attribute must be atomic in the sense that if they are structured, this structure cannot be recognized by the database system [the relations are said to be in *first normal form (1NF)*]. Take as an example the edge defined by vertices v3 and v4. If we write it down as follows:

eid	vertex1	vertex2
e1	(0,0,0)	(2,0,0)

then the values under vertex1 and vertex2 could be stored only as uninterpreted text, and we could never ask the database system to give us the x-coordinates of its vertices. If we wished to do so, we would have to take a more circuitous route using vertex identifiers:

edge		
eid	vertex1	vertex2
●	●	●
●	●	●
e1	v3	v4
●	●	●
●	●	●

The question may now be answered by picking up the vertex identifiers v3 and v4 from row e1 in relation *edge* and looking up the x values in rows v3 and v4 in relation *vertex*.

Which brings us directly to the issue of data manipulation. The example readily demonstrates that, for one, all access to information in the relations is on the basis of values (this is referred to as *associative or value-oriented access*), and at least the following operations are needed:

1. The *selection* of rows from a table on the basis of values in one or more columns. Example: Selection in *vertex* based on values v3 and v4.

2. The selection of column values in a relation (or, as it is called, the *projection* of a relation onto a given set of attributes). Example: The projection of *vertex* on x.

3. The combination (*join*) of two relations into a single one by concatenating the rows that agree in their values under prespecified attributes. Example: We may simulate the traversal from *edge* to *vertex* as discussed above by concatenating the rows in *edge* with those in *vertex* that agree in their values under vertex1 or vertex2, and vid, respectively.

To answer our question, we proceed as follows. Select from *edge* the row with eid='e1', join this row with *vertex* such that the value under vertex1 agrees with those under vid, and do a second join on the basis of attribute vertex2. Note that at this point we have constructed two new relations of identical structure, both combining all attributes from *edge* with those of *vertex* (and, incidentally, each with a single row). Now we merge the two relations into a single one by set union. Finally, since we are interested only in the x-coordinates, we may dispose of the other attributes by projection, with the final result:

result
x
0
2

In fact, there is a language, *SQL* (the subject of current standardization efforts), that relieves the user of the task of constructing the sequence of steps to be performed. One need only state
the desired result relation in terms of properties of the original relations. The remainder—constructing the correct sequence of operations to be performed—is left to the database system. In our
example:

```
select x
from   vertex
where  vid in
          (select vertex1
           from   edge
           where eid='e1'

           union

           select vertex2
           from   edge
           where eid='e1')
```

The expert will notice that a straightforward translation into a
sequence of operations would add two projections before the join.
But both the expert and the novice will start to suspect that, after
all, the relational model may not be such a good choice, considering
that from a robotics point of view an edge should be an object that
is absolutely trivial to handle.

9.4.2 A Database Schema

Now that we have settled on the relational model and understand its
basic concepts, we proceed to cast our example into a database
schema. What we must provide is a list of the relations we wish to
establish, together with their layout in terms of attributes and the
value sets (*domains*) from which to draw the attribute values. As
pointed out in Section 9.2.2, no data may be put into the database
unless and until a schema has been declared for it. Some database
systems require that before starting to build a database, the schema
be declared in its entirety. Others—especially those offering SQL—
permit the introduction of relations in a stepwise manner. For example, to introduce our relation vertex we write:

```
create table vertex
          (vid char(3) NOT NULL,
           x  integer,
           y  integer,
           z  integer)
```

(where NOT NULL indicates that a value for this attribute must always be present).

For complex miniworlds, finding the right database schema (the *database design*) is a science by itself. Ease and naturalness of manipulation, reflection of as much of the miniworld semantics as possible, and maintenance of consistency and best possible performance are but some of the criteria to be observed. This is not the place to go into detail, but a rough outline of one possible design approach follows.

First identify the types of (physical or mental) entities in the miniworld about which you wish to derive statements from the database. Following Section 9.3, such entity types could be robots, arms, grippers, axes, tools, cuboids, faces, edges, vertices, or matrices. Then associate with each entity type a number of properties of interest that apply to all entities of that type. For example, vertices are described by x-, y- and z-coordinates; edges by their two end points and curvature; faces by their edges and surface properties; cuboids perhaps by color, weight, and material; axes by geometric, kinematic, and dynamic properties; robots by precision and load characteristics. In particular, associate with each entity type a property whose values (*key*) uniquely identify the corresponding entity.

One basic premise of the database approach is that entities do not exist in isolation but are related to their environment in a variety of ways. Therefore, as a further step, establish the interrelationships (relationship types) between the entities of various types. For example, robots are *composed* of a base, an arm, and a gripper; different grippers may be *attached* to a robot; an arm is *composed of a sequence of* axes; a cuboid *is composed of* faces; two cuboids are *connected via* a common face. In fact, what we have done is an entity-relationship analysis that was mentioned at the end of Section 9.3 (*composed of* is the converse relationship to *part of* in Fig. 9.4).

The relations of our database schema are now derived in a straightforward manner:

Declare a relation for each entity type, with an attribute for each of its properties. Identify the attribute corresponding to the key property as the *key attribute*.

Declare a relation for each relationship type, with an attribute for each of the interrelated entity types. The assumption is that under these attributes there will appear only values that are keys of tuples within the entity relations.

After a detailed analysis encompassing many properties beyond the ones discussed so far, part of a database schema for our example

(see also Fig. 9.4) may thus look as follows. For reasons of simplicity, instead of writing down each create-table operation, we just
list the relation name and its attributes.

```
robot(rid,type,degrees_of_freedom,no_of_axes,base_frame,
     arm_end_frame,gripper_frame,tcp_max_velocity,
     tcp_max_accel,pos_precision,max_load)
axis(aid,type_degree_of_freedom,max_pos,min_pos,null_pos,z_rot,
     z_trans,x_trans,x_rot,max_velocity,max_accel)
gripper(gid,type,no_of_fingers,max_open,min_open,null_open,
     tcp_init_frame,grip_strategy,max_force,int_range,ext_range)
face(fid,face_type,reflectivity,coarseness)
edge(eid,start_point,end_point,curvature)
vertex(vid,x_coord,y_coord,z_coord)
cuboid(cid,length_x,length_y,length_z)
cylinder(yid,radius,height)
sphere(sid,radius)
cuboid_faces(cid,fid)
face_edges(fid,eid)
robot_axes(rid,aid)
robot_gripper(rid,gid)
```

Usually, a schema derived in this fashion is not optimal under
one or more of the criteria mentioned for the database design. In
cases where a relation derived for a relationship type represents
a function, this relation may be joined with the relation that constitutes the domain of the function. Take, for example, relation
robot_axes, which specifies which robot an axis is part of. The
relation could be dropped by making rid another attribute of relation *axis*. The same applies to relation *robot_gripper*, describing
the gripper currently attached to a robot (this is even a one-to-
one relationship). Relationship relations may also be merged into
an entity relation if the number of entities entering into the relationship is fixed and known beforehand. For example, a relation
edge_vertices has been avoided because the number of end points
of an edge is always known to be 2.

On the other hand, one should also study the interdependencies
among the attributes of a relation. Sometimes it is advisable to
factor these out and thus split a relation into two. An example
will be given below.

One of the drawbacks of the pure set orientation of the relational model also becomes apparent: the axes of a robot are arranged in order, but this order cannot be represented (unless one
adds ordinal numbers or introduces a pointer mechanism based on
the aid's).

9.4.3 Taking Care of Constraints

Choosing the database model and declaring the database schema constituted step 1 of setting up a database. We now turn to step 2, expressing additional consistency constraints. We do not treat the subject exhaustively, just give the flavor of it.

The simplest kind of constraint is the *key constraint*. It signifies that a value under a certain attribute is to uniquely identify a tuple in a given relation. (This is more a formal than a semantic argument: whereas entity identifiers can always be used as keys within entity relations, other attributes may also serve the purpose.) In SQL we notify the database system by writing:

    create unique index key_gripper on gripper(gid)

where *unique* establishes the uniqueness property and, in addition, *index* results in more efficient associative access on the basis of gid values. Since one may also drop an index, the key constraint is a dynamic constraint. Other systems insist that the key constraint be static; in such a case it is included in the declaration of a relation.

The key constraint is a special case of a more general kind of constraint, *functional dependency* (*FD*). An FD, written as follows:

$$A \rightarrow B$$

where A and B are sets of attributes of a given relation R, specifies that each combination of values under A uniquely determines the values under B. Take our relation *gripper*. We observe the following FDs:

    gid → type   (a gripper is of one type)
    type → no_of_fingers   (the type determines all the remaining
                             attributes)
    type → max_open
      ⋮
    type → ext_range

Since SQL does not enforce FDs except for key constraints, one must try to reduce the FDs to key constraints by splitting the relation, factoring out the FDs. In other words, converting the FDs is a subject of database design before running the DBMS rather than of the DBMS itself. Because of a law:

if $A \rightarrow B$ and $A \rightarrow C$ then $A \rightarrow (B, C)$

it suffices to split *gripper* into

```
gripper_instance(gid,type)
gripper_type(type,no_of_fingers,max_open,min_open,null_open,
    tcp_init_frame,grip_strategy,max_force,int_range,ext_range)
```

with keys gid and type, respectively. Incidentally, note that the domain of type in *gripper_instance* is identical to the domain of type in *gripper_type*. Since the latter is a key attribute, the former is referred to as a *foreign key*.

We must consider yet another kind of consistency because certain entities cannot exist without a second entity. Take, for example, a rule that states that the type of a gripper in the database must always be known. Consequently, before introducing a gripper into the database, a tuple with the corresponding type must have been inserted into *gripper_type*. We call such an interdependency *referential integrity* (or also *inclusion dependency* because the set of all type values currently in *gripper_instance* must be a subset of all type values currently kept in *gripper_type*). We note that referential integrity is associated with foreign keys. Other such dependencies are observed for rid and aid in *robot_axes* (a relationship can be established only between known entities), and similarly in *cuboid_faces* and so on. Sad to say, few database systems support referential integrity these days.

Referential integrity is a special case of *existence dependency*. Consider two rules:

An axis may be stored in the database only if it is part of a robot. Grippers may exist independently of robots.

Our database schema as it currently stands treats all entities as mutually independent. Hence the second rule is trivially satisfied. The first one, however, is not. It is a typical example of an entity depending on its existence on a second one. Referential integrity does not solve the problem because it deals only with the existence of relationships. Fortunately, there is way out by means of transactions (Section 9.2.1). A *transaction* is a kind of database procedure consisting of a sequence of database operations that, taken in their entirety, guarantee the consistency of the database. To observe the first rule we introduce a transaction that first checks whether the desired robot exists in the database, then inserts an *axis* tuple, and finally a tuple into *ro-bot_axes*. Provided the database system would check consistency, it would do so only at transaction end (transaction *commit*).

The transaction mechanism would also have to be used to enforce more complicated constraints such as "a cuboid to be rotated in space must not change its shape." Here a (long) sequence of operations would have to be established to determine the vertices of the cuboid and to change their x-, y-, and z-coordinates in the correct manner.

There is another important constraint that in today's systems would have to be checked as part of a transaction, namely *cardinality constraints*. For example, a cuboid has exactly eight vertices; a robot may have between two and six axes.

To summarize, constraints are an important formalism for specifying miniworld semantics that cannot be expressed with the data model. However, since they are often costly to enforce, few of them can directly be communicated to the database system but must be incorporated into the application software or expressed via user transactions.

9.4.4 Using the Database

Step 3, the last step, storing the data in the database, usually is an ongoing exercise. There are generally three update operations, inserting one or more tuples into a relation, deleting one or more tuples from a relation, and modifying one or more tuples in a relation. All three operations are interpreted in the light of the database schema and, unless they are part of a transaction, are rejected if the result does not satisfy the consistency constraints that have been declared to the database system. In case of transactions, the consistency checking is postponed to transaction commit.

Building up a database of several hundred megabytes and more in this way is, of course, a cumbersome and lengthy affair. The robot programmer is more inclined to deal with the natural entities of the miniworld, implying many of the relationships, rather than with a myriad of tuples. For example, a natural entity may be a cuboid, with all its faces, edges, and vertices seen as immediate parts of it. Hence, special tools should be provided as part of the application environment, to permit the entry of larger units and to break these down into tuples of the appropriate relations.

The reverse—collection of tuples into larger units—is true if the database is to be inspected, and information is to be extracted about the various entities. A query such as "give me all about cuboid c5" (meaning, e.g., its faces, surface properties, and vertices) would have to be expressed in SQL as follows:

```
select  fid,reflectivity,coarseness,x_coord,y_coord,z_coord
from    face,vertex
where fid in
        (select  fid
         from    cuboid_faces
         where cid='c5')
and vid in
        (select  start_point
         from    edge
         where eid in
                (select  eid
                 from    face_edges
                 where fid=face.fid)

        union

        select  end_point
        from    edge
        where eid in
                (select  eid
                 from    face_edges
                 where fid=face.fid))
```

The query is a real monster! And so is the result, because it contains a lot of redundancy due to the impossibility of reflecting properly the hierarchical decomposition into faces, each with its own vertices. The query also takes a long time to process because, among other reasons, no fewer than six join operations must be performed. Again, appropriate tools should be provided to the user.

9.4.5 Establishing Individual Views

Views are a mechanism to hide irrelevant portions of the database from a user (e.g., for reasons of convenience or protection) and to let different users approach the database from different angles.

Both purposes are achieved by a facility that lets the user (or the system administrator) declare a view in terms of a select expression. Suppose that a particular user is interested only in the dynamics of axes with a maximum velocity above 3. He then declares:

```
define view axis_dynamics as
        (select  aid,max_velocity,max_accel
         from    axis
         where max_velocity > 3)
```

Consider another user who wishes to deal with cuboids in the way discussed earlier:

```
define view my_cuboid as
        (select  fid,reflectivity,coarseness,x_cooord,y_coord,z_coord
         from   face,vertex
         where vid in

         (select start_point
          from   edge
          where eid in
          (select eid
            from   face_edges
            where fid=face.fid)

          union

          select  end_point
          from   edge
          where eid in
          (select eid
            from   face_edges
            where fid=face.fid)))
```

Views may be interrogated like any other relation. For example, "give me all about cuboid c5":

```
select  *
from    my_cuboid
where fid in
        (select fid
          from    cuboid_faces
          where cid='c5')
```

Views are constructed anew every time they are referred to. Hence, they always reflect the latest state of the database.

9.5 USING OBJECT-ORIENTED DATABASE TECHNOLOGY

Data models underlying database systems currently available in the marketplace (e.g., the relational model as used in the preceding section) have, among others, the following characteristics [17]:

They are *record-oriented*; their basic constructs to organize information in a database are flat collections of attribute values (called

records or tuples) and sets of homogeneously structured records,
that is, instances of the same record type.

They are *value-oriented*; (nearly) all information is expressed in
terms of values (of the attributes of the records), and relationships between records are established by matching the values of
attributes from multiple records (while the network data model
goes a bit beyond mere value-orientation, it is still restricted
enough to fall into this category).

Database record *identification* is primarily by means of attribute values as defined by the user. Since these values may change,
there is no practical way to say whether two records retrieved
at different times are the same.

These characteristics are well in accordance with requirements
found in the classical business and administration types of application that gave rise to the advent of database systems in the 1960s.
Miniworld entities in these areas are rather simple; they typically
comprise a limited number of attributes only and, most important,
do not show any further substructures that would be necessary to
reflect in the database. Furthermore, as (unique) identifiers like
employee numbers, social security numbers, and insurance numbers
have been introduced in the real world anyway, it has been sufficient to offer key attribute facilities to map those into the database,
and they equally lend themselves for expressing relationships.

Looking again at the examples in Sections 9.3 and 9.4, two observations may be made with regard to using these record-oriented
database systems for information management in planning and manufacturing applications:

1. It is indeed possible to represent and to handle a good deal of
   the relevant information within the framework of classical data
   models.

2. However, being forced to do so is rather inconvenient, error
   prone, and most likely very inefficient:

   Since structures for data modeling are rather atomic and homogeneous, complexly structured real-world information must
   be scattered across multiple structures; accordingly, meaningful data retrieval and manipulation operations must be formulated by combining numerous simple queries.

   For these reasons, knowledge about the coherence between
   different small database items, and thus about important miniworld semantics, gets lost within the database. Consequently,
   the knowledge must be incorporated into application programs
   or transactions (which is contrary to the idea of databases),
   making it more difficult to maintain the consistency of the
   database.

Again because of the loss of relevant semantics, the database
management system has no way to take specific properties of
complex structures and the associated operations into account.
It thus can base its optimization measures only on the modeling features it provides. As experience shows, this is usually
rather counterproductive to advanced application requirements
(e.g., those found in planning and manufacturing) and results
in poor overall performance.

In recent years, database research has recognized the problems
mentioned above and has turned its attention toward developing solutions that overcome them. The general idea underlying most approaches is that of an *object-oriented* database system [6,7]: the
data model of such systems provides facilities to define and to deal
with units that allow to represent in a one-to-one way real-world entities of whatever structure and complexity with respect to subunits
and attributes, and of whatever operability. This avoids the necessity of splitting logically related information into multiple, gener-
ically unrelated database units (1:n mapping); thus it incorporates
more semantics within the database and consequently allows for an
efficient implementation by way of exploiting all available structural
and operational knowledge.

Depending on the way and degree of object orientation they provide, the following classes of approaches can be distinguished.

1. *Structurally* object-oriented models support complexly structured objects—that is, objects that may be constructed out of subobjects (which are objects in their own right) beyond incorporating
simple attributes of the usual types. Take as an example our cuboid
which, if one proceeds to more detail, is viewed as the composite of
six faces as subobjects, etc. Structural object orientation also includes a set of appropriate operators to deal with objects. These
operators are generic in the sense that they apply to all object types
that can be defined using structural constructor mechanisms of the
data model; they are thus not type-specific and are predefined in
the data model itself. In our example, a typical operation would retrieve a whole object, or would break it down into its immediate components, and so on, no matter whether the object is a cuboid, a gripper, or a robot.

2. *Behaviorally* object-oriented models provide facilities that allow the definition of new object types including the appropriate type-
specific operators. In the simplest case, these types act as attribute domains only, and their values are represented as byte strings
that are interpreted by the operation code that the user produces
as part of the database design. Matrices could be dealt with in this
way. Behavioral object orientation has been influenced by abstract
data type concepts in the programming language domain.

3. *Fully* object-oriented models combine structural and behavioral object orientation: complexly structured objects may include instances of user-defined types, and user-defined types may use complex object structures as their representation. Take again cuboids, for which translation and rotation have a very special meaning (namely, invariance of all interrelationships between vertices), so that one would associate these operations with all objects of type cuboid; the operations would be realized utilizing matrices and vectors as behavioral domain types.

Furthermore, object-oriented database systems often support *surrogates* to uniquely identify objects in the database. A surrogate is system-generated upon the creation of an object, it never changes during the lifetime of the object, and it is never reused after an object ceases to exist. Primarily for behavioral objects, inheritance mechanisms are offered that allow to establish a taxonomy among object types: objects of a type that is considered to be subordinate to some other type (subtype/supertype) inherit the properties (operations, attributes) defined for the supertype, aside from having their own specific properties.

In the data model aspect, the differences from the more classical systems originate, and probably this is what makes object-oriented database systems attractive for users. However, there are numerous other issues to be rethought or reconsidered for this kind of approach, including transaction management, consistency control, recovery, and user interface. For the purpose of this chapter, though, it should be sufficient to concentrate on object-oriented data models. The following sections therefore present structural and behavioral object orientation with respect to use for planning and manufacturing databases.

9.6 STRUCTURAL OBJECT ORIENTATION

In comparison to relational or similar database systems, structurally object-oriented approaches improve the representability of elaborate real-world information structures by supporting complex objects with appropriate generic operators. Of course, relationships among objects can be expressed and dealt with as well.

There are two main directions to design and implement complex object support. Both attract considerable research in industrial and academic environments, and it may be expected that currently available prototypes will be turned to products within a few years. We first discuss the highlights of both approaches. As the examples from preceding sections suggest, the real world of planning and manufacturing applications includes numerous cases of complexly structured entities that must be dealt with, and thus we may expect that

structurally object-oriented database systems mitigate the data handling problems experienced in this area. This will be demonstrated
by way of examples for one representative of each direction.

9.6.1 Extensions to Relational Technology

Because of the meanwhile undisputed usefulness of the relational
data model for business/administrative applications, one approach
is to keep its basic features intact and incorporate the notion of
complex object into it. There are at least three ways to do so.
    1. Use an additional attribute type *component_of* to establish ex-
plicit links from child tuples (in one relation) to their parent tuple
(in a different relation). These links are readily realized by means
of the surrogate of the respective target tuple. A complex object
is then defined to consist of a tuple together with the transitive
closure of its child tuples. Note that *component_of* attributes essentially capture hierarchical structures; nonhierarchical relationships may be expressed by using reference attributes. For both
kinds of attribute, special semantics are involved that are understood by the DBMS, in contrast to the arbitrariness of the relationships expressed by matching values in the traditional relational way.
It can thus provide appropriate operators to access (copy, delete,
etc.) entire objects, check for the existence of referenced objects,
and the like. The XSQL prototype at the IBM Almaden Research
Center [19] includes complex object support along these lines.
    2. Alternatively, it has been proposed to describe a subobject
by a relational query selecting the appropriate (parts of) tuples from
appropriate relations. This query is then stored as an attribute of
a tuple that represents the superobject. Obviously, arbitrary levels of hierarchical and even nonhierarchical composition can be constructed. When accessing an object, the queries included in its attributes are executed recursively to deliver the required data.
POSTGRES [23] goes this way to provide complex objects; to avoid
tremendous overhead in reexecuting the constructive queries all the
time, it tries to use an object cache that is invalidated and reconstructed upon changes in the subobjects only.
    3. The third approach to include complex objects into the rela-
tional world simply relaxes the requirement that relations be in first
normal form (Section 9.4.1). While this restriction really forms the
baseline of relational theory, the so-called NF^2 model (NF^2 = NFNF =
non-first normal form [21]) does away with it and allows attributes
whose values are relations themselves. An extended theory covering these nested relations is easily defined, mainly because of two
specific operators to nest and unnest relations. A complex object
is considered to be represented by an NF^2-tuple on any level, together with all the relation instances contained in its attributes. As

pure nesting is a hierarchical affair per se, only hierarchical object structures are supported; traditional relational ways must be used to express nonhierarchical relationships. The NF^2 model is, among others, implemented as part of the AIM-P system researched at the IBM Scientific Center at Heidelberg [3]. In fact, AIM-P even extends the basic model by allowing arbitrary tuples, sets, and lists as attribute values. We use this approach in the examples below.

9.6.2 Support of Explicit Structuring Capabilities

While the above-mentioned examples of systems with structural object orientation try to graft complex objects onto flat relations and keep the underlying relational model as much intact as possible, there are also a number of approaches that go the other way round: they start out from providing facilities for explicitly defining the structure of objects to the DBMS, without resorting to an underlying relational structure. If desired, relation like interfaces can be defined on top.

The origins of this kind of data modeling capability go back to the so-called entity–relationship (ER) approach [2]. It presupposes that the real world to be reflected in a database can be perceived as consisting of (concrete or artificial) entities and relationships among them. Consequently, it essentially provides modeling concepts for building blocks of both types. Furthermore, the ER model includes role attributes to distinguish between multiple participations of entity types in a relationship, as well as cardinalities that prescribe or limit how often an entity may enter into a specific relationship.

The original ER approach is restricted to simple entities without any substructure (beyond associating simple attributes with entities). For obtaining structural object orientation, it has been augmented by the concept of molecular aggregation [1]: multiple entities and relationships between them can be aggregated to form new higher level entities in their own right. In its most liberal form, this leads to arbitrarily organized complex objects that may overlap in their subobjects and whose types may be recursive. Of course, a suitable set of generic operators must be defined, along with the object structuring capabilities to make up a real data model (which originally was not the case for the ER approach). We will use the design object data model (DODM) of the DAMOKLES prototype developed at FZI Karlsruhe [8,9] in the examples below.

9.6.3 An Example: Robot Parts Hierarchy

Let us first consider the database representation of the hierarchical arrangement of the individual parts of a robot, that is, its arms (including kinematics and dynamics) and grippers. Remember the

relational world: we had to use numerous relations, one for every
component or subcomponent and additional ones to express the relationships between them. For retrieval, multiple joins had been
necessary.

In the NF^2 nested relations approach, we need only one top-level
relation for robots. For our example, it will be sufficient to include
three attributes, one each for the identifier of a robot, its arms,
and its grippers. The key issue is that both the ARMS attribute
and the GRIPPERS attribute have themselves relations as their values. Figure 9.5 gives the complete relational schema for the RO-
BOTS relation (as an appropriate create-operation). As mentioned
earlier, AIM-P generalizes the pure NF^2 approach, and thus [...]
denotes the construction of tuple types out of attributes, {...} is
the set constructor, and ⟨...⟩ forms ordered lists. A standard relation is a set of tuples; hence it needs [...] brackets for its definition. As can further be seen, the DH matrix to describe the arm
kinematics is represented as an ordered list of four (the FIX prefix
means exactly) tuples consisting of a column number and a four-dimensional vector, which is itself an ordered list of four REAL values.

```
create ROBOTS {
  [ ID: string(20),
    ARMS: {
     [ ID: string(20),
       AXES: ⟨
        [ KINEMATICS:
           [ DH_MATRIX:
              ⟨4 FIX [ COLUMN: integer,
                      VECTOR: ⟨4 FIX real)] ⟩,
              JOINT_ANGLE:
               [ MAX: integer,
                 MIN: integer]
           ],
         DYNAMICS:
           [MASS: real,
            ACCEL: real]
        ] ⟩
     ] },
    GRIPPERS: {
     [ID: string(20),
      FUNCTION: string(20)
     ] }
  ] }
end
```

FIGURE 9.5 NF^2 schema for a robot parts hierarchy.

Figure 9.6 shows a subset of a database adhering to the schema of Figure 9.5. In particular, it includes the data of the left arm of robot Artoo Detoo and of two of its grippers.

Let us see what the (extended) NF² approach offers in terms of dealing with nested relations. The typical select-from-where construction of queries in the flat relational model has been extended to render nested tuples, sets, and lists in the given case. While the desired structure (if it cannot be inferred from the schema) must be specified in the select-clause, the from-part reflects the nested structure of the given relation (when necessary).

For example, to retrieve the robot identifier, the arm identifier, and the DH matrix of the second axis of the left arm of robot Artoo Detoo, the following query might be formulated:

```
select [R.ID,
         select [AR.ID,
                 AR.AXES[2].KINEMATICS.DH_MATRIX
                 ]
         from   AR in R.ARMS
         where  AR.ID = 'left'
         ]
from   R in ROBOTS
where  R.ID = 'Artoo Detoo'
```

{ROBOTS}								
ID	{ARMS}						{GRIPPERS}	
	ID	<AXES>					ID	FUNCTION
		KINEMATICS				DYNAMICS		
		<DH_MATRIX>		JOINT_ANGLE		MASS	ACCEL	
		COLUMN	<VECTOR>	MAX	MIN			
Artoo Detoo	left	1	<1,0,0,0>	-180	180	50	1.0	#200	gripper
		2	<0,0,-1,0>						
		3	<0,1,0,0>					#150	welding
		4	<0,0,100,1>						machine
		1	<1,0,0,0>	-250	60	37.26	2.0		
		2	<0,1,0,0>						
		3	<0,0,1,0>						
		4	<70,0,20,1>						
		1	<0,1,0,0>	-80	250	10.4	6.0		
		2	<0,0,1,0>						
		3	<1,0,0,0>						
		4	<0,40,-10,1>						

FIGURE 9.6 NF² database for a robot part hierarchy.

Note the dot notation for the qualification of attributes down the
hierarchy, and the ⟨variable⟩ in ⟨relation⟩ construct in the from-
clause to obtain a relation variable to be used in the other parts.

To get all information about a specific robot (in the same form as
it is represented in the underlying relation), one simply issues a
query like:

```
select *
from   ROBOTS
where ID = 'Artoo Detoo'
```

which is obviously much simpler than the complicated sequence of
joins that was needed in similar cases for the pure relational model.

Delete and update operations are similar in flavor, and the insertion of tuple, set, and list values is facilitated by constructors for
appropriate values comparable to those used for the schema definition in Figure 9.5. Of course, if the result of a query is to be
used within a programming language for further computation or for
robot and machine control, a query language like the one used for
the examples must be properly embedded into the language of choice
(which includes the provision of cursor concepts or similar to tra-
verse sets in a one-item-at-a-time fashion, as current programming
languages do not usually allow for set-oriented processing). However, the intention of this chapter is to give a feeling for what the
use of databases in planning and manufacturing might be, not to
demonstrate all the details that are necessary to really apply all the
mechanisms we sketch.

Let us now use the DODM data model of DAMOKLES to represent
the same robot information and to formulate the same queries as
above. Figure 9.7 presents the definition of an appropriate schema.

The object type ROBOTS is the equivalent to the ROBOTS nested
relation of Figure 9.5. Since DODM does not offer ordered lists as
a predefined type (which we had used to model the set of robot arms
above), we add a NUMBER attribute to explicitly obtain a way to access the arms by their position.

Note that object types and so on are defined apart from one another for purely syntactic reasons; the rules of composition for complex objects are expressed by means of the structure is-clause,
which specifies what types their subobjects may have. A pictorial
representation of the schema is given in Figure 9.8. Every box
stands for an object type. Its attributes are denoted above the
dashed line; subobject types are drawn below it.

Figure 9.9 depicts the sample database in terms of the DODM
schema.

In a language similar to SQL, a query against this database to
retrieve, as before, the robot identifier, arm identifier, and the

```
object type ROBOTS
   attributes ID: string[20]
   structure is ARMS, GRIPPERS
end ROBOTS

object type ARMS
   attributes ID: string[20]
   structure is AXES
end ARMS

object type GRIPPERS
   attributes ID: string[20];
            FUNCTION: string[20]
end GRIPPERS

object type AXES
   attributes NUMBER: int;
            KINEMATICS: struct
                  DH_MATRIX: DH_M_T array [4];
                  JOINT_ANGLE:
                     struct
                        MAX,MIN: int subr [-360..360];
                     end
                  end;
            DYNAMICS: struct
                  MASS, ACCEL: real;
                  end
end AXES

value_set DH_M_T: struct COLUMN: int subr [1..4];
                     VECTOR: int array [4]
                  end
```

FIGURE 9.7   DODM schema for a robot part hierarchy.

DH matrix of the second axis of the left arm of robot Artoo Detoo would now read:

```
select  ROBOTS.ID,ARMS.ID,AXES.KINEMATICS.DH_MATRIX
from    ROBOTS downto ARMS downto AXES
where   ROBOTS.ID = 'Artoo Detoo' and
        ARMS.ID = 'left' and
        AXES.NUMBER = 2
```

The select-clause contains qualifications through the various parts of compound attributes of object types; the from-clause specifies a

ROBOTS

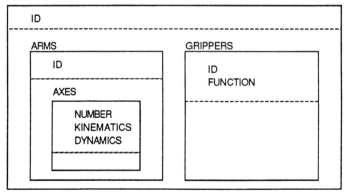

FIGURE 9.8 DODM schema for a robot part hierarchy (graphical representation).

path through the object hierarchy that includes all object types needed in the where-formula or in the select-list.

The retrieval statement for all details of a robot reads like the NF^2 case:

```
select  *
from    ROBOTS
where   ID = 'Artoo Detoo'
```

Let us now look at how to express that two (or more) objects are related to each other (without one being a subobject of the other!). Consider for this example that any two robots may form a sequence in a production line, with a minimal distance to be maintained between them. Figure 9.10 shows schemata and example databases for both the NF^2 and the DODM models.

Obviously, relationships between objects have to be expressed— like anything else—as relations in NF^2. As in the pure relational model, an attribute or a combination of attributes that uniquely identifies the objects to be related (foreign key) is selected for representing them. In DODM, there is a separate construct for relationships. The relates-clause introduces the participating object types by naming them and associating a role attribute with them. From a user standpoint, this should be more natural because one simply considers the whole object to participate in the relationship.

As for querying a relationship, let us assume that we want to find out which robots may follow Artoo Detoo in the sequence. In NF^2, a select-statement resulting in a join must be formulated:

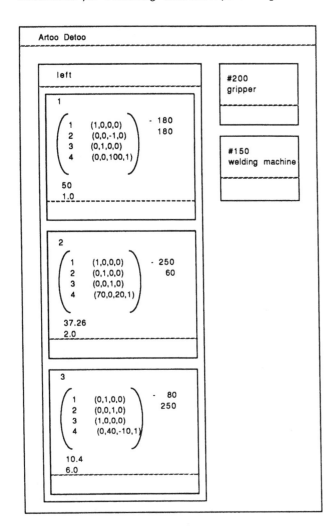

FIGURE 9.9 DODM-database for a robot part hierarchy.

```
select *
from   ROBOTS
where  ID in select SECOND
            from   ROBOT_SEQ
            where  FIRST = 'Artoo Detoo'
```

In DODM, similar to queries exploiting object—subobject relationships, queries may be expressed by explicitly following the relationship

```
create   ROBOT_SEQ{[FIRST:  string(20),          relship  type ROBOT_SEQ
                    SECOND: string(20),              relates    FIRST:   ROBOT,
                    MINDIST:integer]}                           SECOND: ROBOT
end                                                   attributes MINDIST: int
                                                 end ROBOT_SEQ
```

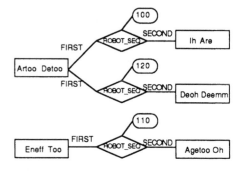

{ROBOT_SEQ}		
FIRST	SECOND	MINDIST
Artoo Detoo	Ih Are	100
Artoo Detoo	Deoh Deemm	120
Eneff Too	Agetoo Oh	110

FIGURE 9.10   Schema and example database for relationships between objects: NF² (left) and DODM (right).

links.   Internally, specific provisions are taken to represent relationships, and thus expensive joins can be avoided.

```
select  *
from    ROBOTS
            via  ROBOT_SEQ.FIRST
            via  ROBOT_SEQ.SECOND
            to   ROBOTS
where ROBOTS.ID = 'Artoo Detoo'
```

Note that in the first case, the nested select is necessary to really find all the information about the robots, not only their identifiers. The from-part of the DODM example specifies the path used for retrieval, from a given object via the roles of a relationship to an object that is finally retrieved. We have used the unabbreviated form, which also works in much more complicated cases; for our example, part of the details might be omitted as long as there are no ambiguities (which depend on the rest of the underlying schema that has not been shown).

9.6.4 An Example: Geometric Modeling

Our next example is taken from modeling geometric solids, more specifically boundary representations of cuboids. This will reveal an inherent difference between the extended NF² and the DODM model:

because of its purely hierarchical approach, the first one does not allow for the direct expression of common subobjects of two or more superobjects; in contrast, the latter does.

In boundary representation, a solid object is specified in terms of its nonoverlapping faces. In turn, each face is modeled by its bounding edges, and an edge by its spanning vertices. Obviously, edges may overlap in their vertices and faces in their edges, thus common subobjects result.

In NF^2, an appropriate nested relation for boundary representations would have to be defined as in Figure 9.11.

An excerpt from a database according to this schema (Fig. 9.12) demonstrates that considerable redundancy must be introduced (and consistently maintained in cases of deletions and updates!) as a result of the overlapping parts. Specialists of relational database design might suggest, for example, that a separate table be kept for the vertex coordinates, thus keeping every vertex only once. However, if one proceeds this way, one has started to flatten the relation in the direction of the pure relational model and most of the advantages of direct complex object representation will be lost.

A query to find the vertices of a given edge and face of a specific cube reads as follows:

```
select {[v.ID,
    (select [l.X,l.Y,l.Z]
    from l in v.LOCATION)
    ]}
from v in e.VERTICES,
     e in f.EDGES,
     f in c.FACES,
     c in BRep
where e.ID = 'el' and
      f.ID = 'fl' and
      c.ID = 3
```

A DODM schema to represent the same information consists of the object types as defined in Figure 9.13.

Principally, there is not much structural difference from the schema of Figure 9.7. Indeed, even the schema for the robot parts hierarchy does not exclude overlaps (e.g., the same arm is made a part of two different robots). There, however, the hierarchy must be enforced either by careful operation of the database or by appropriate consistency constraints. In the case of modeling cuboids in boundary representation, though, the feature of overlapping subobjects is really exploited, as the database subset of Figure 9.14 shows. Overlaps do not show in the schema because a schema always represents type information only, and superobjects with common

```
create BRep {
  [ID: integer,
   FACES: {
    [ID: string(4),
     EDGES: {
      [ID: string(4),
       VERTICES: {
        [ID: string(4),
         LOCATION: [X: real,
                    Y: real,
                    Z: real]
      ]}
     ]}
   ]}
 ]}
end
```

FIGURE 9.11   NF² schema for boundary representations.

subobjects are instances of the same type in our example;   DODM
of course also allows different types with common subobjects.

The query to find information about specific vertices as above
might be formulated in the following way:

```
select *
from   VERTICES upto EDGES upto FACES upto BRep
where  EDGES.ID = 'el' and
       FACES.ID = 'f1' and
       BRep.ID = 3
```

Obviously, redundancy like that in NF² need not be introduced
and maintained in the DODM model of solid objects given in boundary representation.

9.6.5   An Example: Path Planning

In our last example for structural object orientation, we want to
demonstrate one more data model feature that is frequently offered
in these systems, namely object versions [10,13]. Versions are a
means to represent multiple states of the same semantic entity (modeled as a database object), such as different alternatives, revisions,
or the entity's development history.

Some information will remain unchanged over time; other information will vary from time to time. The former is collected into a so-

BRep						
ID	FACES					
	ID	EDGES				
		ID	VERTICES			
			ID	LOCATION		
				X	Y	Z
3	f1	e1	v1	0	0	0
			v2	2	0	0
		e2	v2	2	0	0
			v3	2	3	0
		e3	v3	2	3	0
			v4	0	3	0
		e4	v1	0	0	0
			v4	0	3	0
	f5	e2	v2	2	0	0
			v3	2	3	0
		e5	v3	2	3	0
			v5	2	3	3
		e6	v5	2	3	3
			v6	2	0	3
...	...	...	...	...	...	...

FIGURE 9.12 NF^2 database for boundary representations.

called generic part of a DODM object. The latter results in a version part that consists of a set of versions, each reflecting a particular state of the variable information. Versions are arranged as a version graph whose form (linear, treelike, or acyclic) must be specified in the schema.

Both information in the generic part and in each version may consist of attributes and of subobjects in precisely the same way as it does in nonversioned objects. One restriction must be observed: all versions within the version part of a DODM object must be of identical type. Within one object, versions are sequentially numbered in creation sequence, and of course a version can exist only as part of its generic object. Thus object deletion also deletes all its versions. DODM includes a set of operators to access and manipulate versions. A version may act as an object in its own right, in which case it inherits the invariant attributes and

```
object type B R e p                     object type EDGES
    attributes    ID: int                   attributes    ID: string(4)
    structure is FACES                      structure is VERTICES
end CUBOIDS                             end EDGES


object type FACES                       object type VERTICES
    attributes    ID: string(4)             attributes
    structure is EDGES                          ID: string(4);
end FACES                                       LOCATION: struct
                                                            X: real;
                                                            Y: real;
                                                            Z: real
                                                        end
                                        end VERTICES
```

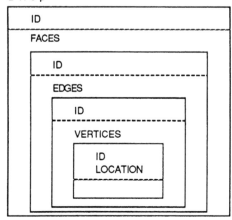

FIGURE 9.13 DODM schema for boundary representations.

subobjects of its generic object. As a consequence, an update of
version data affects a single version, whereas an update of generic
data affects each version. To be more precise, then, generic data
are not necessarily invariant, but rather common to all versions.

For a brief example, consider paths planned for robot movements
in the database. Every path is a sequence of path segments that
are related to one another. Suppose that a number of alternative
paths, and thus segment compositions, are to be explored for a
while. Without going into further details, we assume that several
attributes describe a path as a whole and that no (invariant) subobjects exist on the generic object level.

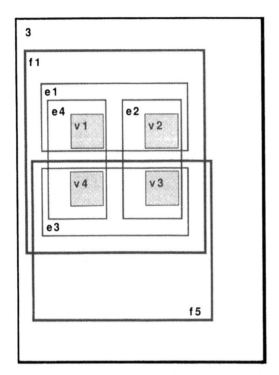

FIGURE 9.14 DODM database (excerpt) for boundary representations.

Figure 9.15 gives an impression of a schema and a small database excerpt with three versions of segment compositions of a specific path.

It has been assumed that versions 2 and 3 have both emerged from version 1. The reader will also note from the example that relationships may also act as members of the structural part of objects and/or versions.

To retrieve a specific version of a given object, a query like

```
select *.3
from   PATHS
where NAME = 'path4'
```

is issued; navigation in the version graph (similar to navigation across relationships) is also available for this purpose.

```
object  type  PATHS
   attributes NAME:  string(10);
               <further generic path data>
   versions  treelike
       (structure  is    SEGMENTS, SEG_SEQ)
end PATHS


object  type  SEGMENTS
       ▪  ▪  ▪
end SEGMENTS


relship  type SEG_SEQ
    relates FROM: SEGMENT,
             TO: SEGMENT
end SEG_SEQ
```

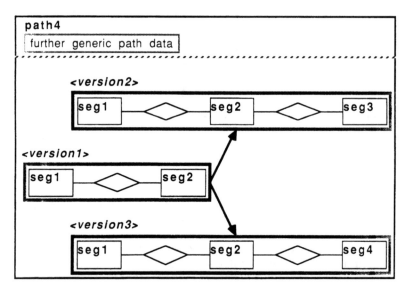

FIGURE 9.15   DODM objects with versions: schema and example database.

9.6.6 Summary

Structural object orientation is primarily based on a single extension to classical data models, namely complex objects in various forms with more or fewer restrictions. Although it is true that the essence of complex objects is no more than a special relationship *is subobject of*, there are overwhelming arguments to make them a special concept:

Complex objects are ubiquitous in engineering applications and are justified as a data modeling construct because they ease database design and use.

Even more important, now that the database system is aware of the special semantics associated with complex objects, it can implement them in the most efficient way. In contrast, if *is subobject of* were a relationship like any other, the system would have no chance to, for example, provide clustering of objects with their subobjects or avoid costly join operations.

Similar arguments hold for versions (relationship *is version of*).

As is true for record-oriented data models, structurally object-oriented approaches allow for the definition of application view to filter out the parts of a database that are needed or allowed for use by a specific application. Also, a number of provisions for model-inherent consistency exist. For example, it is mandatory (as in NF^2) or it is possible to enforce (as in DODM) that subobjects are not created without the existence of their respective superobject, and that upon object deletion, all subobjects are deleted as well. For relationships in DODM, minimal and maximal cardinalities may be defined for participating objects: they determine that a given object must take part in at least a given number of instances of a specific relationship and may be a member of at most a second given number of its instances. Furthermore, referential integrity guarantees that only existing objects participate in relationships (which requires, e.g., automatic deletion of relationship instances upon deletion of an object involved in one of their roles). As usual, key attributes or attribute combinations with the uniqueness property are also provided. All these constraints are automatically maintained by the data model operations or within transactions, respectively.

9.7 BEHAVIORAL OBJECT ORIENTATION

9.7.1 The Notion

In the preceding section we described new structuring concepts offered by advanced data models, such as NF^2 and DODM. As expressive as these concepts might be, they still require a new way of thinking by the engineering user because they are not concepts of his area of expertise; rather they are modeling concepts from the database area.

For the database user (e.g., a programmer of robot programs), it would be much more convenient to deal with concepts that are familiar to this application domain. In the case of robot programming, these concepts would include (models of) physical objects like robots, arms, axes, grippers, or logical objects like frames and assembly plans. With the concept of *objects*, therefore, we associate models of physically or logically existing entities of the application domain. Such an object should then be treated and manipulated as a unit. Thus the object-oriented database paradigm is nothing more or less than the provision of language features at the interface that support the modeling *and* manipulation of each object—in an efficient way.

The object-oriented view of an entity—within a certain context— at the database interface requires the hiding of the information that is not important with respect to the chosen context. For example, for the movement operation of an assembly part, only the current location and the desired location as well as the trajectory along which the assembly part is moved are of interest. The detailed description of the constituent parts (the volume, the material, etc. of the assembly part) are irrelevant in this context. Behavioral object orientation means hiding irrelevant details and thereby abstracting the user's view of an object to the information that is important for the context of the application. The user's perception of an object is therefore restricted to a set of operations that have their counterpart in the respective application domain. Typical operations that are provided for objects of type *assembly part* would be *scaling, translation, rotation, weight, volume*, and so on. All these operations can be applied without having a detailed knowledge of the internal structure of the object's representation.

The behavior-oriented "object view" was first approached in the programming language community with the concept of *abstract data types* (ADTs). The first programming language to provide language constructs for abstract data types was SIMULA 67. The class concept of SIMULA or—today—of Smalltalk allows the programmer to model similar objects as classes. A class consists of a set of functions, forming the interface, and an internal representation. A particular application object is then created as an instantiation of a predefined class.

It is actually a fairly straightforward idea to borrow the ADT concept from programming languages to achieve behavioral object orientation in database systems as well—nevertheless, only in the past couple of years has this been proposed. These approaches are characterized by providing language facilities at the interface level to define an abstract data type for a class of similarly structured and manipulated objects. These objects can then be inserted into the database, manipulated, and retrieved by applying the defined operations. The first research project that introduced the concepts into the realm of DBMSs is ADT-INGRES [24]. Rather than taking a structurally object-oriented database system as a basis for these systems, conventional relational DBMSs were chosen. In R^2D^2, a database system that was developed at the University of Karlsruhe in cooperation with the IBM Scientific Center Heidelberg [14], an abstract data type facility is incorporated in a structurally object-oriented database system.

In the following we first analyze ADT-INGRES as one representative of the systems that integrate objects in a conventional DBMS. Then we give a more detailed description of the R^2D^2 system, to highlight the definition of new data types for engineering applications.

9.7.2 One of the First Approaches: ADT-INGRES

ADT-INGRES was first proposed by Stonebraker and coworkers [22, 24]. Most of the ideas were subsequently incorporated in the database system POSTGRES [23]. ADT-INGRES provides a facility that allows the user to define data types. The representation of the new data type must be specified in the programming language C.

Let us now consider an example. We want to define a relation to store cuboids, basically in terms of its eight vertices. Rather than repeating for each of the vertices that a vertex consists of three decimal numbers (i.e., the x, y, and z coordinates), we specify once and for all an ADT vertex_type and then use it in the definition of the relation cuboids as follows:

```
cuboids (id,
         material: char(10),
         description: char(20),
         V1: ADT: vertex_type,
         V2: ADT: vertex_type,
            . . .
            . . .
         V8: ADT: vertex_type)
```

An example query using the ADT attribute vertex_type would look as follows:

```
range of c is cuboids
retrieve (c.material,c.description,c.V1)
where    c.id=5
```

The output of this query could then—depending on the implementation of the ADT—look like this:

material	description	V1		
		X	Y	Z
copper	massive	1.0	3.5	2.0

And a possible append command could look as follows:

```
append to cuboids
        (id=5,
         material="copper",
         description="massive",
         V1=(1.0,3.5,2.0),
         V2=(...),
         ...
         V8=(...))
```

The user must supply the implementation of such an abstract domain. For our example, this would be:

```
define ADT
        (typename="vertex_type",
         bytesin=9,
         bytesout=9,
         inputfunc-"to_internal_vertex",
         outputfunc="to_external_vertex",
         filename="/usr/ingres/.../vertex")
```

Inputfunc and outputfunc are C subroutines that convert the data type to internal and external representation, respectively. Outputfunc, for example, would extract the X, Y, and Z coordinates from the internal representation and output them in the format shown above. For the implementation of these routines, an intrinsic knowledge of C is required.

An obvious disadvantage of ADT-INGRES is that each abstract data type has to be mapped onto one attribute. In our case this means that the three coordinates are mapped onto an attribute of type string. This is a very unnatural mapping. It would be much more convenient (and natural) to map the coordinates onto three attributes of type float.

The schematic ADT mapping for our data type vertex-type is as follows:

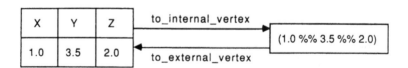

In addition to such a data type, the ADT-INGRES user can define operators on these domains. As an example, let us present the framework of the operator "R_y" which takes as an argument a vertex and an angle. It returns a vertex that is rotated about the y-axis by the given angle. Here we assume that the data type (which is just a numerical type), has been defined previously. The implementation would look as follows:

```
define adtop
      (opname="R_y"
       funcname="rotate_about_y"
       filename="/usr/ingres/.../rotate_y",
       result=ADT: vertex_type,
       arg1=ADT: vertex_type,
       arg2=ADT: angle_type,
       prec like "+")
```

Once again, the file rotate_y must contain a C program implementing this operator.

Similarly one can define the other possible geometric transformations scaling, translation, and rotation about the other two axes.

Let us now demonstrate a database query that rotates our previously inserted copper cuboid:

```
range of c is cuboids
replace   c(V1=R_y(c.V1,PHI),
            V2=R_y(c.V2,PHI),
            ...,
            V8=R_y(c,V8,PHI))
where     c.id=5
```

Discussion

ADT-INGRES provides a novel way of specifying new data types and corresponding operators in a database management system. The advantage of this approach lies in the fact that the operators can be arbitrarily complex. For example, we showed the framework for all the geometric transformations on 3D objects (i.e., scaling, translation, and rotation).

But the additional flexibility of the system also has its penalty. The new data types must be entirely specified in the programming language C. Thus the ADT-INGRES user must be familiar with two quite different systems: the database language QUEL and the programming language C.

Another shortcoming of this approach is inherent in the underlying data model of the database management system INGRES. ADT-INGRES does not allow one to map an ADT onto different tuples (or

relations). It requires each ADT to be mapped completely onto one
attribute. Thus the internal representation of engineering objects
does not reflect the external structure (as the user perceives it)
of the object. This usually results in a fairly tedious transformation process from external to internal representation, and vice
versa. For example, the ADT *vertex_type* had to be mapped into
a character string rather than onto three attributes of type float,
which would have been a much more natural mapping. The reason
for this unnatural mapping is that ADT-INGRES does not provide
any support to handle hierarchical data structures; but precisely
these occur frequently in engineering applications.

ADT-INGRES provides some facilities for behavioral object orientation by allowing the database user to define application-specific
ADT operations. But since the underlying data model is pure relations and thus not structurally object-oriented, these operations
are quite tedious to implement. We now show how abstract data
types can be integrated into one of the structurally object-oriented
data models that extend the relational model, the NF^2 model.

9.7.3 R^2D^2: A Structure-Based Approach

ADT-INGRES is one representative of a behaviorally object-oriented
DBMS based on a conventional data model. To achieve a *fully* object-oriented database system, the structural and the behavioral
paradigms should be combined. This is the objective of the R^2D^2
database system.

Our thesis is that the two approaches to object orientation—that
is, structural and behavioral object orientation—do not contradict
but rather supplement each other. They may enter into a fruitful
symbiosis such that one takes an existing structurally object-oriented database management system, or explicitly develops one, and
augments it by a behaviorally object-oriented user interface. Whereas the user views the database from an abstract level (where the
object structure becomes visible only indirectly via the object operators), the object structure is explicitly established when mapping the behavioral level to the structural level, so that all optimization techniques available on the latter level may be utilized.

Such a two-level approach appears to have three distinct advantages:

Engineering users deal with databases only indirectly via programs
    such as various design tools. Implementation of such tools takes
    place on the level of engineering objects and no longer requires
    a detailed knowledge of database technology.
Implementation of the engineering object types may also be shifted
    from the database expert to the engineering applications specialist,

since mapping the object structures becomes a fairly routine matter.
Database system performance can still be controlled in a manner geared to engineering applications.

We now present the database system R^2D^2 (Relational Robotics Database system with extensible Datatypes). R^2D^2 constitutes such a symbiotic approach to object-oriented database systems. This is achieved by integrating the concept of *abstract data types* (ADT) in the data definition and data manipulation language of a structurally object-oriented DBMS. Thereby the database user can define the data types and operations that correspond to application-specific objects and object manipulations. The behavioral object orientation in R^2D^2 is, again, demonstrated on our specific engineering application introduced in Section 9.3. The structural aspects are given by choosing a particular object-oriented data model, the NF^2 model. The internal representation of an ADT corresponds in R^2D^2 to a (possibly) nested NF^2 relation. Thereby it is guaranteed that the objects defined at the user level are also internally treated as clustered objects. This improves the performance of retrieving an object as a whole over systems in which such an object is segmented over different (flat) relations.

System Architecture

The R^2D^2 database system is based on the *structurally object-oriented* DBMS AIM-P [3], which is a prototype implementation of the nested relational model NF^2. R^2D^2 incorporates the concept of abstract data type in the nested relational model. This allows one to extend the built-in data types by application-specific data types. For this purpose an ADT specification language is provided in which the user defines new data types. Schematically the system architecture is depicted in Figure 9.16.

Assume that applications A and B are both from the same application domain, say robotics. Then both applications could rely on the same collection of application-specific ADTs. One such ADT would be the (model of a) robot *axis*. This ADT *axis* provides a collection of operations that serves as the interface with which the user can access an instance of type *axis*. Among these operations one would certainly provide *translate* and *rotate* to manipulate the kinematics of an axis. Such an ADT is mapped onto its NF^2 representation as defined in the specification language. For implementing the functions, two possibilities exist:

1. The computationally simple operations can be specified directly in an extended database language that is similar to SQL.
2. Computationally more complex operations are programmed in Pascal and are then linked to the DBMS.

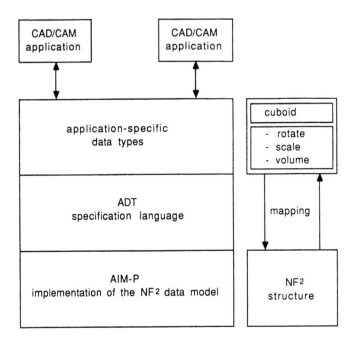

FIGURE 9.16   System architecture of R^2D^2.

In the following we demonstrate the expressive power of R^2D^2 on examples taken from Section 9.3.

Some Basic Data Types

User-Defined Data Types: In Section 9.6 it was demonstrated that querying a deeply nested relation may be quite complex.   The problems associated with the complexity of nested queries can be avoided by using abstract data types, as we demonstrate below. An ADT can be mapped onto a nested NF^2 relation, but the database user need not be aware of the internal representation if only the predefined ADT operations are to be used.

   Language concepts for defining abstract data types have been integrated in programming languages for a long time.   Abstract data types are used for objects that are frequently used and whose internal representation should be hidden from the user.   Here we want to demonstrate the abstract data type facility of R^2D^2 from a user's perspective.   We show how the data types and operations that are essential in robot programming can be integrated into the NF^2 model, thereby making the data model easier to use for storing engineering data.

The general approach to abstract data type definition is as follows. The NF^2 data model has, like most other models, only a limited set of built-in basic data types (character, numeric, Boolean, etc.). Of these basic data types, one can create structured objects, such as lists, tuples, and relations, with nesting of structured objects allowed (e.g., a list could have elements that are lists). This was demonstrated in Section 9.6. In R^2D^2 we allow the user in addition to define data types, which can be much more complex than the basic NF^2 types. In R^2D^2 a user-defined data type can be any "structured" NF^2 object. The following syntax is used to define new data types:

```
create ADT ⟨identifier⟩
is      ⟨NF²_type⟩
with    ⟨operations⟩
end ⟨identifier⟩.
```

The with-clause contains the specification of operations for the abstract data type, as discussed in the next section. The expression ⟨NF^2_type⟩ denotes any NF^2 object. Since we allow an abstract data type to be any NF^2 object, it is, in particular, possible to define abstract data types that are nested NF^2 objects. Furthermore, user-defined data types can be used like any built-in data type; that is, attributes can be of a type that has been defined by the user as an ADT.

Let us now demonstrate how we can support in R^2D^2 the data types that are needed in robotics applications described above. For this purpose we first define the ADTs vector and matrix.

```
create ADT vector
is    < 4 FIX real >     /*exactly length 4*/
with  ......             /*operations on vector*/
end vector.

create ADT matrix
is    < 4 FIX vector >   /*exactly 4 × 4 (a list of a list)*/
with  ......             /*operations on matrix*/
end matrix.
```

In the definition of these data types we use the built-in NF^2 data structure *list*, which is denoted by the pair of brackets "< >". The expression "4 FIX" denotes that the list has exactly four elements. The NF^2 model allows the user to access elements of a list by their position in the list; for example, vector[i] returns the *i*th

component of the list vector. Another built-in function on lists is
INDL, which returns the index range of a list. In our example
INDL(V) returns $1\cdots4$, where V is of type *vector*.

The definition of the ADT *matrix* neatly demonstrates how user-
defined data types can be used like built-in data types. Thus *matrix* consists of a list of vectors, where *vector* is an ADT consisting
of a list of four numeric values. Therefore, the nesting within the
data type *matrix* is actually hidden from the user, who might not
be aware of the internal implementation of a matrix.

Definition of New Operations: So far we have dwelt on the struc-
tual aspects of the ADT concept. For a behavioral orientation, the
definition of an ADT should include user-defined operations. In
R^2D^2 these operations are defined either in an SQL-like language
or in Pascal. An operation can be specified within the ADT definition clause as in the examples below, or it can be added explicitly
to an existing ADT at some later time. Rather than presenting the
formal syntax of this language, let us demonstrate it from a user's
viewpoint by implementing some example operations on the ADT's
vector and *matrix*.

```
create ADT vector
is      < 4 FIX real >
with
operation "+ᵥ" (V,W:vector) returns vector
        return
                select  V[i]+W[i]
                from    i in INDL(V)
        end "+ᵥ";

operation "*ᵥ" (V,W:vector) returns real
        return
                SUM (select V[i]*W[i]
                        from i in INDL(V))
        end "*ᵥ"
end vector.
```

In the last operation "$*_v$" we use the built-in function SUM,
which takes as an argument a list of numeric values and returns
their sum. We also note that we allow for infix operators, which
are denoted by enclosing the operator in quotation marks.

```
create ADT matrix
is      < 4 FIX vector >
with
operation "*ᵥ,ₘ" (V:vector,M:matrix) returns vector
```

```
      return
            select V *ᵥ M[i]
            from i in INDL(M)
      end "*ᵥ,ₘ";
operation row (k:integer,M:matrix) returns vector
      return
            select M[i][j]
            from  i in INDL(M), j in INDL(M[i])
            where j=k
      end row;

operation transpose (M:matrix) returns matrix
      return
            select row(i,M)
            from  i in INDL(M)
end transpose;

operation "*ₘ" (M,N:matrix) returns matrix
      return
            select
                  select transpose(M)[i] *ᵥ,ₘ N[j]
                  from  i in INDL(M)
            from j in INDL(N)
      end "*ₘ"
end matrix.
```

The last operation (query) obviously returns as a result a list of a list. The inner *select* statement loops through all column vectors of transpose(M) (i.e., the rows of the matrix M) and multiplies them with the jth column of N, where j is fixed. The outer *select* loops through all j in INDL(N); that is, it loops through all column vectors of N. The result of this query, therefore, is the multiplication of the two matrices M and N.

More Complex Examples

In addition to these basic types that form the basis of any CAD/ CAM application, one needs to define more complex objects. As a first such example, let us return to our old friend the cuboid, but now in a somewhat different representation. The ADT *cuboid* consists of an internal NF^2 representation and operations to access and manipulate the state of the cuboid. Schematically, the definition of the ADT *cuboid* is shown in Figure 9.17.

The ADT *cuboid* consists of an internal representation (i.e., an NF^2 structure), which—in this case—consists of a tuple with attributes MATERIAL and GEOMETRY. GEOMETRY itself is a composite

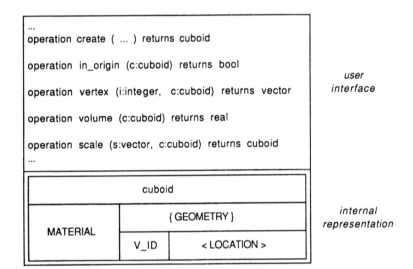

FIGURE 9.17   User interface and internal representation of the ADT cuboid.

type, in this case a list of exactly eight tuples with attributes V_ID and LOCATION representing the vertex identifiers and the x-, y-, z-coordinates of the cuboid's vertices.

The rudimentary user interface to access data objects of type *cuboid* consists of the three functions *in_origin*, which returns true if one of the bounding vertices lies in the origin of the coordinate system; *vertex*, which returns the coordinates of the specified vertex; and *volume*.

R^2D^2 allows the definition of abstract data types by stepwise refinement. Thus the definition of the ADT *cuboid* could be based on the definition of *vector* (we now assume homogeneous coordinates), which was provided earlier.

```
create ADT cuboid is
       [MATERIAL:string(10),
        GEOMETRY:⟨8 FIX [V_ID: string(2),
                          LOCATION: vector ]⟩]
   with
       operation vertex(i:integer,c:cuboid) returns vector
                 return
                 c.GEOMETRY[i].LOCATION
       end vertex;
```

```
     operation in_origin(c:cuboid) returns bool
             return
             exists (v in c): v.LOCATION = ⟨0,0,0,1⟩
     end in_origin;

     operation volume(c:cuboid) returns real
             . . . .

end cuboid.
```

We note that every ADT definition consists of two parts: the internal NF^2 representation and the operations to access this representation. The internal representation of an ADT is specified in the DDL of AIM-P. For the implementation of the operations two possibilities exist. The simple operations, such as *in_origin* and *vertex*, can be specified in the AIM-P query language [14]. Computationally more complex operations, particularly those that are also used in an application program (e.g., *volume*) are specified in Pascal.

Now a database user could state the following query:

```
select vertex(1,c)
from c in CUBOIDS
where in_origin(c)
```

to retrieve the first vertex of (all) cuboids that have one bounding vertex in the origin of the underlying coordinate system. In our example, CUBOIDS is just a relation (set) of objects of type *cuboid*, which could have been declared as follows in R^2D^2:

```
create CUBOIDS
       {cuboid}     —a set of cuboid instances—
end
```

Let us now return to robotics. As mentioned above, a robot axis is described by its dynamics and its kinematics. The position of an axis is described by a 4×4 matrix; that is, we can use the ADT *matrix* to define an ADT *axis*.

```
create ADT axis
is      [ID: string(20),
         KINEMATICS:
                 [DH_MATRIX: matrix],
         DYNAMICS:
                 [MASS: real,
                  ACCEL: real]
        ]
with ......  /*operations on axis*/
end axis.
```

For this ADT we now want to define an operation $rotate_z$, which rotates an axis around its z-coordinate axis by an angle alpha. Rotation of axes is described by multiplying the axis position relative to the robot basis as expressed by the DH matrix with a special 4×4 rotation matrix. The latter is created by an operation $rot\text{-}matrix_z$, which should be part of the ADT *matrix*:

```
operation matrix_rot_z (alpha: real) returns matrix
        return
                ......
end matrix_rot_z
```

Accordingly then, rotation may now be introduced into the ADT *axis* in the form of an additional operation $rotate_z$:

```
operation rotate_z (A: axis, alpha: real) returns axis
        return
                [ID: A.ID,
                 KINEMATICS:
                 [DH_MATRIX: matrix_rot_z(alpha)
                         *_m A.KINEMATICS.DH_MATRIX],
                 DYNAMICS: A.DYNAMICS
                 ]
end rotate_z
```

The abstract data types and operations that were introduced here can be used in the data manipulation constructs as in any built-in data type and operation. Thus, a query that is based on a complex data object becomes much easier to formulate because the user is not required to know the internal representation of the ADT simply to apply the predefined operations. We demonstrate this on an example in which we want to retrieve a specified axis after rotating it by 30° around the z-coordinate axis.

```
select rotate_z (AR.AXES[2],30)
from   AR in R.ARMS,
       R in ROBOTS
where  R.ID = 'Artoo Detoo' and AR.ID = 'left'
```

The abstract data type facilities of the R^2D^2 database system provide powerful constructs for modeling and selecting database objects. The operations associated to an ADT allow the database user to specify application-spectific selection criteria to retrieve particular objects from the database.

9.7.4 Application Programming

But, of course, the development of CAD/CAM applications requires an interface between the database system and a general-purpose programming language. Therefore, one of the main objectives of the R^2D^2 project, aside from supporting the structural modeling of objects, is to provide an engineering programming environment that facilitates the efficient manipulation of objects. This objective is achieved by keeping the objects that are relevant for a particular transaction in a main memory representation that is within the application program's address space. The operations on objects are then performed on this main memory data structure.

In R^2D^2 this capability is provided in the form of an *object cache*. Each engineering transaction possesses a *local* database with an associated object cache. The object cache together with the associated local database form the *private* database of an engineering transaction. Database objects are transferrable into the local database. Objects in the private database are copies of objects stored in the *public* database. Objects that reside in the local database can be fetched into the associated object cache. When an object is resident in the cache, it can be manipulated by operations that are specified in terms of the host programming language (e.g., Pascal). Schematically this is depicted in Figure 9.18.

Interaction Between Global and Local Databases

An engineering transaction may request objects from the public (global) database. Objects thus specified are *checked out* of the public database and stored as a physical copy in the local database—provided no conflict with the checkout specification of another transaction occurred [11]. The local database will typically reside on an engineering workstation. Within the local database, objects are still represented in the format of the underlying data model. Objects are checked out either *for read* or *for update*. In both cases they are appropriately locked, to avoid any uncontrolled concurrency among parallel transactions. In the case of *for update* the objects are *checked* back *into* the public database at the end of the transaction.

Interaction Between the Local Database and the Object Cache

From the local database objects are transferred into the object cache. This can occur via an explicit *prefetch* statement in the program or just by referencing a particular object that exists in the local database. In the latter case it is implicitly *fetched* into the object cache; however, the operation that references an object

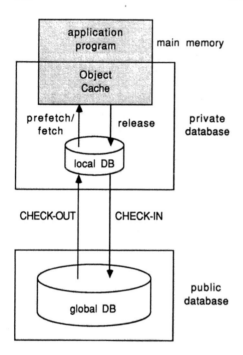

FIGURE 9.18 Overall object cache architecture of R^2D^2.

not resident in the object cache might be rather inefficient because of the additional time required to transform the object. After the first reference, the object remains in the main memory representation until it is either explicitly or implicitly released from the object cache. Thus it is—not considering the performance—transparent to the user whether an object is resident in the cache or in the local database. For implicit release the programmer can—depending on the object size—specify the maximum number of objects of a particular type that may reside in the cache. In case this number is reached, the runtime system will release the least recently used object(s) into the local database.

Transformation of Objects from Their Database
into Main Memory Representation

Pascal has been chosen as a first host programming language of R^2D^2. To provide convenient access from Pascal to the objects resident in the *object cache*, Pascal data structures have been chosen

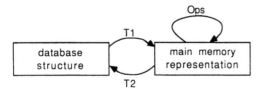

FIGURE 9.19   Transformation steps.

as main memory representation of database objects. This requires the objects to be transformed from their database structure into a Pascal data structure (T1) and vice versa (T2). Operations (Ops) on objects of the underlying data model can be implemented on this Pascal representation. Schematically this is represented in Figure 9.19.

The Pascal representation is—under some user control—automatically generated. Steps T1 and T2 of Figure 9.19 are automatically carried out by the system, meaning in particular that the program code for the transformations is automatically generated.

9.8 CONCLUSIONS

By necessity, this has been a very cursory introduction to a highly complex subject. Assuming that the readers are more interested in using database systems than in developing such systems, we have concentrated on issues that are directly related to designing the database structure, to the creation, manipulation, and retrieval of the data. These issues fall under the heading of application-oriented data structuring of Section 9.2.1. The other topics mentioned there— data integration, consistency, multiuser operation, recovery, transactions, protection, and data independence—are of similar importance. However, since they depend in many of their details on the data structuring facilities chosen, the application-oriented data structuring is the natural topic to start with. This is not to say that the other topics have no effect on usage characteristics—in fact, they do have some effect. Readers interested in these topics are, therefore, referred to the literature mentioned in Section 9.2.4.

One critical issue not discussed in this chapter so far is performance. Database technology remains unattractive for planning and manufacturing applications if response times to update commands or queries remain unacceptably high, or if the throughput of concurrent

requests from within the planning and manufacturing environment
does not satisfy demands. Some of the examples in Section 9.4
seem to indicate that the database systems currently available in
the marketplace may indeed be something of a stopgap measure.
Object-oriented database systems are an attempt to overcome the
weakness. The reasons are very simple: relational systems have
no way of directly reflecting technical objects, whereas object-oriented systems are designed to do precisely this. As a consequence,
the former cannot gear or optimize their internal architecture, storage structure, or access aids to a universe of objects, whereas the
latter can, although—as we have seen—not always perfectly.

To conclude then, the long-range objective of planning and manufacturing applications should be to utilize object-oriented database
technology, in particular because there is a good reason to assume
that products will become commercially available in a few years
hence. In the meantime, classical database technology approaches
should be employed wherever their performance is sufficient because of the many beneficial characteristics they offer beyond application-oriented data structuring.

REFERENCES

1. D. S. Batory and A. P. Buchmann, "Molecular Objects, Abstract
    Data Types, and Data Models: A Framework," *Proceedings of the
    VLDB Conference*, 1984, pp. 172—184.
2. P. Chen, "The Entity—Relationship Model: Toward a Unified
    View of Data," *ACM Trans. Database Syst. 1*:1, 9—36 (March
    1976).
3. P. Dadam et al., "A DBMS Prototype to Support Extended NF^2
    Relations: An Integrated View on Flat Tables and Hierarchies,"
    *Proceedings of the ACM SIGMOD Conference on Management of
    Data*, 1986, pp. 376—387.
4. C. Date, *An Introduction to Database Management Systems*, 4th
    ed. Addison-Wesley, Reading, MA, 1986.
5. R. Dillmann and M. Huck, "A Software System for the Simula-
    tion of Robot-Based Manufacturing Processes," *Robotics 2*:1
    (March 1986).
6. K. R. Dittrich, "Object-Oriented Database Systems," *Proceed-
    ings of the Fifth Entity—Relationship Conference*, North Holland,
    Amsterdam, 1987, pp. 51—66.
7. K. R. Dittrich and U. Dayal, Eds., *Proceedings of the 1986 In-
    ternational Workshop on Object-Oriented Database Systems*. IEEE
    Computer Society Press, New York, 1986.
8. K. R. Dittrich, W. Gotthard, and P. C. Lockemann, "Complex
    Entities for Engineering Applications," *Proceedings of the Fifth*

Entity—Relationship Conference, North Holland, Amsterdam, 1987, pp. 421—440.

9. K. R. Dittrich, W. Gotthard, and P. C. Lockemann, "DAMOKLES: A Database System for Software Engineering Applications," in *Lecture Notes on Computer Science*, Vol. *244*, pp. 353—371. Springer-Verlag, Berlin, 1987.

10. K. R. Dittrich and R. A. Lorie," Version Support for Engineering Database Systems," *IEEE Trans. Software Eng.* (in press).

11. M. Dürr and A. Kemper, "Transaction Control Mechanism for The Object Cache Interface of R^2D^2," *Proceedings of the Third International Conference on Data and Knowledge Bases: Improving Usability and Responsiveness*, Morgan-Kaufman, Jerusalem, pp. 81—89, June 1988.

12. J. Foley and A. van Dam, *Fundamentals of Interactive Computer Graphics*. Addison-Wesley, Reading, MA, 1983.

13. R. H. Katz, *Information Management for Engineering Design: Surveys in Computer Science*. Springer-Verlag, Berlin, 1985.

14. A. Kemper, P. C. Lockemann, and M. Wallrath, "An Object-Oriented Database System for Engineering Applictions," *Proceedings of the ACM SIGMOD Conference on Management of Data*, San Francisco, May 1987, pp. 299—311.

15. A. Kemper and M. Wallrath, "An Analysis of Geometric Modeling in Database Systems," *ACM Comput. Surv.* *19*:1, 47—91 (March 1987).

16. A. Kemper, M. Wallrath, and P. C. Lockemann, "Database Support for Robotics Applications," in U. Rembold and K. Hörmann, Eds., *NATO International Advanced Research Workshop on Languages for Sensor-Based Control in Robotics*, NATO ASI Series, Vol. F29, Springer-Verlag, Berlin, 1987, pp. 283—304.

17. W. Kent, "Limitations of Record-Based Information Models," *ACM Trans. Database Syst.* *4*:1, 107—131 (1979).

18. H. F. Korth and A. Silberschatz, *Database System Concepts*. McGraw-Hill, New York, 1986.

19. R. A. Lorie and W. Plouffe, "Complex Objects and Their Use in Design Transactions," *Proceedings: Engineering Design Applications, Database Week*, 1983, pp. 115—121, San Jose, CA, 1983.

20. A. A. G. Requicha, "Representations for Rigid Solids: Theory, Methods, and Systems," *ACM Comput. Surv.* *12*:4, 437—464 (1980).

21. H. J. Schek and M. Scholl, "The Relational Model with Relation-Valued Attributes," *Inf. Syst.* *11*:2, 137—147 (1986).

22. M. Stonebraker, J. Anton, and E. Hanson, "Extending a Data-Base System with Procedures," *ACM Trans. Database Syst.* *12*:3, 350—376 (1987).

23. M. Stonebraker and L. Rowe, "The Design of POSTGRES,"
     *Proceedings of the ACM SIGMOD Conference on Management of
     Data*, Washington, DC, pp. 340—355, 1986.
24. M. Stonebraker, B. Rubenstein, and A. Guttman, "Application
     of Abstract Data Types and Abstract Indices to CAD Data-
     bases," *Proceedings: Engineering Design Applications, Data-
     base Week*, San Jose, CA, 1983.
25. D. C. Tsichritzis and F. H. Lochovsky, *Data Models*, Prentice-
     Hall, Englewood Cliffs, NJ, 1982.
26. G. Wiederhold, *Database Design*, 2nd ed. McGraw-Hill, New
     York, 1983.

10

Development Toward Autonomous Systems

VOLKER TURAU* *University of Karlsruhe, Karlsruhe, Federal Republic of Germany*

10.1 INTRODUCTION

10.1.1 Why Autonomous Systems?

Up until a few years ago, robots were essentially arms designed to duplicate human movements. Hence, there was usually a rotating base, carrying a column with a telescopic arm, or an arm consisting of revolute joints. The "wrist" could be rotated and bent. Initially, these robots were very costly, so it was uneconomic to use them for simple jobs like handling. One of the earliest applications to find widespread use was in the spot welding of car bodies. Recent developments in sensor and microelectronic technology have spurred a surge of development in robotics.

At present most robots are used by industry for stationary applications. However, there are numerous tasks for which a totally autonomous robot system is required—for example, for repair work under extremely hostile conditions imposed by outer space and deep ocean environments or in a nuclear plant. Other applications are tending machine tools in a flexible manufacturing system or transporting parts in a low piecerate production environment. Discrete parts manufacturing could be substantially improved using autonomous vehicles for material and tool transport, where these vehicles can move independently on the manufacturing floor to any desired location.

Existing transport systems usually consist of a network of roads and guide cables, each adopted to generate an electromagnetic field of low frequency and buried under the surface of roads along the

*Current affiliation: Mathematic and Data Processing Services, Darmstadt, Federal Republic of Germany

routes. By detecting the magnetic field, the vehicles are guided
through the factory. The installation costs are very high, and modification of the network to introduce a new workstation is very expensive and disturbs the normal flow of production. The use of
autonomous vehicles would increase productivity significantly.

Most of the existing experience in implementing autonomous systems has been acquired through exploring moving vehicles, rather
than through autonomous assembly robots. This is probably the
best approach, because autonomous mobility provides a well-defined
task domain where success and failure are easily apparent. For the
first generation of autonomous vehicles, basic issues in vision, planning, and robot control were explored. However, they were all
seriously hampered by primitive sensing and computing hardware
and software. More recent efforts have overcome many of the limitations, and very sophisticated second-generation autonomous vehicle testbeds have evolved.

To support the development of autonomous mobile robot systems,
many industrial countries have started national programs to develop
basic technology for these devices. In the United States, DARPA,
the Defense Advanced Research Projects Agency, has the goal of
building an autonomous land vehicle, which will go on missions over
rugged terrain and attempt to accomplish different assignments [1].
In Europe the programs ESPRIT, RACE, BRITE, and EUREKA are
sponsoring the development of robots and autonomous vehicles [2].
Japan has a national project known as Robots for Critical Work
(RCW), which is designed to guide Japanese robotics research and
development well into the 1990s. RCW was started in 1983, and the
overall theme of the project is to develop robotic systems capable
of operating in hazardous environments.

10.1.2 Some Existing Autonomous Systems

The development of autonomous robot systems started in the early
1970s when the Stanford Research Institute (SRI) built an intelligent vehicle called Shakey [3]. However, it was not until 1980
that this subject elicited a general interest. During this period
much progress was made in developing advanced computer architecture, and sophisticated sensors. This new technology is used to develop more advanced autonomous robot systems. In the following
a short review of existing autonomous robot systems is given.

Autonomous Land Vehicle (ALV)

ALV was developed in connection with DARPA by various universities, research institutes, and companies [4]. The vehicle will
travel on roads, as well as on rough terrain; thus it can examine

the environment, send abstract commands, find out the way, plan
a mission with multiple objectives, and carry out the same. It is
driven by six wheels, which are controlled similar to caterpillar
tracks. ALV is equipped with two cameras, a laser, and an ultrasonic distance measuring unit. The planning activities are done on
four different levels. At the highest level the abstract task description is transformed into a set of geographical objectives and
the various associated constraints. At the next level the global
trajectory is planned and thereby a special route is produced. To
fit the global route to local conditions, a local planning module
(navigator) at the next level details this route. At the lowest level
the reflexive planning (pilot) executes the real-time control of the
vehicle. Apart from this planning hierarchy, there is a multiple-
tier hierarchy for sensor processing.

Autonomous Mobile Robot of the University of Amsterdam

The objective of the Dutch autonomous mobile robot project is to
develop a mobile platform with no connections to the outside world,
which is able to reach a goal position in an office environment [5].
The system is capable of performing "goto(x,y)" tasks, avoiding
obstacles that were unknown to the system beforehand. The tasks
can be defined with the aid of a two-dimensional map of the environment. Obstacles are related with ultrasonic range finders and
a camera. The vehicle has a multiple microprocessor system on-
board. The platform has real-time behavior and can drive with a
velocity of about 0.5 m/s.

Carnegie Mellon University (CMU) Rover

The CMU Mobile Robot Laboratory was started in 1981 by A. M.
Robots in Pittsburgh to pursue research in perception, planning,
and control for autonomously roving robots. The long-term plan is
to develop an accurate, very maneuverable, self-powered vehicle
carrying a small manipulator. The prototype Pluto, has an omnidirectional drive system for accurate control of robot motion in 3
independent degrees of freedom. The design uses three complex
wheel assemblies, each with two motors to independently drive and
steer its own wheel. Pluto is equipped with a collection of sensors
including cameras, sonar, and bump detectors. It was used in a
series of experiments in vision, navigation, and planning. The
planning is rule-based, and the implementation is based on a blackboard system. So-called expert modules, which control the operation of sensors and actuators, are distributed over the processor network. An executive level for each processor is responsible
for process scheduling. The robot system is controlled by a set

of primitives, which provide process handling, message-based inter-
process communication, and access to the blackboard.

KAMRO: Karlsruhe Autonomous Mobile Assembly Robot

At the Institute for Real-Time Computer Control Systems and Robotics of the University of Karlsruhe, an autonomous mobile robot
has been under development since 1985 to be used for transport
and assembly operations in a flexible manufacturing cell [6]. It is
provided with two Puma 260 robots, which are based on a mobile
platform. The vehicle has four wheels with an individual drive for
each and is provided with diagonally placed passive rollers, so that
it can move in any direction. It is equipped with a camera for two-
dimensional part identification. The assembly operations are supported
by two force—torque sensors and two cameras located in the end-effectors. The navigator is based on an ultrasonic sensor system and
four cameras. It has a hierarchically structured control system. The
executive and the sensor modules will be implemented as blackboard
systems, which are coupled through a special communication interface.
The planning module produces a plan, which consists of a sequence
of elementary operations. The elementary operations are executed
by individual elementary operation modules. These modules comprise the expert knowledge necessary to execute an elementary operation.   Therefore, they are able to carry out autonomously complex, sensor-guided tasks. Thus, the planning module specifies
only the desired goal, whereas the elementary operation modules
specify the way this goal is actually achieved.

MIT AI/Lab Mobile Robot

The intention of the mobile robot project at the Artificial Intelligence Laboratory at MIT is to build a mobile robot, that wanders
around in an unconstrained laboratory area and in computer machine rooms [7]. Currently installed sensors are a ring of twelve
Polaroid sonar time-of-flight range sensors and two Sony CCD cameras. The sonars are arranged symmetrically around the rotating
body of the robot, and the cameras are on a tilt head. The system is based on a control architecture that differs substantially
from the structures described so far. Layers of the control system are built to let the robot operate at increasing levels of competence. Layers are made up of asynchronous modules that communicate over low bandwidth channels. Higher level layers can
subsume the roles of lower levels by suppressing their outputs.
However, lower levels continue to function as higher levels are
added.

Microbe

Development of the experimental autonomous system MICROBE
was begun at the Institute of Control Techniques of the Technical
University in Munich in 1979 [8]. This project aims at achieving
the automation of transportation problems employing sensors and
knowledge-controlled processes. MICROBE is a four-wheeled vehicle with step motor driven wheels and rolling supports. It is
equipped with a rotating ultrasonic sensor for distance measurements.

Yamabiko

Yamabiko is a self-contained robot built for artificial intelligence
research. The development started in 1978 at the University of
Tsukuba (Japan). The robot consists of two microcomputers, ultrasonic sensors in four directions, and two independently driven
wheels. The robot can understand its simple world with ultrasonic
sensors while moving around. Mobility is the most important feature
of Yamabiko. It has no global planning system. Echoes from the
ultrasonic source are evaluated directly to detect the presence of
walls and obstructions.

10.1.3 Demands on Autonomous Systems

The majority of current industrial robot applications are performed
using position control alone without significant external sensing. Instead, the environment is engineered to eliminate all significant
sources of uncertainty. Special-purpose devices are designed to
compensate for uncertainty in each grasping or assembly operation.
In the absence of any form of sensing, a fixed sequence of operations is the only possible type of robot program. This model is
certainly not powerful enough to support an autonomous system.
Therefore, the first criterion for an autonomous system is the ability to explore the environment with sensors. Sensing enables robots to perform tasks in the presence of significant environmental
uncertainties, without special-purpose tools. Sensors can be used
to identify the position of parts, to inspect parts, to detect errors
during manufacturing operations, and to accommodate to unknown
surfaces.

The second criterion for an autonomous system is the ability to
monitor its activities and to react to unexpected events. Robots
fail whenever they encounter an unexpected event, no matter how
minor. Even with complete models of robot kinematics and dynamics, it is impossible to anticipate all possible errors. Hence, methods

of reasoning about tasks and events are needed to enable the robot
to work reliably without human intervention. Knowledge is needed
to detect unexpected situations at execution time and to handle them.
After detecting an unexpected event, the robot must be able to analyze the situation. This requires that a complete description of the
current status of its world is available at any time. Furthermore,
the system needs to know how to perform a diagnosis with the help
of the sensor system and to reason automatically.

To be fully autonomous, a system needs to be able to recover
after unexpected events. This in turn calls for automatic plan generation ability. The generation of a plan includes task decomposition, task scheduling, resource management, and geometric planning (manipulator transfer movements, route planning for mobile
robots, etc.). For example, planning of transfer movements requires the ability to plan a path for the manipulator to avoid collisions with objects in the workspace and the ability to choose safe
grasp points on objects.

The above-mentioned tasks can be accomplished only with the
help of a common database. The database contains all available information: geometrical data about robots and objects, the task description, information about the accuracy of sensors, and so on.
When the environment is not known a priori, however, some mechanism must be available for presenting the positions of objects and
their features. Some of these positions are fixed throughout the
task; others must be determined from sensor information, and still
others bear a fixed relation to variable positions.

Often the robot will be trying to achieve multiple goals, some of
them conflicting. It may be trying to reach a certain point ahead
of it, while avoiding local obstacles. Often the relative importance
of goals is context-dependent. The control system must be respon-
sive to high priority goals, while still servicing necessary "low level"
goals. In many cases, an assembly process cannot be considered to
be a linear succession of commands to be executed by the robot. A
parallel description of the task allows the control system to choose
among several possibilities according to a decision criterion. Thus,
an autonomous robot system needs the ability to perform parallel actions.

The robot ought to be robust. When some sensors fail, it should
be able to adapt and continue its work by relying on those still functional. When the environment changes drastically, the robot should
be able to achieve a modicum of sensible behavior.

These requirements are difficult to obtain because, to achieve
real-time response, large amounts of processing power are necessary. One way of achieving this is to apply several processors to
the problem. All this, however, brings the need to develop new
and adequate distributed control and problem-solving mechanisms.

In summary, for a robot system to be autonomous, it must have the ability to:

Sense the environment
Monitor its execution
Reason about tasks and events
Recover after unexpected events
Generate plans automatically
Achieve multiple goals
Perform parallel actions
Respond in real time

10.2 ARCHITECTURE OF AN AUTONOMOUS MOBILE ROBOT

10.2.1 Overall Structure

The primary goal of the architecture of an autonomous robot system should be flexibility. Flexibility is required because the precise characteristics of an autonomous system are unknown. To face the requirements enumerated in the preceding section, a distributed software control structure seems to be appropriate. In such an architecture, expert modules run as independent processes and exchange information over a common memory. This could be realized by a blackboard structure. To simplify system integration, general specifications for several independent modules must be established. The modules are distributed over a processor network and communicate through messages.

Expert modules are specialized subsystems used to control the operation of the actuators and sensors, interpret sensory, and feedback data, to plan strategies to accomplish tasks, and to supervise the execution of the plan. Figure 10.1 illustrates the basic structure of an autonomous robot system.

A control system for a completely autonomous mobile robot must perform many complex information processing tasks in real time. It operates in an environment where the boundary conditions are changing rapidly, and these boundary conditions are determined over very noisy channels, since there is no straightforward mapping between the information acquired through sensors (e.g., ultrasonic, TV camers) and the form required for boundary conditions. Hence data must be transformed into suitable forms. This must be done on different levels. There are several possibilities to realize such a system.

Albus [9] has advocated a hierarchical control system for sensing, interpreting, and planning, as well as for execution. In this architecture, high level goals are decomposed through a succession

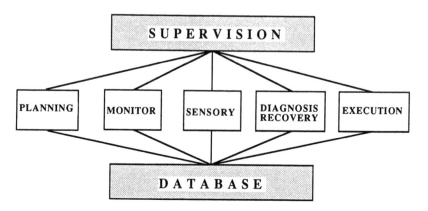

FIGURE 10.1   Software structure.

of levels, each producing strings of simpler commands to the next lower level. The bottom level generates the signals for the robot actuators. Each control level is a separate process with a limited scope of responsibility. Each performs the generic control function of sampling its input and generating appropriate outputs. An important aspect of a control structure is the distribution of knowledge through the system. In a hierarchical model, much information about the current status is hidden in the different layers of the system. This conflicts with the goal of centralized information. An instance of each layer must be built to run the robot system at all. Later changes to a particular piece (to improve it or extend its functionality) must be done in such a way that the interfaces to adjacent pieces do not change; if this is not possible, the effects of change must be propagated to neighboring pieces, changing their functionality too.

Brooks [7] describes an architecture based on subsuming levels. Layers of the control system are built to let the robot operate at increasing levels of competence. The key idea of levels of competence is that it is possible to build layers of a control system corresponding to each level of competence and simply add a new layer to an existing set to move to the next higher level of overall competence. Brooks [18] suggested the following levels of competence for an autonomous mobile robot:

0. Avoid contact with objects (whether the objects move or are stationary).
1. Wander aimlessly around without hitting things.

2. "Explore" the world by seeing places in the distance that look reachable and heading for them.
3. Build a map of the environment and plan routes from one place to another.
4. Notice changes in the "static" environment.
5. Reason about the world in terms of identifiable objects and perform tasks related to certain objects.
6. Formulate and execute plans that involve changing the state of the world in some desirable way.
7. Reason about the behavior of objects in the world and modify plans accordingly.

Since a level of competence defines a class of valid behavior, it can be seen that higher levels of competence provide additional constraints on that class.

To realize such an architecture, a complete robot control system that achieves level 0 competence is built. This system is thoroughly debugged and it is never altered. Next, another control layer is built, one that is able to examine data from the level 0 system and is also permitted to inject data into the internal interfaces of level 0, suppressing the normal data flow. The zeroth layer continues to run, unaware of the layer above it, which sometimes interferes with its data paths. The same process is repeated to achieve higher levels of cometence.

The advantage of this architecture is that individual layers can be working on individual goals concurrently. The suppression mechanism then mediates the actions that are taken. So, there is no need to make an early decision on which goal should be pursued. Another advantage is that individual layers can be realized on different, loosely coupled processors.

A drawback of this architecture is that the system can be extended only by adding a new layer on top of the rest. But some extensions (e.g., adding a new sensor) are bound to have an effect on all levels of competence.

Another approach to achieving a control structure is used in the development of an autonomous mobile assembly robot (KAMRO) at the University of Karlsruhe [6]. The main features of that approach are modularity and the introduction of parallelism between operations. The modular aspect of the system is very important. Many operations are frequently used and are the same in many assembly problems. Therefore, operations are described independently, so that they can be used in several tasks. In such a system, every operation is a composition of a set of elementary operations. An elementary operation is the "smallest" executable directive. The elementary operations are executed by special modules, each dedicated to its own elementary operation. These modules comprise the expert knowledge necessary to execute such an operation. The key idea is that the planning module specifies only the desired goal, whereas the elementary operation modules specify the way this goal is actually achieved.

The second feature of this approach is the introduction of parallelism. In many cases, the assembly process cannot be considered to be a linear succession of commands to be executed by the robot. A parallel description of the task permits the central system to choose among several possibilities, according to a decision criterion. Elementary operations can be executed in parallel. For example, the assembly of a workpiece will usually be carried out by two robots.

10.2.2 KAMRO

This section presents the overall structure of the autonomous mobile robot KAMRO developed at the University of Karlsruhe. The task of the robot is to perform assembly and transport operations in a flexible manufacturing cell (Fig. 10.2) that contains various assembly stations and a material storage system. From a given set of assembly instructions for a product, the computer system produces an executable plan. The robot selects parts from a material storage system and transports them to a workstation. After docking at the station, the two robot arms will assemble the parts to a complete workpiece under the supervision of a multisensor system.

A complete assembly operation has the following pattern. The robot system receives the order to assemble a complex workpiece. The planning system obtains information about the workpiece from a CAD system and determines the necessary elementary assembly operations and their order. The robot collects the different components of the workpiece from the material storage system and transports them to an assembly station. A binary vision system identifies the parts and determines their relative position and orientation. The assembly is accomplished by the robot arms. The operation is supported by the force—torque sensors, which are integrated into the end effectors. These sensors allow the measurement of the forces between the workpiece and the environment. For example, the information is used to detect collisions and to initiate corresponding corrections.

A second vision system monitors the assembly and informs the assembly system about eventual problems. In case of a disturbance, the system accomplishes the necessary corrections. After the system has finished the assembly, it inspects the functionality of the workpiece and moves to the next workstation. Here the robot detects possible obstacles with ultrasonic sensors and determines a

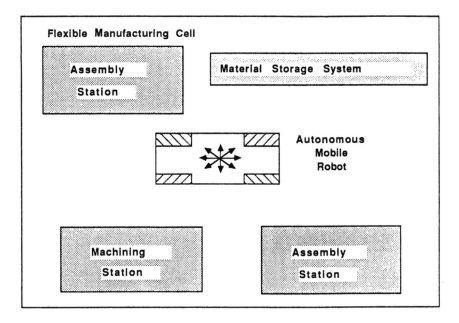

FIGURE 10.2   Flexible manufacturing cell.

safe path around the obstacles. The docking procedure is supported
by two cameras, which are integrated into the end-effectors of the
manipulators. Figure 10.3 shows the robot system KAMRO.

Overall Structure of KAMRO

To realize an autonomous system, various technologies must be
integrated. To simplify this integration, several independent subsystems were designed. Each of these subsystems is a self-contained module performing a well-defined set of operations. Figure
10.4 shows the different modules and how they communicate.

The central unit of the system is the supervision module (SM).
It is an active, superior module, which coordinates and controls the
activation of the entire system and the interaction between the modules. The SM has an interface to the user. The user presents a
task description in a suitable form to the system. This task description includes information about the objects and the layout of
the workcell. The SM activates the planning module (PM). This
module produces an executable plan, which consists of a sequence
of elementary operations (EO). An EO is the "smallest" directive

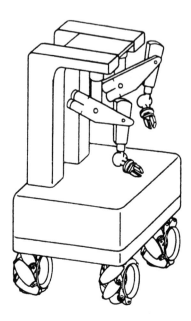

FIGURE 10.3   KAMRO.

the PM is able to plan; hence it cannot be decomposed any further. The EOs are executed by elementary operation modules (EOM). Each of these modules is dedicated to its own EO.   The EOMs comprise the expert knowledge necessary to execute an EO.   Therefore they are able to carry out automatically complex, sensor-guided tasks. Thus, the PM specifies only the desired goal, whereas the EOMs specify the way this goal is actually achieved.

This principle replaces the conventional concept of operators and operands.   The new concept consists of messages and objects.   Objects correspond to operands and are represented by the EOMs, a message is a request to an object to accomplish an operation (operator).   It is a characteristic of the message—object concept that the responsibility for executing an operation lies with the object.   This concept allows a strong decoupling of the abstract task and the actual realization of the solution.   For example, there is an EOM that is responsible for insertion.   This module should realize the insertion of a pin even when the parts are slightly displaced.   Moreover, it should detect defective parts and send a corresponding message to the SM.

For every EO, the PM determines preconditions that must be satisfied at the beginning of the execution. Moreover, the PM establishes

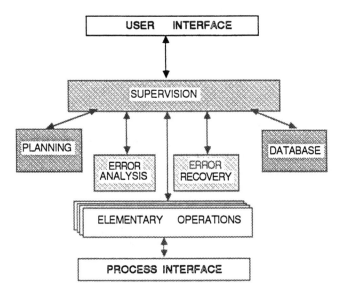

FIGURE 10.4 Overall structure of the robot system.

the scope of independence for the EOMs during execution. Further-more, for each EOM there are criteria that state when the execution of an EO must be stopped (in case of collision, end-effector error, time exceeded, etc.).

All information about the robot and the manufacturing cell is kept in the database. All modules have access to the database via the SM. To realize this centralization of information, special techniques for knowledge representation must be used to ensure the inclusion of new information and to allow modifications or retrieval of existing information quickly [11]. Keeping the information centralized makes it easier to assure data integrity.

There are two types of information. Static information includes the task description and the models of objects and the robot, which do not change during the execution. The dynamic information rep-resents the current state of the cell and must be updated whenever the state of the system changes. At any time during execution, the information kept in the database should be sufficient to reconstruct the present state of the workcell. Only the expert knowledge is stored in the relevant modules.

The execution of the plan is directed and monitored by the supervision module, which has expert knowledge about execution strategies in general. The actual state of the system arises out of the status reports of the different modules. The reports are kept in the database, and the supervision module has access to all information stored in the database. Depending on the status reports, the SM stops or initiates processes and updates the model about the current state of the cell. The initiation of a process includes the handing over of a detailed specification of the task. In the case of initiating an EOM, the information handed over includes parameters that specify the goal and the scope of independence and parameters for the criteria that decide when to stop a task.

If an EOM cannot carry out its task, the SM activates the error analysis module (EAM). The task of the EAM is to find the reason for the failure and the precise state of the workcell after the failure (e.g., location of displaced part, distance between two assembly parts). This information forms the basis for the error recovery module (ERM), which decides how to proceed after a failure.

The following sections discuss in detail the elementary operation modules, the error analysis module, and the error recovery module.

Elementary Operation Modules

The elementary operation modules, located at the lowest level of the control system hierarchy, have an interface to the robot system and the sensor system. Each EOM is responsible for the execution of a fixed EO. Every task presented to the robot system is decomposed into a sequence of EOs. Because of the hierarchical design of the system, it is always possible to include a new elementary operation. This increases the flexibility of the system. The following elementary operations are used in a prototype of the system:

Approach	Detach	Dock
Depart	Grasp	Move
Transfer	Measure	Special functions
Insert		

With these EOs it is possible to accomplish fundamental assembly tasks. For example, the assembly of the Cranfield Benchmark [12] was realized with these EOs. The EOs "moving" and "docking" are needed to navigate the robot.

To achieve a high degree of autonomy, the robot must be able to react quickly to changes in the workcell. This necessitates an intelligent execution of the plan. Therefore, the EOMs can execute a task, even if the conditions are not in accordance with the plan. The scope in which these modules can make their own decisions is determined by the PM. An EOM should be able to carry out:

Goal-oriented motions to achieve wanted relations
Corrective motions to improve position accuracy
Motions to measure positions and relations

These motions are controlled by fast sensory feedback loops. At the beginning of the execution of an EO, the corresponding module checks whether the preconditions are satisfied. During execution, the EOM monitors sensor values that are specfic to the EO and, if necessary, it stops execution. In the latter case the module sends a message containing the corresponding sensor values to the SM, which in turn activates the error analysis module.

In the example that follows, the EO "insert" is explained and examples of preconditions, monitoring rules, and feedback loops are given (Fig. 10.5a).

EXAMPLE:
Elementary operation "insert"
Preconditions: the axis of the pin and the axis of the hole are aligned except for a small angle α (Fig. 10.5b).
the absolute value of the difference of the distance between the gripper fingers and the diameter of the pin is less than E_1.
Monitoring rules: if the absolute value of the force vector is greater than E_2, then stop all motions.

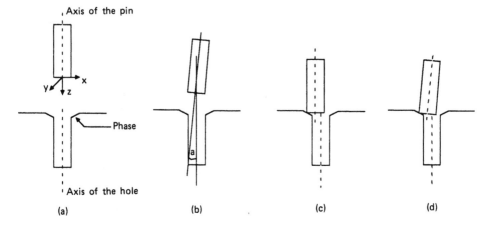

(a)   (b)   (c)   (d)

FIGURE 10.5 Elementary operation "insert."

Here α, E_1, and E_2 are constants that depend on the robot, the workpieces, and the accuracy of the sensor system.

Feedback loops: if the pin is in contact with the chamfer, transfer in direction d by distance l (Fig. 10.5c, 10.5d).

Here d is equal to the projection of the force vector on the xy-plane and l is equal to half the width of the chamfer.

Error Analysis Module

The error analysis module (EAM) is activated if an EOM cannot accomplish its assigned task according to the execution plan. The causes for the failure can be manifold:

The preconditions for an EO are not in accordance with the actual conditions.
An unexpected event forces the EOM to stop execution.
The EOM is not able to terminate the assigned task within a given period of time (e.g., because of inaccurate data about the layout of the workcell).

The purpose of the EAM is to establish the exact reason for the failure of an EO. The result of this analysis will serve as a basis for the generation of a new plan. The EAM makes use of several knowledge sources, both active and passive. Active knowledge sources are the sensors integrated into the robot system (vision system, tactile sensors, distance sensors, etc.). Passive knowledge sources are the world model, information about the sensors (availability, accuracy, reliability, etc.), and the execution plan. Furthermore, there must be knolwedge about the mode of operation of the EOs (tasks, strategies, potential errors, etc.).

The error analysis is based on the following models (Fig. 10.6):

First, there are a geometrical model and a functional model. The geometrical model represents the general layout and the geometrical relations between the components of the robot. The functional model describes the behavior of the robot and the feeders. These two models enable the system to predict the result of a robot action under the assumption that the workcell is laid out as planned and that the robot is functioning correctly. The functional model is represented in form of rules and facts and the geometrical model is represented by CAD data. Both models are kept in the database.

Second, there is a causal model. The causal model contains properties of the robot that are causally related, in that the value of one property is determined by the values of other properties. Most of

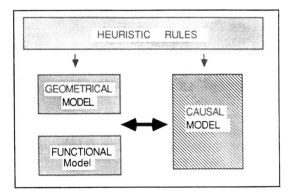

FIGURE 10.6 Error analysis model.

the properties are observables and the values are measured by the sensor system. The causal relations are represented as rules.

EXAMPLE:
If the distance between the gripper fingers is greater than the diameter of object A and if object A has the shape of a cylinder, then the robot is not holding object A.

Furthermore the causal model contains rules about relations between different EOs. This is particularly effective if a failure has occurred because of an undetected side effect of an earlier EO.

EXAMPLE:
In the case of a collision during the execution of an "approach" EO, assembly parts may be displaced.

The rules themselves are organized hierarchically. There are domain rules, analysis rules and solution rules. The domain rules accomplish a first classification of the error. They focus on the analysis on a particular domain. On the second level there are the analysis rules, which are applicable within a domain. They select a property to be measured and a sensor that executes this measurement. The solution rules interpret the results of the preceding rules and associate a solution with these results. This association will employ positive and negative (e.g., absence of a certain symptom) evidence.

The activation of the EAM is the result of the failure of an EO. The corresponding EOM assigns the information about the failure of the task to the EAM. Based on this information and with the help of the passive knowledge sources, the module tries to infer the reason for the failure. In many cases it will not be possible to do this unambiguously. But it will be practicable to narrow down the possible reasons for the type of failure. The module chooses one of these as a provisional hypothesis. This selection is done using heuristical rules derived during an experimental phase.

Now the module utilizes its causal model to select a sensor that helps to verify this hypothesis. The information this sensor supplies is quantitative and must be transformed into qualitative information. Then the module checks to ensure that this result does not contradict any information contained in the knowledge sources or information gathered by the sensor system. In case of a contradiction this procedure will be repeated, or a different sensor will be selected. Based on this new information the module tries to verify the hypothesis or rejects it. Thereafter, it selects a new one and the process starts again. The process terminates either if a reason for the failure is found or if none of the hypotheses can be verified. In the first case the geometrical model is adjusted to the changes of states of the real world and the correctness of the analysis is checked with the help of the functional model. In the latter case the module was not able to explore the reason for the failure.

Error Recovery Module

The task of the error recovery module (ERM) is to decide how to proceed with the execution after an error has been detected. This is done on the basis of the analysis accomplished by the EAM and the context of the present EO within the task. This means that the plan produced by the PM must contain more information than actually is needed for normal execution. The ERM is able to group errors into classes of errors such that for problems within one class, the same recovery routines can be applied. The efficiency of the ERM depends on the available knowledge about the context within an EO. To a certain extent, this knowledge can be derived from the plan produced by the PM. For sections of the assembly process, where errors are likely to occur, special recovery routines can be stored. In a given situation the ERM must decide whether one of the available recovery routines is applicable. In complex situations for which no such routines are available, the ERM must activate the PM to produce a new plan. However, since the generation of a new plan is usually very time-consuming, it is desirable to have many recovery routines.

The main characteristics of the robot system KAMRO are the following:

1. Knowledge about execution is located at the lowest level of the hierarchy: the elementary operation modules.
2. The plan produced by the planning module contains a limited amount of freedom to adapt to diverging situations.
3. The execution of the plan is event-driven.
4. The message—object concept allows a strong decoupling of the abstract task and the actual realization of the solution.

10.3 COMPONENTS OF AN AUTONOMOUS MOBILE ROBOT

10.3.1 Planning System

The basic task of a robot planning system is to convert the user's specification of a task into a robot-level program to carry out the task. The main role of the task planner is to plan the robot-specific motion and sensing commands necessary to achieve the goal of the task. The complexity of a planning system depends on the level of the task description. For example, it is assumed that the order in which the different parts are to be assembled is already contained in the task description. One method of doing this is to use a precedence graph. Alternatively, another program produces such a precedence graph from the user's description.

The development of robot planning systems is supported by the use of CAD/CAM and artificial intelligence tools. To convert a task-level specification to a robot-level specification, a planning system must solve the following problems.

Layout planner: The planner must choose where in the workspace each operation is to take place, in ways that minimize sensing and positioning errors as well as reduce the time for the completion of the task.

Strategic planner: This part of the planner is responsible for the more global decisions of the planning process, such as finding a feasible sequence of elementary actions for the specified task, assigning resources on the basis of heuristics, and specifying the frame parameters for the selected move and grasp operations, respectively.

Route planner: The route planner must produce a collision-free path among possible obstacles, such that the robot can move along that path from start to goal. The navigation of an autonomous mobile robot can be done on-line, such that moving and unknown obstacles can be detected and avoided.

Gross motion planner: The gross motion planner must choose efficient, collision-free paths for the manipulator and the parts it carries. Furthermore, the gross motion planner must synchronize the movements of the different manipulators.

Fine motion planner: The system must plan a strategy of sensing and motion that guarantees that parts-mating operations will be reliable despite errors in control and sensing. Moreover, this part of the system is responsible for planning how to grasp, such that the grasp is stable, avoiding collisions with local obstacles while grasping the part or placing it.

Many of the decisions made by the planner involve complex calculations. These decisions manifest themselves in the numeric values of positions—that is, in the position of parts and workstations, dimensions of grippers and mobile platforms, directions of compliant motions, and so on. These decisions are tightly interrelated and may propagate across succeeding operations. On the surface, they may appear to be independent operations. For example, the choice of a grasp point is affected by the way the part will be mated or placed.

Planning is one of the main techniques of artificial intelligence. Therefore, techniques of artificial intelligence are widely used in planning systems. For a review of planning techniques in artificial intelligence, see Hertzberg [13]. Early research was devoted to very abstract problems and artificial environments (e.g., block world problems). Other researchers look at non-domain-specific planning aspects. Only recently real-world problems have found increasing attention. There are basically three different categories of planning.

Planning in a single abstraction space: This is the simplest form of planning, where all operators and world states are on the same level of abstraction. One type of planning of this form is logical inference, which tries to prove the goal from the initial situation using a set of operators. The plan is constructed from the sequence of operators. Another type of planning of this form is heuristic search within a state-space representation. A classical system utilizing this technique is STRIPS [14].

Planning in multiple abstraction spaces: The basic idea of this technique is to avoid details in an early state of planning to reduce the complexity of the state space. If a plan is found on a high level of abstraction, it is planned in more detail on the next lower level. A classical representative is ABSTRIS [15]. There are two principal ways to abstract the problem space: situation abstraction and operator abstraction.

Meta-planning: Meta-planning can be described as "planning how to plan." This means that there is an additional level of planning at which strategic decisions are made. These decisions include choosing which planning method should be used in which situation. These planning strategies can be thought of as the operators of the meta-planning level. The first system to use this technique was MOLGEN [16].

Besides this AI-oriented research, there has been much work on real-world robot planning. Popplestone, Lozano-Pérez, and Taylor and their coworkers [17–19] proposed an approach to synthesize sensor-based programs written in the robot programming language AL from task-level specifications. This method relies on representing prototypical motion strategies for particular tasks as parametrized robot programs, known as procedure skeletons. Such a skeleton contains all the motions, error tests, and computational capacity needed to carry out a task, but many of the parameters necessary to speify motions and tests remain to be given. The applicability of a particular skeleton to a task depends on the presence of certain features in the model and the values of parameters of uncertainties. A similar approach based on procedure skeletons is taken in the LAMA system [20]. LAMA formulated the relationship of task specification, obstacle avoidance, grasping, skeleton-based strategy synthesis, and error detection in one system. The deficiencies of existing methods for geometric reasoning and sensory planning have prevented implementation of a complete task-level robot planning system.

The difficulties in motion planning led most of the researchers to solve more delimited subproblems, mainly grasp planning, transfer motion planning, and parts-mating planning:

1. Grasp planning consists of computing the grasping position of a gripper together with the approach and depart trajectories [21, 22].
2. Transfer planning consists of computing a collision-free trajectory for moving a part from its initial position to a goal position [23].
3. Parts-mating planning consists of synthesizing a sensor-based program for controlling the motions of the robot when achieving a given relation between two parts [24,25].

Lozano-Pérez and Brooks [18] suggested the following approach to a planning system:

1. The use of a small number of powerful planning modules to identify the range of possible values of the parameters needed for grasping, gross motion, and fine motion.
2. The use of constraint propagation to choose feasible values for parameters that affect more than one operation.

3. The use of skeleton programs to indicate stereotyped sequences of operations required to execute common tasks.
4. The use of configuration space to reason about legal robot motions for grasping, fine motion, and gross motion.

Robot programs must tolerate a certain degree of uncertainty if they are to be successful, because the actual positions of objects at task execution time will differ from those in the model. The principal sources of errors are part variation, robot position errors, and modeling errors. It is not possible to produce programs that guarantee success under worst-case error assumptions because there are too many uncertainties to be included in such a program. To ensure the autonomy of the system, the planner must use expectations of the uncertainty to choose motions and sensing strategies that are efficient and robust.

A robot programming system requires a complete world model and a complete task specification. The world model for a task must contain at least the following information:

Geometric descriptions of all objects and robots in the task environment
Physical descriptions of all objects (e.g., mass and center of gravity)
Kinematic descriptions of manipulators
Relative positions of robots and objects
Sensor capabilities and uncertainties
Joint limits and acceleration bounds of the manipulators and the mobile platform

The problem of decomposing and successfully executing a specified robot task involves reasoning about constraints from two distinct sources: (a) constraints arising from task and object characteristics and (b) constraints arising from the specific configuration of the robot in use. The planning system transforms a task-oriented problem description into a plan that describes how the given problem can be solved by the robot. Frommherz [26] proposed the presentation of the solution of the specified assembly problem in robot-level code, called explicit solution language (ESL). An ESL program can be translated into a program written in any other explicit programming language, which can directly be executed by the physical or simulated robot system.

The plan contains a sequence of such action elements as operations for pick and place, push, and turn with assigned resources (robot, gripper). To generate the plan, it is necessary to select and order the plan elements. The next step is to detail these plan elements. Therefore, the planning system consists of the submodules

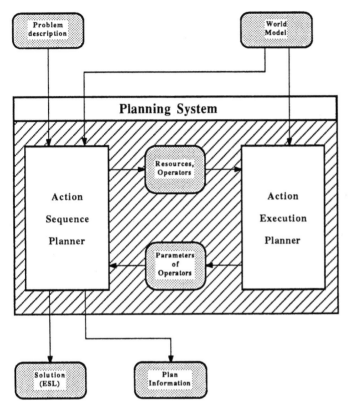

FIGURE 10.7 The components of the planning system.

ASP (action sequence planner) and AEP (action execution planner); see Figure 10.7.

The ASP is responsible for the global decisions of the planning process; these include finding a feasible sequence of elementary actions for the specified task, assigning resources on the basis of heuristics, and specifying the frame parameters for the selected move and grasp operations, respectively [27].

The AEP is responsible for the detail planning of the selected actions. For the geometrical planning of motions, the AEP considers possible collisions of obstacles in the world with the robot, the gripper, and the payload. The AEP is integrated into the planning process of the ASP. This means that a plan element proposed by the ASP is modified when it turns out that it cannot be executed because of geometrical constraints. For example, if obstacles prevent the execution of a pick-and-place operation, either the proposed resources or the proposed operation must be changed. In some situations there will be several different operations to realize a task. Then the ASP should propose to the AEP the operation that has the highest probability of being executed. This is done because the geometric planning is very time-consuming.

Next we discuss in more detail the action sequence planner, the action execution planner, and the route planner as they are under development at the University of Karlsruhe. For a review of path planning methods for robot motion, see Chapter 8 of this book.

Action Sequence Planner

To appreciate the structure, submodules, and interfaces of the action sequence planner (ASP) as it was designed at the University of Karlsruhe [27], and to understand its functionality, it is important to know how problems are presented to the ASP. This is shown using the Cranfield Benchmark [12] as an example. Figure 10.8 shows the 17 parts of the Cranfield Benchmark in their initial positions on the fixture and the fully assembled workpiece.

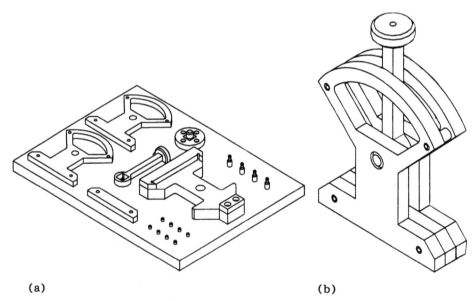

(a)   (b)

FIGURE 10.8   The Cranfield Benchmark assembly kit: (a) assembly kit   and (b) final assembly.

The problem description, which is an input of the ASP, is an elaborated form of task specification. In the case of an assembly task, the specification contains a set of pick-and-place operations and a set of constraints related to their execution sequence. These constraints depend only on the task, not on the layout, components, etc. This information may be represented by a precedence graph. The nodes represent the single pick-and-place operations and the edges the precedence relations. The precedence graph of the Cranfield Benchmark is shown in Figure 10.9. For example, it contains

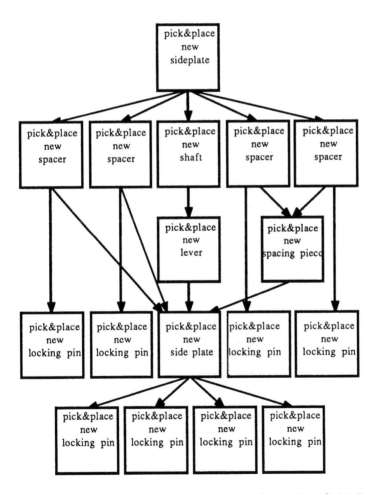

FIGURE 10.9 The precedence graph of the Cranfield Benchmark.

the information that a workpiece of type "shaft" must be placed at position pos04 before a workpiece of type "lever" can be placed at pos07.

In the task description, only the type of the workpiece to be assembled is specified. For example, "pick&place new spacer at pos02" means that a new object of type "spacer" is to be picked and placed at the position pos02. The assignment to a special item of that type of workpiece is done by the ASP depending on the robot types, the gripper types, and the layout of the assembly cell.

In the following, the submodules of the ASP are described in more detail (see Fig. 10.10). These submodules communicate via a common working memory. The ASP analyzes the task specification and proposes a sequence of actions and resources, which are refined by the

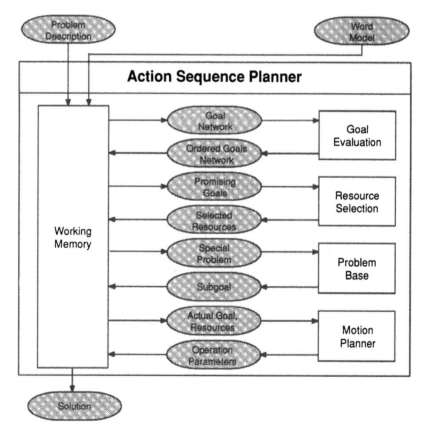

FIGURE 10.10 The structure of the assembly planning system (ASP).

subplanners. Since the geometrical planning of motions and other details is very time-consuming, the ASP utilizes heuristic knowledge to propose only promising plans to the subplanners. For different problems, domain-specific knowledge bases are provided for:

The evaluation of goals
The selection of the resources
The handling of obstacles
The planning of robot action details

In the case of pick-and-place operations, the subplanner for the refinement is called motion planner.

The goal evaluation examines the possible sequences of the different subgoals given by the precedence graph. Because it contains rules that favor promising goals, the probability that problems will arise during the detail planning phase or during the execution is reduced.

The following rule gives an example for a heuristic:

"If a workpiece of type x is to be assembled and there are different objects of type x available, use the nearest one first."

This strategy has the effect of ensuring that workpieces that might be obstacles for later pick-and-place operations are used first. Moreover, the different paths should be without any intersections, and the sum of their length should be a minimum. This example shows that the evaluation depends on the components and the layout of the robot cell.

If there are several pick-and-place operations that can be executed, the following heuristic chooses one to be executed first:

"If there are several pick-and-place operations that can be executed, choose the one for which the distance of the gripper and the center of the assembly task is minimal."

This heuristic reduces the danger of collision of the gripper with parts already assembled. Heuristics like these are simple and may produce the wrong result in special cases. Therefore, the system does not destroy possible solutions but orders them according to the heuristic rules.

The submodules for resource selection analyzes which gripper available in the robot cell may grip the workpiece in the start and goal positions of a pick-and-place operation. The special grip for the operation is selected such that the direction of the grip and the mating direction are equal. This means that the gripper will not grip the workpiece on the side where it probably would contact other objects.

After the selection of a gripper and the determination of the grip transformation, the robot configuration for the start and the goal positions is analyzed. If several robots in the assembly cell can execute the necessary motion, the robot that is able to do the job best will be selected to perform the pick-and-place operation. The arm selection uses the distances of the start and goal positions to the border of the joint space and to the Cartesian working space of the robot as criteria. This choice depends on the kinematics of the available robots.

To provide the information necessary for resource selection, different subroutines for geometrical calculations are needed. However, they do not take any possible collision among objects into consideration. This problem is treated in the submodule "motion planner."

The problem base contains rules for the special case that a goal of the task description cannot be achieved directly (i.e., by executing one pick-and-place operation). For example, the goal "pick&place new spacer to pos02" cannot be achieved directly if a spacer at position pos02 would collide with another object X. A similar problem is given if all objects of type "spacer" are hidden by an object X. The solution of the problem is to find both a location for X and resources that can pick and place X to that location. Then the subgoal to place X at that location is added to the task description. The "findspace" operation must take into account that X should not be an obstacle for later pick-and-place operations. Therefore, the entire task must be considered to find a suitable location for X. This problem is known as the interaction problem.

Action Execution Planner

This section presents an algorithm for generating collision-free paths for a robot with 6 degrees of freedom and revolute joints. The algorithm is part of the action execution planner under development at the University of Karlsruhe [28]. For an assembly robot, many motions are gross motions, which don't need the support of a sensor system (e.g., transfer parts from one place to another). Thus the ability to generate collision-free paths is very important for an autonomous robot system. The algorithm assumes that the obstacles for the robot are convex and polyhedral.

The approach taken by Hörmann treats the arm (consisting of a shoulder, an upper arm, and a forearm) and the hand (with the payload) separately. His approach utilizes the configuration space (C-space) method. A point in the configuration space describes the position and the orientation of the robot (e.g., the joint space of a manipulator is a configuration space representation). Here the configuration space is the Cartesian C space for the first three joints of the arm. This method allows the decoupling of the links of the robot such that the C-space obstacles for each link can be computed separately. This is done to decrease the execution time of the algorithm.

In the first step of the algorithm, the explicit free space representation in the configuration space of the arm is calculated. The configuration space is the Cartesian space relative to a reference point in the wrist of the manipulator. Since a point in this configuration space does not unambiguously specify the configuration of the robot, kinematic states of the robot are introduced. For each state, the different C-space obstacles are calculated. The C-space obstacles are determined by a contact analysis between a real obstacle and the links of the robot. The representation of this space uses a space grid consisting of cubes of the same size. The obstacles correspond to connected sets of cubes in this space.

The C-space obstacles and the real obstacles are mapped into this cube space. The space outside these obstacles is called free configuration space (i.e., with respect to free space). A configuration for the arm is safe if the reference point of the arm is contained in the free C-space. In the second step, the orientation range of the hand is divided into a finite number of subsets. For each discrete orientation the occupied cubes are calculated. Similarly, a pattern of cubes for each rotation between two adjacent orientations is calculated.

It is now possible to describe the configuration of the entire robot by specifying the following values:

The coordinates of the cube, in which the reference point of the arm is located
The kinematic state
The coordinates of the hand orientation

Here it is assumed that the reference point of the arm is located at the center of a cube.

To prove that such a configuration is safe, it is necessary to prove that the cube is within free C space for this particular kinematic state and that the space occupancy pattern of the hand is within free space. Furthermore, the movement between two adjacent cubes is safe if the corresponding orientation patterns are safe. A rotation between two adjacent orientations is safe if the corresponding rotation pattern is safe.

The third step of this algorithm is to produce a map. This map is realized in the form of a graph, where the arcs represent safe translations or safe rotations, and the nodes represent safe configurations. To find a path from a start to a goal position, a search algorithm is applied to this state-space graph. The paths found by this algorithm consist of a chain of basic motions that connect the start and goal configurations.

The foregoing algorithm is used for gross motion planning. But grasping and fine motion planning also involve gross motion planning at the same time, since these motions are impossible without moving the robot manipulator. Hence, the manipulator C-space obstacles must be taken into account in some way. The nature of the algorithm allows the integration of the Cartesian C-space obstacles into fine motion and grasp planning.

Route Planner

This section presents methods for route planning for a navigation system of an autonomous mobile robot. In many cases the problem is reduced to "finding a minimum path among polygonal obstacles in two dimensions," which is a problem of computational geometry. A commonly used part in most of the knowledge algorithms involves graph search algorithms (e.g., A*) to find a minimum distance path in the "visibility graph" of obstacles vertices. A drawback of these methods is that a minimum Euclidean distance path in such a space will pass through a subset of the vertices of the obstacles. This is not a desirable solution, because the danger that the robot will collide with an obstacle is very high. Furthermore, many of these methods assume that rotation of the robot can be ignored and consider translation only. The computational expense of these algorithms would probably increase dramatically if applied to the extra degrees of freedom in three dimensions.

To develop an autonomous mobile robot moving in a cluttered environment, it is necessary to use fast, three-dimensional, collision-free motion planning. Several methods have been proposed. Many of the approaches taken utilize the configuration space technique. This method is computationally very expensive. It requires, first, mapping a world description into a configuration space (i.e., generating the C-space obstacles). In general, this step consumes a great deal of time and memory. Furthermore, the search must be performed in a high dimensional space. Another approach is to use an octree representation in a three-dimensional Cartesian space. An octree is a recursive decomposition of a cubic space into subcubes. To find a collision-free path, special search techniques are applied to the octree.

The potential field approach [29] offers very good possibilities for fast obstacle avoidance. This method can be extended to moving obstacles, since stability of mechanism persists with a continuously time-varying potential field. The only drawback of this method is that the algorithm gets stuck at local minima in the potential field.

A possible solution could be a combination of a general path planner and a potential field planner.

Herman [30] suggested a structure for a module that automatically generates collision-free motions for robots. The major components are as follows.

World representation: An adequate representation of the robot and its environment is required. Common forms for such a representation are surface-based CAD models, swept volumes, cellular arrays, octrees, and analytic surface equations.

World description acquisition: The world description may be obtained from various sources. These include databases and sensors (TV camera, ultrasonic system).

Search space representation: The search for collision-free path occurs in a search space. The representation might be the same as the world representation, but it is often different. Examples of search space representations that are usually different from world representations are configuration spaces, Voronoi-based spaces, generalized cylinder-free spaces, and medial axis-free spaces.

Mapping of the world-space to the search space: If the search space representation is different from the world representation, a procedure that maps the world to the search space is needed.

High level task planner: In this component, the overall task of the robot is planned. Information about specific position constraints on the robot are sent to the path planner, which finds a collision-free path satisfying these constraints.

Path planner: This component uses a set of search techniques to find collision-free path in the search space.

Trajectory planner: Here the path obtained by the search process is converted into a trajectory that can be executed by the robot. Typically, the path-planning process is concerned only with collision-free configurations in space, not with velocity, acceleration, smoothness of motion, and so on. These factors are handled by the trajectory planner. The output of this component forms the motion commands to the servo mechanism of the robot.

In the following, different methods for route planning are described. Herman [30] describes a fast path-planning system developed at National Bureau of Standards (NBS). In this system the path planning occurs in a three-dimensional Cartesian space represented as an octree. The search algorithm used is a combination of different standard algorithms (e.g., A^*, hill climbing, and test). Primitive shapes are used to approximate the robot and its swept volume paths. The search algorithm does not explicitly handle rotations. Rotations could be incorporated by decoupling

them from translations. Thus, the result would be a sequence of
pure rotations and pure translations.

Khatib [29] presents a real-time obstacle avoidance approach
based on the artificial potential field concept. The philosophy of
the artificial potential concept can be described as follows:

"The robot moves in a field of forces. The position to be reached
is an attractive pole for the robot, and obstacles are repulsive
surfaces for the robot."

The basis for the application of the potential field approach is the
operational space formulation. The operational space approach has
been formalized by constructing its basic tool, the equations of motion in the operational space of the robot.

The approach taken by Soetadji [31] permits stationary obstacles
to be convex or nonconvex, polyhedral, or overlapping. A cube-
based model is used, which contains the three-dimensional representation of the objects. The objects have different characteristics,
which must be considered for the mapping procedure. The obstacles
can be classified as stationary but static, such as a wall; stationary
but dynamic, such as a door (open or closed); and moving obstacles
and objects, such as a mobile robot or a human being.

A perception system is responsible for the recognition of moving
obstacles; it operates during the robot motion phase. A cube-based
model is used because it meets the following requirements on a geometrical model:

Three-dimensional object representation
Acceptable approximation error
Conformity between physical and modeled objects
No restriction on the number, size, or shape of objects
Existence of powerful mapping algorithms

The internal data structure of the cube-based model is an octree
structure. The environment is decomposed into cubes of different
sizes. The global cube, which contains the entire robot workspace,
is subdivided into smaller cubes until a base cube of predetermined
size is reached. A base cube signifies the presence of an obstacle
in a fixed unit volume. The subdivision of a cube into eight subcubes starts when the mapping algorithm detects an object within
the parent cube. A cube that is not found to contain any object
is termed empty, whereas a partly occupied cube contains some object and therefore undergoes further subdivision until the detected
object is completely described by a sequence of base cubes.

The general idea of this planning algorithm is to produce a map of passable routes off-line. This map is derived from the free space representation with the help of robot geometry. The actual route planning is done on-line by searching a route from the start to the goal position in this map. The output consists of high level "move" commands (e.g., translate, rotate, circle) for the vehicle.

To determine possible routes within the free space, information about the ground condition, on which the mobile robot operates, must be taken into consideration. Map generation is done in three successive steps. In the first step, the cubes are investigated to determine whether they intersect with the ground. Such an intersection is called a tile. In the next step, the set of tiles is separated into passable and nonpassable units. A tile is passable if all configurations in which the robot is on that tile do not lead to collision with any obstacle. Thus, in this step, the height, breadth, gradient of ground, and orientation of the robot are taken into consideration. In the last step, the connection beween neighboring, passable tiles are established.

For the second step, algorithms are presented only for rotational symmetric robots and robots that are symmetric relative to a plane. In the latter case a declaration is made as to whether the tile is completely passable, not passable, or passable only within a definite orientation. This declaration is valid for each point of the tile; that is, a tile is passable only when the mobile robot may rotate about each point of the tile in each direction without a collision with obstacles.

For finding the collision-free route, a search module is needed. The map produced serves as a database and contains information on possible routes. A search method in this context should always find a route if one exists, with a minimum of storage and search time. Furthermore, the selected route should be the shortest. To meet these requirements, the breadth-first search and A* algorithms are used.

An example of a three-dimensional workspace is shown in Figure 10.11. The workspace has a size of 256 × 256 × 256 × unit, where a unit is equal to 4 cm. The sample environment contains a bridge (A) with two ramps (B and C), three blocks (D, E, and F), and a cylinder (G), to which the robot can move using ramps H, I, J, and K. In this example 1646 cubes, 6031 tiles, and 22,124 connections are generated. There were 700 kilobytes of memory needed to store this information. Figure 10.12 shows three examples of sequences of tiles found by the search module. The examples are taken from Soetadji [31].

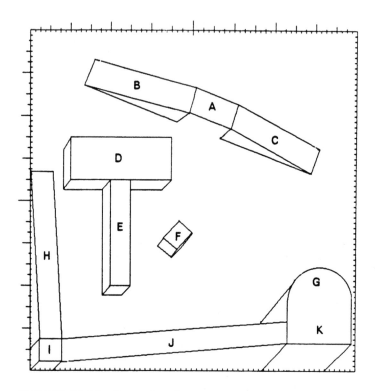

FIGURE 10.11   Example of a three dimensional work-space.

10.3.2   Navigation Systems

There are three different applications of navigation systems.   The
first is used when a mobile robot moves within an environment that
is unknown a priori.   An example is an autonomous land vehicle
with the task of moving through open countryside for a one-time
trip, taking into account terrain elevations or other constraints,
but not knowing beforehand where individual obstacles are located.
Many of these applications are based on a "discover—move" loop.
The first step is to determine the free space in the environment
of the robot with the help of a sensor system.   Such a system
usually includes a TV camera and ultrasonic sensors.   The second
step is to produce a map of the free space and to select goal positions in this free space. The map is produced to avoid deadlocks.
The third step is to choose the "best" goal position among the possible goal positions and to plan a route to reach that position. The
fourth step is to move to this goal position.   If this position is not
the final position, the system starts with step 1 again.

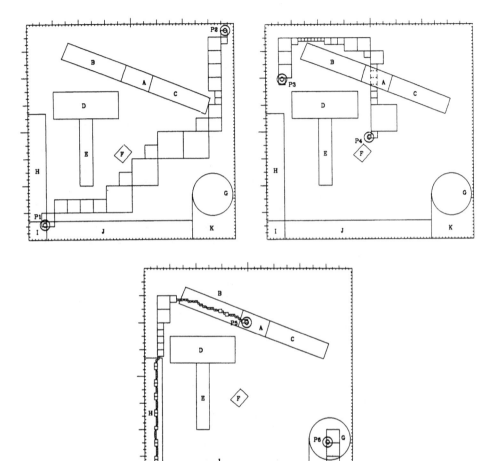

FIGURE 10.12 Three examples of the sequences of tiles found by the search module.

If during this movement new sensor data report that an object is inside an area that was previously free space, one of the following explanations is valid:

Previous free space reporting is erroneous.
The new object report is erroneous.
There are moving objects in the environment.

The most dangerous response to such an inconsistency is to ignore the new object, because this could lead to a collision.

The second application is used in case a robot moves within a known environment (i.e., the location of the obstacles is a priori known). Examples are robots moving through an office area, or in a flexible manufacturing cell. In that case the navigation system tries to match the sensor values with the map. An example of such a matching is the determination of the position and orientation of the robot in a manufacturing cell, or the detection of an obstacle that is not shown on the map initially given to the robot. Environment recognition is performed by searching for estimated perspective information, which is equivalent to the real perspective information from an original image and by obtaining the robot orientation and position necessary for generating this estimated perspective information. The approximate position and orientation of the mobile robot on the map can be estimated from the number of revolutions of the wheels of the robot, if the initial robot position and orientation are previously known. The problem is that the errors are accumulating. Thus the estimate perspective information will match the real perspective information, only with a difference. If matching between the two sets of perspective information is confirmed, the true position of the robot might be obtained from the difference between these two sets of information. Moreover, obstacles might be detected through the observation that an object doesn't match the map at all.

A mixture of the first two applictions is the problem of "road following." Navigation of roadways at speeds up to 30 km/h was demonstrated before 1979 in Japan, when a passive stereo vision system was used to steer a small car around obstacles [32]. The system drove vehicles on a roadway, provided it could see a guardrail, which served as a continuous "obstacle." A road following system should be able to accomplish the following tasks:

Execution of a prespecified user mission over a mapped network of
    sidewalks, including turning at intersections.
Recognition of landmarks, road intersections, obstacles, etc.
Avoiding obstacles.
Making decisions in real time.

10.3.3 Sensor System

Autonomous mobile robots require adaptive motion control based on feedback from the work environment via visual, tactile, and other types of sensor. The integration of sensors allows intelligent robots to accommodate changes in the workspace, such as changes in position and/or orientation of workpieces and to perform complex operations such as automated assembly and sorting. A completely environmentally interactive robot will require the integration of many sources of sensor input. There has been a growing interest in recent years in the development of multisensor systems. Henderson [33] introduced the concepts of "logic sensor" and "multisensor kernel system." Kak [11] presented a concept of knowledge-based robotic assembly cells, using multiple sensors.

Robot-mounted sensors extend robot capabilities and can be grouped into three categories based on the way the sensor information supports the robot action.

1. *Sensors to support local motion*: Compliant motion is the typical example of this kind of use. Compliant motion meets external constraints by specifying how the robot's motion should be modified in response to the forces generated when the constraints are violated. Contact with a surface, for example, can be guaranteed by moving so that a small force normal to the surface is maintained. Using this technique, the robot can achieve and retain contact with a surface that may vary significantly in shape and orientation from the programmer's expectations. Another application of compliant motion is the insertion of a peg in a hole.

2. *Sensors to determine precise positions and orientations of objects*: Vision systems and ultrasonic sensors are typical examples of this category. In the simplest case the information is required only for the purpose of one motion—for example, the determination of the orientation of an object in order to grasp it. A more complex example is given in the case of all parts of a workpiece lying on a table, with the sensor system having to recognize the parts and their positions.

3. *Sensors to monitor the execution of an action*: Consider the case of an assembly operation in a cluttered environment. During the assembly, every motion of the robot arm should be monitored and if there is a significant increase in force, all motions should be stopped to avoid any damage. A similar example is the monitoring of the movements of a mobile robot. In that case contact sensors mounted in the bumpers could monitor the movement of the robot.

Sensors for Assembly

Assembly seems to be a difficult job for a robot. Because assembly requires great dexterity, much research effort has gone into visual and tactile sensing to guide and supervise the work. If robots were expected to assemble in the same way as human beings, very complex visual and tactile sensing would be needed. In practice, however, many assembly jobs are simple, and robots can tackle these immediately. Frequently, jobs can also be tackled by robots as long as some modification to the procedure is made. On the other hand, many offer little hope for robot assembly, however dexterous the robot, and however good its sensing may be.

The main operations in assembly are to pick something up with a vertical movement, move it horizontally, and then move it down vertically for insertion. These operations need to be performed quickly and smoothly.

Robot sensing systems fall into two fundamental categories: noncontact sensing and contact sensing. A noncontact sensor measures the response to light or sound or magnetic or electromagnetic energy in the environment surrounding a robot. On the other hand, a contact sensor measures the response to some form of physical contact between the robot and its environment.

Contact Sensing: There is a need for contact sensing in assembly operations, since vision systems are very often impractical for real-time control of a robot. Furthermore, contact sensing allows a robot to analyze environments that cannot be seen because of inadequate illumination or object obstruction. Other disadvantages of vision systems are that large amounts of memory are required to store visual information as opposed to the information of contact sensors, and that the required amounts of processing time are very high. The function of a tactile sensor system is to measure, if possible, the spatial differences between the tactile images observed during gripping and those expected. An additional function of the tactile sensor system is to check for slippage of the object within the gripper during transfer. There are three fundamental tactile sensing operations:

Joint forces: sensing the force applied to the robot end effector.
Touch: sensing the pressures applied to various points on the gripper surface.
Slip: sensing any movement of the object while it is being grasped or transported.

Harmon [34] compiled the following list of desirable tactile sensor properties:

1. Resolution of detail, 1—2 mm.
2. Threshold sensitivity, 1 gram; dynamic range, 1000:1.
3. Response time, 1—10 ms, enabling rapid assembly and recovery.
4. Robust and durable skin to withstand harsh industrial environments.
5. Low hysteresis: the same signal for increasing and decreasing force increments.
6. Local processing of raw data.

Assembly tasks require a measurement precision of the order of the repeatability of the robot used. This is typically 0.1 mm. Close-fitting parts may require better than 0.1 mm accuracy, and in these cases remote center compliance tools can be of assistance. For a review of tactile sensors for robots, see Yardley [35].

Noncontact Sensing: Among the noncontact sensors are vision systems, laser scanners, and ultrasonic sensors. Vision is probably the single most important sensing ability that an intelligent robot can possess. Important applications are parts recognition, parts sorting, precision assembly operations, and determination of spatial relationships. Any robot would be more intelligent if it could acquire information about its environment through its own vision system, rather than being limited by a knowledge base provided by its programmer. On the other hand, a robot vision system can actually be used to build the knowledge base of the robot.

Many existing computer vision systems utilize a TV camera, photodiode array camera, or charge-coupled device (CCD) camera. As the camera scans a scene, its output is converted to a digital code by an analog-to-digital converter and stored in memory as a digital image. The computer must then analyze the digital images and apply some degree of artificial intelligence to understand the scene. The computer extracts such information as object edges, regions, boundaries, and color. In some systems, the digital images are compared to image templates to classify and recognize objects in the scene. The image templates must have been stored in the computer memory for the comparison task. Using this technique, the computer can recognize distinct, well-defined patterns.

To handle tasks that require flexible positioning of a monocular camera, an arm-mounted camera is very useful. Possible applications of such a camera include tracking, current world model verification, and other supervisory work requiring camera repositioning. In the absence of range data, such an eye-in-hand camera may possess unique advantages over a static overhead camera. An eye-in-hand camera doesn't suffer from parallax errors in the calculation of locations and orientations for an object in the scene. Moreover, such a camera can get by with lower resolution, hence can produce results in shorter time.

Sensors for Navigation

To support the navigation system of a mobile robot, several different sensors can be used. Among those are shaft encoders, contact bumpers, sonar devices, and stereo vision. The main objective of the sensor system is to produce a map of the environment. The

second objective is to avoid collision with obstacles. Once a map
has been produced, either by the robot or by the user, the sensor system can be used to determine the position of the robot in
its environment. After the initial position has been determined,
the information of the shaft encoders can be used to determine a
path on the map. Hence, at any time it is theoretically possible to
determine the position of the robot. All sensing devices possess
some degree of uncertainty in measurement. It is important to consider this uncertainty in producing a map, because the inferred location of an object sensed from different locations would be different without considering uncertainty.

To get a three-dimensional image of the environment, range data
must be combined with two-dimensional images. The main techniques
for ranging are triangulation and time-of-flight measurements. There
are two different types of triangulation: active and passive. Passive
triangulation involves merging two images taken from stationary devices to produce stereo images. The difficulty with passive triangulation is the correspondence problem. Active triangulation uses
a stationary camera and a moving light source. The scene is scanned
by a moving spot or strip of light. The reflected image points are
detected, and their displacement is used to calculate range.

Time-of-flight ranging involves the calculation of the time it takes
for a signal to reach and return from an object. Different signal
sources are used, including electromagnetic, light, and sound. The
speed of the signal determines the depth resolution, and the amount
of object resolution is determined by the width of the signal. Therefore laser ranging results in good object resolution and ultrasonic
ranging provides good depth resolution. Depending on the application, simple position and proximity sensors can be used to complement or replace a sophisticated vision system.

Another type of sensor to support navigation consists of bumpers to which pressure switches are attached. In addition to protection when the robot accidentally crashes, the bumpers provide
very definite information about the presence and location of an object. Usually bumpers are arranged in a circular fashion around
the robot (Fig. 10.13). Under the assumption that the robot contacts only one object at a time, the geometry of the bumpers allows
interpretation of the contact. For example, this information can be
used to guide a robot through doorways.

10.3.4 World Model

The global database contains the known set of facts associated with
a particular domain and is the model of the "world." At any in-
stant the information contained in the database must be sufficient
to describe the current state of the cell in detail. This is necessary

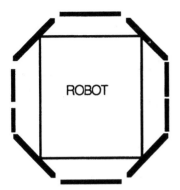

FIGURE 10.13 Bumpers arranged around a robot.

for planning, diagnosis, and error recovery of robot actions. The state of the workcell or the robot environment includes the locations of all objects and their features, as well as the position and the kinematics of the robot. For an assembly robot, the following information is necessary:

Part geometry
Part location
Features of parts
Mass characteristics of the part
Geometric tolerances
Grasping conditions
Symmetry axes and planes

Entries in the world model are changed whenever a new object is identified, whenever an object is moved or assembled, whenever an object is manipulated, and whenever the robot or vehicle is moved. During the execution of a task, different modules need information about the status of the workcell. Since access to the global database is asynchronous, care must be taken to make sure that at any given time the entries are valid. One way to ensure data integrity is to associate with each data item a time stamp. This allows comparison of different entries in the database.

Different descriptions of objects are needed for different purposes. For example, consider the case of planning a collision-free trajectory for an object. Here the existence of holes in the object is not relevant. But for planning the insertion of a pin into a hole in the object, the geometry of the hole is very important. A hierarchically structured description of objects could meet this

requirement.   At each level of description more details are added.
There are several possibilities for representing rigid solid objects:

Boundary representation: Figure 10.14 illustrates the solid repre-
    sentation scheme that is widely used in CAD systems.   Solids are
    represented indirectly, in terms of boundaries.   In this scheme
    boundaries are represented in terms of such primitive entities as
    unbounded surfaces, curves, and points, which together may be
    used to define "faces."
Constructive solid geometry: Solids are represented as compositions,
    via the regularized set operations, of other solids that must be,
    at the lowest level, primitive solids.   Examples of operations are
    union, intersection, and difference; examples of primitives are
    cubes, spheres, and cylinders (Fig. 10.15).

FIGURE 10.14   Boundary representation.

Sweep representations: **Figure 10.16** illustrates a "translational sweep" scheme, where a solid is represented as the volume swept by a two-dimensional set when it is translated along a line. Generally solids are defined by sweeping surfaces or solids, either of which may vary parametrically, along curves.

A representation scheme's "descriptive power" (i.e., the class of objects that may be described) depends on the richness of the scheme's set of representational primitives—for example, the available surfaces for boundary representations, or the available primitive solids for constructive solid geometry representation. For a

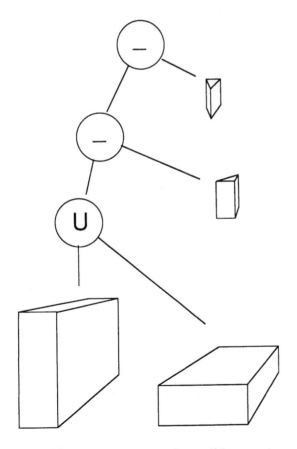

FIGURE 10.15 Constructive solid geometry.

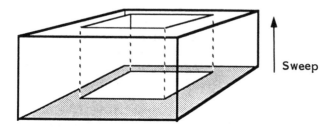

FIGURE 10.16   Sweep representations.

survey on methods of geometric modeling see Baer et al. [36]. The above-mentioned representation scheme can be used only to describe static objects.   To model a robot, different aspects must be considered (e.g., kinematics and dynamics).

The kinematics of a robot are defined by number, types and mutual positions of the axis.   The description is usually in the form of kinematic chains, where each axis has its own coordinate system, whose position is determined relative to the coordinate system of the preceding axis.

The dynamic model describes the gravitational forces and torques influencing the robot and caused by intrinsic dynamics, as well as the forces and torques caused by the robot motion.   The dynamical model is a mathematical model consisting of higher order differential equations.   Its parameters can be obtained from the geometric and kinematic models of the robot [37,38].

10.3.5   Knowledge Representation

The representation of knowledge about the robot and its environment in an internal model is crucial to both the processing of sensory data and the decomposition of tasks and goals.   In the first case the knowledge supports the sensory processing modules in selecting processing algorithms appropriate to the expected incoming sensory data.   The sensor module can thereby detect the absence of expected events and measure deviations between what is observed and what is expected.   The role of knowledge in planning was explained above.   Knowledge is very often complex and vague; frequently it is incomplete, and it can contain contradictions. Most knowledge is a mixture of information and various levels of instructions, which control the use of the information.

Knowledge may be taught (i.e., entered by storing parameters during a training session using a sample part), or it may be generated

from a CAD database, which contains a geometrical representation
of expected parts. Bowen and Kowalski [39] described the concept of knowledge assimilation in four steps.

1. If input data can be proved from the current database, the new
    database is identical to the current one.
2. If one item of information in the current database can be proven
    from the rest of the database together with input data, the new
    database is the database obtained by assimilating input into the
    rest of the database.
3. If addition of input to the current database results in a con-
    tradiction, input data must not be added to the current data-
    base. The new database is identical to the current one.
4. If input is independent of the current database, the new data-
    base represents the database obtained by adding input to the
    current database.

These rules suggest that knowledge assimilation can be performed
in four steps:

Provability check
Redundancy check
Contradiction check
Independence check

Both the provability and the contradiction check are intended to
judge whether input data should be added to the database. This
knowledge is called assimilable. The redundancy check is applied
to knowledge already contained in the knowledge base. The independence check should take place after the input data have been
checked as to assimilability, since this is a prerequisite for the
check. Miyachi et al. [40] presented a Prolog-based program for
knowledge assimilation of rules and facts.

There are several representation schemes available, but no single representation scheme will suffice for all situations and objects
a robot normally will be asked to deal with. Therefore, a collection
of different representation schemes is needed, and eventually a top-
level expert system will have to be developed, to invoke automatically the correct representation strategy on the basis of partial evidence about the object under examination.

There are basically two approaches to knowledge representation:
declarative and procedural. Declarative representation is a neutral
way to represent knowledge, independent of its use. Control is
achieved at a higher level, by general-purpose knowledge processing strategies, which are applied uniformly. An example of this approach is the implementation of the language Prolog. The knowledge

base is formed from associations between items of information and rules of inference operating on the associations. The control is achieved through the backtracking mechanism of the interpreter. A variation of this approach is to include control elements in the knowledge base as distinct items. An implementation of a rule-based system in LISP is an example of this approach, since LISP doesn't provide an inference mechanism.

The proceduralistic approach does not distinguish between knowledge and control. The representation of knowledge doesn't differ syntactically from the representation of the control mechanism. When the knowledge base is processed, the context of a particular process determines whether information is taken as a value or used for control purposes. A Pascal program is an example of procedural knowledge representation. Several such representation schemes are presented in the subsections that follow.

Semantic Networks

The principle of a semantic network is to regard knowledge as a set of associations between concepts. A network generally consists of nodes and arcs. In this case the nodes represent the concepts and the arcs are associations linking the nodes. Unlike a Prolog program, which lists each association separately, a semantic network can be very complex, with many arcs referring to the same node.

Figure 10.17 illustrates a semantic network to represent the following associations:

Sideplate_2 is a sideplate and part of the benchmark and is to be assembled before the lever.
Sideplate_1 is a sideplate and part of the benchmark.
Lever is part of the benchmark and is to be assembled before sideplate_1.

In a semantic network usually nodes are nouns and the associations are verbs. The nouns include specific individuals (e.g., sideplate_1) and general classes (e.g., <sideplate>). Figure 10.17 describes a static situation, but it is also possible to construct dynamic semantic networks.

Frames

The above-mentioned prototype concept leads to the method of knowledge representation by frames. A frame is a structure that is instantiated as many times as needed. A frame has a name and consists of named slots. Each slot may be filled with a value or a reference to another frame. Thus, a frame in its basic form is

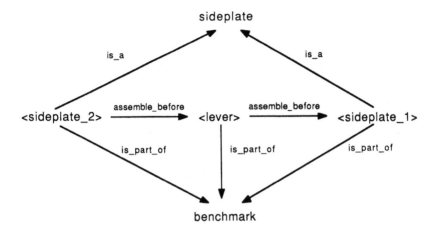

FIGURE 10.17   A semantic network.

rather like a record in Pascal. Figure 10.18 shows a frame describing the workpiece "Lever." The last slot refers to another
frame describing the workpiece benchmark, which is defined elsewhere.

Lever

   area: 5645

   center: (256,256)

   number_of_holes: 1

   area_of_holes: 857

   coordinates_of_holes: (256,380,19)

   number_of_straight_edges: 11

   number_of_curved_edges: 1

   mass: 350

   part_of: benchmark

FIGURE 10.18   Frame "Lever."

Slots of frames may also contain calls to processing procedures. The foregoing example might be extended to include a call to a procedure that calculates the weight of the workpiece benchmark. Frames are more formally structured than semantic networks, and thus more suited to describe static situations.

Object-Oriented Programming

In the most basic form, object-oriented programming views computation as consisting of the passing of messages between entities called "objects," which beyond the capacity to send and receive messages, possess local procedures whereby they can do work, including deciding how to react upon receipt of a message, executing methods, and sending further messages. Hence "objects" encapsulate both data and methods, which operate on the representation much like instances of an abstract data type in a more traditional language like Ada. "Classes" are templates from which objects having properties specified by the templates may be created. The collection of objects of a class have common operations and therefore uniform behavior.

Another important aspect of object-oriented programming is that classes can have any number of subclasses. Subclasses share the representation of the superclass and the operations applicable to all objects. This property is called inheritance. It is possible to override an operation provided by a superclass by allowing a redefinition of the operation locally. Object-oriented languages can be characterized as follows:

object-oriented = objects + object-classes + class inheritance

One of the first object-oriented programming languages was Smalltalk [41]. Being object-oriented, Smalltalk uniformly treats everything as an object: system components (compiler, window management, etc), primitive data types (integers, Booleans, etc.), and graphic elements (rectangular areas, arcs, etc.). The definition of objects includes a definition of the operations that may be applied to the objects. For example, the operation "+" applied to an integer is just ordinary addition of integers, but the operation "+" applied to an object of the class "three-dimensional-path for manipulator" is the concatenation of the two paths. Figure 10.19 shows an example describing the class "manipulator."

The ability of Smalltalk to reason about objects in the same manner as the "real-life" physical objects leads to intuitive, understandable systems. This makes object-oriented programming very attractive for application in robotics.

class name:	manipulator
superclass:	robot
instance variables:	configuration is a set of six angles describing the configuration of the manipulator
methods:	grasp
	transfer
	rotate
	insert

FIGURE 10.19 Definition of the class "manipulator."

REFERENCES

1. M. Stefik, "Strategic Computing at DARPA: Overviews and Assessment," *Commun. ACM 28*:7, 690—704 (1985).
2. H. J. Warnecke and H. Linder, "Trends in Robotics Research in the European Community," *Proceedings of ICAR*, Tokyo, 1985, pp. 7—13.
3. A. M. Thompson, "The Navigation System of the JPL Robot," in *Proc. 5th Int. Joint Conf. Artificial Intell.*, Massachusetts Inst. Technol., 1977.
4. T. Linden et al., "Architecture and Early Experiment with Planning for the ALV," *Proceedings of the IEEE Conference on Robotics and Automation*, 1986, pp. 1615—1621.
5. F. Tuijnman, M. Beemster, W. Duinker, L. O. Hartzberger, E. Kuijpers, and H. Muller, "A Model for Control Software and Sensor Algorithms for an Autonomous Mobile Robot," *International Conference on Intelligent Autonomous Systems*, Amsterdam, 1986, pp. 617—622.
6. U. Rembold and R. Dillmann, "Autonomous Robot of the University of Karlsruhe," *Proceedings of the 15th ISIR*, Tokyo, 1985, pp. 11—13.
7. R. A. Brooks, "A Robust Layered Control System for a Mobile Robot," *IEEE J. Robotics Autom.* RA-2:1, 14—23 (1986).
8. P. Kampmann, E. Freyberger, G. Karl, and G. Schmidt, "Real-Time Knowledge Acquisition and Control of an Experimental Autonomous Vehicle," *International Conference on Intelligent Autonomous Systems*, Amsterdam, 1986, pp. 294—307.
9. J. S. Albus et al., "Hierarchical Control for Robots in an Automated Factory," *Proceedings of the 13th ISIR*, Chicago, 1983, pp. 29—43.

10. R. A. Brooks, "Aspects of Mobile Robot Visual Map Making," in H. Hanafusa and H. Inoue, Eds., *Robotics Research 2*, MIT Press, Cambridge, MA, 1984, pp. 369—375.

11. A. C. Kak, K. L. Boyer, C. H. Chen, R. J. Safranek, and H. S. Yong, "A Knowledge-Based Robotic Assembly Cell," *IEEE Expert*, Spring 1986, pp. 63—83.

12. K. Collins, A. J. Palmer, and K. Rathmill, "The Development of an European Benchmark for the Comparison of Assembly Robot Programming Systems," *Robot Technology and Applications, Proceedings of the First Robotics Europe Conference*, Brussels, 1984.

13. J. Hertzberg, "Planerstellungs-Methoden der Künstlichen Intelligenz," *Informatik-Spektrum*, 9, 149—161 (1986).

14. R. E. Fikes and N. J. Nilsson, "STRIPS: A New Approach to the Application of Theorem Proofing to Problem Solving," *Artif. Intell.* 2, 189—208 (1971).

15. E. D. Sacerdoti, "Planning in a Hierarchy of Abstraction Spaces," *Artif. Intell.* 5, 115—135 (1974).

16. M. J. Stefik, "Planning With Constraints (Molgen)," *Artif. Intell.* 16, 111—141 (1981).

17. R. J. Popplestone, A. P. Ambler, and I. Bellos, "RAPT, A Language for Describing Assemblies," *Ind. Robot.* 5:3, 131—137 (1978).

18. T. Lozano-Pérez and R. A. Brooks, "An Approach to Automatic Robot Programming," in J. W. Boyse and M. S. Picket, Eds., *Solid Modeling by Applications*, Plenum Press, New York, 1984.

19. R. H. Taylor, "The Synthesis of Manipulator Control Programs from Task-Level Specifications," Ph.D. dissertation, Stanford University, Report AIM-282, July 1976.

20. T. Lozano-Pérez and P. H. Winston, "LAMA: A Language for Automatic Mechanical Assembly," *Proceedings of the Fifth IJCAI*, Cambridge, MA, August 1977.

21. T. Lozano-Pérez, "Automatic Planning of Manipulator Transfer Movements," *IEEE Trans. Syst., Man., Cybern.*, Vol. SMC-11, pp. 681—698, Oct. 1981.

22. C. Laugier, "A Program for Automatic Grasping of Objects with a Robot Arm," *Proceedings of the 11th International Symposium on Industrial Robots*, Tokyo, 1981.

23. R. A. Brooks, "Planning Collision-Free Motions for Pick-and-Place Operations," *First International Symposium on Robotics Research*, Bretton Woods, NH, 1983.

24. B. Dufay and J. C. Latombe, "An Approach to Automatic Robot Programming Based on Inductive Learning," *First International Symposium on Robotics Research*. Bretton Woods, NH, 1983.

25. T. Lozano-Pérez, M. T. Mason, and R. H. Taylor, "Automatic Synthesis of Fine-Motion Strategies for Robots," *Int. J. Robotics Res.* 3:1 (1984).

26. B. Frommherz, "Robot Action Planning," *CIM Europe 1987 Conference*, Tatton Hall, Cheshire, England, May 1987.

27. B. Frommherz and K. Hörmann, "A Concept for a Robot Action Planning System," NATO Advanced Research Workshop on Languages for Sensor-Based Control in Robotics, Italy, September 1986.

28. K. Hörmann, "A Cartesian Approach to Findpath for Industrial Robots," NATO Advanced Research Workshop on Languages for Sensor-Based Control in Robotics, Italy, 1986.

29. O. Khatib, "Real-Time Obstacle Avoidance for Manipulators and Mobile Robots," *Int. J. Robotics Res.* 5:1, 90—98 (1986).

30. M. Herman, "Fast, Three-Dimensional, Collision-Free Motion Planning," *IEEE Int. Conf. on Robots and Automation*, pp. 1046-1049 (1986).

31. T. Soetadji, "Cube-Based Representation of Free Space for the Navigation of an Autonomous Mobile Robot," *International Conference on Intelligent Autonomous Systems*, Amsterdam, 1986, pp. 546—561.

32. S. Tsugawa, T. Yatabe, T. Hirose, and S. Matsumoto, "An Automobile with Artificial Intelligence," *Proceedings of IJCAI-79*, 1979.

33. T. Henderson and E. Shilcrat, "Logical Sensor Systems," *J. Robotic Syst.* 1:2, 169—193 (1984).

34. L. D. Harmon, "Automated Tactile Sensing," *Robotics Res.*, Vol. 1, No. 2, pp. 3—32, Summer 1982.

35. A. M. M. Yardley and K. D. Baker, "Tactile Sensors for Robots: A Review," in P. Scott, Ed., *The World Yearbook of Robotics Research and Development*, Kogan Page, 1986.

36. A. Baer, C. Eastman, and M. Henrion, "Geometric Modelling: A Survey," *Computer-Aided Des.* 11:5, 253—272 (September 1979).

37. R. Dillmann and C. Blume, *Frei programmierbare Manipulatoren*. Vogel-Verlag, Würzburg, 1981.

38. M. Kircanski and M. Vukobratovic, "Computer-Aided Generation of Manipulator Kinematic Models in Symbolic Form," *Proceedings of 15th International Symposium on Industrial Robots*, Tokyo, 1985, pp. 1043—1049.

39. K. A. Bowen and R. A. Kowalski, "Amalgamating Language and Meta-Language in Logic Programming," in K. L. Clark and S. A. Tärlund, Eds., *Logic Programming*. Academic Press, New York, 1981.

40. T. Miyachi, S. Kunifuji, H. Kitakami, K. Furukawa, A. Takeuchi, and H. Yokota, "A Knowledge Assimilation Method

for Logic Databases," *New Generation Comput.* 2, 385—404 (1984).

41. A. Goldberg and D. Robson, *Smalltalk 80: The Language and its Implementation.* Addison Wesley, Reading, MA, 1986.

11

Some Examples of Advanced Robot Systems

Compiled and Translated by ULRICH REMBOLD   *University of Karlsruhe, Karlsruhe, Federal Republic of Germany*

11.1 INTRODUCTION

Literature describing robot applications is abundant. Readers interested in obtaining an overview of typical jobs robots are performing in manufacturing systems should read the proceedings of conferences on robots and advanced manufacturing. One will find that to solve a manufacturing problem with the help of a robot, it does not suffice to purchase the device and expect it to do meaningful work on the factory floor. Usually, there is a considerable amount of experimentation involved to set up the robot in a workcell and to teach it the work it is to perform. An additional task is the design of the gripper and the peripheral equipment (part feeders, fixtures, etc.). When sensors are needed in a rather unstructured manufacturing environment, the process of finding a solution becomes even more involved. As a matter of fact if vision or other complex sensors are needed, often a practical solution cannot be obtained.

This chapter shows with the use of practical examples how complex useful robot applications are. The first contribution discusses the handling of limb material whereby radiators for cars are assembled involving the handling of rubber hoses. The cooperation of two robots to insert cable trees is shown in the second example. The third example shows the use of various vision systems to aid robots doing their assignments. In the last example, the operation of an autonomous mobile robot is discussed.

11.2 THE AUTOMATIC ASSEMBLY OF HOSES
USING INDUSTRIAL ROBOTS

HANS-JÜRGEN WARNECKE   *Fraunhofer Institute of Technical Production and Automation, Stuttgart, Federal Republic of Germany*

BRUNO FRANKENHAUSER   *Robert Bosch GmbH, Reutlingen, Federal Republic of Germany*

11.2.1 Assembly of Limb Materials

Assembly is the part of production that still offers the greatest op-
portunity for automation [1]. In some sectors, assembly accounts
for more than 50% of the total production cost. In addition, assembly increases in complexity as more product variants and types
are produced to meet customer demands. A further advantage of
automating the assembly is the possibility of creating a clearly
structured flow of material.

The importance of the use of robots in assembly becomes evi-
dent from statistics. In December 1988 there were 18,000 industrial robots installed in the Federal Republic of Germany. Of these,
about 34% were used for assembly work. By comparison, in 1984
only 59 out of the 4800 industrial robots were applied for assembly
work.

The assembly of nonrigid or limb parts (hoses, seals, wrapping
materials, etc.) is very difficult to automate. This contribution
describes a method for the assembly of hoses and the construction
of a pilot cell to perform the work.

The basic problems involved in hose assembly are [2]:

The need to create high joining forces and torques by the robot,
     the gripper, or the tool to perform the assembly process.
The inaccuracy of the robots, grippers, and tools involved in the
     joining process.
The high workpiece tolerances.
A high level of deformation of the workpiece as a result of the join-
     ing and gripping operations.
The possibility that materials of the parts to be joined will have non-
     linear mechanical properties and undesirable temperature coeffi-
     cients, which adversely effect the assembly operation.
Small gripping surfaces of the workpieces.
The possibility that gripping can be done only at a great distance
     from the joining area (the joining section is the beginning of
     the hose).

When designing automatic assembly stations using industrial robots, these problems result in a highly complex system (cost and

time intensive). To be able to quickly and efficiently carry out the planning and design of an assembly station, all working parameters must be known. To simulate the overall joining process and to establish the assembly parameters for the work described here, the finite element method and the software package MARC were used [2].

Thus, the following problems could be solved.

Calculation of the relevant assembly parameters: in particular, the joining force, the required gripping force, the maximum and minimum reach, and studies on the effect of tolerances on the assembly process.

Finding a design for the assembly of limb parts with base parts by simulating the joining process.

Study of subproblems of the assembly station (e.g., necessary closing force of the gripper, joining force to be applied by the industrial robot).

Selection and definition of the necessary joining strategies (simulation of the joining process).

Determination of constraints of the equipment and simulation of failures (e.g., folding of the hose because too long a reach was chosen).

Investigation of additional parameters that may benefit the joining process (e.g., support of the joining process by heating the hose, the use of lubricants).

Using this program, it is a simple matter to compute the behavior of incompressible materials with a lateral contraction coefficient $r = 0.5$ (rubber). There exists for the calculation of hoses a large choice of materials parameters, both for rubber and thermoplastics. In addition, there is the possibility of defining the properties of other materials. For the study of the maximum allowable tolerances between joining part and the base part (maximum possible eccentricity and/or angular error), it is possible to generate a three-dimensional model of both mating components.

Figure 11.1 gives a summarized version of the software structure of the input and output data. Input is achieved via a preprocessor, which generates and checks the element data and the node data. The preprocessor is the program system MENTAT. The results stored from the generation of the finite element model are adapted to MARC using a transformation program. The data are then edited using the postprocessor MENTAT, both graphically and in the form of lists.

The calculations take a lot of computer time; therefore a powerful Siemens VP200 mainframe was used.

In the calculations, the base part was modeled as a rigid body, both for the symmetrical and asymmetrical joining processes. The

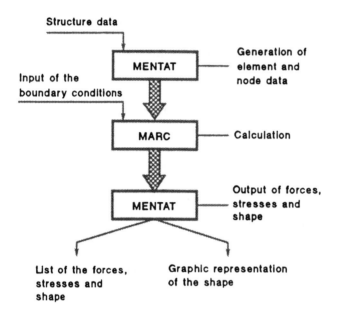

Structure data

Input of the
boundary conditions

MENTAT — Generation of element and node data

MARC — Calculation

MENTAT — Output of forces, stresses and shape

List of the forces,
stresses and
shape

Graphic representation
of the shape

FIGURE 11.1 Structure of the input and output to the finite element modeler.

separation of the joining part into element structures has a strong influence on the precision of the calculations or of the simulation. An imprecise separation produces unreliable results, while a very precise separation increases the calculating time considerably. The MARC element type 82 was used for all symmetrical calculations. This is an axially symmetrical, freely definable ring with a four-sided sectional area. There are five nodes per element, four of which are on the corners; the fifth has only one degree of freedom with regard to the force (hydrostatic pressure) and cannot be shared with other elements. Therefore, this element is not easy to use in the analysis of incompressible materials. Each corner node has 2 degrees of freedom, one axial displacement (z-direction), and one radial displacement (r-direction). Each element has a total of 9 degrees of freedom. The contact between the base part and the joining part, and the resulting friction, are studied by observing the position and orientation of specially developed constructs, the so-called GAP elements.

For the state of the individual elements the corresponding frictional force parts are taken into account. The GAP elements have two nodes, but they have neither a true length nor other physical

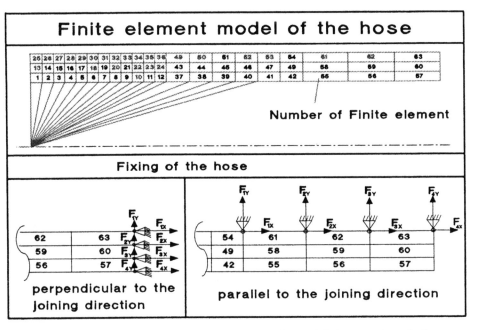

FIGURE 11.2 Finite element structure of the joining part and the base part.

properties (Fig. 11.2). As previously mentioned, MARC has at its disposal a wide range of materials. Depending on the available material, the so-called Mooney—Rivlin material law with three definable parameters or the four-parameter deformation law was used for modeling (Fig. 11.3).

For modeling of assembly robots doing parallel work, the base parts are assumed to be completely rigid. The joining parts are mapped by a three-layered model into the radial direction, having at any time 21 elements in the axial direction (this applies to symmetrical calculations only).

Figure 11.4 compares the calculated values of the joining forces in a round base part configuration with the values obtained during testing. The individual phases of the joining process can also be seen. During the joining process the joining part may fold inward. Half of all joining trials tested subsequently failed by curling upward. By simulating the assembly process, however, it is possible to find suitable means of dealing with these problems. They are:

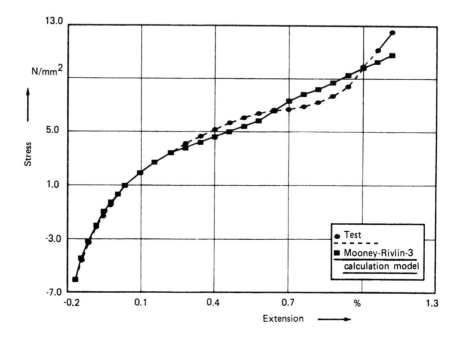

FIGURE 11.3 Stress–elongation diagram of rubber; comparisons of the measured material behavior and the ideal model (Mooney–Rivlin).

The use of lubricants
The selection of a design suitable for assembly for joining parts and
  base parts
The choice of the best base and joining part materials
The selection of the best joining strategy

It is also possible to calculate the stresses produced in the joining part during the joining process, and to portray them graphiclly with the help of the MENTAT program [2]. The comparative stress is then calculated according to the Cauchy method (Fig. 11.5).

Knowing the stress distribution and its magnitude at the respective points, the following aspects can be considered:

The quality of the sealing effect between the joining part and the
  base part (possible optimization of the base part geometry)
The optimization of the compression ratio with respect to material
  failure at these points

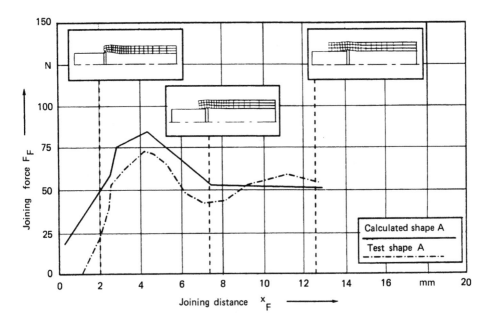

FIGURE 11.4 Comparisons of the calculated and experimentally measured joining force for a rubber hose.

The recognition of the locations at which a failure of the joining
    process might occur (e.g., sharp edges or transitions on the
    base part, where curling up is possible)
The selection of the base part material having the best friction co-
    efficient
The best choice of joining part materials (with respect to friction,
    shaping behavior, etc.)
The correct geometrical design of the base part

As a result of the joining force and torque, a deformation of the
joining part occurs during the joining process (Fig. 11.6). From
the degree of deformation, the following can be determined:

Unstable hose zones (possible failure during the joining process)
Type of motions the industrial robot must perform
Statement concerning success of the joining operation

During joining, the part located between the base part and the
gripper is compressed like a spring. This stores energy in the
hose. The industrial robot travels along its preprogrammed path
during the joining process. If the joining process is finished (in

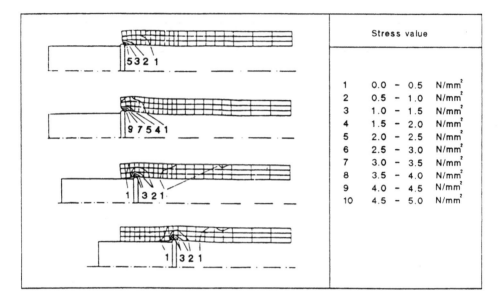

FIGURE 11.5 Stress calculated according to the Cauchy method.

accordance with the programming), then the gripper is opened. The energy stored in the hose is released. As a result of the compression, the preprogrammed path and the actual joining path are no longer the same. From the size of the calculated compression, a correction value can be determined for the joining path to be traversed by the industrial robot.

An additional assembly parameter, which is necessary for the success of the assembly process, is the maximum permissible tolerance (eccentricity and angular error) between the axes of the base part and joining part. Figure 11.7 shows the calculation for a selected assembly configuration. It can be seen that an eccentricity of 3 mm or more gives rise to a failure due to folding and the resultant slipping on the base part. This can be seen in Figure 11.7 by noting that with a joining displacement of 3 mm or more, the joining force rapidly diminishes.

11.2.2 Simulation of the Assembly Process

Knowledge of the assembly parameters and of the results of the simulation of the assembly process (e.g., of pushing a hose over a

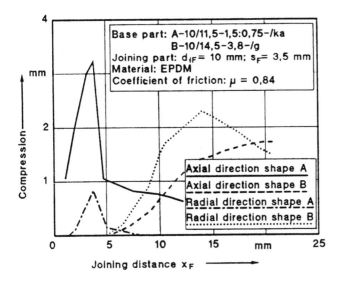

FIGURE 11.6 Deformation of the joining part due to the joining force: d_{iF} = internal diameter, s_F = wall thickness; EPDM = reinforced ethyl-propylene rubber.

radiator brace) are important for the success of the assembly operation. From knowledge of the joining sequence, conclusions can be drawn regarding:

The feasibility of the planned joining sequence
The design of the base part geometry (the goal here is the optimization of the joining sequence)
The choice of the base part material (friction ratios)
The choice of the joining part materials (friction ratios, shapeability of the joining materials)
The design of the assembly cell (in particular the incorporation of joining aids)
The decision support for the selection of the suitable joining strategy (optimization of the joining movements)

Figure 11.4 showed three phases in which a rubber part is pushed over a base part. It can be seen that for a base part with a round geometry and without a centering section, a folding inward of the joining part may occur during the first contact between the joining and base part (placing the hose on the base part). When moving the base part in increments, the joining part suddenly "snaps" over the base part.

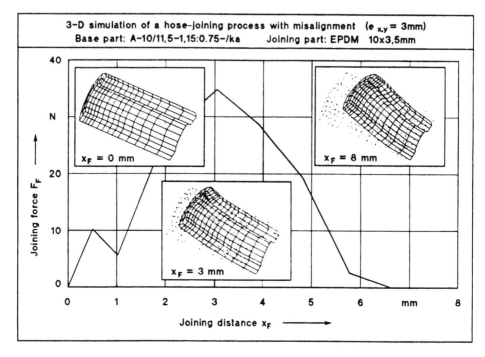

FIGURE 11.7 Deformation of the effect of misalignment on the joining force and distance.

During tests, a failure rate of 50% due to curling inward was observed. This can be completely eliminated by selecting a suitable joining strategy (optimum choice of the movement between joining and base parts). Furthermore, this phenomenon can be prevented by a suitable arrangement of the base part surface and by using lubricants.

When comparing the tests with the results from the calculation and the finite elements simulation, a good agreement of the results can be seen. The errors, which occur in the calculation of the assembly parameters (order of magnitude between 15 and 20%), result primarily from imprecise information about the real materials, which is entered into the ideal material model. In addition, the following factors are important:

The tolerances between the joining and modeled parts (difference between the real part and its ideal model), particularly in regard to the roundness and the measuring tolerances of the joining part, must be determined.

The friction ratios can be only roughly estimated, since effects such
   as adhesion cannot be incorporated into the calculation.
Material mixtures (particularly of thermoplastics) are unpredictable
   variables.
Often, the assembly of the parts is very inconsistent.

The simulation of the joining process is very easy with the finite
element method. It is possible to statically simulate the individual
joining steps separately from each other and to calculate the parameters effecting them. In addition, the whole joining process can be
simulated dynamically over a preset joining displacement; that is, it
is possible to present a continuous assembly operation of a hose
over a radiator brace on the screen.

By varying the geometric and technical parameters affecting the
assembly process, it is possible to arrive at an optimum solution in
regard to functional or design requirements. By using the finite
element method, it is possible to do rapid planning and designing of
the part assembly system using flexible automated assembly stations.
Thus, the assembly of a hose can be observed without the need for
experiments.

11.2.3 Compensation of Tolerances Using Various Joining Strategies

When automating the assembly of hoses, problems are associated
with the tolerances that arise in the entire assembly system. Passive and active joining strategies can be used to compensate for
tolerances. This can be done by performing suitable movements
with the joining part by the industrial robot about the base part.
The degree of freedom of the industrial robot should actually be
greater than that of the actual joining strategy employed.

Figure 11.8 shows four basic forms of the joining strategies used
together with their corresponding joining phases, namely:

The orientation phase
The contact phase
The stabilization phase
The concluding phase

The basic joining operation, the purely translational movement, is
called strategy I. Here the joining part and the base part are
pushed together along their center axes. However, this strategy
often does not render good results. The workpiece must be given
an additional twist, which is a rotation about the joining axis. The
superposition of the rotational with the translational movements leads
to strategy II.

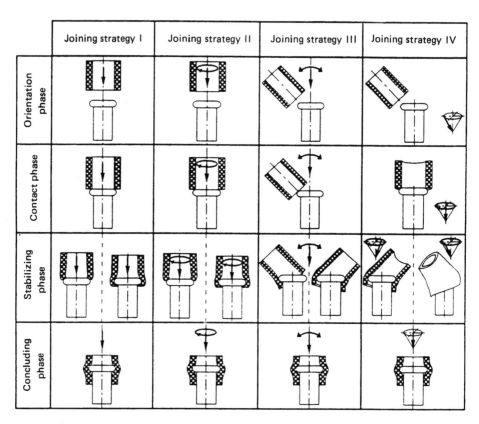

FIGURE 11.8   Four joining strategies and joining phases.

    The additional rotation of a joining part causes an increase in the relative speed between the mating surfaces, which at an appropriate angular speed may result in reduced friction.

    Because of the symmetrical properties of the joining part, rotations about the second or third principal axis are identical.   The superposition of the translational movement with a rotational movement (pitch and yaw) about one of these axes is described as strategy III.   In manual assembly this is the most common joining motion. A combination of a translational movement and a rotational movement about all principal axes (wobble) is strategy IV.   Irregular movements are characterized by overlapping wobble movements.

    The significance of the boundary constraints of the joining process was described before.   The kinematic boundary constraints are particularly important for the choice of a strategy.   The existence

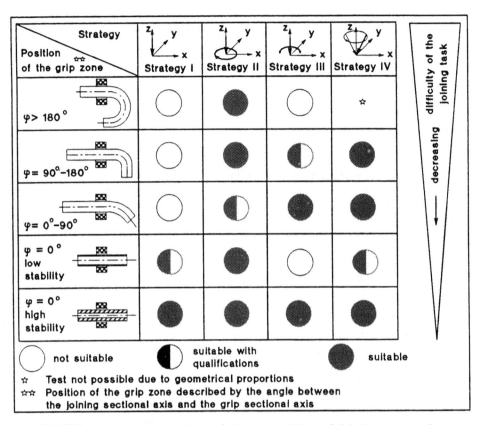

FIGURE 11.9 Determination of the suitability of joining strategies for various grip zones.

of a suitable free joining space, which enables the unhindered completion of joining, is essential.

The required gripping space about the workpiece depends on the geometry of the robot end-effector, the fixtures, and the workpiece. Figure 11.9 shows how the geometry of a tubular part influences mating strategies. It shows, also, joining strategies for parts with large internal diameters and small wall thicknesses and for parts with small internal diameters and large wall thicknesses. For restricted gripping space, joining strategies I and II may lead to problems (especially for hoses 1 and 2), as failures arise due to arching, folding, curling, and so on. With an increase in the distance of the gripping space from the joining sectional area, and an

increase in lift length and the angle between the axis of the joining
sectional area and the axis of gripping sectional area, joining strategy
III is of limited value. In addition, the reach was studied in relation
to the gripping space (for straight hoses only). It was found that
an increase of the reach from 30 to 50 mm leads to a decrease of the
compensation of tolerances.

The adjustable tolerances decrease from 3 mm at a reach of 30 mm
to 2 mm at a reach of 50 mm. The same tendency was observed with
the joining strategy II, III, and IV.

The relationship between the external diameter of the joining part
and its internal diameter is a measurement of stability during assembly. A study was made of a joining part with a large internal diameter and small wall thickness (D_{af}/D_{if} = 1.2) and one with a small internal diameter and large wall thickness (D_{af}/D_{if} = 1.9). Joining
strategies II and IV were found to be basically interchangeable. For
both diameter ratios, joining strategy II showed a circular field within which tolerance could be compensated for. From here on this field
is called the tolerance field.

The effect of tolerance compensation for different joining part materials was studied for all four strategies using polyvinyl chloride
(PVC), PVC with textile reinforcement, and rubber with textile reinforcement. The results obtained are summarized in Figure 11.10. The
tolerance field for joining strategy II is largely independent of the material, while that for joining strategies I, III, and IV is very dependent on the material used. Basically, rubber with textile reinforcement has the smallest possibility for tolerance compensation. The
tolerance field for PVC and PVC with textile reinforcement using
joining strategies III and IV are only slightly different with respect
to size and shape. This illustrates the greater tolerance compensation capability of thermoplastics compared with elastomers. The asymmetrical shape of the tolerance field, particularly for joining strategy
IV, is caused by the inhomogeneity of the material and geometrical
tolerances of the joining parts and base parts.

The size of the tolerance field is in essence determined by the
joining time, the programmed joining speed, and acceleration of the
industrial robot. The joining time, however, has no influence on
the shape of the tolerance field.

For the experiment, the joining time in relation to the joining
speed was measured. The joining time was varied between 2 and
200 seconds. As can be seen from Figure 11.11, the maximum adjustable tolerance field for parts with a small internal diameter and
large wall thickness is smaller than for parts with a large internal
diameter and small wall thickness. For all strategies it could be noticed that only in exceptional cases did a joining time exceeding 10
seconds cause an increase in the compensation capability. It further
could be shown that joining parts with a wide stability range the

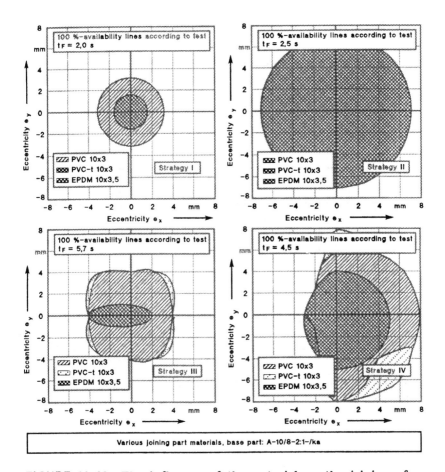

FIGURE 11.10 The influence of the material on the joining of a part.

joining strategies I, II and IV correlate very well, while joining strategy III is very dependent on the stability of the joining part.

A number of assembly parameters and boundary conditions influence the choice of joining strategies; the most important are the minimal adjustable eccentricity and angular error.

For all four joining strategies it was possible to show that tolerance adjustment assembly is possible in a hose. The advantages of this method are high economic efficiency and low engineering cost. The results of the study are summarized in Figure 11.12. The best joining strategy depends on the particular joining case and the related assembly parameters.

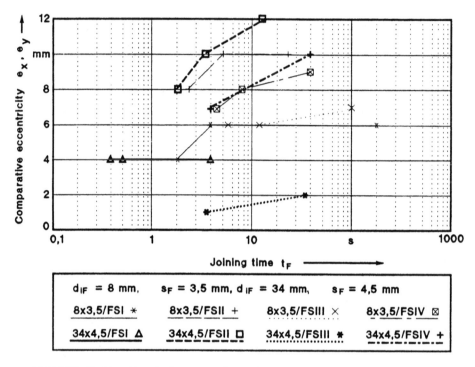

FIGURE 11.11   The influence of the joining time on the adjustable
eccentricity; abbreviations as in Figure 11.12.

11.2.4   Constructions of the Pilot Operation

To test the methods developed, an automatic assembly station was
constructed.   Commercially available products were chosen as joining parts and base parts.   The joining part was a cooling water
hose with one inlet and two outlets.   The base parts were a radiator, a cooling fluid compensation tank, and a tubular connector.   The
overall construction of the test cell used for the automated assembly
of hoses is shown in Figure 11.13.

Joining strategy / Evaluation criterion	Observations		FS I	FS II	FS III	FS IV
Minimal comparable eccentricity	mm	d_{aF}/d_{iF} <1,5	4	8	1	7
		d_{aF}/d_{iF} >1,5	4	8	6	7
Minimal comparable angular error	o	d_{aF}/d_{iF} <1,5	<5	<15	<5	<10
		d_{aF}/d_{iF} >1,5	<10	<20	<5	<15
Shape tolerance compensation field	○ ellipt. / ● circ.f.	d_{aF}/d_{iF} <1,5	●	●	○	◐
		d_{aF}/d_{iF} >1,5	●	●	●	◐
Influence of the stifness of the joining part	○ strong / ● weak	d_{aF}/d_{iF} <1,5	◐	●	○	◐
		d_{aF}/d_{iF} >1,5	●	●	●	●
Influence of various base part geometrical configurations	○ strong / ● weak	d_{aF}/d_{iF} <1,5	◐	●	○	●
		d_{aF}/d_{iF} >1,5	●	●	◐	●
Influence of the position of the grip zone	○ strong ● weak		○	●	○	●
Influence of the choice of material	○ negative / ● positive	EPDM-T	◐	●	○	○
		PVC	●	●	●	◐
		PVC-T	◐	●	●	●
Relative joining force effort	%	d_{aF}/d_{iF} <1,5	100	50	75	100
		d_{aF}/d_{iF} >1,5	100	50	50	50
Necessary joining time	s		>1	>2	>4	>4
Number of joining parts to be assembled >1	●possible ○not poss		●	○	●	◐
Relative joining space	%		20	20	75	100
Necessary number of IR degrees of freedom	-		1	2	3	6
Manual programming effort	h		0,5-1	2-4	6-8	6-8

EPDM-t	... Ethyl-Propylene-Rubber reinforced	d_{aF} ... Joining part external diameter
PVC	... Polyvinylchloride	d_{iF} ... Joining part internal diameter
PVC-t	... Polyvinylchloride reinforced	IR ... Industrial robot
FS	... Joining strategy	unless otherwise stated the test results refer to EPDM-T

FIGURE 11.12 Comparison of joining time and adjustable eccentricity.

The test cell consists of the following components:

An industrial robot with its controls

A six-component force—torque sensor to monitor the joining forces and torques

A handling device for bringing the workpieces to the work area

A gripper with its control

A laser scanner to recognize the conduits

A gripper jaw change system

A computer to process the sensor information

FIGURE 11.13   The pilot assembly cell for the assembly of hoses using an industrial robot.

A programmable gripper was developed for feeding and handling of the part to be joined during assembly. Additional elements to carry out the following functions were integrated into the gripper:

A position controller
A gripping force controller
A slip sensor to assure that the robot has a firm grip of the work-
    piece

Control of the gripping force is needed to make certain that gripping is done with high precision. Since the hoses are subject to
strong tolerance variations of the external diameter, it is not possible to guarantee a secure grip if a nominal diameter value is
used. In addition, because of unnecessary travel of the gripper
jaws, an unacceptable preforming of the hose can take place, which
can lead to a failure of the assembly process. Since hoses with various external diameters must be gripped, it is necessary to change

either the gripper or the gripper jaws. Since available change systems for gripper jaws needed between 6 and 8 seconds for a changeover, a new system with the following characteristics was developed:

An interface for measuring signals
A changeover time of less than 2—3 seconds

Shorter changeover times can be achieved only by using a so-called flying change of the gripper jaws: the changer is mounted directly in the robot hand and the gripper is brought into the work position by a rotating or sliding change mechanism. The device developed is shown in Figure 11.14.

During operation, the robot moves the gripper into the jaw change system. Cam followers 1 and 2 are brought simultaneously into contact with the two guides (links). Through the tapered shape of the guide, cam follower 1 is pressed upward and moves a spring-loaded piston up. As a result, the gripper jaws are separated from the gripper jaw holder. The jaws are received by pins and held in position by preloaded steel balls. The still open piston now receives the next gripper jaws and travels with the new gripper jaws out of the change system. Signals (e.g., from the slip sensor) are transferred via precision-engineered contacts to the measuring system. They have a useful life of more than a million load cycles.

To recognize the hoses, an image processing system was provided that had the following features:

Performed the distance measurements, since the hose can be shifted in any rotational position.
Measured the external diameter of the joining part or the center of the cross section of the joining part.
Assured a high availability, since this system is a vital component of the cell.
Performed an acceptance test of the joining operation.

A laser scanner was chosen for solving this problem. Figure 11.15 shows the information flow between the laser scanner and the industrial robot. The sensor data are brought via a preprocessing unit and an IEC bus to a computer; the latter unit serves as sensor processor and protocol converter. The data are evaluated and edited. From here they are sent via a serial interface to the industrial robot control. The communication between the personal computer and the industrial robot control takes place via a common software protocol.

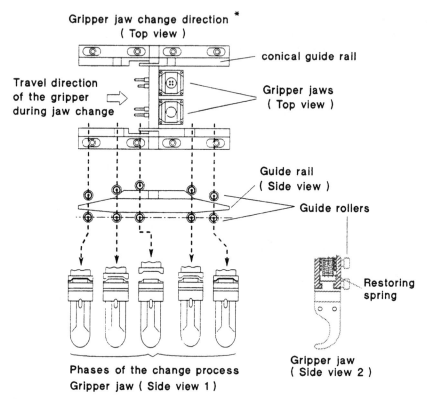

Gripper jaw change direction *
(Top view)

conical guide rail

Travel direction
of the gripper
during jaw change

Gripper jaws
(Top view)

Guide rail
(Side view)

Guide rollers

Restoring
spring

Phases of the change process
Gripper jaw (Side view 1)

Gripper jaw
(Side view 2)

* Drawing not to scale

FIGURE 11.14 Principle of the "flying" changer for grippers.

With the system configuration shown, it is possible to automate
the entire hose assembly process. The availability of the assembly cell was considerably increased as a result of the modular design.

An essential precondition for the automation of a hose assembly
is a product designed for assembly by robot.

A great number of good assembly procedures and general rules
of product design for automatic assembly are listed in Reference
3.

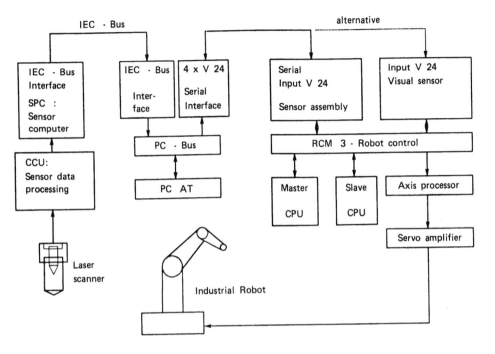

FIGURE 11.15 Schema of the control system of the experimental hose assembly robot.

The most important results of this work, which will assist in the automatic assembly of hoses or will make the assembly economically feasible, are listed in Figure 11.16. Figure 11.16a shows design features of the hose, and Figure 11.16b presents features of the base part.

It must be remembered that the design of the base parts is very much dependent on and influenced by boundary conditions such as the closeness of the contact and the capabilities of robot.

It is not possible to achieve a well-automated system by the proper arrangement of the joining parts and base parts alone. A product designed for assembly is extremely important. Until now, wide-scale automation of assembly operations was not possible because the necessary joining prerequisites have almost never existed. A further difficulty is the hitherto strict adherence of manufacturing tolerances that were too close for most components and subassemblies involved in the assembly process.

a) Base part

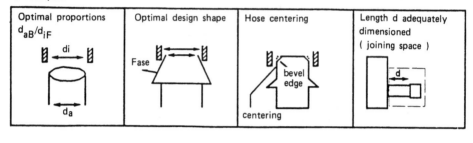

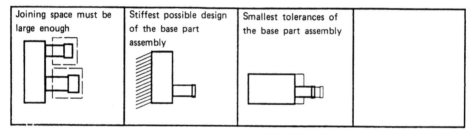

FIGURE 11.16 Proposed design elements for mating hoses with tubular sections: (a) base part and (b) joining part.

11.3 A DUAL-ARMED ROBOT SYSTEM FOR ASSEMBLY TASKS

KLAUS FELDMANN   *Friedrich Alexander University, Erlangen, Federal Republic of Germany*

DETLEF CLASSE   *Robert Bosch GmbH, Nürnberg, Federal Republic of Germany*

11.3.1 Introduction

In the design of modern assembly systems, an attempt is made to reduce assembly costs by replacing handwork by the use of automatic machines. A distinction is made between the rigid transfer line as an efficient assembly tool for large batch products and flexible assembly cells, which are able to assemble a given product spectrum automatically. Small and medium-sized lots are the main application field for these cells. For reasons of economy, retooling and setup times must be minimized. These devices can assemble automatically parts of a certain similarity without having to reconfigure the whole equipment. It is not always possible to adjust workpieces, grippers, and auxiliary devices in a way that permits an

b) Joining part

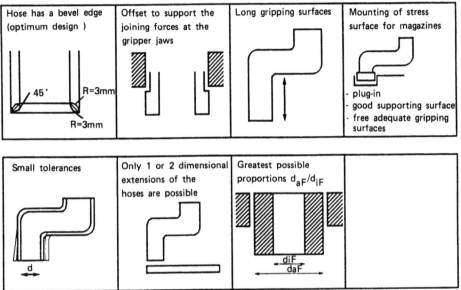

Hose has a bevel edge (optimum design)	Offset to support the joining forces at the gripper jaws	Long gripping surfaces	Mounting of stress surface for magazines

45° R=3mm
R=3mm

- plug-in
- good supporting surface
- free adequate gripping surfaces

Small tolerances	Only 1 or 2 dimensional extensions of the hoses are possible	Greatest possible proportions d_{aF}/d_{iF}	

d

$\dfrac{d_{iF}}{d_{aF}}$

automatic adaption to a new task to be made. Mostly, it is necessary to change the handling equipment, grippers, or fixtures, or to have the adaption done by sensors or program instructions.

Good results have been reached with the assembly of electromechanic and electronic devices or with special products in the automotive industry. Oil pump and wheel assembly are well-known examples. Flexible devices for assembly are industrial robots or, for more elementary applications, numerically controlled (NC) drives for basic assembly operations. Despite a trend to use standardized devices, the need for grippers and fixtures may be very expensive. For numerous assembly tasks it is inevitable to see totally newly designed, special-purpose machines using NC axis.

11.3.2 Two-Armed Assembly

Man mostly uses both arms to assemble parts. Either the workpiece is of complex design, difficult to handle, and has to be held with both hands, or the second arm is operating as a joining tool (e.g., a screwdriver) that makes mating easier or possible at all. With successfully automated assembly processes it is often possible to replace the second arm by a special fixture or gripper. Thus, the purpose of the remaining arm is to guide a tool or gripper. The situation is different for limb parts including hoses, seals, cable trees, and springs [4]. These parts yield to forces and cause particular difficulties during handling and joining. Sometimes they can be assembled automatically by a specially designed gripper [4,5]. An attempt to solve the problems mentioned is the use of two arms. The system discussed here was designed to increase the flexibility of an assembly operation by the coordination of two robots.

The proposed assembly method can be used for the following tasks:

Joining hoses
Laying cable trees
Inserting large seals
Mounting springs

11.3.3 The Arrangement of the Assembly Cell

Figure 11.17 shows the assembly cell used for practical tests. It contains two Cartesian robots having three axes each, which can be coordinated by a common control system. The control is designed such that coordinated and uncoordinated operations are possible. The latter operating mode allows conventional handling operations. A gripper exchange system supports the use of both operation modes.

The programming is done by a problem-oriented explicit programming language whose structure is outlined in the next section. Its practical usefulness is enhanced by a teach panel (Fig. 11.18). A joystick containing strain gages is assigned to each robot to perform simultaneous motions. The axes of the individual robots are controlled in a way that ensures that the handling speed is proportional to the force exerted on the joystick [6]. Thus teach-in of a point can be easily done by guiding the effector and storing the coordinates by operating a push-button control. The teach points are entered into two lists, and variables are assigned to them which can be referenced by the user program. The user can examine in detail the position of an axes on a display terminal. Thus it is possible to take measurements of the assembly object and the gripper.

The central control element is the terminal; here programs are developed, corrected, completed, and described with the help of an editor. For the teach-in operation, control is handed over to the teach panel. The control unit consists of two 16-bit computers; each has an own arithmetic processor and is assigned to one robot. The synchronization of both robots and the collision monitoring during program execution is done via a common data field in a dual-

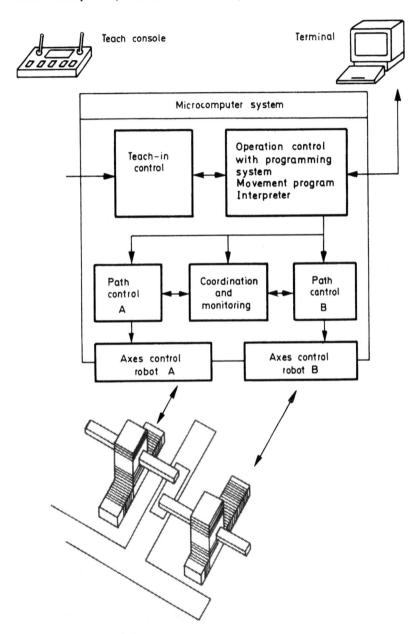

Teach console

Terminal

Microcomputer system

Teach-in control

Operation control with programming system
Movement program Interpreter

Path control A

Coordination and monitoring

Path control B

Axes control robot A

Axes control robot B

FIGURE 11.17   The flexible assembly cell and the controller.

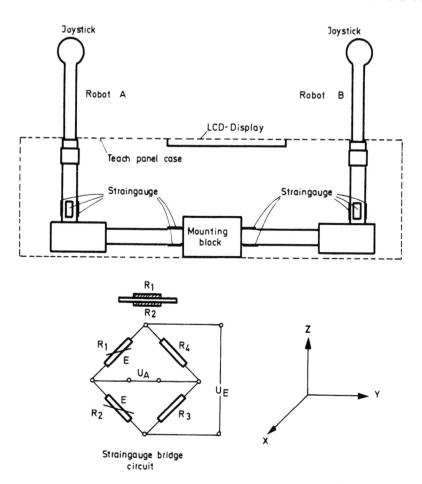

FIGURE 11.18   The teach panel.

port RAM. The control system can be subdivided into several hierarchical levels, which are realized separately for each robot (Fig.
11.19). The common motion program, developed for the assembly
task, is split up into two separate programs before execution. This
is done off-line because of the long computing times involved. Thus,
the microcomputers can be fully employed with the execution of the
robot programs, tasks synchronization, and collision control avoidance.   The lowest control functions of speed and current control
are done by hardware. The position control and all higher tasks
of the control hierarchy are realized in software.

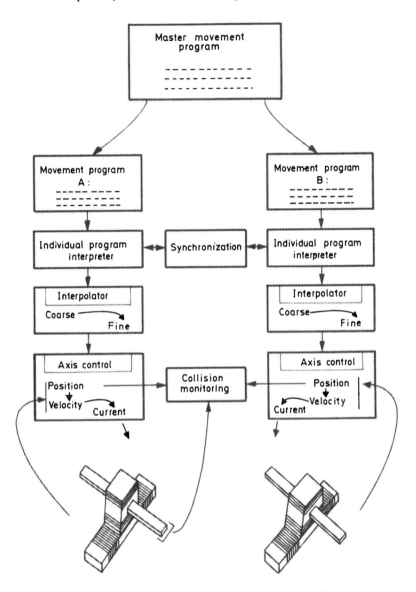

FIGURE 11.19 Schema of the hierarchical controller.

11.3.4 Programming

The assembly cell serves as an experimental testbed for special assembly tasks. To carry out the experiments efficiently, it was

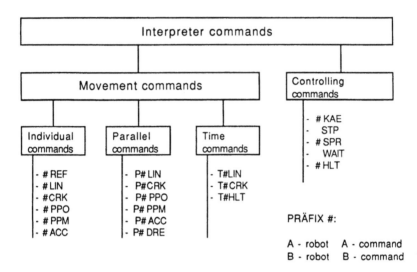

FIGURE 11.20   The interpreter commands.

necessary to set up a programming and control system that satisfies
the special requirements of two-armed assembly operations. The
textual programming language developed for this purpose is problem-oriented. The language is based on interpreter commands of
two types: "motion" and "control" (Fig. 11.20). The control commands perform the execution of the program, while the motion commands cause the explicit movements of the robots.

Fundamentally, the motion commands consist of an abbreviated description of the type of motion desired between a start and a goal
point of a path. For an economical assembly, both robots must be
able to carry out individual tasks either completely independently or
in a joint operation mode. Accordingly, the language commands can
describe the motion of either separate or joint devices. To the latter group belong the so-called time commands, by which simultaneous
motion cycles are realized for several paths. The starting point of
the motion program is the reference mark (# REF).

Individual Commands

A typical set of motion commands was designed because it is available for every comfortable industrial robot (Fig. 11.20). Point-to-
point (PPO), circular path (CRK), and straight-line motions (LIN) in
space were implemented. The point-to-point motion commands are
supplemented with smooth transition sections for neighboring path

segments (PPM) and an accelerated linear motion (ACC). The latter allows one to increase the speed from a given initial value to a final speed. The acceleration distance may be the whole length of a path segment. All the individual commands have a prefix marking the selected robot.

Parallel Commands

A parallel command, characterized by a prefix P, is used to move robots synchronously. For this type of motion one robot is determined to be a master by a second prefix, and the other is declared automatically to be the slave. Both end positions of the robots, which had been reached by the execution of a command, can be defined by a vector and form an origin of a Cartesian coordinate system. The positions can be transferred into another location by a shift vector a. On account of the kinematics of the robots, they cannot be rotated with regard to each other. The shift vector a is kept constant during a motion (Fig. 11.21). The master robot executes the individual motion described by the command following the second prefix. The corresponding motion of the slave robot is synchronized with that of the master. The slave command is generated at the program compilation time.

A special case of a parallel motion command is the rotation P # DRE, which generates for the slave robot a circular motion with the same rotation angle and center that had been programmed for the master robot. What remains constant is not the shift vector a along the motion but its value $|a|$. Generally, the slave robot moves along a circle with another radius in another rotational plane.

To make the calculation easier, a new coordinate system is first created at the turning point that is axially parallel to the original coordinate system. For the following considerations, robot A is taken as master and robot B as slave. From the start vector s_A and end vector z_A of the motion of the master robot, the turning axis g is obtained by forming the cross product (Fig. 11.22):

$$\vec{g} = \vec{s_A} \times \vec{z_A} \tag{11.1}$$

From the scalar product of the vectors s_A and z_A, the rotational angle α is obtained:

$$\alpha = \arccos \frac{\vec{s_A} \times \vec{z_A}}{\vec{s_A} \times \vec{z_A}} \tag{11.2}$$

The rotation of the master robot about angle α in the rotational plane can be described by the rotation matrix M. When M is applied to the start vectors s_A the goal vector z_A is obtained.

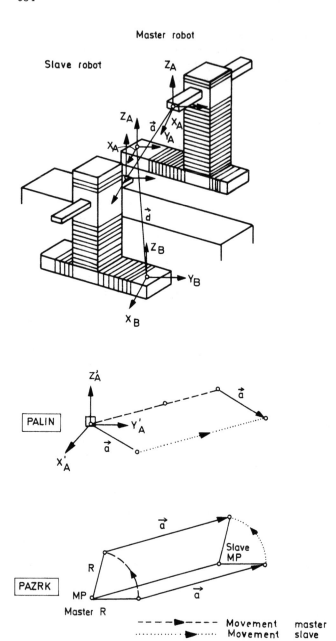

FIGURE 11.21 The parallel commands and constant shift vector.

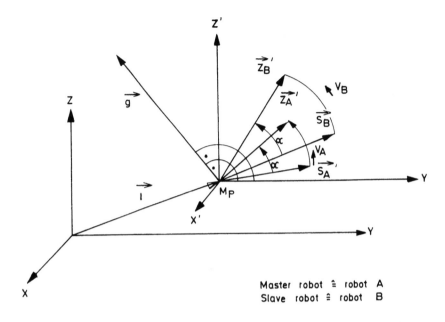

FIGURE 11.22   The motion caused by the parallel command P#DRE.

$$\vec{z_A'} = \vec{M} \times \vec{s_A'} \tag{11.3}$$

By projecting the start vector z_B of the slave robot into the rotational plane of the master robot, the robation matrix M can be used to describe the motion process:

$$\vec{z_{BT}'} = \vec{M} \times \vec{s_{BT}'} \tag{11.4}$$

After retransformation, the goal vector that was sought is obtained.

The relation of speeds can be derived from the equal motion times of both robots and correspond to the radio of the radii.

$$\frac{v_A}{v_B} = \frac{r_A}{r_B} \tag{11.5}$$

Time Commands

Another possibility for synchronizing motion process is to pass single motion commands at equal time segments.   Commands that

generate this type of motion have the prefix T (Fig. 11.20). Time
commands create a flowing motion of both devices, although the distance vector between them is not constant. Since both devices reach
the actual position at the same time, waiting is not necessary, even
with succeeding parallel commands. Thus, the commands are capable of achieving minimal clock times for assembly tasks where objects are handled jointly.

Control Commands

In addition to the motion commands, some helpful commands for
the sequence control exist. The command #DAE enables a change
of the constant field that specifies the transition behavior between
two path sections (Section 11.3.5). The command #HLT causes the
manipulator to maintain the end point of the last motion for a given
time. With the jump command #SPR, a jump is envoked; the destination is a motion command at another location of the program.
The command STP causes the termination of the running program.

Essential preconditions for two-armed robot assembly are mechanisms for synchronized and unsynchronized motions. If parts of the
assembly task require individual and coordinated motions, the robot
completes the program part to be executed first and then waits for
the second one. Status words characterizing the wait and enable
states of both robots are communicated by a global data field. They
represent the synchronization mechanisms of motion (Fig. 11.19). A
corresponding status byte describes the preparatory stage for the
program execution. The user has access to the synchronization
mechanisms by inserting the control command WAIT. Coordinate
commands generate these mechanisms automatically.

Assembly Task "Assemble a Spring"

The successive steps in mounting a spring are outlined in Figure 11.23. For the various assembly situations the find phase is
shown. The associated commands are indicated below the individual diagrams. Phase 1 is preceded by the preparation of the spring
and the grip operation. In phases 2—4 the steps are shown to manipulate the spring through a clearance. The spring is oriented in
phase 5, and the assembly is completed in phase 6. The fine motions needed to insert the ends of the spring through the eyes of
the gripper are not shown.

Obstacles and work clearances are typical conditions that compli-
cate the automatic assembly of electromechanical gears. Often they
require the construction of expensive grippers and increase the cost
of an application.

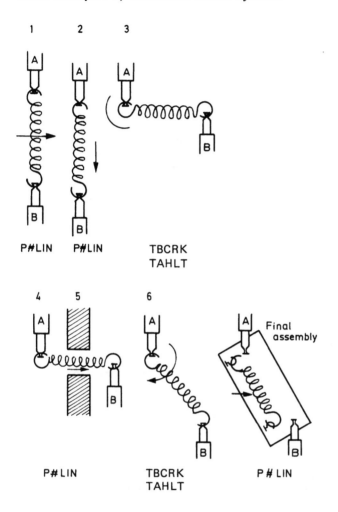

FIGURE 11.23 Six phases of a spring assembly.

11.3.5 Control Features

Figure 11.19 showed that two identical controllers are used, one for each robot arm. To perform path control, a two-stage inter-polator with coarse and fine interpolation is used [7]. Recursive al-gorithms help to save computing time. They require the partitioning

of the coarse interpolator into a real-time and a background component to assure fast synchronization.

Smoothing Transitions Between Path Segments

The trajectory of a robot is made up of single path segments.
In general there would be discontinuities in speed and acceleration
between two neighboring path segments if a robot tried to follow
the path exactly. The control loops of the axes are equipped with
P controllers, the behavior of which can be approximated by a first-
order proportional time delay. However, discontinuities in speed
behavior cause a deviation from the programmed trajectory. Since
robots are generally made up of different axes, the deviations may
be unpredictable.

This behavior causes problems with one-arm activities, and it is
intolerable for two-arm assembly operations: sensitive workpieces
can be destroyed or damaged by inaccuracies of the trajectory.
Higher precision can be gained only by slowing down the trajectory motion speed. However, this is contradictory to the requirement of short cycle times in the assembly of electromechanical parts.
This conflict can be eliminated with a polynomial approximation of
the path [8]. The positioning commands must be used accordingly.

For parallel commands, the transitions for both robots are adjustable to assure that the shift vector between the effectors remains constant. The only exception is the P#DRE command. The
trajectory along which the robot joints travel and their speed may
differ considerably. Since speed is used to determine the coefficients of the polynomials, different transition behaviors for both
robots may be set. However, for parts with small dimensions the
trajectories are very similar and the resulting change of the shift
vector is very small. Thus, changes of length can be absorbed by
a flexible gripper mount. The transition within the polynomial is
done at the time $t = -T_a$ when leaving the old segment and $t = +T_a$
when entering the new segment, where T_a is the time needed to accelerate the axis considered from standstill to its maximum speed (Fig.
11.24). With the control command KAE, the user can set the acceleration times T_a, hence the transition behavior, within a wide
range. The transition process as a function of time is described
by a fifth-degree polynomial. The coefficients are determined from
continuity considerations for position, speed, and acceleration [7]:

$$x(-T_a) = x_1 \qquad x(+T_a) = x_2$$
$$x(-T_a) = v_{1x} \qquad x(+T_a) = v_{2x}$$
$$x(-T_a) = a_{1x} \qquad x(+T_a) = a_{2x}$$

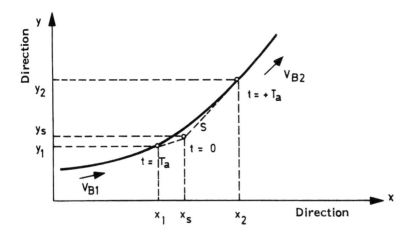

FIGURE 11.24   Merging of two path segments.

Collision Control

Industrial robots have a relatively large workspace that often overlaps the workspace of other devices. An assembly cell with cooperating robots has an overlapping common workspace where the coordinated motions take place. Thus the danger of collision exists and possible causes must be considered. Several methods to avoid collision have been investigated. According to the time of their execution, they can be divided into two groups:

Off-line procedures to simulate the motion sequences with a model; a collision can be either recognized on a screen [9] or calculated analytically [10]. Such procedures are suitable for the detection of errors during the layout of an assembly cell and programming of the robots.

On-line procedures to detect a collision with the help of sensors during the execution of the program. They also may detect device faults and other unexpected influences.

The on-line procedure offers greater safety for the operation of a laboratory system. For the system discussed here, an indirect method using an angular decoder and a direct method using a sensitive protective hood were used (Fig. 11.19). With the indirect method, the signal to indicate danger of collision is obtained from robot internal state variables. Since the precondition for this method is a sufficiently precise kinematic model of the robot, the calculated

effort for the Cartesian design of the robot is minimal. The user
is able to define a locally variable protective space, corresponding
to the motions of the master robot.

11.3.6 Cable Tree Assembly

Video recorders are consumer electronic devices that are produced
cheaply in large quantities. One problem area with their assembly
is the installation of cables. Although considerable efforts have
been made to replace cables with printed circuits, it is not possible
to replace them all. Cables are limp and cannot be handled well
in automatic assembly. Often single cables are combined in cable
trees or flat ribbon cables.

Figure 11.25 shows four steps to install a flat cable. The cables
are terminated at both ends with plugs. In the first step each robot grips a plug and then together they carry the cable to the actual assembly place. At the beginning of the second step the first
plug is turned by 90° and inserted. Then the cable is straightened
out and the second plug is inserted.

11.4 VISION-GUIDED ROBOT APPLICATIONS

ULRICH REMBOLD   *University of Karlsruhe, Karlsruhe, Federal
Republic of Germany*

Vision techniques are used for a variety of robot applications. The
most important ones are locating and inspection of parts. These
tasks may be done with 2D or 3D vision systems.

The 2D vision technology has matured over the past decade, and
there are numerous well-designed binary and gray scale vision systems available. The latter are more difficult to apply, and quite a
lot of experimentation is necessary to provide good lighting conditions. In many applications light is the most crucial point of the
vision process.

There are also numerous 3D vision systems available. However,
they are very difficult to use, and complex algorithms must be developed to interpret pictures of a 3D scene; for example, an object
can be presented to the camera from different views. To be able
to positively identify this object, there must be a sensor hypothesis available for every view. In simple cases it often is possible to
identify an object in a picture from key features without knowing
too much about its 3D shape. However, this situation is rather the
exception.

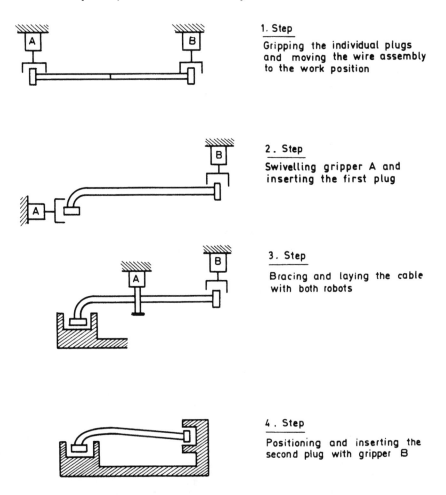

1. Step

Gripping the individual plugs and moving the wire assembly to the work position

2. Step

Swivelling gripper A and inserting the first plug

3. Step

Bracing and laying the cable with both robots

4. Step

Positioning and inserting the second plug with gripper B

FIGURE 11.25 Wire assembly for a video cassette recorder.

Another problem involves connecting the vision system to the robot. When a part has to be handled, it is necessary to locate and identify it and then to determine a suitable grasp position. Frequently, this work is done on a moving conveyor. Both the camera and the robot have unique coordinate systems. The camera identifies the object in regard to its own coordinate system. For grasping, however, the robot must know where the object lies in reference to its own coordinate system. This requires the transfer of the object's location from the camera to the robot coordinate system. There are two ways of performing this coordinate transformation. The first one uses the frame concept of a high level programming language, whereby calibration is done with the help of a round disc that is shown to the robot and to the camera, in reference to their respective coordinate systems. The coordinate system of the camera is made known to the robot via a frame inversion. With the second method, the camera position in reference to the robot coordinate system is obtained by a similar calibration procedure; however, merging of the two coordinate systems is done via complex trigonometric calculations, considering the kinematic structure of the robot.

11.4.1 Two-Dimensional Vision

The examples given below were obtained through the courtesy of ASEA Brown Boveri Industrie-Roboter Friedberg, Federal Republic of Germany.

Figure 11.26 shows an example of an automatic assembly station for car doors. The station consists of a bin-picking robot, two

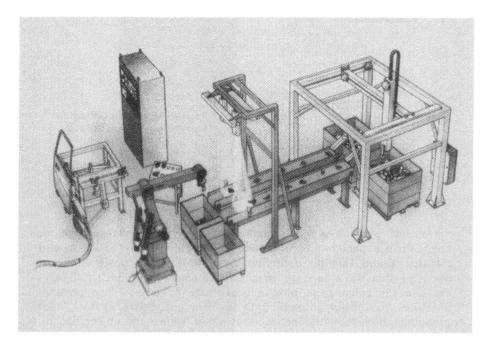

FIGURE 11.26 An automatic assembly station for car doors.

conveyors, a camera system, an assembly robot, and a fixture for the door being assembled. With an electromagnet, the first robot selects different types of hinges from two bins and places them at random on the conveyor belt. The parts travel down the belt and are recognized by the vision system, which identifies them and determines their position and orientation. Thereafter, the second robot is guided to the pickup position to grasp the proper part. After buffering, the hinges are clamped into a fixture and bolted to the door.

The second example depicts a flexible deburring station for castings (Fig. 11.27). The workcell consists of two roller conveyors, a robot, a vision system, and a grinder. Parts are brought via the upper roller conveyor to the vision station for identification. The robot picks them up in the proper position and brings them to the grinder. The parts are rotated along the surface of the grinding wheel to remove the burrs. After completion of deburring, parts are placed on the second roller conveyor for transportation.

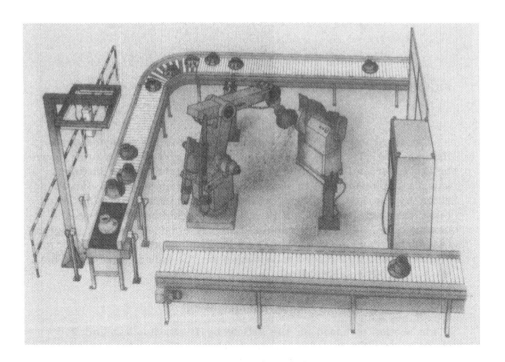

FIGURE 11.27 A flexible robot-operated deburring station.

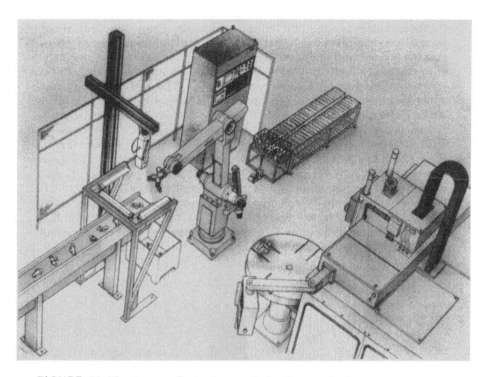

FIGURE 11.28   A manufacturing cell for brass fittings.

   The third example shows a flexible machining center for brass
fittings (Fig. 11.28).   The center consists of a vision system, a robot, a roller and belt conveyor, an indexing table with fixtures,
and a machine tool.   The brass parts are brought to the pickup position by the belt conveyor, where they are identified.   The robot
takes the parts and inserts them into a machining fixture on the indexing table.   After the machine tool has obtained its part program,
the workpiece is processed.   Upon completion, the part is lifted out
of the fixture and placed in a basket on the roller conveyor. Thereafter, the cycle is repeated.

11.4.2   Stereo Vision-Guided Robotics

(This section is based on the article by D. E. B. Lees and P.
Trepagurer, "Stereo vision guided robotics."   Courtesy of Auto-
matix, Inc. Billerica, Massachusetts.)

The use of two photographic images of the same scene to give depth information has been known since at least the mid-nineteenth century, when the stereoscope (still popularly sold under the trademark "View Master") was invented by Wheatstone. The stereo approach to robotics vision is essentially the same, although the two images need not be in the same plane, as they are in a stereoscope. Figure 11.29 illustrates the general two-part stereo approach; the second part is entirely general and mathematical, while the first part is application-dependent.

The mathematical part of the algorithm essentially uses trigonometry to locate in space the feature that corresponds to the two binocular images. The vision algorithms for this stereo system assume a "pinhole" model of the camera optics. This means that all rays reaching the camera focal plane have traveled through a common point referred to as the optics "pinhole." Since any light ray striking the focal plane has gone through this single, fixed point in space (the pinhole), a focal plane location together with the pinhole location determine a unique line in space. A point is imaged by a pair of lines, intersecting in space at the original object point. Physically this can be thought of as triangulation of the object's position, as illustrated in Figure 11.30. The stereo code takes as input certain camera calibration coefficients, which essentially characterize the cameras in space, together with the focal plane coordinates, and uses them to locate the corresponding space point.

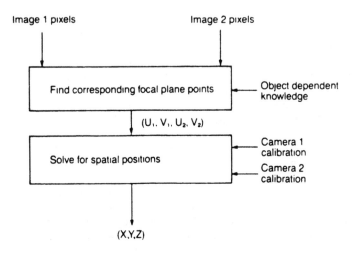

FIGURE 11.29 Stereo vision processing consists of two parts.

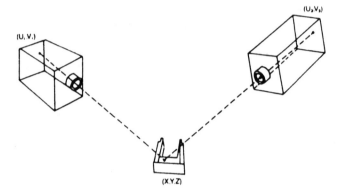

FIGURE 11.30   Stereo projection of a point into focal planes.

In practical situations, the lines in space from a common point
will not appear to intersect because of noise and deviations from
pinhole optics in a real camera.   Thus, point location is done in a
least-squares sense.   That is, the point that is chosen will minimize the sum of the squares of the normal distances to each of the
triangulation lines.

As mentioned above, to perform stereo reconstruction, it is necessary to know the characteristics of the camera in space—the position and orientation of the effective pinhole—as well as other cam-
era-dependent parameters such as the focal length of the lens and
the size of a pixel in the focal plane.

One characteristic of the camera calibration procedure is that it
is not necessary to attempt to measure any of these parameters directly.   Instead, the cameras take pictures of an object called "calibration target," with distinctive features on it in known positions.
When the apparent positions of these features in the image plane
are compared with their known positions on the calibration target
in space, all the necessary information about camera calibration can
be inferred without any explicit measurement.   This technique
greatly simplifies the calibration process.

The camera images then must be processed to correctly extract
the relevant features.   A pair of digitized images of the same object from different camera positions are processed to find features
that are to be located in space.   The image pair may be generated
by two rigidly fixtured cameras, by two cameras mounted on a robot arm, or by a single camera mounted on a robot arm and moved
to two positions.   For stereo reconstruction to proceed, features of

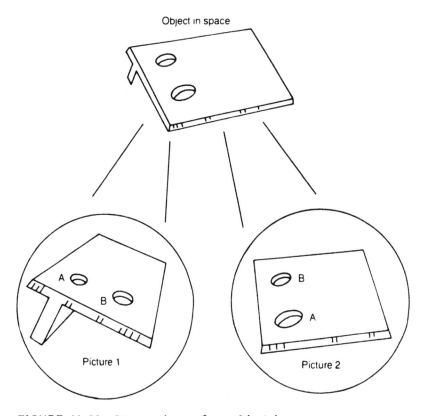

FIGURE 11.31 Stereo views of an object in space.

interest must first be extracted from the digitized images. The system provides techniques that apply to this task and are embedded in the RAIL programming language.

After the features have been extracted, the correspondence between the same features in the two images must be found. For example, the object in Figure 11.31 has two holes. The application-specific portion of the software must recognize that hole A in picture 1 corresponds to hole B in picture 2. Similarly hole B in picture 1 corresponds to hole A in picture 2. The user must write application-specific code to solve this correspondence problem for each application. This problem is the same one faced by a photo-interpreter performing stereo mapping from photographs. However, the process can be automated in the factory because the objects of interest are extremely simple relative to those found in cartography.

When the apparatus to perform stereo reconstruction in space is in place, the resulting information must be presented to the robot in usable form. During setup, the robot is manually taught to move a tool along a 3D path over a part to perform a production operation (e.g., welding, spraying, drilling and cutting). Then stereo vision is used to determine the position of three distinctive features on that part.

When the production line is started, poorly fixtured parts will be presented to the system with position and orientation somewhat different from the part on which the robot tool operations were taught. Part tracking finds the same three distinctive features on each part.

Three points (if they are noncollinear) are sufficient to determine the position and orientation of a rigid body in space. Thus, the full six-dimensional offset transformation (three translations and three rotations) found from the use of stereo vision on the part's three reference features can be applied to the tool path previously taught, transforming it for the current part's orientation. Special situations may simplify the process. For instance, if for some reason it is known that the angles of a part will remain the same, and the part can only translate, then a single reference point will suffice.

This technique assumes, however, that the part can be considered as a rigid body. This means that all parts, although they may be very poorly fixtured, must be identical to within the tolerance required by the manufacturing operation being performed by the robot. Other techniques, such as a continuous tracking system, take many pictures very rapidly to accommodate continuous variation in the part. Additional hardware is required for this approach, which is also computationally more expensive. It was found that in many applications, the rigid body approximation was quite adequate.

Stereo vision can be applied to a great many applications (Figs. 11.32—11.35). Figure 11.32 shows the location of parts that have been stacked on a pallet in a partially ordered manner. The location of the part on the top of the stack is determined using stereo vision, and a robot is directed to pick it up. This process is repeated until the whole tiering rack is unloaded. The determination of which part is on top makes use of a priori knowledge about part shape and size.

Figure 11.33 shows a car body on an assembly line. When the car halts at a station on the line, stereo vision is used on gage holes in each of the four wheel wells, to determine the body location and orientation so that robots can spray sealant on wheel well seams. Large positioning errors may be present, but the robot can still accurately direct a narrow spray to the desired portion of the automobile.

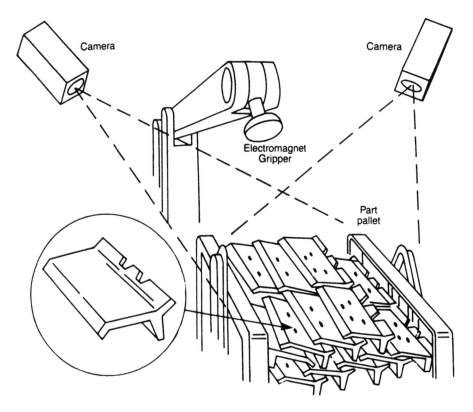

FIGURE 11.32 Tiering rack bin picking.

Another example is shown in Figure 11.34. In this automotive application, cameras on the end of a robot arm are used to locate a sheet metal part that has been tack welded onto the automobile frame. The holes in the sheet metal part are the features that are reconstructed. The cameras are placed on the robot's wrist, so that after the part has been located, they pivot out of the way as the seams are arc welded with the MIG torch positioned above them. Placement of the cameras on the robot arm, rather than floor-mounted fixturing as in Figure 11.33, can greatly reduce the number of cameras required. Another advantage of robot-mounted cameras is the saving in fixturing costs. Furthermore, in some cases space considerations on a line may make fixtures for cameras difficult to use.

Figure 11.35 illustrates an application in which slotted plates are welded to a larger assembly. Stereo vision is used to locate the slot

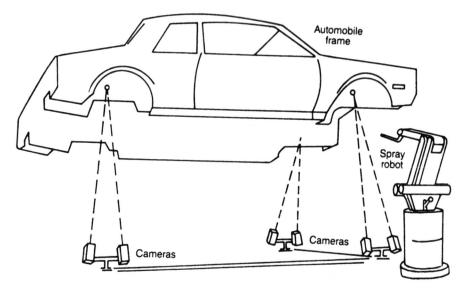

FIGURE 11.33 Stereolocation of an automobile to direct a robot.

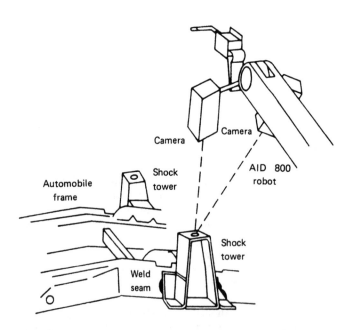

FIGURE 11.34 Welding shock towers to automobile frames.

in each plate so that a welding robot can automatically weld the assembly, even though the parts are imprecisely fixtured and the plate positions may vary relative to the part. This is an example of being able to use fewer than three features in a special case. The case is confined to the plane of the large assembly, so that only the slot needs to be reconstructed—it gives both translation information and, because it is asymmetric, the rotation angle normal to the plane.

The common feature of all four examples is that objects whose location is not accurately known, or objects that are imprecisely fixtured, can be located with high accuracy using stereo vision to generate part locations for robots. This can save money, compared with precise part fixturing.

11.5 MOBILE AUTONOMOUS ROBOT IPAMAR PERFORMS FREE-RANGING AGV OPERATIONS

HANS-JÜRGEN WARNECKE and GERHARD DRUNK *Fraunhofer Institute of Technical Production and Automation, Stuttgart, Federal Republic of Germany*

11.5.1 Introduction

Free-ranging operations of automated guided vehicles (AGV) are regarded as essential capabilities of transportation systems. The development of a sensor and control system for the navigation of transportation vehicles is an important part of an autonomous mobile robot.

The goal of the research discussed in this section was to find a new system control architecture that on one hand is powerful enough to fulfill the requirements of autonomous operation in a real-world environment and on the other hand, is straightforward enough to be acceptable for industrial use with regard to technological as well as financial considerations.

The specification of the control structure for the autonomous system was based on the requirements of AGV operations in an industrial environment. An attempt was made to overcome the disadvantages of using inductive guide systems.

The work was divided into two phases. First, the requirements of an autonomous AGV were analyzed. Second, the functions and solution alternatives for the sensor components and control components were established [13]. The results of the work are discussed as follows: Section 11.5.2 briefly surveys the state of the art. Section 11.5.3 describes integrated sensor action planning as the synthesis of a new control structure for mobile autonomous robots. Section 11.5.4 discusses the feasibility demonstration performed with the

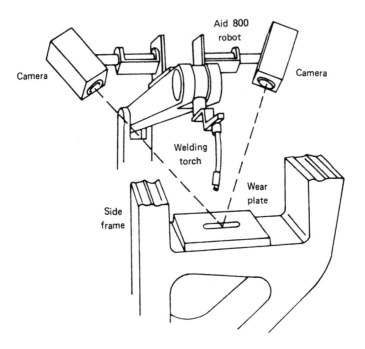

FIGURE 11.35   Locating the part with binocular vision.

research platform IPAMAR. And finally, Sections 11.5.5 and 11.5.6 offer a detailed description of the principles of AGV operation and the most important experimental results obtained.

11.5.2   State of the Art

Today, most of the automated guided vehicles for transportation tasks use the inductive guidance principle. The vehicles follow the magnetic field generated by embedded cables carrying ac signals with a frequency of typically 3—15 kHz. Inductive guidance has reached a high standard, looking at further advances, however, the technological limits of the principle seem to be reached.

Other passive tracking methods use reflective or fluorescent strips bonded to the floor. These strips can be installed easily, but problems arise from the mechanical wear incurred under the harsh conditions of the factory environment.

The drawbacks of wire-guided AGV systems are their poor flexibility with respect to the layout of the transportation routes and the high cost of the ground installation in an operating production system. Here, free-ranging navigation without guide paths, landmarks, or beacons would constitute essential progress for the application of AGV systems.

The development of navigation systems for autonomous mobile robots is an important research topic. Several institutions in Japan, North America, and Europe are concerned with the development of sensors, controls, and artificial intelligence modules for the navigation of experimental vehicles. General hierarchical control architectures for mobile sensor-guided robots are described in References 14—16. Several structures use the blackboard concept as interface between different control modules [17,18]. There are also approaches organized in terms of behavioral aspects of the vehicle [19,20].

Although the control structures of these concepts differ in several apsects, generalized components for planning, task execution, perception, and motion control can be identified (Fig. 11.36). In some cases an additional component for learning is implemented.

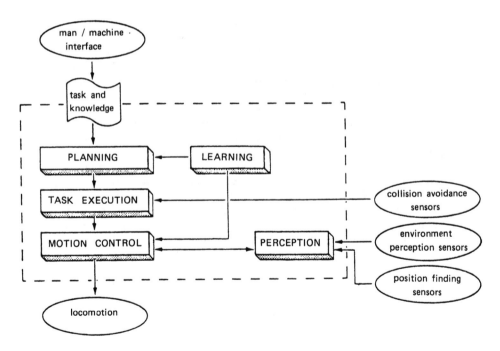

FIGURE 11.36 Components of a control system of autonomous mobile robots.

The first mobile robot projects were conducted in the United States with the robot Shakey at Stanford Research Institute [21], the Cart at Stanford University[22], and the JPL Rover at California Institute of Technology. At present important projects are conducted at MIT [23] and the Carnegie-Mellon University [24].

In Japan, too, extensive research work is done in the field of mobile autonomous robots. Yamabiko [25], at the University of Tsukuba, and MELDOG [26] are well-known examples. The work in Europe started in France [27] with the LAAS vehicle, where pioneer work was performed. Other important projects are conducted in Germany [16,28].

At present, the application of these control structures for mobile autonomous robots still is accompanied by enormous problems involving lack of computational power, inability to understand the real world, low reliability, and high cost.

With these problems in mind, a new sensor and control structure was developed for the navigation of free-ranging AGVs.

11.5.3 The Control Architecture of Integrated Sensor Action Planning

The synthesis of the sensor and control system is based on an analysis of functions as well as on existing solutions for sensor and control components. By minimizing the sensor requirements to the fundamental operational demands, odometry was chosen for local navigation and an arrangement of ultrasonic range finders for environment perception and collision avoidance.

For efficient control architecture of an autonomous mobile robot, computation time should be kept to a minimum and the operational environment should be as simple as possible. The perception components usually need the largest amount of computation.

In most cases a local world model that stores and updates sensor data is maintained. Operations with sensor data include data preprocessing, feature extraction, coordinate transformation, data fusion, correspondence matching, position error computation, and model updating. Access to the world model by motion control is performed by a set of interface procedures. The output of the sensor data processing operations is used for:

Collision avoidance; the selected path is free or not free
Position estimation; the position error at the present location
On-line motion guidance (i.e., the distance to the wall is maintained
    when traveling along a wall)

The basic idea is to find a control architecture that performs these three tasks without using a local world model. This goal was achieved by a new control structure named ISAP (integrated sensor action planning). This architecture makes the direct processing of sensor results possible by defining the application of sensors at the task planning level.

With the help of the global world model, all sensor actions for the collision avoidance and correction of position error are generated as sensor commands, which are executed on the motion control level. The functions include transducer selection initiation of time measurements, coordinate transformation for sensor parameters, and data to retrieve the correct world model. This information can be considered to be the condensed version of the local world model.

For collision avoidance, different sensors based on the ultrasonic principle are used. The shape of their surveillance area actually depends on the situation and cannot be preplanned in the global world model. For this reason, the sensor functions are implemented in the controller. The generated collision avoidance commands are used by the execution monitor. Position estimates are updated in the actual industrial environment, aided by the selection of the most suitable locations for referencing. Thus, only parts of the world must be known exactly for the operation. Parallel or perpendicular walls are chosen as references to obtain defined conditions while sensing.

An on-line motion guidance principle is used with obstacle avoidance strategies. This is initiated by local replanning. The sensor-controlled functions for the trajectory generation do not require model knowledge.

Thus, collision avoidance, position estimation, and on-line motion guidance can be done with the direct processing of the sensor results. For this, the necessary model knowledge is already implied in the sensor action modules. The overall control structure is shown in Figure 11.37.

Planning

The global world model is constructed with the aid of a menu-driven CAD workstation. Edges, areas, attributes, and positions are the input elements. The results can be inspected on a graphic color display.

The transportation tasks requested by an operator are first decomposed into more global locomotion and load handling tasks. Then a detailed route planning, including the sensor application planning, results in a sequence of motion and sensor commands.

Task Execution

The interpreter reads commands generated by the planner and starts the related motion control procedures. Status information from

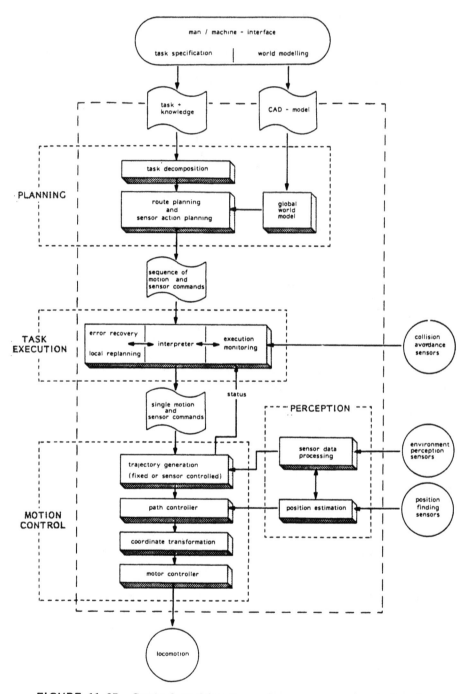

FIGURE 11.37 Control architecture of a sensor-supported action planning system of an autonomous mobile robot.

motion control and warnings from the collision avoidance sensors are supervised by the execution monitor. When an error occurs, especially one resulting from an unknown obstacle, local replanning is performed by the error recovery module. It initiates a sensor-guided recovery strategy.

Perception

The perception module determines position errors by comparing the results of the sensors with the expected values. The difference is used to update the position estimation, supplying the vehicle's position to the path controller. The results of the ultrasonic perception sensors also can be applied directly for on-line motion guidance.

Motion Control

The robot uses a very powerful motion control module with an accurate guidance system to follow continuous paths. Other features are trajectory interpolation, feedback control, and coordinate transformation from world coordinates to steering and driving angles.

11.5.4 Physical Description of the Vehicle IPAMAR

For practical demonstration and testing, the research platform IPA-MAR (IPA Mobile Autonomous Robot) was developed (Fig. 11.38). The overall structure of the physical components is shown in Figure 11.39. Figure 11.40 gives a view into the interior of IPAMAR. The main parts of the vehicle are locomotion components, perception sensors, vehicle computer, technical equipment, load handling module, and man—machine interface.

Locomotion

The vehicle uses a three-wheel-drive system. The active front drive wheel is steered and the two wheels on the rear axle are passive. Robust dc motors are applied. The total weight of the vehicle is 260 kg.

Perception

Local navigation is performed by an odometry principle using front wheel encoders. Two groups of sonic range finders are installed. The first group, with eight pairs of transducers, is arranged in front of the vehicle for obstacle detection. An additional tactile safety bumper is mounted in front and on both sides of the vehicle. Perception of the walls is done by the second group of ultrasonic sensors consisting of five precisely focused units. The docking

FIGURE 11.38 The autonomous mobile robot IPAMAR performing free-ranging AGV operations (IPA, Stuttgart).

process at the load handling station is supported by two optical range finders, one on each side. The optical and the sonic sensors can be seen in Figure 11.41.

Vehicle Computer

The vehicle computer is a multiprocessor system with two 32-bit Motorola 68020 processors with 68881 floating-point coprocessors and a common static 512 kb CMOS-RAM. The peripheral devices are connected by two serial ports, six 8-bit parallel interfaces, and four D/A and eight A/D converters. There is enough computation capacity available for complex control algorithms. This computer executes the complete vehicle control software.

Technical Equipment

The technical equipment consists of a 24 V power supply, the electrical control gear, and the vehicle controller. A complete security system is implemented to satisfy safety requirements. Communication between vehicle and host computer is performed by an

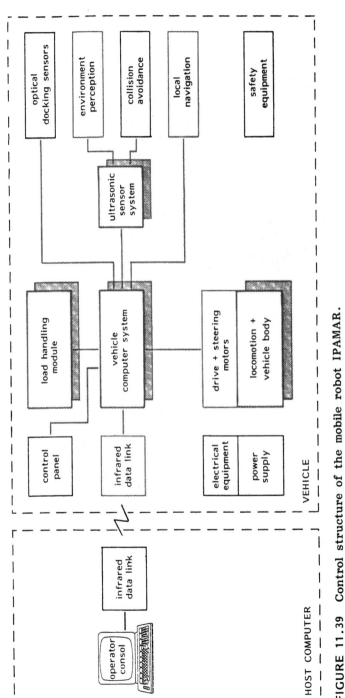

FIGURE 11.39 Control structure of the mobile robot IPAMAR.

FIGURE 11.40   The interior of IPAMAR (IPA, Stuttgart).

infrared communication link with a range of up to 30 m.   The data
transmission is done with the help of a communication protocol.

Load Handling Module

For demonstrating the AGV's operation, the vehicle is equipped
with a container handling device.   With a lift attachment and a telescopic platform, simple passive load/unload stations can be served.

Man—Machine Interface

For the input of the commands of the transportation tasks as well
as for the CAD data of the global world model, an IBM PC/AT is
used as a host computer.   The construction of the global world
model is menu-driven with a mouse as input device and a color
graphic display.   At the moment, the route and sensor action planning is done on the host computer and the complete program is then
transmitted to the vehicle computer via the infrared communication
link.

Program development for the vehicle control software is directly
performed on the vehicle computer.   For this purpose, terminals,

FIGURE 11.41   The arrangement of optical and ultrasonic sensors of IPAMAR (IPA, Stuttgart).

an external Winchester/floppy disk unit, and a printer can be connected to the vehicle computer.

11.5.5 Operation of IPAMAR as a Free-Ranging AGV

The sensor and control system developed allows free-ranging navigation of the IPAMAR vehicle in the factory environment. The basic operation is described below.

Course Description

For the description of the course, the floor plan entered into the CAD system includes the desired routes, the position of the load/unload stations, and the initial position of the vehicle. Then the control program for the complete course is generated automatically and transmitted to the vehicle by the infrared communication link. As soon as the vehicle has completed its self-calibration at the starting position, the system is ready for operation. Changes of the course are executed in the same way without any changes in hardware. An example of the CAD model is given in Figure 11.42.

Transportation Tasks

Commands for the transportation of a container from one station to another are fed into the host computer by an operator. The task is transmitted to the vehicle by the infrared command link, where it is decomposed into a sequence of motion and handling operations.

Navigation

The local estimation of position that is performed with the aid of the odometer is updated by predefined sensor actions of the five ultrasonic range finders. Figure 11.43 gives an example of position

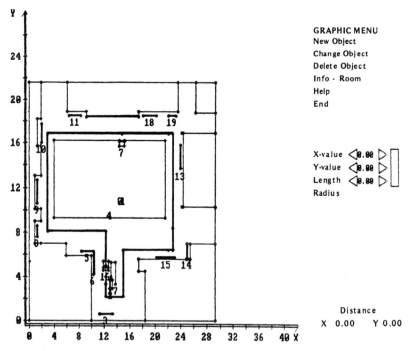

FIGURE 11.42   CAD presentation of the global world model of a factory.

updating. In this case the sensor action planner has chosen a parallel wall on the left side of the vehicle for correcting the y-direction and the orientation with the help of two sensors on that side. Correction for the x-direction is done by the sensor in front of the vehicle.

Collision Avoidance

When an unknown obstacle or person suddenly blocks the vehicle's path, the belt of ultrasonic sensors recognizes the situation and stops the vehicle under computer control. In this case the emergency brake is not applied (Fig. 11.44). The vehicle waits until the obstacle disappears and then resumes its operation. The next stage of the development will include avoidance strategies for obstacles that partially block the path.

Docking

Docking requires high accuracy, and for this purpose optical range finders are applied. By recording the measured distances

FIGURE 11.43 Perception of obstacles in front and to the right of IPAMAR (IPA, Stuttgart).

FIGURE 11.44 Collision avoidance system using a belt of ultrasonic sensors (IPA, Stuttgart).

to an edge of the docking station during the approach, the distance and orientation are obtained. These can be corrected during the final approach to the load/unload station (Fig. 11.45).

11.5.6 Experimental Results

The complete mobile robot system was developed in the extremely short time of one year. The system structure was successfully tested and proved to be robust against disturbances.

At the moment, accuracy for local navigation is still rather low in a dead reckoning mode. Thus, position update must be performed every 5—10 m. Local navigation errors result from the simplification of the kinematic model. In particular, the load-dependent radius of the rubber wheel and the undefined contact points in curved trajectories are reasons for position errors. By developing a refined model of the mechanical system, it will be possible to improve the accuracy of dead reckoning operations.

The next step will be the implementation of local obstacle avoidance strategies and the application of a sonic array sensor.

FIGURE 11.45 Optical docking sensor for correcting the position of the robot (IPA, Stuttgart).

11.5.7 Conclusions

A new control architecture for mobile autonomous robots was presented. Based on the analysis of existing control structures and the requirements for operating in the "natural" environment of a factory floor, the architecture of integrated sensor planning—ISAP—was derived.

Whereas most control systems for mobile robots require a local world model, in this approach sensor actions for navigation and collision avoidance are preplanned from a partly known global world model at the path-planning level. The planning module produces complete sequences of motion and sensor application commands. The results of the sensors are directly evaluated by the motion control module.

The mobile autonomous robot IPAMAR serves as a testbed for the control structure. Equipped with a load handling module, IPAMAR performs AGV tasks along an experimental test course installed at the institute's laboratory. The efficiency of the integrated sensor action planning architecture was proved by the successful operation of IPAMAR.

REFERENCES

1. E. Abele et al., "Einsatzmöglichkeiten von flexiblen automat-
    isierten Montagesystemen in der industriellen Produktion,"
    Montagestudie/Arbeitsgemeinschaft Handhabungssysteme (ARGE-
    HHS) unter Beteilgung des Fraunhofer-Institutes (IPA), Stutt-
    gart, VDI-Verlag, Dusseldorf, 1987.
2. H. J. Warnecke and B. Frankenhauser, "Fem als Planungshilfs-
    mittel für Roboter-Montagestationen für biegeschlaffe Teile,"
    CAE-J. 3, 50-59 (1987).
3. H. J. Warnecke and R. Bäßler, "Vorgehensweise zur montage-
    gerechten Produktgestaltung," *Robotersysteme* 3:1, 37—45
    (1987).
4. H.-J. Warnecke and B. Frankenhauser, "Montage biegeschlaffer
    Teile mit Industrierobotern," *wt-Z. ind. Fertig.* 76:1 (1986).
5. F.-W. Nolting, "Flexible Automatisierung von Fügeprozessen in
    der Feinwerktechnic—Einhaken von Zugfedern," in *Feinwerk-
    technik & Meßtechnik*, Carl Hanser Verlag, München (will be
    published soon).
6. J. Leinert, "Entwurf und Realisierung einer Industrieroboter-
    steuerung mit Programmiersystem," Diploma thesis, Lehrstuhl
    für Technische Elektronik, Universität Erlangen-Nürnberg,
    1985.
7. H. F. Schütz, "Entwurf und Realisierung leistungsfähiger
    Interpolatoren für Industrieroboter-Bahnsteuerungen auf einem
    Mikrorechner," Diploma thesis, Lehrstuhl für Elektrische
    Antriebe und Steuerungen, Universität Erlangen-Nürnberg,
    1985.
8. R. P. Paul, "Robot Manipulators," in *Mathematics, Program-
    ming and Control*, MIT Press, Cambridge, MA, 1983.
9. R. Dillman, U. Rembold, C. Blume, et al., "Grafische Simu-
    lation von Handhabungs- und Montage-einrichtungen," *Tech-
    nische Rundschau* 38, 1983.
10. T. Lozano-Pérez and M. A. Wesley, "An Algorithm for Plan-
    ning Collision-Free Paths Among Polyhedral Obstacles," *Com-
    mun. ACM* 22:10 (October 1979).
11. "To see or not to see. . . Flexible Manufacturing with Ma-
    chine vision," ASEA Robotics, Västerås, Sweden.
12. D. E. B. Lees and P. Trepagnier, "Stereo Vision Guided Ro-
    botics," in *Electronic Imaging*, Morgan-Grampion Publishing,
    1984.
13. G. Drunk "Sensors for Mobile Robots," in P. Dario et al.,
    Eds., *Sensors and Sensory Systems for Advanced Robots*,
    Springer-Verlag, Berlin, 1987.

14. J. S. Albus, C. R. McLean, A. J. Barbera, and M. L. Fitzgerald, "Hierarchical Control for Robots in an Automated Factory," *Proceedings of the 13th ISIR*, Chicago, 1983.

15. A. Meystel, "Nested Hierarchical Controller for Intelligent Mobile Autonomous Systems," *Proceedings of the Conference on Intelligent Autonomous Systems*, Amsterdam, December 1986, pp. 416–458.

16. R. Dillmann and U. Rembold, "Autonomous Robot of the University of Karlsruhe," *Proceedings of the 15th ISIR*, Tokyo, 1985, pp. 91–102.

17. S. Y. Harmon, "Coordination of Intelligent Subsystems in Complex Robots," *Proceedings of the First Conference on Artificial Intelligence Applications*, 1984, pp. 64–69.

18. C. Thorpe and T. Kanade, "Vision and Navigation for the CMU Navlab," *Proceedings of the SPIE Conference on Mobile Robots*, Cambridge, MA, October 1986, Vol. 727, pp. 261–266.

19. J. L. Crowley, "Representation and Maintenance of a Composite Surface Model," *Third IEEE Conference on Robotics and Automation*, San Francisco, March 1986.

20. Y. Y. Huang, Z. L. Cao, S. J. Oh, E. U. Kaltan, and E. L. Hall, "Automatic Operation for a Robot Lawn-Mower," *Proceedings of the SPIE Conference on Mobile Robots*," Cambridge, MA, October 1986, Vol. 727, pp. 344–354.

21. A. J. Nilsson, "A Mobile Automation: An Application of Artificial Intelligence Techniques," *Proceedings of the First IJCAI* 1969.

22. H. P. Moravec, "The Stanford Cart and the CMU Rover," *Proc. IEEE 71*:7 (July 1983).

23. P. W. Cudhea and R. A. Brooks, "Coordinating Multiple Goals for a Mobile Robot," *Proceedings of the Conference on Intelligent Autonomous Systems*, Amsterdam, December 1986, pp. 168–174.

24. J. L. Crowley, "Navigation for an Intelligent Mobile Robot," *IEEE J. Robotics Autom. RA-1*:1 (March 1985).

25. Y. Kanayama, J. Iijima, H. Ochiai, H. Watarei, and K. Ohkowa, "A Self-Constrained Robot: Yamabiko," *Third USA–Japan Computer Conference*, 1978, pp. 246–250.

26. S. Tachi and K. Komoriya, "Guide Dog Robot," *Second International Symposium of Robotics Research*, Kyoto, 20–23 August 1984.

27. G. Geralt, "Mobile Robots," in M. Brady et al., Eds., *Robotics and Artificial Intelligence*, Springer-Verlag, Berlin, 1984, pp. 168–174.

28. F. Freyberger, P. Kampmann, and G. Schmidt, "Ein wissensgestütztes Navigationsverfahren für autonome mobile Roboter," *Robotersysteme 2*:2, 149–161 (1986).

Index